# Elliptic and Modular Functions from Gauss to Dedekind to Hecke

This book presents the fundamental results of modular function theory as developed during the nineteenth and early twentieth centuries, focussing particularly on those interesting methods and techniques that appear to have been overlooked or are not generally well known. Of particular note are Jacobi's derivation of the infinite products for his elliptic functions based on his transformation theory; his first proof of the triple product identity; Hermite's derivation of elliptic functions, establishing the conditions for the ratio of two periodic entire functions to be doubly periodic; Mordell's proof of Ramanujan's conjectures on the Euler products of certain modular forms, based on the work of Hurwitz, Kiepert and Klein on the multiplier equation; and Hecke's work on the representation of integers as sums of squares using Dirichlet series of signature $(2, k, 1)$.

RANJAN ROY is the Huffer Professor of Mathematics and Astronomy at Beloit College and has published papers in differential equations, fluid mechanics, complex analysis, and the development of mathematics. He has received the Allendoerfer Prize, the Wisconsin MAA teaching award, and the MAA Haimo Award for Distinguished Mathematics Teaching; he was twice named Teacher of the Year at Beloit College. He is a co-author of three chapters in the NIST *Handbook of Mathematical Functions*, the co-author of *Special Functions* with Andrews and Askey, and the author of *Sources in the Development of Mathematics*.

# Elliptic and Modular Functions from Gauss to Dedekind to Hecke

RANJAN ROY

*Beloit College*

**CAMBRIDGE**
UNIVERSITY PRESS

# CAMBRIDGE
## UNIVERSITY PRESS

University Printing House, Cambridge CB2 8BS, United Kingdom

One Liberty Plaza, 20th Floor, New York, NY 10006, USA

477 Williamstown Road, Port Melbourne, VIC 3207, Australia

4843/24, 2nd Floor, Ansari Road, Daryaganj, Delhi - 110002, India

79 Anson Road, #06-04/06, Singapore 079906

Cambridge University Press is part of the University of Cambridge.

It furthers the University's mission by disseminating knowledge in the pursuit of
education, learning, and research at the highest international levels of excellence.

www.cambridge.org
Information on this title: www.cambridge.org/9781107159389

© Ranjan Roy 2017

First published 2017

*A catalogue record for this publication is available from the British Library.*

*Library of Congress Cataloging-in-Publication Data*
Names: Roy, Ranjan, 1948–
Title: Elliptic and modular functions from Gauss to Dedekind to Hecke / Ranjan Roy, Beloit College,
Department of Mathematics.
Description: Cambridge : Cambridge University Press, 2017. | Includes bibliographical references and index.
Identifiers: LCCN 2016036250 | ISBN 9781107159389 (hardback : alk. paper)
Subjects: LCSH: Elliptic functions. | Modular functions. | Functions.
Classification: LCC QA343 .R69 2017 | DDC 515/.983 – dc23 LC record available at https://lccn.loc.gov/
2016036250

ISBN 978-1-107-15938-9 Hardback

# Contents

# *Preface*

Modular function theory has developed a great deal during the past two hundred years, and its foundations have been reworked several times. During the course of this process, some earlier approaches and methods have naturally fallen into disuse or obscurity. The purpose of this work is to present the fundamental results of modular function theory as developed during the nineteenth and early twentieth centuries, focusing particularly on those interesting methods and techniques that appear to have been overlooked or are not generally well known. While the study of these methods has intrinsic value, it may also be useful to contemporary researchers in modular function theory. The study of nineteenth-century mathematics is often believed to be quite cumbersome, but a little familiarity with its mode of presentation and context readily dissolves much of this obstacle. To illustrate this point, I have remained very close to the original expositions and have provided some details about the creators of modular function theory. It is my hope that substantial portions of this book will be accessible to beginning graduate students, and to undergraduates with a solid knowledge of complex analysis.

The triple product identity is the expression of a theta series as an infinite product, and by 1808 this identity was known to Gauss; Chapter 2 is devoted to Gauss's work on theta functions. A special case of the triple product identity, known as the pentagonal number theorem, was conjectured by Euler in 1741 and proved by him in 1750. This theorem is so named because the powers of the variable in the series are pentagonal numbers. Euler proved this result by a very interesting and fruitful method, and Gauss, perhaps around 1796–97, showed that the method could also be applied to the situation in which the powers were squares or triangular numbers. This work of Gauss has received hardly any attention. In 1882, Cayley applied this method to reprove an important identity of Sylvester. Again, in 1919, Ramanujan employed it to prove the Rogers-Ramanujan identity. More recently, Andrews has shown that Euler's method can be applied to verify several significant $q$-series identities.

Jacobi's fundamental results on infinite products for the Jacobi elliptic functions are discussed in Chapter 3. These infinite products may be obtained in several ways, one due to Abel; most expositions use the method of theta functions, found by Jacobi. However, we present Jacobi's first method, given in his *Fundamenta Nova* of 1829. This method employs the transformation of elliptic functions, a method that has hardly appeared in textbooks since 1900. We also present Jacobi's three proofs of the triple product identity, especially Jacobi's little-known and ingenious first proof.

Eisenstein's method of constructing elliptic and modular functions and his proofs of important basic results have been given in Chapter 4, closely following Eisenstein's 1847 monograph in some detail. We remark that Weil's excellent book[1] gives a concise presentation of this material. Chapter 4 also contains Hurwitz's first 1881 proof of the fact that Weierstrass's discriminant function $\Delta(\omega)$ is expressible as an infinite product in $q = e^{i\pi\,\omega}$, Im $\omega > 0$. Hurwitz's elegant and more familiar second proof is presented in Chapter 12.

Hermite's noteworthy and useful 1858 work on the transformation of theta functions, dealt with in Chapter 5, led him to his most significant contributions to modular function theory. Hermite employed a powerful new theta functions notation, depending upon two indices. We also discuss Smith's 1866 paper, in which he used this helpful notation to prove Jacobi's formula for the product of four theta functions, a formula from which many other results in elliptic and theta function theory can be easily derived. Jacobi briefly mentioned this formula without any details in a letter to Hermite, included by Jacobi in his 1847 *Opuscula Mathematica*. He wrote that he had derived this formula in his lectures on theta series, delivered in the mid-1830s. These lectures were not available at the time Smith read the *Opuscula*, though they were later published by Borchardt. These lectures have been treated in textbooks, including Stalker's outstanding 1998 book,[2] but Smith's work, initiated by Jacobi's passing remark, has received little attention.

After the initial work by Abel and Jacobi on elliptic and modular functions, mathematicians realized that further advances in this theory required the use of the then-novel methods of complex analysis; we discuss these matters in Chapter 6. Also included is a treatment of the much earlier work of Cotes, of about 1715, and de Moivre, of the 1720s, on the factorization of $x^n - a^n$, and Simpson's 1759 dissection of an infinite series, since these results were applied by Sohnke in 1837 and others to determine the modular equation. Hermite's method of 1848 for construction of elliptic functions, also included in Chapter 6, is a topic now fallen into some obscurity. He started with the observation that a periodic entire function could not have a second period without becoming a constant. He then considered the ratio of two periodic entire functions, each with the same period, and enquired as to the conditions under which this function could have a second period. A particular case of his solution produced the Jacobi elliptic functions as the ratios of theta functions.

Hypergeometric functions, the topic of Chapter 7, encompass many important functions as special or limiting cases. As such, they were a major focus of nineteenth-century mathematical research, on which Gauss, Kummer, and Riemann wrote important papers. Gauss's early interest in hypergeometric functions was aroused by his observation that the complete elliptic integral $K$ could be expressed as a hypergeometric function of $k^2$. Then Riemann, in his 1856–57 lectures on hypergeometric functions, was able to express them as contour integrals; this insight led him to a proof that the

[1] Weil (1976).    [2] Stalker (1998).

modulus $k^2$ was invariant with respect to the principal congruence subgroup of level 2. Although it is a striking example of the way an integral changes as the contour moves around singular points, Riemann's fascinating proof of this theorem does not seem to have made its way into our textbooks or discussions.

Dedekind's impressive effort to provide a foundation for the theory of modular functions independent of elliptic functions is the main topic of Chapters 8 and 9. He described the fundamental domain of the modular group and defined the modular invariant by means of a conformal mapping of the fundamental domain onto the complex plane. He denoted this mapping function by $v$; a year later, Klein named it $J$, so that it has since been called Klein's $J$ invariant. Dedekind also introduced his famous $\eta$ function, defined as

$$\eta(\omega) = cv^{-\frac{1}{6}}(\omega)(1-v)^{-\frac{1}{8}}\left(\frac{dv}{d\omega}\right)^{\frac{1}{4}},$$

where $c$ denotes a constant. He then defined the elliptic modular functions $k^2$, $K$, $k'^2$, and $K'$ in terms of the $\eta$ function. Dedekind regretted the fact that, in the end, he was forced to employ theta functions to show that the $\eta$ function could be expressed as an infinite product. Hurwitz used Eisenstein series to overcome this difficulty.

Modern mathematicians define the $\eta$ function immediately as an infinite product, and, in fact, Dedekind himself had taken this step in his comments on some fragmentary results of Riemann. In these remarks, Dedekind proved, without the use of integrals, that the modular transformation of the $\eta$ function was expressible in terms of a finite arithmetical sum. Rademacher called this a Dedekind sum and developed its properties independent of its $\eta$ function origins. For example, he proved that the Dedekind sum could be expressed as a finite sum of values of the cotangent function. Chapter 9 includes Rademacher's first proof of this result, utilizing integrals; the proof given by Rademacher in his well-known book on Dedekind sums avoids this device.

Weierstrass and Klein elucidated the relationship of algebraic invariants and modular forms, the topic of Chapter 10. Klein gave the modern definition of the $J$ invariant:

$$J(\omega) = \frac{g_2^3(\omega)}{\Delta(\omega)}.$$

He then showed that his invariant-theoretic approach allowed him to apply Jacobi's results to obtain the infinite product for $\Delta$ and the Lambert series expansion for $g_2$.

Chapter 11 discusses modular and multiplier equations, connecting up Jacobi's work with that of Dedekind, Klein, Kiepert, and Hurwitz, and leading to Mordell's resolution of Ramanujan's conjectures. Even before he discovered elliptic functions, Jacobi used a trial-and-error method to find modular equations of orders 3 and 5. Upon discovering elliptic functions, Jacobi was able to establish the $p+1$ roots of the modular equation of prime order $p$. A few years later, Sohnke made use of these roots to construct modular equations of orders 7, 11, 13, 17, and 19. Jacobi also found the fifth-order multiplier equation. Three decades later, Joubert showed that methods analogous to those of Sohnke could be employed to develop multiplier equations of small prime order. Chapter 11 gives an expanded treatment of the methods of Sohnke and Joubert, who were mentioned briefly by Borwein and Borwein.[3]

Kiepert and Klein saw that $\sqrt[12]{\Delta(\omega)}$ could be interpreted as a multiplier in the transformation of an elliptic integral. Klein came to this conclusion by way of his invariant-theoretic approach, whereas Kiepert arrived at this idea through his use of Weierstrass elliptic functions to solve algebraic equations of the fifth degree. As Kiepert's teacher, Weierstrass had suggested that such a method might lead to a solution of a general equation of the fifth degree without its reduction to an equation of a special form. Although this suggestion did not pan out, it led Kiepert to the consideration of the transformation equation satisfied by $\sqrt[12]{\Delta(p\omega)}$. Klein gave the roots of the general equation of order $p$ up to a factor of a twelfth root of unity. His student, Hurwitz, determined these twelfth roots exactly in his 1881 thesis. It was this work of Hurwitz that Mordell employed to resolve the conjectures of Ramanujan on Euler products associated with modular forms.

The ideas of Hurwitz's innovative 1904 paper are well known and have appeared in many textbooks, perhaps because Serre discussed them in a 1957 seminar and soon thereafter presented them in his *Cours d'arithmétique*. Previous chapters concentrated primarily on methods now little known; by contrast, Chapter 12 presents familiar results and methods within modular function theory, but as Hurwitz first thought them out. Such mathematical results, as originally conceived and presented even one hundred or more years ago, may well reveal completely new or lost insights and avenues. A translation of Hurwitz's original 1904 paper is included as an appendix to this book, by which the reader may see that the ideas contained in old mathematical papers are often very accessible.

Ramanujan may be seen as the founder of an essential aspect of the theory of modular forms: the theory of Euler products, the topic of Chapter 13. It is generally well known that Ramanujan conjectured that the Dirichlet series associated with $\Delta(\omega)$ had an Euler product. However, his work in this area was much more extensive, including his statement that similar Euler products must exist for $\sqrt{\Delta}$, $g_2\Delta$, $g_3\Delta$, $g_2^2\Delta$, $g_2 g_3\Delta$, $g_2^2 g_3\Delta$. Moreover, he correctly surmised that linear combinations of modular forms of a given level had Euler products even when the original modular forms did not. This led Birch to remark in 1975[4] that Ramanujan's insight into the arithmetic of modular forms was greater than initially realized. It is often mentioned that Mordell proved Ramanujan's conjecture on $\Delta(\omega)$, usually in the context of Hecke's work. But again, as we discuss in Chapter 13, Mordell also proved Ramanujan's conjectures that other related modular forms had Euler products.

Hecke's work on Dirichlet series and modular forms, that is, series and functions of signature $\{\lambda, k, \gamma\}$, has been thoroughly treated in Berndt and Knopp's recent book.[5] Consequently, I have confined Chapter 14 to the background to and foundation for Hecke's work and a key result of Hecke on the dimension of the space of functions of signature $\{2, k, 1\}$. Chapter 15 is devoted to Hecke's little-known work on representations of integers as sums of squares, with additional mention of Glaisher's contributions in this area. In lectures at Princeton and Michigan in 1938, Hecke showed how

---

[3] Borwein and Borwein (1987).    [4] Birch (1975).    [5] Berndt and Knopp (2008).

Dirichlet series of signature $\{2, k, 1\}$, with $k$ a positive integer, could be applied to the problem of the number of representations of an integer as a sum of an even number of squares. Hecke is not mentioned in Grosswald's encyclopedic work of 1983[6] on sums of squares; Berndt and Knopp mention this work of Hecke without too much detail.

Hecke operators on modular forms for the full modular group is the topic of Chapter 16, with the exposition closely following Hecke's own treatment of the topic. Thus, this chapter completes the narrative of the previous three chapters on the connection between modular forms and Dirichlet series. Because Hecke was unaware that his operators were Hermitian, his work is somewhat labored and incomplete; still, it is of interest to note how much he could do under this constraint. Hecke's student Petersson showed that the operators were Hermitian, thus greatly simplifying the theory.

In this book we can see in some detail that a study of the relevant works of past centuries can greatly expand one's mathematical perspective and tool box. In addition, as one studies the development of the theory of modular functions, one encounters acknowledged masters, such as Gauss, Dedekind, Hecke, and others. Yet neglecting the work of lesser-known mathematicians, such as Joubert, Kiepert, and Sohnke, would be regrettable, since they added significant insights and advances. Clearly, the rate of mathematical output has increased at least thirty- or forty-fold since the early twentieth century. Thus, there is a relatively small body of work dating from the nineteenth or earlier centuries, and of this body, very few papers are of lasting interest within a given topic. Reading those few papers is not excessively demanding, and the benefits of so doing, especially for one's teaching, far outweigh any inconvenience. With increasing electronic availability, study of the old works has been made even easier.

My next volume will focus on the developments in modular forms after Hecke's 1935–37 papers, discussing connections with topics such as quadratic forms, elliptic curves, and Ramanujan's conjectures on partitions. There exist many introductory works on such topics, so I will attempt to elucidate important and interesting aspects of the more advanced topics within a manageable number of pages.

I first thank my wife for typesetting and editing this work and for her valuable help in translating; note here that unattributed translations are mine. I am indebted to NFN Kalyan for creating the wonderful portrait of Gauss for the cover. Kalyan's work is constructed of ten layers of etched glass, illuminated by colored LED light. I also owe a great debt to Kieran Donaghue for his expert assistance in German translations. I am grateful for Bruce Atwood's and Zhitai Li's skillful construction and corrections of diagrams. Cindy Cooley and Chris Nelson provided indispensable help with library materials; Sarah Arnsmeier assisted me with secretarial work. Thanks to Paul Campbell for guidance on making the bibliography. Finally, I thank Ann Davies and Beloit College for supporting me during the process of preparing this book.

---

[6] Grosswald (1983).

# 1

# *The Basic Modular Forms of the Nineteenth Century*

## 1.1 The Modular Group

Nineteenth-century mathematicians did not always give precise definitions in their work, though from the context and examples, their meaning is usually quite discernible. We here offer some examples of the basic concepts created and studied by these mathematicians, such as modular forms and the multiplier system, and discuss how these ideas would now be classified and defined.

The modular group $SL_2(\mathbb{Z})$ consists of all $2 \times 2$ matrices with integer entries and determinant 1:

$$SL_2(\mathbb{Z}) = \left\{ \begin{pmatrix} a & b \\ c & d \end{pmatrix} \,\middle|\, a, b, c, d \in \mathbb{Z}, \ ad - bc = 1 \right\}. \tag{1.1}$$

It can be deduced from results proved in 1775 by Lagrange,[1] in connection with the reduction of quadratic forms, that the modular group can generated by

$$S = \begin{pmatrix} 1 & 1 \\ 0 & 1 \end{pmatrix} \quad \text{and} \quad T = \begin{pmatrix} 0 & -1 \\ 1 & 0 \end{pmatrix}. \tag{1.2}$$

Here the notation for $S$ and $T$ follows early researchers such as Mordell and Rademacher; some recent works have interchanged $S$ and $T$.

Observe that $S$ and $T$ generate $SL_2(\mathbb{Z})$ by an application of the Euclidean algorithm. Given a $2 \times 2$ matrix $\begin{pmatrix} a & b \\ c & d \end{pmatrix}$, suppose $c \neq 0$ and $|a| \geq |c|$. In that case, there exists an integer $n$ such that

$$S^n \begin{pmatrix} a & b \\ c & d \end{pmatrix} = \begin{pmatrix} a + cn & b + dn \\ c & d \end{pmatrix} \equiv \begin{pmatrix} a_1 & b_1 \\ c & d \end{pmatrix}, \tag{1.3}$$

where $0 \leq a_1 < |c|$. By multiplying the matrix on the far right-hand side of (1.3) by $T$, we can move the entries $c$ and $d$ to the first row:

$$T \begin{pmatrix} a_1 & b_1 \\ c & d \end{pmatrix} = \begin{pmatrix} -c & -d \\ a_1 & b_1 \end{pmatrix}. \tag{1.4}$$

---

[1] See Weil (1983) pp. 216–217 and 321.

Now if $a_1 \neq 0$, then the calculation indicated in (1.3) and (1.4) can be repeated until we obtain a matrix in which the entry in the first column and second row is 0. Since the determinant of all matrices in the modular group is 1, the matrix containing this entry must take the form

$$\begin{pmatrix} \pm 1 & \pm k \\ 0 & \pm 1 \end{pmatrix}.$$

Next, note that

$$S^{-k}\begin{pmatrix} \pm 1 & \pm k \\ 0 & \pm 1 \end{pmatrix} = \begin{pmatrix} \pm 1 & 0 \\ 0 & \pm 1 \end{pmatrix}$$

and

$$T^2 \begin{pmatrix} -1 & 0 \\ 0 & -1 \end{pmatrix} = \begin{pmatrix} 1 & 0 \\ 0 & 1 \end{pmatrix},$$

proving that $S$ and $T$ as given in (1.2) are the generators of $SL_2(\mathbb{Z})$.

Let $\mathbb{H}$ denote the set of all complex numbers $x + iy$, where $x$ is a real number and $y > 0$. The set $\mathbb{H}$ is called the upper half-plane. The elements of $SL_2(\mathbb{Z})$ act upon the elements of $\mathbb{C} \cup \{\infty\}$, on the complex plane, and on the point at infinity by means of the equation

$$\begin{pmatrix} a & b \\ c & d \end{pmatrix}(z) = \frac{az + b}{cz + d}, \quad \text{where } z \in \mathbb{C} \cup \{\infty\}. \tag{1.5}$$

Note that if $z \in \mathbb{H}$, then

$$\operatorname{Im} \frac{az + b}{cz + d} = \frac{\operatorname{Im} z}{|cz + d|^2} > 0. \tag{1.6}$$

Thus, (1.5) serves also to define the action of $SL_2(\mathbb{Z})$ on $\mathbb{H}$.

The modular group and some of its subgroups were used in unpublished works of Gauss in his theory of binary quadratic forms and in his work on the arithmetic-geometric mean of two complex numbers. However, it seems that Dedekind was the first mathematician to explicitly define the modular group and its fundamental domain. He did this in a paper of 1877, and two years later, Felix Klein singled out the congruence subgroups of the modular group as the concept capable of bringing order and focus to the theory of modular functions.

Let $\Gamma = \Gamma(1)$ designate the modular group $SL_2(\mathbb{Z})$ and, for a positive integer $N$, let

$$\Gamma(N) := \left\{ \begin{pmatrix} a & b \\ c & d \end{pmatrix} \in \Gamma \ \middle| \ \begin{pmatrix} a & b \\ c & d \end{pmatrix} \equiv \begin{pmatrix} 1 & 0 \\ 0 & 1 \end{pmatrix} \bmod N \right\}. \tag{1.7}$$

$\Gamma(N)$ is called the principal congruence subgroup of level $N$. Any subgroup of $\Gamma$ containing $\Gamma(N)$ for some $N$ is called a congruence subgroup of level $N$, or simply a congruence subgroup. Adolf Hurwitz, in his 1881 doctoral dissertation, explicitly defined the subgroup

$$\Gamma^0(N) := \left\{ \begin{pmatrix} a & b \\ c & d \end{pmatrix} \in \Gamma \ \middle| \ b \equiv 0 \bmod N \right\}. \tag{1.8}$$

Hurwitz also proved that the index of $\Gamma(N)$ in $\Gamma$ was given by

$$N^3 \prod_{p|N} \left( 1 - \frac{1}{p^2} \right), \tag{1.9}$$

and that the index of $\Gamma^0(N)$ in $\Gamma$ was

$$N \prod_{p|N} \left( 1 + \frac{1}{p} \right). \tag{1.10}$$

We observe that (1.10) also expresses the index of the subgroup

$$\Gamma_0(N) = \left\{ \begin{pmatrix} a & b \\ c & d \end{pmatrix} \in \Gamma \ \middle| \ c \equiv 0 \bmod N \right\}. \tag{1.11}$$

The subgroup (1.11) is called the Hecke subgroup of the modular group, even though it was used by Hurwitz, because Hecke employed it extensively in his researches. Hurwitz considered another useful congruence subgroup:

$$\Gamma_1(N) = \left\{ \begin{pmatrix} a & b \\ c & d \end{pmatrix} \ \middle| \ \begin{pmatrix} a & b \\ c & d \end{pmatrix} \equiv \begin{pmatrix} 1 & * \\ 0 & 1 \end{pmatrix} \bmod N \right\}. \tag{1.12}$$

The $*$ notation indicates that the entry is arbitrary and does not satisfy any congruence relation mod $N$. Note that the index of $\Gamma_1(N)$ in $\Gamma$ is

$$N^2 \prod_{p|N} \left( 1 - \frac{1}{p^2} \right). \tag{1.13}$$

Let $\Gamma'$ be a congruence subgroup of $\Gamma = SL_2(\mathbb{Z})$. Two points $z_1, z_2 \in \mathbb{H}$ are called $\Gamma'$-equivalent if there exists a $\gamma \in \Gamma'$ such that $\gamma(z_1) = z_2$. A fundamental domain for $\Gamma'$ is a closed region $F$ in $\mathbb{H}$ such that every point in $\mathbb{H}$ is $\Gamma'$-equivalent to a point in $F$ and no two points in the interior of $F$ are $\Gamma'$-equivalent. Sometimes the interior of $F$ is called the fundamental domain.

Dedekind found a fundamental domain for the modular group $\Gamma$; in his 1877 paper, he gave a description but not a diagram. A year later, Klein drew a picture of this, as well as pictures of fundamental domains of several congruence subgroups. The fundamental domain of the modular group as described by Dedekind and diagramed by Klein is represented in Figure 1.1, where the fundamental domain comprises the region including and enclosed by the two dark vertical lines and the arc of the circle between them. Dedekind and Klein discussed the subgroup $\Gamma(2)$; Klein's diagram of its fundamental domain is given in Figure 1.2, in which the fundamental domain is composed of the region including and enclosed by the two dark vertical lines and the two semicircles between them. Interestingly, Gauss had earlier drawn a diagram for a fundamental domain of $\Gamma(2)$ in a manuscript he did not publish.

An important subgroup of $\Gamma$, connected with a theta constant, given later in (1.40), is the group generated by $S^2$ and $T$. This is a congruence subgroup of level 2 and index 3. Denoted in Hecke's notation by $G(2)$, this subgroup can be shown to be a conjugate of $\Gamma_0(2)$:

$$G(2) = S^{-1}T \, \Gamma_0(2)T^{-1}S. \tag{1.14}$$

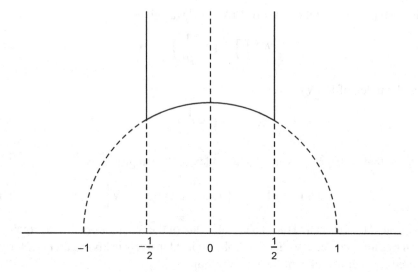

Figure 1.1. Fundamental Domain.

In their famous 1890–92 work on modular forms, Klein and R. Fricke denoted (1.14) by $\Gamma_3$, since it was a subgroup of $\Gamma$ of index 3. A fundamental domain for $G(2)$, or $\Gamma_3$, is shown in Figure 1.3; this fundamental domain consists of the region including and enclosed by the two vertical lines at $+1$ and $-1$ and the semicircle between them.

The points corresponding to $\{\infty\} \cup Q$, where $Q$ is the set of rational numbers, are called cusps. Observe that for any rational number $\frac{a}{c}$ in its reduced form, there exist integers $b$ and $d$ such that $ad - bc = 1$. Since

$$\frac{a\infty + b}{c\infty + d} = \frac{a}{c},$$

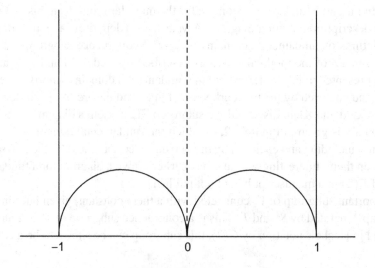

Figure 1.2. Fundamental Domain of $\Gamma(2)$.

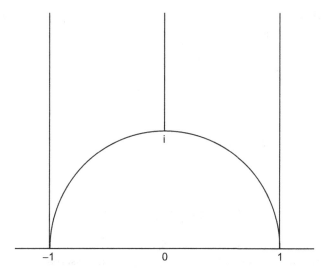

Figure 1.3. Fundamental Domain of $G(2)$.

it follows that a rational point is $\Gamma$-equivalent to infinity. Thus, the modular group permutes the cusps transitively. This may also be understood from the fact that the fundamental domain for $\Gamma$ has only one cusp, that is, the cusp at infinity. A subgroup $\Gamma'$ of $\Gamma$ does not necessarily act transitively on the cusps. This means that for a given $\Gamma'$, there is usually more than one $\Gamma'$-equivalence class among the cusps $\{\infty\} \cup Q$. These equivalence classes are called cusps of $\Gamma'$. Figure 1.2 illustrates that there are three inequivalent cusps of $\Gamma(2)$: $0, -1, \infty$. Observe that $-1$ is equivalent to 1, since $S^2 \in \Gamma(2)$ and $S^2(-1) = 1$. Note that $G(2)$ has two inequivalent cusps, one at $\infty$ and one at $-1$.

## 1.2 Modular Forms

Eisenstein, in his theory of elliptic functions, considered the series

$$G_k(\omega_1, \omega_2) = \sideset{}{'}\sum_{m,n} \frac{1}{(m\,\omega_1 + n\,\omega_2)^k}, \tag{1.15}$$

where $k \geq 2$ was an integer and $\omega_1, \omega_2$ were complex numbers such that $\text{Im } \frac{\omega_1}{\omega_2} > 0$. The summation was to be taken over all integers $m, n$ except $m = n = 0$. He showed that the series converged absolutely when $k \geq 3$. This implied that, with integers $a, b, c, d$ such that $ad - bc = 1$, if

$$\omega_1' = a\,\omega_1 + b\,\omega_2$$
$$\omega_2' = c\,\omega_1 + d\,\omega_2,$$

and if $k \geq 3$, then

$$G_k(\omega_1', \omega_2') = G_k(\omega_1, \omega_2). \tag{1.16}$$

Eisenstein noted that $G_k(\omega_1, \omega_2) \equiv 0$ when $k \geq 3$, with $k$ odd. Now if we write $\omega = \frac{\omega_1}{\omega_2}$ and $G_k(\omega) \equiv G_k(\omega, 1)$, then

$$G_k(\omega) = \sideset{}{'}\sum_{m,n} \frac{1}{(m\omega + n)^k} \qquad (1.17)$$

and 1.16 can be rewritten as

$$G_k(\omega') = (c\omega + d)^k G_k(\omega), \qquad (1.18)$$

where

$$\omega' = \frac{a\omega + b}{c\omega + d}, \quad ad - bc = 1. \qquad (1.19)$$

Since $G_{2k}(\omega + 1) = G_{2k}(\omega)$, $G_{2k}(\omega)$ has a Fourier series expansion with period one. This Fourier series was found by Hurwitz in his doctoral dissertation:

$$\begin{aligned}
G_{2k}(\omega) &= 2\zeta(2k) + 2\frac{(2\pi i)^{2k}}{(2k-1)!} \sum_{n=1}^{\infty} \sigma_{2k-1}(n)e^{2\pi i n \omega} \\
&= 2\zeta(2k)\left(1 + \frac{(-1)^k 4k}{B_{2k}} \sum_{n=1}^{\infty} \sigma_{2k-1}(n) e^{2\pi i n \omega}\right), \qquad (1.20)
\end{aligned}$$

where $\sigma_{2k-1}(n) = \sum_{d|n} d^{2k-1}$ and where $B_{2k}$ were Bernoulli numbers defined by

$$\frac{x}{e^x - 1} = \sum_{n=0}^{\infty} B_n \frac{x^n}{n!}.$$

Note that $B_1 = -\frac{1}{2}, B_{2m+1} = 0, m \geq 1$. The series $G_{2k}(\omega)$ is called an Eisenstein series. The discriminant function

$$\Delta(\omega) = \left(60G_4(\omega)\right)^3 - 27\left(140G_6(\omega)\right)^2 \qquad (1.21)$$

played an important role in Weierstrass's theory of elliptic functions. Weierstrass employed his theory of sigma functions (equivalent to theta functions) to show that

$$\Delta(\omega) = (2\pi)^{12} e^{2\pi i \omega} \prod_{n=1}^{\infty} (1 - e^{2\pi i n \omega})^{24}. \qquad (1.22)$$

Observe that (1.18) implies that

$$\Delta\left(\frac{a\omega + b}{c\omega + d}\right) = (c\omega + d)^{12} \Delta(\omega), \quad ad - bc = 1. \qquad (1.23)$$

As a consequence of (1.18) and (1.22), we can see that

$$j(\omega) = \frac{1728\left(60G_4(\omega)\right)^3}{\Delta(\omega)} \qquad (1.24)$$

satisfies, for integers $a, b, c, d$ with $ad - bc = 1$, the relation

$$j\left(\frac{a\omega + b}{c\omega + d}\right) = j(\omega).$$ (1.25)

The function (1.24) was considered by Hermite in a 1858 paper on modular equations. In 1877, Dedekind defined $v(\omega) = \frac{1}{1728}j(\omega)$ by means of a conformal mapping of the fundamental domain of $\Gamma(1)$. Klein published a paper in 1878 in which he independently rediscovered this function, (1.24), through invariant theory. From that time onward, the function became known as Klein's $J$ function, where $J(\omega) = v(\omega) = \frac{1}{1728}j(\omega)$. It also appears that Gauss was moving toward this function, in the context of his work in the theory of quadratic forms, though he did not explicitly define it. Hermite gave the Fourier expansion

$$j(\omega) = e^{-2\pi i\omega} - 744 + 196884e^{2\pi i\omega} + \cdots.$$ (1.26)

Another function, $k^2$, arose naturally within the theory of elliptic functions when the elliptic function was regarded as an elliptic integral. Hermite gave the relation between $k^2$ and $j$:

$$j = 256\frac{(k^4 - k^2 + 1)^3}{k^2(1 - k^2)^2}.$$ (1.27)

In a paper of 1828, Jacobi proved that $k^2$ satisfied an equation similar to (1.25), provided that

$$\begin{pmatrix} a & b \\ c & d \end{pmatrix} \in \Gamma(2).$$ (1.28)

Next note that we may define $\sqrt[12]{\Delta(\omega)}$, a function studied by Kiepert, Hurwitz, and Klein, from (1.22) by using the equation

$$\sqrt[12]{\Delta(\omega)} = 2\pi\, e^{\frac{\pi i\omega}{6}} \prod_{n=1}^{\infty}(1 - e^{2\pi in\omega})^2.$$ (1.29)

By writing $U = \begin{pmatrix} a & b \\ c & d \end{pmatrix} \in \Gamma(1)$, (1.23) will imply that

$$\sqrt[12]{\Delta(U\,\omega)} = v(U)(c\,\omega + d)\,\sqrt[12]{\Delta(\omega)},$$ (1.30)

where $v(U)$ denotes a suitable twelfth root of unity. Now it is clear from (1.29) that

$$v(S) = e^{\frac{\pi i}{6}},$$ (1.31)

where $S$ is as in (1.2), because

$$\sqrt[12]{\Delta(\omega + 1)} = e^{\frac{\pi i}{6}}\,\sqrt[12]{\Delta(\omega)}.$$ (1.32)

Similarly, with $T$ as in (1.2),

$$v(T) = e^{\frac{3\pi i}{2}};$$ (1.33)

this follows from (1.30), since

$$\sqrt[12]{\Delta\left(-\frac{1}{\omega}\right)} = v(T)\,\omega\,\sqrt[12]{\Delta(\omega)}. \tag{1.34}$$

Taking $\omega = i$ in (1.34) yields the value of $v(T)$ in (1.33). Note that if $U$ and $V$ belong to the modular group $\Gamma(1)$, then (1.30) implies that

$$v(UV) = v(U)v(V). \tag{1.35}$$

Now since $\Gamma(1)$ is generated by $S$ and $T$, it is possible in principle to find the value of $v(U)$ for any $U \in T(1)$. In fact, in his doctoral thesis of 1881, Hurwitz proved the transformation formula:

For any integers $a, b, c, d$ with $ad - bc = 1$,

$$\sqrt[12]{\Delta(a\,\omega_1 + b\,\omega_2, c\,\omega_1 + d\,\omega_2)} = e^{\left((1-c^2)(db+3d(c-1)+c+3)+c(d+a-3)\right)\frac{i\pi}{6}}\sqrt[12]{\Delta(\omega_1, \omega_2)}. \tag{1.36}$$

Four years earlier, Dedekind gave the transformation formula for

$$\eta(\omega) = e^{\frac{\pi i \omega}{12}} \prod_{n=1}^{\infty} (1 - e^{2\pi i n \omega}), \tag{1.37}$$

except that he stated his result in terms of Dedekind sums, to be defined later. A famous particular case of Dedekind's result is

$$\eta\left(-\frac{1}{\omega}\right) = \sqrt{\frac{\omega}{i}}\,\eta(\omega). \tag{1.38}$$

The function in (1.37) is now known as the Dedekind $\eta$ function. The theta constants can be written in terms of $\eta$ functions; for example,

$$\theta_3(\omega) = \prod_{n=1}^{\infty} \left(1 - e^{i\pi i n \omega}\right)\left(1 + e^{(2n-1)\pi i \omega}\right)^2 \tag{1.39}$$

$$= \frac{\eta^2\left(\frac{\omega+1}{2}\right)}{\eta^2(\omega+1)}.$$

Note that a theta function is a function of two variables: $x$ and $q = e^{\pi i \omega}$, as given in (3.189). When $x = 0$, we sometimes refer to the function as a theta constant. Some powers of theta constants and their ratios are types of modular forms or modular functions.

The infinite product for $\theta_3(\omega)$ can also be expressed as a series:

$$\theta_3(\omega) = \sum_{n=-\infty}^{\infty} e^{\pi i n^2 \omega} = 1 + 2\sum_{n=1}^{\infty} e^{\pi i n^2 \omega}. \tag{1.40}$$

By the year 1808, Gauss knew that the infinite product (1.39) equalled the series (1.40), but he did not publish his results along these lines. In 1828, Jacobi rediscovered this fact. Gauss applied Fourier series to prove that

$$\theta_3\left(-\frac{1}{\omega}\right) = \sqrt{\frac{\omega}{i}}\,\theta_3(\omega). \tag{1.41}$$

It is not known exactly when Gauss proved (1.41). However, on the basis of the position of this work in his notebooks and other factors, it appears likely that he did this work around 1808. In 1817, Cauchy published (1.41) in a paper on waves using Fourier transforms. In 1828, Jacobi proved (1.41) based on his theory of elliptic functions. Note that

$$\theta_3(\omega + 2) = \theta_3(\omega). \tag{1.42}$$

Gauss and Jacobi defined other theta functions:

$$\theta_0(\omega) = \sum_{n=-\infty}^{\infty} (-1)^n\, e^{\pi i n^2 \omega}, \tag{1.43}$$

$$\theta_1(\omega) = \sum_{n=-\infty}^{\infty} (-1)^n\, e^{2\pi i \left(\frac{2n+1}{2}\right)^2 \omega}, \tag{1.44}$$

$$\theta_2(\omega) = \sum_{n=-\infty}^{\infty} e^{2\pi i \left(\frac{2n+1}{2}\right)^2 \omega}. \tag{1.45}$$

They showed that, with $S$ defined by (1.2),

$$\begin{aligned}
\theta_0(S\omega) &= \theta_0\left(-\frac{1}{\omega}\right) = \sqrt{\frac{\omega}{i}}\,\theta_2(\omega), \\
\theta_2(S\omega) &= \theta_2\left(-\frac{1}{\omega}\right) = \sqrt{\frac{\omega}{i}}\,\theta_0(\omega), \\
\theta_1(S\omega) &= \theta_1\left(-\frac{1}{\omega}\right) = -\sqrt{i\omega}\,\theta_1(\omega).
\end{aligned} \tag{1.46}$$

In 1828, Jacobi proved a result that can be used to show that $k^2(\omega)$ is invariant with respect to $\Gamma(2)$:

$$k^2(\omega) = \frac{\theta_2^4(\omega)}{\theta_3^4(\omega)} = e^{i\pi\omega} \prod_{n=1}^{\infty} \left(\frac{1 + e^{2n\pi i\omega}}{1 + e^{(2n-1)\pi i\omega}}\right)^8. \tag{1.47}$$

Key definitions serve to categorize functions such as the $\theta$, $\eta$, $\Delta$, $k^2$, and so on: A function is denoted a meromorphic modular form of integer weight $k$, belonging to $\Gamma(1)$ or with respect to $\Gamma(1)$, if: $f$ is meromorphic on the upper half-plane $\mathbb{H}$; and

$$f\left(\frac{a\omega + b}{c\omega + d}\right) = (c\omega + d)^k f(\omega), \tag{1.48}$$

for $\omega \in \mathbb{H}$ with $\begin{pmatrix} a & b \\ c & d \end{pmatrix} \in \Gamma(1)$; and $f(\omega)$ has a Fourier expansion

$$f(\omega) = \sum_{n=-m}^{\infty} a_n e^{2\pi i n \omega}. \tag{1.49}$$

In the case $k = 0$, $f(\omega)$ is called a modular function belonging to $\Gamma(1)$. Note that the function $z = e^{2\pi i \omega}$ maps the vertical strip of the fundamental domain of $\Gamma(1)$ into the unit disc $|z| < 1$; also note that

$$f(\omega) = \sum_{n=-m}^{\infty} a_n z^n.$$

When $m > 0$, we say that $f$ has a pole of order $m$ at $\infty$ or $i\infty$; when $-m \geq 0$, we say that $f$ is holomorphic at $\infty$. Now when $f$ is holomorphic in the upper half-plane and holomorphic at $\infty$ (or $i\infty$), then $f$ is called a modular form of weight $k$ belonging to $\Gamma(1)$. And if $f$ is a modular form of weight $k$ and $a_0 = 0$, then $f$ is known as a cusp form of weight $k$.

Denoting the vector space of modular forms of weight $k$ by $M_k(\Gamma)$ and the space of cusp forms of weight $k$ by $S_k(\Gamma)$, we see that $S_k(\Gamma)$ is a subspace of $M_k(\Gamma)$. Observe that $j(\omega)$ is a modular function and that

$$G_k(\omega) \in M_k(\Gamma),$$
$$\Delta(\omega) \in S_{12}(\Gamma),$$
$$G_{k-12}(\omega)\Delta(\omega) \in S_k(\Gamma) \text{ for } k > 12.$$

Examples of functions in (1.36) and (1.38) lead us to an extension of modular forms. We denote $v$ as a multiplier system of weight $k$ for $\Gamma(1)$ if

$$\text{for } i = 1, 2, 3, \quad U_i = \begin{pmatrix} a_i & b_i \\ c_i & d_i \end{pmatrix}, \text{ and } U_3 = U_1 U_2,$$

with $U_1$, $U_2$ arbitrary elements of $\Gamma(1)$, we have $|v(U_i)| = 1$ and

$$v(U_3)(c_3\,\omega + d_3)^k = v(U_1)v(U_2)(c_1 U_2\,\omega + d_1)^k (c_2\,\omega + d_2)^k. \tag{1.50}$$

In the case when $k$ is an integer, (1.50) reduces to

$$v(U_3) = v(U_1)v(U_2). \tag{1.51}$$

Next, let $v(S) = e^{2\pi i t}$, where $t$ is a real number and $S$ is given by (1.2). A function $f$, meromorphic on $\mathbb{H}$ and with the series expansion

$$f(\omega) = \sum_{n=-m}^{\infty} a_n e^{2\pi i (n+t)\omega}, \tag{1.52}$$

is designated a meromorphic modular form of weight $k$ with multiplier system $v$ if

$$f(U\,\omega) = v(U)(c\,\omega + d)^k f(\omega), \tag{1.53}$$

where

$$U = \begin{pmatrix} a & b \\ c & d \end{pmatrix} \in \Gamma(1). \tag{1.54}$$

If $-m \geq 0$ in (1.52) and if $f$ is holomorphic in $\mathbb{H}$, then $f$ is called a modular form of weight $k$ with a multiplier system $v$; but if $-m \geq 1$ in (1.52) and if $f$ is holomorphic in $\mathbb{H}$, then $f$ is called a cusp form with a multiplier system $v$.

The function $\sqrt[12]{\Delta(\omega)}$ is an example of a cusp form of weight 1 with multiplier system $v$ such that $v(S) = e^{\frac{\pi i}{6}}$ and $v(T) = e^{\frac{3\pi i}{2}}$. It was proved by Rademacher that $\eta(\omega)$ is an example of a cusp form of weight $\frac{1}{2}$ with multiplier system $v$ such that $v(S) = e^{\frac{\pi i}{12}}$ and $v(T) = e^{-\frac{i\pi}{4}}$.

The function $k^2(\omega)$ is invariant under the action of $\Gamma(2)$. The fundamental domain of $\Gamma(2)$ has three inequivalent cusps at $-1, 0$, and $\infty$. In order to define a meromorphic modular form for $\Gamma(2)$, conditions similar to (1.49) should hold at these cusps. Even more generally, we may let $G$ be a group of level $N$ so that $G$ may have more than one cusp. Apart from a condition analogous to (1.48), we will require that for any $U$, as in (1.54),

$$f(U\omega)(c\omega + d)^{-k} = \sum_{n=-m}^{\infty} a_n e^{\frac{2\pi i n \omega}{N}}. \tag{1.55}$$

Then, as before, we can define modular functions, modular forms, and cusp forms of weight $k$ with respect to the group $G$. By checking the behavior at the cusps, it can be shown that $k^2(\omega)$ is a modular function for $\Gamma(2)$. See the exercise at the end of this chapter.

Hecke defined another class of functions $f$, and he designated them as having signature $(\lambda, k, \gamma)$:

$$f(\omega) = \sum_{n=0}^{\infty} a_n e^{\frac{2\pi i n \omega}{\lambda}}, \tag{1.56}$$

$$\frac{f\left(-\frac{1}{\omega}\right)}{(-i\omega)^k} = \lambda f(\omega), \tag{1.57}$$

$$f(x + iy) = O(y^{-c}) \tag{1.58}$$

as $y \to 0^+$ with $c$ a constant.

It can be seen that when $k$ is even and $\geq 4$, the vector space of functions of signature $(1, k, (-1)^{\frac{k}{2}})$ is identical with the space $M_k(\Gamma(1))$. An example of a function of signature $(2, k, 1)$ would be $\theta_3^k(\omega)$. Note also that $\sqrt{\Delta(\omega)}$ is an example of a function of signature $(2, 6, 1)$.

## 1.3 Exercises

Check the behavior of the function $k^2(\omega)$, defined by (1.47), at the three cusps $0, -1$, and $\infty$:

1. Use (1.47) to show that $\lambda(\omega) \to 0$ as $\omega \to i\infty$.

2. Use the first equation in (1.47) along with (1.41), (1.46), and (3.197) to show that

$$k^2\left(-\frac{1}{\omega}\right) = 1 - k^2(\omega).$$

Deduce that $k^2 \to 1$ as $\omega \to 0$ in the fundamental domain.

3. Prove that

$$k^2(\omega + 1) = \frac{k^2(\omega)}{k^2(\omega) - 1}.$$

Deduce that as $\omega \to -1$ in the fundamental domain, $k^2(\omega) \to \infty$. See (5.31).

# 2

---

# *Gauss's Contributions to Modular Forms*

## 2.1 Early Work on Elliptic Integrals

The early study of modular functions and forms is linked with the theory of elliptic functions, and these functions initially arose from the study of elliptic integrals, that is, integrals of the form

$$\int \frac{dx}{\sqrt{p(x)}}, \tag{2.1}$$

where $p(x)$ is a polynomial of degree 3 or 4, without repeated roots. An important example of an elliptic integral is the lemniscatic integral:

$$\int \frac{dx}{\sqrt{1 - x^4}}. \tag{2.2}$$

This integral is related to the lemniscate, defined by the equation

$$(x^2 + y^2)^2 = x^2 - y^2 \quad \text{or} \quad r^2 = \cos 2\theta. \tag{2.3}$$

Suppose $P$ is a point on the lemniscate (Figure 2.1) with polar coordinates $(r, \theta)$. Then the length of the arc $OP$ is given by the integral (2.2). This follows from the fact that, in polar coordinates, the infinitesimal arclength $ds$ is given by

$$ds^2 = dr^2 + r^2 \, d\theta^2.$$

Differentiating the polar equation (2.3) with respect to $\theta$, we get

$$2r \, dr = -2 \sin 2\theta \, d\theta$$

and therefore

$$ds^2 = dr^2 + \frac{r^4}{1 - \cos^2 2\theta} \, dr^2 = \frac{dr^2}{1 - r^4};$$

$$\text{arc } OP = \int_0^r ds = \int_0^r \frac{dt}{\sqrt{1 - t^4}}. \tag{2.4}$$

Thus, the length of the arc of the lemniscate is given by the lemniscatic integral (2.4).

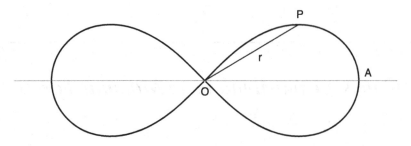

Figure 2.1. Leminiscate.

In 1691, Jacob Bernoulli (1654–1705) discovered integral (2.2) in connection with a problem dealing with an elastic band. In 1694, he found the lemniscatic curve and its arclength, but he was unable to evaluate it in terms of the elementary functions. He therefore evaluated this integral as an infinite series, first expanding $(1 - x^4)^{-\frac{1}{2}}$ as an infinite series by use of the binomial theorem, and then integrating term-by-term. He thus had for $|k| \le 1$

$$\int_0^k (1 - x^4)^{-\frac{1}{2}} \, dx = \int_0^k \left( 1 + \sum_{n=1}^{\infty} \frac{\frac{1}{2} \cdot \frac{3}{2} \cdots \frac{2n-1}{2}}{n!} x^{4n} \right) dx$$

$$= k + \sum_{n=1}^{\infty} \frac{1 \cdot 3 \cdots (2n-1) \, k^{4n+1}}{n! \, 2^n \, (4n+1)}.$$

In 1718, the noted amateur mathematician Giulio Carlo Fagnano (1682–1766) published a paper explaining how to divide the length of the arc $OA$ into 2, 3, or 5 equal parts using only straightedge and compass. He accomplished this by showing that if

$$r = \frac{2u \sqrt{1 - u^4}}{1 + u^4}, \tag{2.5}$$

then

$$\int_0^r \frac{dt}{\sqrt{1 - t^4}} = 2 \int_0^u \frac{dt}{\sqrt{1 - t^4}}. \tag{2.6}$$

Thus, the arclength corresponding to the radius vector $r$ is twice the arclength corresponding to the radius vector $u$, where $u$ and $r$ are related by (2.5).[1] Note that $u$ and $r$ may be obtained from one another by successively solving only quadratic equations. This allows for geometric constructibility, meaning that a lemniscate can be halved or doubled by straight-edge and compass. For example, to find the point $Q$ on the curve corresponding to the radius vector $u$ such that $2 \operatorname{arc} OQ = \operatorname{arc} OA$, we take $r = 1$ and solve equation (2.5). Trisection and quinsection are also possible but require a little more work.

Fagnano's papers on the lemniscate went unnoticed until December 1751, when Euler (1707–1783) read them in Fagnano's collected works, *Produzioni Mathematiche*.

---

[1] Fagnano (1750) vol. 2, pp. 304–313. See also pp. 293–297.

Euler immediately perceived the significance of Fagnano's results; he himself had done some work on elliptic integrals in the 1730s and knew that it was unlikely that an elliptic integral could be evaluated in terms of elementary functions. Euler saw that the particular cases in Fagnano's work could give a new direction to the study of elliptic integrals. He observed that (2.5) could be seen as a generalization of the double-angle formula for the sine function

$$s = 2t \sqrt{1 - t^2}, \tag{2.7}$$

where $s = \sin 2\phi$ and $t = \sin \phi$. Note that $\phi$ gives the arclength of the unit circle, and that

$$\phi = \int_0^{\sin \phi} \frac{dt}{\sqrt{1 - t^2}} \tag{2.8}$$

is the analog of (2.4). Now note that the double-angle formula is a particular case of the addition formula for the sine function

$$\sin(\phi + \psi) = \sin \phi \, \cos \psi + \cos \phi \, \sin \psi. \tag{2.9}$$

Moreover, the integral in (2.8) is actually $\phi = \arcsin t$; by taking $\psi = \arcsin s$, Euler could rewrite (2.9) as

$$\phi + \psi = \int_0^s \frac{du}{\sqrt{1 - u^2}} + \int_0^t \frac{du}{\sqrt{1 - u^2}} = \int_0^z \frac{du}{\sqrt{1 - u^2}} = \arcsin z,$$

where

$$z = s\sqrt{1 - t^2} + t\sqrt{1 - s^2} = \sin \psi \, \cos \phi + \sin \phi \, \cos \psi.$$

In 1752, soon after reading Fagnano's papers, Euler found the addition formula for the lemniscatic integral. In a paper of 1753,[2] "De integratione aequationis differentialis $\frac{mdx}{\sqrt{1-x^4}} = \frac{ndy}{\sqrt{1-y^4}}$," he presented the addition formula in the form:

$$\int_0^u \frac{dt}{\sqrt{1 - t^4}} + \int_0^v \frac{dt}{\sqrt{1 - t^4}} = \int_0^r \frac{dt}{\sqrt{1 - t^4}}, \tag{2.10}$$

where

$$r = \frac{u\sqrt{1 - v^4} + v\sqrt{1 - u^4}}{1 + u^2 \, v^2}. \tag{2.11}$$

Note that by setting $u = v$ in the addition formula (2.11), Euler showed how to double the lemniscatic arclength (2.4). We do not know how he discovered this amazing result, because his proof took the form of a mere verification. Euler also viewed (2.11) as a solution of the differential equation

$$\frac{dy}{\sqrt{1 - y^4}} = \frac{dx}{\sqrt{1 - x^4}}. \tag{2.12}$$

---

[2] Eu.I-20, pp. 58–79, especially 65–66.

In a May 30, 1752 letter to his friend Goldbach,[3] Euler wrote that, whereas the equation

$$\frac{dy}{\sqrt{1-y^2}} = \frac{dx}{\sqrt{1-x^2}}$$

had the integral

$$y^2 + x^2 = c^2 + 2xy\sqrt{1-c^2},$$

with $c$ an arbitrary constant, the equation (2.12) had the integral

$$y^2 + x^2 = c^2 + 2xy\sqrt{1-c^4} - c^2x^2y^2. \qquad (2.13)$$

And when we solve this equation as a quadratic in $y$, the result is

$$y = \frac{x\sqrt{1-c^4} \pm c\sqrt{1-x^4}}{1+x^2c^2}, \qquad (2.14)$$

the same as (2.11), except for the $\pm$.

In the same paper of 1753,[4] Euler considered the more general differential equation

$$\frac{dy}{\sqrt{1+my^2+ny^4}} = \frac{dx}{\sqrt{1+mx^2+nx^4}}. \qquad (2.15)$$

He proved that the integral of this differential equation was given by

$$c^2 - x^2 - y^2 + nc^2x^2y^2 + 2xy\sqrt{1+mc^2+nc^4} = 0. \qquad (2.16)$$

Euler then obtained the addition formula by solving the quadratic in $y$ to get

$$y = \frac{x\sqrt{P(c)} \pm c\sqrt{P(x)}}{1-nc^2x^2}, \qquad (2.17)$$

where $P(x) = 1 + mx^2 + nx^4$. Though Euler did not say so explicitly, it is then an immediate generalization of (2.10) that if

$$\int_0^x \frac{dt}{\sqrt{P(t)}} \pm \int_0^c \frac{dt}{\sqrt{P(t)}} = \int_0^y \frac{dt}{\sqrt{P(t)}},$$

then $y$ would be given by (2.17).

He also found the general integral of the differential equation (2.15) when $P(x)$ was replaced by

$$p(x) = A + 2Bx + Cx^2 + 2Dx^3 + Ex^4.$$

In a paper of 1765,[5] Euler attempted to prove that by means of a fractional linear transformation, $p(x)$ could always be reduced to the particular case

$$P(x) = 1 + mx^2 + nx^4.$$

---

[3] Fuss (1968) vol. 1, pp. 564–568.    [4] Eu.I-20, pp. 63–67.
[5] Eu.I-20, pp. 302–317, especially pp. 303–304. The paper is entitled, "Integratio aequationis $\dfrac{dx}{\sqrt{A+Bx+Cx^2+Dx^3+Ex^4}} = \dfrac{dy}{\sqrt{A+By+Cy^2+Dy^3+Ey^4}}$."

There was a gap in his reasoning: it was possible within his reduction for $m$ and $n$ to be complex, whereas Euler intended them to be real. This defect was finally removed by Legendre in a work of 1792.[6] In another work, Legendre showed that any elliptic integral

$$\int \frac{A(x)\,dx}{\sqrt{p(x)}},$$

where $p(x)$ was a fourth-degree polynomial in $x$ and $A(x)$ was a rational function in $x$, could be reduced to the consideration of integrals of the first, second, and third kinds, respectively[7]:

$$F(k, x) = \int_0^x \frac{dt}{\sqrt{(1 - t^2)(1 - k^2 t^2)}}, \tag{2.18}$$

$$E(k, x) = \int_0^x \frac{\sqrt{1 - k^2 t^2}}{\sqrt{1 - t^2}}\,dt, \tag{2.19}$$

$$\prod(n,\, k,\, x) = \int_0^x \frac{dt}{(1 + nt^2)\sqrt{(1 - t^2)(1 - k^2 t^2)}}. \tag{2.20}$$

## 2.2 Landen and Legendre's Quadratic Transformation

As noted by Mittag-Leffler,[8] the eighteenth-century work on elliptic integrals stood on two pillars: the addition theorem of Euler and the quadratic transformation of elliptic integrals, due to John Landen (1719–1790). Landen published this transformation in 1771 and then elaborated upon it in 1775.[9] Legendre gave a nice formulation of Landen's transformation[10]:

$$\text{If} \quad \sin(2\phi - \theta) = \lambda \sin\theta, \quad \text{then}$$

$$(1 + \lambda) \int_0^\theta (1 - \lambda^2 \sin^2 u)^{-\frac{1}{2}}\,du = 2 \int_0^\phi (1 - k^2 \sin^2 u)^{-\frac{1}{2}}\,du,$$

$$\text{where} \quad k = \frac{2\sqrt{\lambda}}{1 + \lambda}.$$

Now if we set

$$t = \sin u, \quad \frac{dt}{\sqrt{1 - t^2}} = du, \quad x = \sin\phi, \quad y = \sin\theta, \quad k' = \sqrt{1 - k^2},$$

the theorem takes the form:

$$\text{If} \quad y = (1 + k') \frac{x\sqrt{1 - x^2}}{\sqrt{1 - k^2 x^2}} \quad \text{and} \quad k = \frac{2\sqrt{\lambda}}{1 + \lambda}, \quad \text{then}$$

$$(1 + \lambda) \int_0^y \frac{dt}{\sqrt{(1 - t^2)(1 - \lambda^2 t^2)}} = 2 \int_0^x \frac{dt}{\sqrt{(1 - t^2)(1 - k^2 t^2)}}.$$

---

[6] Legendre (1792) pp. 9–10.    [7] Legendre (1811–1817) vol. 1, p. 19.
[8] Mittag-Leffler (1923), an English translation of Mittag-Leffler's 1876 paper written in Swedish.
[9] Landen (1775).    [10] Legendre (1811–1817) vol. 1, p. 81.

This result may also be formulated in terms of a differential equation similar to (2.12) and (2.15):

$$\text{If } y = \frac{(1 + k')x\sqrt{1 - x^2}}{\sqrt{1 - k^2x^2}}, \quad k' = \sqrt{1 - k^2}, \quad k = \frac{2\sqrt{\lambda}}{1 + \lambda}, \tag{2.21}$$

$$\text{then } \frac{(1 + \lambda)\,dy}{\sqrt{(1 - y^2)(1 - \lambda^2 y^2)}} = \frac{2\,dx}{\sqrt{(1 - x^2)(1 - k^2 x^2)}}. \tag{2.22}$$

Thus, the addition theorem for elliptic integrals and the quadratic transformation (2.21) may be viewed as a particular cases of the problem:

Suppose $p(x)$ and $q(x)$ are polynomials of degree 3 or 4 and without repeated roots. Find an algebraic relation $F(x, y) = 0$ that effects a transformation of the elliptic differential $\frac{dx}{\sqrt{p(x)}}$ into the elliptic differential $\frac{dy}{\sqrt{q(y)}}$.

Note that $F(x, y) = 0$ is an integral of the differential equation $\frac{dx}{\sqrt{p(x)}} = \frac{dy}{\sqrt{q(y)}}$.

Once the transformation (2.21) is known, it is not difficult to show that (2.22) will also be satisfied. In fact, this involves only elementary algebra and verifying the formulas:

$$1 - y^2 = \frac{(1 - (1 + k')x^2)^2}{1 - k^2x^2},$$

$$1 - \lambda^2 y^2 = \frac{(1 - (1 - k')x^2)^2}{1 - k^2x^2},$$

$$\sqrt{(1 - y^2)(1 - \lambda^2 y^2)} = \frac{1 - 2x^2 + k^2x^4}{1 - k^2x^2}, \tag{2.23}$$

$$(1 + \lambda)\,dy = \frac{2(1 - 2x^2 + k^2x^4)\,dx}{(1 - k^2x^2)\sqrt{(1 - x^2)(1 - k^2x^2)}}. \tag{2.24}$$

Now (2.22) follows from (2.23) and (2.24).

### 2.3 Lagrange's Arithmetic-Geometric Mean

In a January 3, 1777 letter[11] to his friend Condorcet, Joseph-Louis Lagrange (1735–1813) wrote that he was intrigued by Landen's result, though he had not completely verified it. Then in a 1784–1785 paper, "Sur novelle méthode de calcul intégral," Lagrange applied Landen's transformation to the approximate numerical evaluation of an elliptic integral. It was in this connection that he discovered the concept of the arithmetic-geometric mean of two numbers. Lagrange expressed Landen's transformation in a remarkable form[12]:

If $p > q > 0$, with

$$p' = p + \sqrt{p^2 - q^2}, \quad q' = p - \sqrt{p^2 - q^2} \tag{2.25}$$

---

[11]  Lagrange (1867–1892) vol. 14, p. 41.
[12]  Lagrange (1867–1892) vol. 2, pp. 253–312, especially pp. 264–270.

and if

$$R = R(p, q, y) = \sqrt{(1 \pm p^2 y^2)(1 \pm q^2 y^2)}, \quad y' = \frac{yR}{1 \pm q^2 y^2}, \qquad (2.26)$$

then

$$\frac{dy}{R} = \frac{dy'}{R'}, \qquad (2.27)$$

where $R' = R(p', q', y')$.

Lagrange also noted that $p$ and $q$ were, respectively, the arithmetic mean and the geometric mean of $p'$ and $q'$, that is, $p = \frac{p'+q'}{2}$ and $q = \sqrt{p'q'}$. For the purpose of approximate evaluation of an elliptic integral, Lagrange defined two sequences: Let $p_0 = p$, $q_0 = q$ with $p > q$, and for any positive or negative integer $n$, set

$$p_n = p_{n-1} + \sqrt{p_{n-1}^2 - q_{n-1}^2}, \quad q_n = p_{n-1} - \sqrt{p_{n-1}^2 - q_{n-1}^2}, \qquad (2.28)$$

or

$$p_n = \frac{p_{n+1} + q_{n+1}}{2}, \quad q_n = \sqrt{p_{n+1}\, q_{n+1}}. \qquad (2.29)$$

The two bilateral sequences were given by

$$\cdots p_{-n},\ p_{-n+1}, \ldots, p_0, \ldots, p_n,\ p_{n+1} \cdots$$
$$\cdots q_{-n},\ q_{-n+1}, \ldots, q_0, \ldots, q_n,\ q_{n+1} \cdots.$$

In the positive direction of increasing $n$, the sequence $p_n$ would be increasing to infinity and the sequence $q_n$ would be decreasing to zero. So Lagrange took $q_n = 0$ for sufficiently large $n$ to approximately evaluate the integral

$$\int \frac{dy}{\sqrt{(1 \pm p_n\, y^2)(1 \pm q_n\, y^2)}} \approx \int \frac{dy}{\sqrt{1 \pm p_n\, y^2}}. \qquad (2.30)$$

The integral on the right-hand side of (2.30) can be computed in terms of a logarithm. In the negative direction, Lagrange noted that $p_{-n} - q_{-n} \to 0$ as $n \to \infty$. Observe that this follows from the facts that the sequence $p_{-n}$ is decreasing, the sequence $q_{-n}$ is increasing, and also that

$$p_{-n} - q_{-n} \leq \frac{p_{-n+1} - q_{-n+1}}{2} \leq \frac{p - q}{2^n}.$$

The common value to which the sequences $p_n$ and $q_n$ converge, as $n \to \infty$, is called the arithmetic-geometric mean, or agM, of $p = p_0$ and $q = q_0$. Thus, for sufficiently large $n$, the elliptic integral could be approximately found from the exactly computable integral

$$\int \frac{dy}{1 \pm p_{-n} y^2}.$$

### 2.4 Gauss on the Arithmetic-Geometric Mean

In an April, 1816 letter[13] to his friend and former student, Schumacher, Gauss wrote that he had been studying the arithmetic-geometric mean since 1791, when he was fourteen years old. This certainly suggests an independent discovery, since Gauss did not have access to Lagrange's work at that time. In a paper dated by Schering as written before 1800,[14] Gauss denoted the agM of two positive numbers $a \geq b$ by $M(a, b)$. To show that the subsequences of $a_n$ and $b_n$ converged to the same value when they were defined by

$$a_1 = \frac{a+b}{2}, \quad b_1 = \sqrt{ab}, \quad a_2 = \frac{a_1 + b_1}{2}, \quad b_2 = \sqrt{a_1 b_1}, \quad \text{etc.,} \tag{2.31}$$

he observed that

$$b \leq b_1 \leq b_2 \leq \cdots \leq b_n \leq \cdots \leq a_n \leq \cdots \leq a_1 \leq a.$$

He then noted that if $a = b$, then $a_n = b_n$ and $M(a, b) = a$. On the other hand, if $a > b$, then

$$4(a_n^2 - b_n^2) = (a_{n-1} + b_{n-1})^2 - 4a_{n-1} b_{n-1} = (a_{n-1} - b_{n-1})^2$$

and

$$\frac{a_n - b_n}{a_{n-1} - b_{n-1}} = \frac{a_{n-1} - b_{n-1}}{4(a_n + b_n)} = \frac{a_{n-1} - b_{n-1}}{2(a_{n-1} + b_{n-1}) + 4b_n} < \frac{1}{2}.$$

Gauss hence concluded that

$$a_n - b_n < \frac{a_{n-1} - b_{n-1}}{2} < \frac{a-b}{2^n}.$$

This meant that the increasing sequence $b_n$ and the decreasing sequence $a_n$ converged to the same number, the arithmetic-geometric mean, $M(a, b)$.

It is reported that Gauss stated that in 1794 he knew the relation between the arithmetic-geometric mean and the theta series.[15] Examples of these series are

$$P(x) = \sum_{n=-\infty}^{\infty} x^{n^2}, \qquad Q(x) = \sum_{n=-\infty}^{\infty} (-1)^n x^{n^2}.$$

Gauss probably became interested in the study of such series after reading Euler's *Introductio in Analysin Infinitorum*, where he found Euler's famous identity, known as the pentagonal number theorem:

$$\prod_{n=1}^{\infty} (1 - x^n) = \sum_{n=-\infty}^{\infty} (-1)^n x^{\frac{n(3n-1)}{2}}. \tag{2.32}$$

---

[13] Peters (1860–65) vol. 1, p. 125.
[14] Gauss (1863–1927) vol. 3, pp. 361–365; for Schering's remark, see pp. 492–3.
[15] Gauss (1863–1927) vol. 10/2, p. 18 of L. Schlesinger's article on Gauss's function theoretic work.

The connection of the series $P(x)$ and $Q(x)$ with the agM may be perceived through the identities

$$P(x) + Q(x) = 2P(x^4), \tag{2.33}$$

$$P^2(x) + Q^2(x) = 2P^2(x^2), \tag{2.34}$$

$$P(x)Q(x) = Q^2(x^2). \tag{2.35}$$

It follows from (2.34) that $P^2(x^2)$ is the arithmetic mean of $P^2(x)$ and $Q^2(x)$; (2.35) implies that $Q^2(x^2)$ is the geometric mean of $P^2(x)$ and $Q^2(x)$. Note that we require $|x| < 1$ for the absolute convergence of the series. Thus, the agM of $P^2(x)$ and $Q^2(x)$ must be

$$P(0) = P(\lim_{n \to \infty} x^{2^n}) = Q(\lim_{n \to \infty} x^{2^n}) = Q(0) = 1.$$

Therefore, if two numbers are expressible as the series

$$\sum_{n=-\infty}^{\infty} x^{n^2} \text{ and } \sum_{n=-\infty}^{\infty} (-1)^n x^{n^2},$$

then their arithmetic-geometric mean is 1.

In an undated manuscript,[16] "Hundert Theoreme über die neuen Transcendenten," Gauss wrote a proof of formulas (2.33) through (2.35). Since this proof involves only addition and multiplication of series, it is probable that this is the proof he mentioned that he found in 1794. To prove (2.33), observe that

$$P(x) + Q(x) = \sum_{n=-\infty}^{\infty} x^{n^2} + \sum_{n=-\infty}^{\infty} (-1)^n x^{n^2} = 2 \sum_{n=-\infty}^{\infty} x^{(2n)^2}$$

$$= 2 \sum_{n=-\infty}^{\infty} x^{4n^2} = 2P(x^4).$$

Gauss's proof of (2.34) is much more interesting. He noted that

$$P(x) + iQ(x) = (1 + i)Q(ix) \tag{2.36}$$

$$P(x) - iQ(x) = (1 - i)P(ix). \tag{2.37}$$

Both (2.36) and (2.37) can be checked in a similar fashion; for (2.36), we need only show that the coefficient of $x^{n^2}$ is the same on each side of (2.36), or that

$$1 + i(-1)^n = (-1)^n i^{n^2}(1 + i). \tag{2.38}$$

When $n$ is even, (2.38) is correct, because $i^{n^2} = 1$, and when $n$ is odd, $i^{n^2} = i$, and $1 - i = -i(1 + i)$. By multiplying (2.36) and (2.37), Gauss obtained a formula for the sum of the squares:

$$P^2(x) + Q^2(x) = 2P(ix)Q(ix). \tag{2.39}$$

---

[16] Gauss (1863–1927) vol. 3, pp. 466–469, especially pp. 466–467.

In his "Hundert Theoreme" manuscript, Gauss showed that

$$P(x)Q(x) = \left(Q(x^2)\right)^2 \tag{2.40}$$

followed easily from the product expansions for $P(x)$ and $Q(x)$:

$$P(x) = \prod_{n=1}^{\infty}(1 - x^{2n})(1 + x^{2n-1})^2 \tag{2.41}$$

$$Q(x) = \prod_{n=1}^{\infty}(1 - x^{2n})(1 - x^{2n-1})^2. \tag{2.42}$$

Observe that (2.41) and (2.42) imply that

$$
\begin{aligned}
P(x)Q(x) &= \prod_{n=1}^{\infty}(1 - x^{2n})^2\,(1 - x^{4n-2})^2 \\
&= \prod_{n=1}^{\infty}(1 - x^{4n})^2\,(1 - x^{4n-2})^2\,(1 - x^{4n-2})^2 \\
&= \prod_{n=1}^{\infty}(1 - x^{4n})^2\,(1 - x^{4n-2})^4 = Q^2(x^2).
\end{aligned}
$$

Note that (2.39) and (2.40) combined produce (2.34).

How might Gauss have discovered (2.41) and (2.42)? Observe that each implies the other by changing $x$ to $-x$. Of course, both formulas are special cases of the triple product identity, but Gauss did not discover this identity until approximately 1800. In his unpublished and undated manuscript, "Zur Theorie der transcendenten Functionen gehörig," he presented a method resembling Euler's proof of the pentagonal number theorem

$$\prod_{n=1}^{\infty}(1 - x^n) = \sum_{n=-\infty}^{\infty}(-1)^n x^{\frac{n(3n-1)}{2}}. \tag{2.43}$$

It is possible that Gauss's 1794 proof followed the lines of Euler's proof.[17] For convenient comparison of the two, we give an outline of Euler's reasoning: Let $P_0$ denote the infinite product on the left-hand side of (2.43). Euler showed inductively that

$$P_{n-1} = 1 - x^{2n-1} - x^{3n-1}P_n, \quad n \geq 1, \tag{2.44}$$

where

$$P_n = \sum_{k=0}^{\infty} x^{kn}(1 - x^n)(1 - x^{n+1})\cdots(1 - x^{n+k}), \quad n \geq 1. \tag{2.45}$$

In the manner of his time, Euler gave only the first few steps of the induction so that his intention became clear. The basic idea of his proof was to break up the $k$th term in

---

[17] Eu.I-2, pp. 390–398 or Fuss (1969) vol. 1, pp. 522–524.

$P_n$ into two parts

$$x^{kn}(1 - x^{n+1}) \cdots (1 - x^{n+k}) - x^{(k+1)n}(1 - x^{n+1}) \cdots (1 - x^{n+k}), \qquad (2.46)$$

and then to add the negative part of the $k$th term to the positive part of the $k + 1$th term:

$$-x^{(k+1)n}(1 - x^{n+1}) \cdots (1 - x^{n+k}) + x^{(k+1)n}(1 - x^{n+1}) \cdots (1 - x^{n+k+1})$$
$$= -x^{(k+2)n+k+1}(1 - x^{n+1}) \cdots (1 - x^{n+k}). \qquad (2.47)$$

After applying this process to each of the terms of $P_n$, it is clear that

$$P_n = 1 - x^{2n+1} - x^{3n+2} P_{n+1}.$$

To obtain the series on the right-hand side of (2.43), Euler repeatedly applied (2.44).
  Now Gauss proved $(2.42)^{18}$ in the form

$$\prod_{n=1}^{\infty} \frac{1 - x^n}{1 + x^n} = 1 + 2 \sum_{n=1}^{\infty} (-1)^n x^{n^2}. \qquad (2.48)$$

For this purpose, he defined the series:

$$P = 1 + \sum_{k=1}^{\infty} \frac{x^{kn}}{1 + x^n} \cdot \frac{(1 - x^{2n+k})(1 - x^{n+1})(1 - x^{n+2}) \cdots (1 - x^{n+k-1})}{(1 + x^{n+1})(1 + x^{n+2})(1 + x^{n+3}) \cdots (1 + x^{n+k})}, \qquad (2.49)$$

$$Q = \frac{x^n}{1 + x^n} + \sum_{k=1}^{\infty} \frac{x^{(k+1)n}}{1 + x^n} \cdot \frac{(1 - x^{n+1})(1 - x^{n+2}) \cdots (1 - x^{n+k})}{(1 + x^{n+1})(1 + x^{n+2}) \cdots (1 + x^{n+k})}, \qquad (2.50)$$

$$R = P - Q, \qquad (2.51)$$

where $n \geq 1$ and $|x| < 1$ for the series to converge absolutely. He first calculated $R$ by subtracting the $k$th term of $Q$ from the $k$th term of $P$ to get

$$R = \frac{1}{1 + x^n} + \sum_{k=1}^{\infty} \frac{x^{kn}}{1 + x^n} \cdot \frac{(1 - x^n)(1 - x^{n+1}) \cdots (1 - x^{n+k-1})}{(1 + x^{n+1})(1 + x^{n+2}) \cdots (1 + x^{n+k})}, \qquad (2.52)$$

and denoted the series in (2.52) by $\phi(x, n)$. Gauss again computed $R$, this time by subtracting the $k$th term in $Q$ from the $(k + 1)$th term in $P$ to obtain

$$R = 1 - \frac{x^{2n+1}}{1 + x^{n+1}} - \frac{x^{3n+2}}{1 + x^{n+1}} \cdot \frac{1 - x^{n+1}}{1 + x^{n+2}} - \frac{x^{4n+3}}{1 + x^{n+1}} \cdot \frac{1 - x^{n+1}}{1 + x^{n+2}} \cdot \frac{1 - x^{n+2}}{1 + x^{n+3}} - \cdots. \qquad (2.53)$$

Comparing the series for $R$ produced the relation

$$\phi(x, n) = 1 - x^{2n+1} \phi(x, n + 1) \quad \text{for } n \geq 1. \qquad (2.54)$$

A repeated application of (2.54) then yielded

$$\phi(x, n) = 1 - x^{2n+1} + x^{4n+4} - x^{6n+9} + x^{8n+16} - \cdots, \quad n \geq 1. \qquad (2.55)$$

---

[18] Gauss (1863–1927) vol. 3, pp. 437–439.

Comparing (2.55) with the right-hand side of the result we wish to prove, (2.48), it is clear that the case $n = 0$ is of primary interest. However, the series $P$ and $Q$ are divergent when $n = 0$, although (2.52) produces the value, as noted by Gauss:

$$\phi(x, 0) = \frac{1}{2}. \tag{2.56}$$

Care is needed here because of the divergence of $P$ and $Q$, of which Gauss was well aware. We thus employ a notation not used by Gauss in his informal notes. Let $p_1, p_2, p_3, \ldots$ and $q_1, q_2, q_3, \ldots$ denote the successive terms of the series $P$ and $Q$ when $n = 0$. Then

$$\phi(x, 0) = \lim_{m \to \infty} \left( (p_1 - q_1) + (p_2 - q_2) + \cdots + (p_m - q_m) \right)$$
$$= \lim_{m \to \infty} \left( (p_1 + (p_2 - q_1) + (p_3 - q_2) + \cdots + (p_m - q_{m-1}) \right) - \lim_{m \to \infty} q_m. \tag{2.57}$$

Gauss noted that $\lim_{m \to \infty} q_m = \frac{T}{2}$, where $T$ denoted the product on the left-hand side of (2.48), and that

$$\lim_{m \to \infty} \left( p_1 + (p_2 - q_1) + \cdots + (p_m - q_{m-1}) \right) = 1 - x\phi(x, 1).$$

We remark that Gauss himself referred to $T$ as the last term of the series $Q$ when $n = 0$. By (2.55),

$$1 - x\phi(x, 1) = 1 - x + x^4 - x^9 + x^{16} - \cdots .$$

Thus, (2.57) could be rewritten as

$$\frac{1}{2} = 1 - x + x^4 - x^9 + x^{16} - \cdots - \frac{T}{2}, \tag{2.58}$$

or, to complete the proof,

$$T = \prod_{n=1}^{\infty} \frac{1 - x^n}{1 + x^n} = 1 + 2 \sum_{n=1}^{\infty} (-1)^n x^{n^2}. \tag{2.59}$$

Gauss gave several proofs of (2.42). As we have shown, one used the idea Euler had applied to prove the pentagonal number theorem. Another proof was based upon the triple product identity that Gauss discovered between 1798 and 1808, though it went unpublished:

$$\sum_{n=-\infty}^{\infty} q^{n^2} x^n = \prod_{n=1}^{\infty} \left( 1 - q^{2n} \right) \left( 1 + q^{2n-1} x \right) \left( 1 + \frac{q^{2n-1}}{x} \right). \tag{2.60}$$

The product formulas for $P(x)$ and $Q(x)$, given by (2.41) and (2.42), can be obtained from (2.60) by taking $x = 1$ and $x = -1$, respectively. In fact, the pentagonal number theorem is also a consequence of (2.60) when we take $x = -q^{\frac{1}{2}}$ and replace $q$ with $q^{\frac{3}{2}}$.

The product on the right-hand side of (2.60) then becomes

$$\prod_{n=1}^{\infty}(1-q^{3n})\left(1-q^{3n-\frac{3}{2}+\frac{1}{2}}\right)\left(1-q^{3n-\frac{3}{2}-\frac{1}{2}}\right)$$

$$=\prod_{n=1}^{\infty}(1-q^{3n})\left(1-q^{3n-1}\right)\left(1-q^{3n-2}\right)=\prod_{n=1}^{\infty}(1-q^{n}).$$

The sum on the left-hand side of (2.60) is then given by

$$\sum_{n=-\infty}^{\infty}(-1)^{n}q^{\frac{3n^2}{2}+\frac{n}{2}}=\sum_{n=-\infty}^{\infty}(-1)^{n}q^{\frac{n(3n-1)}{2}},$$

where the last equality follows when one changes the index of summation from $n$ to $-n$. This proves the pentagonal number theorem.

Gauss related the arithmetic-geometric mean $M(a, b)$ of two positive numbers, $a >$ $b > 0$, to the elliptic integral:

$$\frac{1}{M(a, b)}=\frac{2}{\pi}\int_{0}^{\frac{\pi}{2}}\frac{d\theta}{\sqrt{a^2\cos^2\theta+b^2\sin^2\theta}}\equiv I(a, b).\qquad(2.61)$$

This can be verified using Lagrange's result, contained in (2.25) through (2.27). Let $I(a, b)$ be the integral defined by (2.61) and let $x=\frac{\cot\theta}{b}$ to obtain

$$I(a, b)=\frac{2}{\pi}\int_{0}^{\infty}\frac{dx}{\sqrt{(1+a^2x^2)(1+b^2x^2)}}.\qquad(2.62)$$

Now by (2.26) and (2.27), we know that if

$$x=y\sqrt{\frac{1+a_1^2y^2}{1+b_1^2y^2}},\quad\text{with }a_1=\frac{a+b}{2},\quad b_1=\sqrt{ab},\qquad(2.63)$$

then

$$\frac{dx}{\sqrt{(1+a^2x^2)(1+b^2x^2)}}=\frac{dy}{\sqrt{(1+a_1y^2)(1+b_1y^2)}}.\qquad(2.64)$$

Note that (2.62) and (2.64) imply that $I(a, b)=I(a_1, b_1)$; using the notation in (2.31) and letting $c=M(a, b)$, we can see that

$$I(a, b)=I(a_1, b_1)=I(a_2, b_2)=\cdots=I(c, c)=\frac{2}{\pi}\int_{0}^{\infty}\frac{dx}{1+x^2c^2}=\frac{1}{c}.$$

This proves Gauss's formula (2.61). In his 1818 paper on an astronomical topic,[19] Gauss provided an alternative proof by applying a different transformation. He set

$$\sin\theta=\frac{2a\sin\theta'}{(a+b)\cos^2\theta'+2a\sin^2\theta'}=\frac{2a\sin\theta'}{a+b+(a-b)\sin^2\theta'}.\qquad(2.65)$$

---

[19] Gauss (1863–1927) vol. 3, pp. 333–355, especially pp. 352–355.

Streamlining Gauss's proof, set $y = \sin\theta$, $x = \sin\theta'$, and $\lambda = \frac{a-b}{a+b}$; then (2.65) takes the form

$$y = \frac{(1+\lambda)x}{1 + \lambda x^2}.\tag{2.66}$$

Moreover, we have

$$\sqrt{a^2 \cos^2\theta + b^2 \sin^2\theta} = a\sqrt{1 - \frac{a^2 - b^2}{a^2}\sin^2\theta}$$
$$= a\sqrt{1 - k^2 \sin^2\theta},$$

where $k^2 = \frac{4\lambda}{(1+\lambda)^2}$ or $k = \frac{2\sqrt{\lambda}}{1+\lambda}$, and we know that

$$\sqrt{a_1^2 \cos^2\theta + b_1^2 \sin^2\theta} = a_1\sqrt{1 - \frac{a_1^2 - b_1^2}{a_1^2}\sin^2\theta}$$
$$= \frac{a+b}{2}\sqrt{1 - \lambda^2 \sin^2\theta}.$$

Thus, the relation $I(a, b) = I(a_1, b_1)$ becomes

$$\frac{dy}{\sqrt{(1-y^2)(1-k^2y^2)}} = \frac{(1+\lambda)dx}{\sqrt{(1-x^2)(1-\lambda^2x^2)}},\tag{2.67}$$

where $y$ is given by (2.66) and $k = \frac{2\sqrt{\lambda}}{1+\lambda}$. This last step can be proved in the same way that (2.22) was proved by using (2.21). We call (2.66) and (2.67) Gauss's quadratic transformation.

It is of interest to note that (2.21), due to Landen, and (2.66), due to Gauss, are the two basic quadratic transformations of elliptic integrals and that one followed by the other gives the duplication of the elliptic integral. Writing

$$z = \frac{(1+\lambda)y}{1 + \lambda y^2} \quad \text{with} \quad \lambda = \frac{2\sqrt{k}}{1+k}$$

and

$$y = \frac{(1+k')x\sqrt{1-x^2}}{\sqrt{1 - k^2 x^2}}, \quad \text{where} \quad k'^2 = 1 - k^2,$$

we have, from (2.22) and (2.67),

$$\frac{dy}{\sqrt{(1-y^2)(1-\lambda^2 y^2)}} = \frac{2}{1+\lambda}\frac{dx}{\sqrt{(1-x^2)(1-k^2 x^2)}}$$
$$= \frac{1}{1+\lambda}\frac{dz}{\sqrt{(1-z^2)(1-k^2 z^2)}}.\tag{2.68}$$

Note that $\lambda$ and $k$ were called moduli of the corresponding elliptic integrals and the relation $\lambda(1+k) = 2\sqrt{k}$ was called a quadratic modular equation, since it related the moduli by means of a quadratic transformation. Now equation (2.68) implies the duplication

$$\frac{dz}{\sqrt{(1 - z^2)(1 - k^2 z^2)}} = \frac{2\,dx}{\sqrt{(1 - x^2)(1 - k^2 x^2)}}.$$

As Lagrange had observed, the convergence of $a_n$, $b_n$ to $M(a, b)$ is very rapid. In fact, this convergence can be shown to be quadratic, that is

$$|a_n - M(a, b)| \le c\,|a_{n-1} - M(a, b)|^2,$$

where $c$ is a constant independent of $n$. Roughly, this means that the number of digits of agreement between $a_n$ and $M(a, b)$ doubles after each iterate. Gauss made some numerical calculations to verify this; he took

$$a = 1, \quad b = 0.2; \quad a = 1, \quad b = 0.6; \quad a = 1, b = 0.8; \quad a = \sqrt{2}, \quad b = 1.$$

Note that for the case $a = \sqrt{2}$, $b = 1$, the arithmetic-geometric mean is given by

$$\frac{1}{M(\sqrt{2}, 1)} = \frac{2}{\pi} \int_0^{\frac{\pi}{2}} \frac{d\theta}{\sqrt{2 \cos^2 \theta + \sin^2 \theta}} = \frac{2}{\pi} \int_0^1 \frac{dt}{\sqrt{1 - t^4}}.$$

Gauss computed four iterates of the agM,[20] each to twenty-one decimal places:

$$a\phantom{_4} = 1.41421356237309504\,8802, \quad b\phantom{_4} = 1.00000000000000000000,$$
$$a_1 = 1.20710678118654752\,4401, \quad b_1 = 1.18920711500272106\,6717,$$
$$a_2 = 1.19815694809463429\,5559, \quad b_2 = 1.19812352149312012\,2607,$$
$$a_3 = 1.19814023473559220\,7441, \quad b_3 = 1.19814023467730720\,5798,$$
$$a_4 = 1.19814023473559220\,7441, \quad b_4 = 1.19814023473559220\,7439.$$

Since $a_4$ agrees with $b_4$ to nineteen decimal places and since $M(\sqrt{2}, 1)$ lies between $a_4$ and $b_4$, it follows that $a_4$ and $b_4$ yield the correct value of $M(\sqrt{2}, 1)$ up to nineteen decimal places. Note that $a_2$ and $b_2$ agree up to four decimal places and $a_3$ and $b_3$ agree up to nine decimal places. This rapid convergence was employed by Y. Kanada and his collaborators in their efforts to compute the value of $\pi$ to billions of digits. We also mention that other methods for computing $\pi$ were devised by J. M. and P. B. Borwein and D. V. and G. V. Chudnovsky, based partly on Ramanjuan's work on modular equations and series expansions of $\pi$.[21]

## 2.5  Gauss on Elliptic Functions

According to Gauss's mathematical diary,[22] on September 9, 1796, he found the inverse of the integral $\int (1 - x^3)^{-\frac{1}{2}}\,dx$ as a series. He accomplished this by means of Newton's reversion of series method. A few days later, he found the series for the inverse of

---

[20] Gauss (1863–1927) vol. 3, p. 364.   [21] See Berggren et al. (1997), Borwein and Borwein (1987).
[22] See Dunnington (2004) pp. 469–484, a translation of Gauss's diary by J. Gray.

$\int (1 - x^n)^{-\frac{1}{2}} dx$. He then noted on January 7, 1797 that

$$\int \sqrt{\sin x}\, dx = 2 \int \frac{yy\, dy}{\sqrt{1 - y^4}}, \quad \int \sqrt{\frac{1}{\sin x}}\, dx = 2 \int \frac{dy}{\sqrt{1 - y^4}}, \quad yy = \frac{\sin}{\cos} x,$$

where he indicated that $yy$ (or $y^2$) signified either $\sin x$ or $\cos x$. Clearly, Gauss was developing an interest in the lemniscatic integral and looking into defining its inverse. By March 1797, he had indeed defined the inverse of the lemniscatic integral; he wrote,

$$\sin \operatorname{lemn} \left( \int_0^x \frac{dt}{\sqrt{1 - t^4}} \right) = x, \quad \cos \operatorname{lemn} \left( \frac{1}{2}\omega - \int_0^x \frac{dt}{\sqrt{1 - t^4}} \right) = x, \qquad (2.69)$$

$$\text{where} \quad \frac{\omega}{2} = \int_0^1 \frac{dt}{\sqrt{1 - t^4}}\, dt. \qquad (2.70)$$

In his notes, Gauss sometimes abbreviated sin lemn and cos lemn as $s$ and $c$, respectively. For convenience, we shall denote these functions as sl and cl. In a diary entry of March 19, 1797, Gauss noted that in order to divide the lemniscate into $n$ equal parts, one had to solve an equation of degree $n^2$. This follows from a repeated application of the addition formula for the lemniscatic integral or the lemniscatic function. As we can see from Euler's addition formula (2.14), the addition formula, in terms of sl and cl, is given by[23]

$$\operatorname{sl}(a \pm b) = \frac{\operatorname{sl} a \sqrt{1 - \operatorname{sl}^4 b} \pm \operatorname{sl} b \sqrt{1 - \operatorname{sl}^4 a}}{1 + \operatorname{sl}^2 a \operatorname{sl}^2 b} = \frac{\operatorname{sl} a \operatorname{cl} b \pm \operatorname{cl} a \operatorname{sl} b}{1 \mp \operatorname{sl} a \operatorname{sl} b \operatorname{cl} a \operatorname{cl} b}. \qquad (2.71)$$

Note that if we take $x = \operatorname{sl} u$ and $y = \operatorname{cl} u$, then $c = 1$ in (2.13) and the resulting relation is useful in the derivation of the second equation in (2.71). Now an equation of degree $n^2$ must have $n^2$ roots, of which $n$ correspond to the division points on the real lemniscate. This implies that the majority of the roots must be complex. Perhaps this was the impetus for Gauss to extend the lemniscatic functions sl and cl to functions of complex variables. A formal change of variables $t = iu$ gave him

$$i \int_0^x (1 - t^4)^{-\frac{1}{2}} dt = \int_0^{ix} (1 - u^4)^{-\frac{1}{2}} du;$$

so that, in terms of the lemniscatic function, he had

$$\operatorname{sl}(iy) = i \operatorname{sl} y. \qquad (2.72)$$

Gauss also showed that

$$\operatorname{cl}(iy) = \frac{1}{\operatorname{cl} y}. \qquad (2.73)$$

Now observe that the function $y = \sin x$ is the inverse of the integral

$$x = \int_0^y \frac{dt}{\sqrt{1 - t^2}}$$

---

[23] Gauss (1863–1927) vol. 10/1, pp. 147–154.

and has a period $2\pi$, where $\pi$ is given by the integral

$$\frac{\pi}{2} = \int_0^1 \frac{dt}{\sqrt{1-t^2}}.$$

Analogously, the function $\mathrm{sl}\, x$ has a period $2\omega$, where $\omega$ is given by the integral (2.70). Gauss's extension of sl to a function of a complex variable showed that by use of (2.71) through (2.73), he could obtain the formula

$$\mathrm{sl}(a+ib) = \frac{\mathrm{sl}(a) + i\,\mathrm{sl}(b)\,\mathrm{cl}(a)\,\mathrm{cl}(b)}{\mathrm{cl}(b) - i\,\mathrm{sl}(a)\,\mathrm{sl}(b)\,\mathrm{cl}(a)}. \qquad (2.74)$$

Now $\mathrm{sl}(0) = 0$ and $\mathrm{cl}(0) = 1$; hence, by periodicity, $\mathrm{sl}(2\omega) = 0$ and $\mathrm{cl}(2\omega) = 1$. Thus, by (2.72),

$$\mathrm{sl}(a + i\,2\omega) = \mathrm{sl}(a).$$

In this way, Gauss found that $\mathrm{sl}\, x$ was a doubly periodic function with two independent periods, $2\omega$ and $2i\,\omega$.

In a diary entry written between March 19 and 21, 1797, Gauss stated that the lemniscatic curve was geometrically divisible into five equal parts. In notes on the lemniscatic function, with $\mathrm{s} = \mathrm{sl}\, x$ and $\mathrm{c} = \mathrm{cl}\, x$, he wrote the formulas[24]

$$\mathrm{sl}\,(2x) = \mathrm{s}\,\mathrm{c}\,(1+\mathrm{s}^2)\frac{2}{1+\mathrm{s}^4} = \mathrm{s}\,\mathrm{c}\,(1+\mathrm{c}^2)\frac{2}{1+\mathrm{c}^4}, \qquad (2.75)$$

$$\mathrm{sl}\,(3x) = \mathrm{s}\,\frac{3 - 6\mathrm{s}^4 - \mathrm{s}^8}{1 + 6\mathrm{s}^4 - 3\mathrm{s}^8}, \qquad (2.76)$$

$$\mathrm{sl}\,(5x) = \mathrm{s}\,\frac{\mathrm{s}^{24} + 50\mathrm{s}^{20} - 125\mathrm{s}^{16} + 300\mathrm{s}^{12} - 105\mathrm{s}^8 - 62\mathrm{s}^4 + 5}{1 + 50\mathrm{s}^4 - 125\mathrm{s}^8 + 300\mathrm{s}^{12} - 105\mathrm{s}^{16} - 62\mathrm{s}^{20} + 5\mathrm{s}^{24}} \qquad (2.77)$$

These relations can be proved by use of the addition formula (2.71) combined with a the particular case of (2.71) or (2.13):

$$\mathrm{s}^2 + \mathrm{c}^2 + \mathrm{s}^2\,\mathrm{c}^2 = 1 \quad \text{or} \quad (1+\mathrm{s}^2)(1+\mathrm{c}^2) = 2 \qquad (2.78)$$

$$\text{or} \quad \mathrm{s} = \sqrt{\frac{1-\mathrm{c}^2}{1+\mathrm{c}^2}} \quad \text{or} \quad \mathrm{c} = \sqrt{\frac{1-\mathrm{s}^2}{1+\mathrm{s}^2}}. \qquad (2.79)$$

Note that (2.78) follows from (2.71) by taking $a = \frac{\omega}{2}$ and choosing a minus sign for $b$.

If we wish to trisect the complete arc *OPAO* of the lemniscate, it is sufficient to construct lengths

$$r = \mathrm{sl}\left(\frac{\omega}{3}\right) \quad \text{and} \quad r = \mathrm{sl}\left(\frac{2\omega}{3}\right).$$

Observe that the left-hand side in (2.76) is zero when $x = \frac{\omega}{3}$ or $\frac{2\omega}{3}$ and that this makes the numerator of the right-hand side, a polynomial of degree nine, equal to zero for these values of $x$. This is an example of Gauss's assertion that to divide a lemniscate into $n$ equal parts, one must solve an equation of degree $n^2$. This equation may be reduced to

---

[24] See Gauss (1863–1927) vol. 3, pp. 404–405 and vol. 10/1, pp. 147 and 161–62.

an equation of degree $n^2 - 1$ because $\text{sl}\,x$ is a factor of the polynomial of degree $n^2$. In the case $n = 3$, (2.76) shows that the equation to be solved is $s^8 + 6s^4 - 3 = 0$. This is a quadratic equation in $y = s^4$; hence, $\text{sl}(\frac{\omega}{3})$ and $\text{sl}(\frac{2\omega}{3})$ are constructible by straight-edge and compass. In a similar way, (2.77) implies that $\text{sl}(\frac{\omega}{5})$ is a solution of an equation of degree 24. In fact, it is an equation of degree 6 in $\text{sl}^4$:

$$y^6 + 50y^5 - 125y^4 + 300y^3 - 105y^2 - 62y + 5 = 0. \qquad (2.80)$$

Gauss factored this polynomial as[25]

$$(y^2 - 2y + 5)(y^4 + 52y^3 - 26y^2 - 12y + 1).$$

The quartic polynomial can be further factored:

$$\left(y^2 + \left(26 + 12\sqrt{5}\right)y + 9 + 4\sqrt{5}\right)\left(y^2 + \left(26 - 12\sqrt{5}\right)y + 9 - 4\sqrt{5}\right).$$

The coefficients of this polynomial are constructible numbers, and hence the solutions of (2.80) are constructible.

In this effort, Gauss was clearly attempting to extend his cyclotomic theory to the division of a lemniscate into $n$ equal parts. In Section 7 of his famous *Disquisitiones Arithmeticae*, he showed that a circle could be divided into $n$ equal parts if $n$ were a prime of the form $2^m + 1$. Gauss showed that $e^{\frac{2\pi i}{n}}$ satisfied an irreducible equation of degree $n - 1$, so that when $n - 1$ was of the form $2^m$, then the equation could be solved by a sequence of quadratic equations whose coefficients were constructible numbers. This then implied that $\cos\frac{2\pi}{n}$ and $\sin\frac{2\pi}{n}$ were constructible numbers for such $n$. Although Gauss developed the corresponding theory for the lemniscatic function, he did not publish it. Still, he gave a clear indication of this work on the lemniscatic function in the same section of his *Disquisitiones*:[26]

> The principles of the theory which we are going to explain actually extend much farther than we will indicate. For they can be applied not only to circular functions but just as well to other transcendental functions, e.g. to those which depend on the integral $\int (1 - x^4)^{-\frac{1}{2}}\,dx$ and also to various types of congruences. Since, however, we are preparing a substantial work on transcendental functions and since we will treat congruences at length as we continue our discussion of arithmetic, we have decided to consider only circular functions here.

In 1827, the Norwegian mathematician Niels Abel (1802–1829) published his researches on elliptic functions with an application to the division of the lemniscate. He showed that if the number of divisions was a prime of the form $2^m + 1$, then the division could be done in a geometric way. Note that the proof of this theorem would not be quite analogous to the proof of such a theorem for the circle, since the algebraic equation obtained in this case would be of degree $n^2 - 1$, not a power of 2. Thus, in this case, one is not assured of obtaining a sequence of quadratic equations. To achieve his end, Abel had to first define complex multiplication of elliptic functions. The passage from the *Disquisitiones* clearly indicates that Gauss had also discovered complex multiplication of elliptic functions. However, he did not elaborate on this topic in the

---

[25] Gauss (1863–1927) vol. 3, pp. 404–405 and vol. 10/1, p. 162.    [26] Gauss (1965) p. 407.

manuscripts he left behind, writing only that Abel had followed the same road he had taken. Thus, we here follow Abel to describe Gauss's ideas.

Suppose $\phi$ is an elliptic function with periods $\omega_1$ and $\omega_2$. Then $\phi(\alpha x)$ must also have periods $\omega_1$ and $\omega_2$ when $\alpha = n$, an integer. An elliptic function admits of complex multiplication by a complex number $\alpha$ if $\phi(\alpha x)$ has periods $\omega_1$ and $\omega_2$, that is if $\alpha\omega_1$ and $\alpha\omega_2$ are periods of $\phi(x)$. This will occur when there are integers $m_1$, $n_1$, $m_2$, $n_2$ such that

$$\alpha\,\omega_1 = m_1\,\omega_1 + n_1\,\omega_2, \tag{2.81}$$

$$\alpha\,\omega_2 = m_2\,\omega_1 + n_2\,\omega_2. \tag{2.82}$$

Dividing (2.81) by (2.82) and setting $\tau = \frac{\omega_1}{\omega_2}$, we see that the ratio of the periods must satisfy the equation

$$\tau = \frac{m_1\tau + n_1}{m_2\tau + n_2}.$$

This means that $\tau$ must satisfy the quadratic equation

$$m_2\,\tau^2 + (n_2 - m_1)\,\tau - n_1 = 0.$$

Therefore, $\tau$ is of the form $\frac{a + \sqrt{b}}{c}$, where $a$, $b$, $c$ are integers.

Abel considered elliptic functions that were the inverses of elliptic integrals; he showed that the the ratio $\tau$ of the two periods of each function must be a complex number. Thus, he proved that $b$ must be negative and (2.82) could be rewritten as

$$\alpha = m_2\,\tau + n_2, \quad \tau = \frac{\alpha - n_2}{m_2}.$$

Thus, the ratio of the two periods had to take the form

$$\tau = p + q\sqrt{-m}, \tag{2.83}$$

where $p$ and $q$ were rational numbers and $m$ was a square-free positive integer.

Abel was also able to show,[27] using methods similar to those used to prove (2.82) and (2.83), that an elliptic function $\phi(x)$ would admit of complex multiplication by $\alpha$ if the ratio of its periods $\tau$ was of the form (2.83) and $\alpha$ also took that form. Moreover, he showed that $\phi(\alpha x)$ was a rational function of $\phi(x)$. Note that in the case of the lemniscatic function, $\tau = \sqrt{-1}$ clearly takes the form (2.83).

Abel applied complex multiplication to reduce an algebraic equation of degree $n^2$, satisfied by $\mathrm{sl}(\frac{\omega}{n})$, to two equations each of degree $n$ and each containing $\mathrm{sl}\,x$ as a factor.[28] The degrees of the equations would thereby be reduced to $n - 1 = 2^m$ and he could solve these equations by successively solving of quadratic equations. Abel himself illustrated this technique for $n = 5$; to show the constructibility of $\mathrm{sl}(\frac{\omega}{5})$, he first had to factorize 5 as $5 = (2 + i)(2 - i)$ and then prove that when $y = \mathrm{sl}\,x$

$$\mathrm{sl}((2 + i)x) = y\,i\,\frac{1 - 2i - y^4}{1 - (1 - 2i)\,y^4} \equiv z \tag{2.84}$$

---

[27] Abel (1965) vol. 1, pp. 377–383.    [28] Ibid. pp. 352–362.

and

$$\mathrm{sl}(5x) = \mathrm{sl}((2-i)(2+i)x) = -z\,i\,\frac{1+2i-z^4}{1-(1+2i)z^4}. \tag{2.85}$$

Now to solve the equation $\mathrm{sl}(5x) = 0$, Abel had first to solve $z^4 = 1 + 2i$ and then

$$y\,i\,\frac{1-2i-y^4}{1-(1-2i)y^4} = (1+2i)^{\frac{1}{4}}. \tag{2.86}$$

Note that (2.86) could be solved by dividing it by the conjugate equation

$$-y\,i\,\frac{1+2i-y^4}{1-(1+2i)y^4} = (1-2i)^{\frac{1}{4}},$$

leading to a quadratic equation in $y^4$. Thus, Abel had the constructibility of $\mathrm{sl}(\frac{k\omega}{5})$ for $k = 1, 2, 3, 4$. These results can be tied up with Gauss's factorization of (2.80) by observing that

$$\left(y^4 - (1-2i)\right)\left(y^4 - (1+2i)\right) = y^8 - 2y^4 + 5.$$

### 2.6   Gauss: Theta Functions and Modular Forms

By 1808, Gauss had discovered the theta functions in their series as well as in their product forms. He defined these functions by their series:

$$P(t) = 1 + 2\sum_{n=1}^{\infty} e^{-n^2\pi t}, \tag{2.87}$$

$$Q(t) = 1 + 2\sum_{n=1}^{\infty} (-1)^n e^{-n^2\pi t}, \tag{2.88}$$

$$R(t) = 2\sum_{n=1}^{\infty} e^{-(n+\frac{1}{2})^2\pi t} \tag{2.89}$$

where, for convergence, the real part of $t$ was taken as positive. He gave the transformation formulas

$$P(t) = \frac{1}{\sqrt{t}} P\left(\frac{1}{t}\right), \quad Q(t) = \frac{1}{\sqrt{t}} R\left(\frac{1}{t}\right), \quad R(t) = \frac{1}{\sqrt{t}} Q\left(\frac{1}{t}\right). \tag{2.90}$$

In fact, he found the more general theorem:[29] Let $a, b, c, d$ be integers, $ad - bc = 1$, and $t' = \frac{at-bi}{d+cti}$. Consider the six cases modulo 2:

$$
\begin{aligned}
a &\equiv 1 \quad 1 \quad 1 \quad 0 \quad 1 \quad 0, \\
b &\equiv 0 \quad 1 \quad 0 \quad 1 \quad 1 \quad 1, \\
c &\equiv 0 \quad 0 \quad 1 \quad 1 \quad 1 \quad 1, \\
d &\equiv 1 \quad 1 \quad 1 \quad 1 \quad 0 \quad 0.
\end{aligned}
$$

---

[29] Gauss (1863–1927) vol. 3, p. 386.

Then, for $h = \sqrt{i^\lambda(d + cti)}$, where the value of $\lambda$ varies, we have

$$hP(t') = P(t)\ Q(t)\ R(t)\ R(t)\ Q(t)\ P(t), \tag{2.91}$$

$$hQ(t') = Q(t)\ P(t)\ Q(t)\ P(t)\ R(t)\ R(t), \tag{2.92}$$

$$hR(t') = R(t)\ R(t)\ P(t)\ Q(t)\ P(t)\ Q(t). \tag{2.93}$$

Observe that Re $t > 0$, so that Re $t' > 0$. In this theorem, Gauss determined the action of the modular group $\Gamma = \Gamma(1)$ on the three functions $P, Q, R$. In particular, he had the results in (2.90). These follow from case 6, where $\lambda = 3$.

Since we usually define a modular form as a function on the upper half-plane, we take $t = -i\omega$, so that the condition Re $t > 0$ translates to Im $\omega > 0$. Set

$$P_1(\omega) = 1 + 2\sum_{n=1}^{\infty} e^{\pi in^2\omega} = 1 + 2\sum_{n=1}^{\infty} q^{n^2} \equiv \theta_3(\omega), \tag{2.94}$$

$$Q_1(\omega) = 1 + 2\sum_{n=1}^{\infty} (-1)^n e^{\pi in^2\omega} = 1 + 2\sum_{n=1}^{\infty} (-1)^n q^{n^2} \equiv \theta_0(\omega), \tag{2.95}$$

$$R_1(\omega) = 2\sum_{n=1}^{\infty} e^{\pi i(n-\frac{1}{2})^2\omega} = 2\sum_{n=1}^{\infty} q^{(n-\frac{1}{2})^2} \equiv \theta_2(\omega), \tag{2.96}$$

where $q = e^{\pi i\omega}$. Equations (2.90) can now be rewritten

$$P_1\left(-\frac{1}{\omega}\right) = \sqrt{\frac{\omega}{i}}\,P_1(\omega), \quad Q_1\left(-\frac{1}{\omega}\right) = \sqrt{\frac{\omega}{i}}\,R_1(\omega), \quad R_1\left(-\frac{1}{\omega}\right) = \sqrt{\frac{\omega}{i}}\,Q_1(\omega).$$

$$\tag{2.97}$$

On the basis of Gauss's general theorem, or directly from (2.94) through (2.96), we may write

$$P_1(\omega + 1) = Q_1(\omega), \quad Q_1(\omega + 1) = P_1(\omega), \quad R_1(\omega + 1) = \sqrt{i}\,R_1(\omega). \tag{2.98}$$

Observe that all the cases of Gauss's theorem follow from (2.97) and (2.98), since $\Gamma(1)$ is generated by

$$\begin{pmatrix} 1 & 1 \\ 0 & 1 \end{pmatrix} \quad \text{and} \quad \begin{pmatrix} 0 & -1 \\ 1 & 0 \end{pmatrix}.$$

Gauss's proof[30] of (2.97) began with the consideration of the periodic function

$$T = \sum_{k=-\infty}^{\infty} e^{-\alpha(k+\omega)^2}$$

of period 1. Note that Re $\alpha > 0$ for the series to be absolutely convergent. He expanded this periodic function as a Fourier series. Now in 1777, Euler had expanded a periodic function as a series in cosines and used the orthogonality of the cosines to derive the coefficients of the series as integrals.[31] This paper was published in 1794, and it is

---

[30] Gauss (1863–1927) vol. 3, pp. 436–437.    [31] See Eu.I-16/2, pp. 333–341.

possible that Gauss had seen it. In any case, Gauss applied a method that yielded a special case of the summation formula now named after Poisson. In doing this, he set

$$T = A + 2B \cos \omega P + 2C \cos 2\omega P + 2D \cos 3\omega P + \cdots ;$$

today we might write this as $T = A_0 + 2 \sum_{n=1}^{\infty} A_n \cos 2n\pi\omega$. Skipping steps (2.100) through (2.103), Gauss obtained (2.104):

$$A_n = \int_0^1 T \cos 2n\pi\omega \, d\omega \tag{2.99}$$

$$= \int_0^1 \left( \sum_{k=-\infty}^{\infty} e^{-\alpha(k+\omega)^2} \right) \cos 2n\pi\omega \, d\omega \tag{2.100}$$

$$= \sum_{k=-\infty}^{\infty} \int_0^1 e^{-\alpha(k+\omega)^2} \cos 2n\pi\omega \, d\omega \tag{2.101}$$

$$= \sum_{k=-\infty}^{\infty} \int_k^{k+1} e^{-\alpha\omega^2} \cos 2n\pi(\omega - k) \, d\omega \tag{2.102}$$

$$= \sum_{k=-\infty}^{\infty} \int_k^{k+1} e^{-\alpha\omega^2} \cos 2n\pi\omega \, d\omega \tag{2.103}$$

$$= \int_{-\infty}^{\infty} e^{-\alpha\omega^2} \cos 2n\pi\omega \, d\omega. \tag{2.104}$$

He next evaluated (2.104) by observing that

$$e^{-\alpha\omega^2} \cos 2n\pi\omega = \frac{1}{2} e^{-\alpha\omega^2 + i2\pi n\omega} + \frac{1}{2} e^{-\alpha\omega^2 - i2\pi n\omega}$$

$$= \frac{1}{2} e^{\frac{-n^2\pi^2}{\alpha}} \left( e^{-\alpha\left(\omega - \frac{in\pi}{\alpha}\right)^2} + e^{-\alpha\left(\omega + \frac{in\pi}{\alpha}\right)^2} \right). \tag{2.105}$$

In 1730, Euler had used Wallis's formula for $\pi$ to determine that[32]

$$\int_0^1 (-\log x)^{\frac{1}{2}} \, dx = \frac{\sqrt{\pi}}{2}.$$

A change of variables $x = e^{-\alpha\omega^2}$ and then integration by parts yields

$$\int_0^{\infty} e^{-\alpha\omega^2} \, d\omega = \frac{1}{2} \sqrt{\frac{\pi}{\alpha}}$$

or

$$\int_{-\infty}^{\infty} e^{-\alpha\omega^2} \, d\omega = \sqrt{\frac{\pi}{\alpha}}.$$

---

[32] Eu.I-14, pp. 1–24, especially p. 12.

Gauss knew this and remarked that it easily followed that

$$\int_{-\infty}^{\infty} e^{-\alpha(\omega - \frac{in\pi}{\alpha})^2} \, d\omega = \sqrt{\frac{\pi}{\alpha}}. \tag{2.106}$$

Now in order to rigorously deduce (2.106), Gauss required Cauchy's integral theorem. We may wonder whether Gauss was aware of this theorem in 1808, or perhaps he made a mere formal change of variables with a complex parameter, as Euler and Laplace had done, unaware of the required underpinnings. Indeed, three years later, in a letter to Bessel, Gauss wrote about his discovery of what is now known as Cauchy's theorem. It is possible that Gauss knew the result in 1808. Anyway, he had the value of the integral

$$A_n = e^{-\frac{n^2\pi^2}{\alpha}} \sqrt{\frac{\pi}{\alpha}}, \tag{2.107}$$

and also the formula

$$T = \sqrt{\frac{\pi}{\alpha}} \{1 + 2e^{-\frac{\pi^2}{\alpha}} \cos 2\pi\omega + 2e^{-\frac{4\pi^2}{\alpha}} \cos 4\pi\omega + \cdots\},$$

or, in another form,

$$\sum_{k=-\infty}^{\infty} e^{-\alpha(k+\omega)^2} = \sqrt{\frac{\pi}{\alpha}} e^{-\alpha\omega^2} \sum_{k=-\infty}^{\infty} e^{-\frac{\pi^2}{\alpha}(k+\frac{\alpha\omega i}{\pi})^2}. \tag{2.108}$$

Of course, Gauss could have arrived at (2.107) directly, by several methods; this is an interesting exercise. More generally, Gauss stated the formula

$$\sum_{k=-\infty}^{\infty} e^{-\alpha(k+\omega)^2} [\cos 2(k+\omega)\alpha\psi - i \sin 2(k+\omega)\alpha\psi]$$

$$= \sqrt{\frac{\pi}{\alpha}} \sum_{k=-\infty}^{\infty} e^{-(k-\frac{\alpha\psi}{\pi})^2 \frac{\pi^2}{\alpha}} [\cos 2k\omega\pi - i \sin 2k\omega\pi],$$

or, if $\alpha\alpha' = \pi^2$ and $-\alpha\psi = \omega'\pi$,

$$\sum_{k=-\infty}^{\infty} e^{-\alpha(k+\omega)^2} \left[\cos(2k+2\omega)\omega'\pi - i \sin(2k+2\omega)\omega'\pi\right]$$

$$= \sqrt{\frac{\pi}{\alpha}} \sum_{k=-\infty}^{\infty} e^{-\alpha'(k+\omega')^2} [\cos 2k\omega\pi - i \sin 2k\omega\pi].$$

As a corollary, Gauss obtained

$$\sum_{k=-\infty}^{\infty} e^{-\alpha(k+\omega-\frac{1}{4})^2} - \sum_{k=-\infty}^{\infty} e^{-\alpha(k+\omega+\frac{1}{4})^2}$$

$$= 4\sqrt{\frac{\pi}{\alpha}} \sum_{n=1}^{\infty} (-1)^{n-1} (2n-1) e^{-(2n-1)^2 \frac{\pi^2}{\alpha}} \sin(4n-2)\omega\pi;$$

as another corollary, he stated that the expression

$$\sum_{k=-\infty}^{\infty} t e^{-\pi t^4 k^2 - 2\pi t^2 k \sqrt{u} - \frac{\pi u}{2}}$$

did not change its value when $t$ was replaced by $\frac{1}{t}$ and $u$ was replaced by $-u$.

To obtain the first result in (2.90), we take $\alpha = \pi t$ and $\omega = 0$ in (2.108), and to obtain the second result in (2.90), we take $\alpha = \pi t$ and $\omega = \frac{1}{2}$. The third result follows from the second by changing $t$ to $\frac{1}{t}$. Gauss did not publish (2.90); the first formula in (2.90) was rediscovered by Cauchy.[33] He found this formula in the course of his work on the theory of waves. He also independently determined the reciprocity of the Fourier cosine transform given by:

If $f(x) = \sqrt{\dfrac{2}{\pi}} \displaystyle\int_0^\infty \phi(t) \cos tx \, dt$, then $\phi(x) = \sqrt{\dfrac{2}{\pi}} \displaystyle\int_0^\infty f(t) \cos tx \, dt$,

calling $f$ and $\phi$ reciprocal functions. This work of Cauchy contained the first published version of the Poisson summation formula, stated as

$$\sqrt{\alpha} \sum_{n=\infty}^{\infty} f(n\alpha) = \sqrt{\beta} \sum_{n=-\infty}^{\infty} \phi(n\beta), \tag{2.109}$$

where $\alpha\beta = 2\pi$ and $\phi$ denotes the Fourier cosine transform of $f$. Cauchy set $f(x) = e^{-\frac{x^2}{2}}$ and found the cosine transform to be $\phi(x) = e^{-\frac{x^2}{2}}$; thus, he designated $e^{-\frac{x^2}{2}}$ a self-reciprocal function. When this is substituted in (2.109), the result is the first equation in (2.90). In 1828, Jacobi proved this case by elliptic functions, and later in the same year, he proved the other cases.[34]

## 2.7  Exercises

1.  Prove Euler's formula

$$\prod_{n=1}^{\infty} (1 + q^n x) = 1 + \sum_{m=1}^{\infty} \frac{q^{\frac{m(m+1)}{2}} x^m}{(1-q)(1-q^2) \cdots (1-q^m)} \tag{2.110}$$

by denoting the product by $f(q, x)$ and observing that

$$f(q, x) = (1 + qx) f(q, qx). \tag{2.111}$$

Assume $f(q, x) = 1 + \sum_{m=1}^{\infty} a_m(q) x^m$ and use (2.111) to get

$$a_m(q) = \frac{q^m a_{m-1}(q)}{1 - q^m}.$$

See Euler (1988) pp. 261–262.

---

[33] Cauchy (1882–1974) ser 1, vol 1, pp. 5–318, especially pp. 300–303.

[34] See Jacobi (1965) vol. 1, pp. 260 and 264.

2. By the same method used in exercise 1, prove another formula of Euler:

$$\prod_{n=1}^{\infty} \frac{1}{(1-q^n x)} = 1 + \sum_{m=1}^{\infty} \frac{q^m x^m}{(1-q)(1-q^2)\cdots(1-q^m)}.$$

See Euler (1988) pp. 265–266.

3. Deduce from the identity in exercise 1 Euler's combinatorial theorem:

> The number of different ways in which the integer $n$ can be expressed as a sum of $m$ different positive integers is the same as the number of different ways in which $n - \frac{m(m+1)}{2}$ can be expressed as the sum of the numbers $1, 2, 3, \ldots, m$.

4. Using the identity in exercise 2, derive another combinatorial theorem of Euler:

> The number of different ways in which the integer $n$ can be expressed as a sum of $m$ positive integers is the same as the number of different ways in which $n - m$ can be expressed as the sums of the integers $1, 2, 3, \ldots, m$.

The theorems presented in exercises 3 and 4 were first stated and proved by Euler in response to a September, 1740 letter from Phillipe Naudé, a mathematician of French origin then working in Berlin. The letter posed the two combinatorial problems: Find the number of ways in which a given positive integer can be expressed as a sum of a fixed number, first: of distinct positive integers, and secondly: of integers not necessarily distinct. Euler replied to Naudé within two weeks, outlining the solutions, and soon afterward he presented the detailed solutions to the Petersburg Academy. His paper to the Academy is reproduced in Eu.I-2, pp. 163–193. See also Euler (1988) p. 267.

5. Note Euler's proof of the pentagonal number theorem:

$$\prod_{n=1}^{\infty}(1-x^n) = \sum_{n=-\infty}^{\infty} (-1)^n x^{\frac{n(3n-1)}{2}}$$

$$= 1 + \sum_{n=1}^{\infty} (-1)^n \left( x^{\frac{n(3n-1)}{2}} + x^{\frac{n(3n+1)}{2}} \right).$$

(a) Observe that

$$(1-\alpha)(1-\beta)(1-\gamma)\cdots$$
$$= 1 - \alpha - \beta(1-\alpha) - \gamma(1-\alpha)(1-\beta) - \delta(1-\alpha)$$
$$\times (1-\beta)(1-\gamma) - \cdots.$$

(b) Apply the algebraic identity in (a) to show that

$$P_0 \equiv \prod_{n=1}^{\infty}(1-x^n)$$
$$= 1 - x - x^2(1-x) - x^3(1-x)(1-x^2) - x^4(1-x)$$
$$\times (1-x^2)(1-x^3) - \cdots$$
$$= 1 - x - x^2 P_1,$$

where

$$P_1 = 1 - x + x(1-x)(1-x^2) + x^2(1-x)(1-x^2)(1-x^3) + \cdots.$$

(2.112)

(c) In (2.112), multiply out each term by the $(1-x)$ it contains to prove that

$$\begin{aligned} P_1 &= 1 - x + x(1-x^2) - x^2(1-x^2) + x^2(1-x^2)(1-x^3) \\ &\quad - x^3(1-x^2)(1-x^3) + \cdots \\ &= 1 - x^3 - x^5 P_2, \end{aligned}$$

where

$$P_2 = 1 - x^2 + x^2(1-x^2)(1-x^3) + x^4(1-x^2)(1-x^3)(1-x^4) + \cdots.$$

(d) Prove inductively that

$$P_{n-1} = 1 - x^{2n-1} - x^{3n-1} P_n,$$

(2.113)

where

$$P_n = \sum_{k=0}^{\infty} x^{kn} (1-x^n)(1-x^{n+1}) \cdots (1-x^{n+k}).$$

(2.114)

(e) Show that a repeated application of (2.113) and (2.114) proves the pentagonal number theorem.

  The pentagonal number theorem was conjectured by Euler in 1741. He found a proof nine years later; see Eu.I-2, 254–294. Euler explained his proof in a letter to his mathematical confidant Christian Goldbach; see Fuss (1968), vol. 1, pp. 522–524.

6. Prove Euler's statement that the complete integral of the equation

$$\frac{dx}{\sqrt{1-x^3}} = \frac{dy}{\sqrt{1-y^3}}$$

is given by

$$x^2 + y^2 + c^2 x^2 y^2 = 4c - 4c^2(x+y) + 2xy - 2cxy(x+y).$$

See Fuss (1968) vol. 1, p. 568. Also see Eu.1.20, pp. 74–75.

7. This exercise sketches Gauss's proof of the identity

$$\sum_{n=0}^{\infty} x^{\frac{n(n+1)}{2}} = \prod_{n=1}^{\infty} \frac{1 - x^{2n}}{1 - x^{2n-1}},$$

(2.115)

modeled on Euler's proof of the pentagonal number theorem. Let

$$P = \sum_{k=0}^{\infty} \frac{x^{kn}(1 - x^{2n+2(k+1)}) \prod_{j=1}^{k} (1 - x^{n+2j})}{\prod_{j=0}^{k} (1 - x^{n+2j+1})}$$

$$Q = \sum_{k=0}^{\infty} x^{(k+1)n} \prod_{j=0}^{k} \frac{1 - x^{n+2j+2}}{1 - x^{n+2j+1}}.$$

Show that

(a) $\psi(x, n) \equiv P - Q$

$$= \sum_{k=0}^{\infty} x^{kn} \prod_{j=0}^{k} \frac{1 - x^{n+2j}}{1 - x^{n+2j+1}}.$$

(b) $\psi(x, n) = 1 + x^{n+1} + x^{2n+3}\psi(x, n + 2), \quad n \geq 1.$

Hint: Subtract the $k$th term of $Q$ from the $(k + 1)$th term of $P$.

(c) $\psi(x, n) = 1 + x^{n+1} + x^{2n+3} + x^{3n+6} + x^{4n+10} + \cdots, \quad n \geq 1,$

$\psi(x, 0) = 0.$

(d) $\psi(x, 0) = 1 + x + x^3\psi(x, 2) - \prod_{j=1}^{\infty} \frac{1 - x^{2j}}{1 - x^{2j-1}}.$

Deduce (2.115) from (d). See Gauss (1863–1927) vol. 3, pp. 439–440.

8. We present the outline of one of Gauss's proofs of the triple product identity:

$$\sum_{n=-\infty}^{\infty} x^{n^2} y^n = \prod_{n=1}^{\infty} (1 - x^{2n})(1 + x^{2n-1}y) \left(1 + \frac{x^{2n-1}}{y}\right). \tag{2.116}$$

Let $P \equiv \prod_{n=1}^{\infty} (1 + x^{2n-1} y) \left(1 + \frac{x^{2n-1}}{y}\right) = \sum_{m=-\infty}^{\infty} A_m y^m.$

(a) Show that changing $y$ to $x^2 y$ changes the product to

$$\frac{1 + \frac{1}{xy}}{1 + xy} P = \frac{1}{xy} P,$$

and thus $\sum_{m=-\infty}^{\infty} A_m y^m = xy \sum_{m=-\infty}^{\infty} A_m x^{2m} y^m.$

(b) Deduce that

$$A_m(x) = x^{2m-1} A_{m-1}(x) = \cdots = x^{m^2} A_0(x), \quad m \in \mathbb{Z},$$

and hence

$$P = A_0(x) \sum_{m=-\infty}^{\infty} x^{m^2} y^m. \qquad (2.117)$$

(c) Let $[x] = \prod_{n=1}^{\infty} (1 - x^n)$ and let $y = -1$ in (2.117) to show that

$$A_0(x) = \frac{\prod_{n=1}^{\infty} (1 - x^{2n-1})^2}{\sum_{n=-\infty}^{\infty} (-1)^n x^{n^2}} = \frac{[x]^2}{[x^2]^2} \cdot \frac{1}{\sum_{n=-\infty}^{\infty} (-1)^n x^{n^2}}. \qquad (2.118)$$

(d) Let $y = i$ in (2.117) to prove that

$$A_0(x) = \frac{\prod_{n=1}^{\infty} (1 + x^{4n-2})}{\sum_{n=-\infty}^{\infty} (-1)^n x^{4n^2}} = \frac{[x^4]^2}{[x^2][x^8]} \cdot \frac{1}{\sum_{n=-\infty}^{\infty} (-1)^n x^{4n^2}}. \qquad (2.119)$$

(e) Now use (2.118) and (2.119) to show that

$$[x^2] A_0(x) = [x^8] A_0(x^4) = [x^{32}] A_0(x^{16}) = \cdots = 1,$$

and hence $\quad A_0(x) = \dfrac{1}{\prod_{n=1}^{\infty} (1 - x^{2n})}.$

For the proofs of (a) and (b), see Gauss (1863–1927) vol.3, p. 434 and for the proofs of (c), (d), and (e), see pp. 446–447 of the same volume.

9. Show that (2.16) is an integral of (2.15).
10. Prove Gauss's quadratic transformation formula by showing that (2.66) implies (2.67).
11. Prove equation (2.17), Euler's addition formula.
12. (a) Show that $\text{sl}'x = \frac{d}{dx} \text{sl}\,x = \sqrt{1 - \text{sl}^4 x}$ and then prove that

$$\text{sl}(x+y) + \text{sl}(x-y) = \frac{2\text{sl}\,x\,\text{sl}'\,y}{1 + \text{sl}^2 x\,\text{sl}^2 y}.$$

   (b) Show that

$$\text{sl}(3x) + \text{sl}\,x = \frac{2\text{sl}(2x)\,\text{sl}'\,x}{1 + \text{sl}^2(2x)\,\text{sl}^2 x}.$$

   (c) Prove (2.76).
   (d) Show that

$$\text{sl}(5x) + \text{sl}\,x = \frac{2\text{sl}(3x)\,\text{sl}'(2x)}{1 + \text{sl}^2(3x)\,\text{sl}^2(2x)}.$$

   (e) Prove (2.77).
13. Demonstrate that the formula $s^2 + c^2 + s^2 c^2 = 1$, where $s = \text{sl}x$, $c = \text{cl}x$, follows from (2.13). Deduce that

$$\text{sl}(x+y) = \frac{sc' + s'c}{1 - scs'c'}, \qquad (2.120)$$

where $s' = \text{sl}\,y$, $c' = \text{cl}\,y$.
   For these formulas, see Gauss (1863–1927) vol. 3, p. 404.

14. Verify:

$$\mathrm{sl}(x+y) = \frac{\mathrm{sl}^2 x - \mathrm{sl}^2 y}{\mathrm{sl}\, x\sqrt{1 - \mathrm{sl}^4 y} - \mathrm{sl}\, y\sqrt{1 - \mathrm{sl}^4 x}};$$

$$\sqrt{1 + \mathrm{sl}^2(x+y)} = \frac{\sqrt{(1 + \mathrm{sl}^2 x)(1 + \mathrm{sl}^2 y)} + \mathrm{sl}\, x\, \mathrm{sl}\, y\, \sqrt{(1 - \mathrm{sl}^2 x)(1 - \mathrm{sl}^2 y)}}{1 + \mathrm{sl}^2 x\, \mathrm{sl}^2 y}.$$

For these and similar formulas, see Gauss (1863–1927) vol. 3, p. 406.

15. Prove that

$$\mathrm{sl}(x+y) = \frac{\mathrm{sl}\, x\sqrt{1 - \mathrm{sl}^4 y} + \mathrm{sl}\, y\sqrt{1 - \mathrm{sl}^4 x}}{1 + \mathrm{sl}^2 x\, \mathrm{sl}^2 y} \qquad (2.121)$$

is equivalent to (2.120).

# 3

## Abel and Jacobi on Elliptic Functions

### 3.1 Preliminary Remarks

We have seen that Gauss discovered elliptic functions and their double periodicity in the late 1790s, but he did not publish any of his work on this topic. In the mid-1820s, he apparently considered publishing these results, but abandoned the idea after he read Abel's first paper: "Recherches sur la fonctions elliptiques" of 1827. This was a monograph of more than 120 pages, in which Abel inverted the elliptic integral

$$\alpha = \int_0^x \frac{dt}{\sqrt{(1 - c^2 t^2)(1 + e^2 t^2)}}, \quad 0 \leq x \leq \frac{1}{c},$$

to define an elliptic function $x = \phi(\alpha)$. He then employed an imaginary change of variables, a change he failed to justify,[1] to obtain $xi = \phi(\beta i)$, where

$$\beta = \int_0^x \frac{dt}{\sqrt{(1 + c^2 t^2)(1 - e^2 t^2)}}, \quad 0 \leq x \leq \frac{1}{e}.$$

We note that Abel may have been aware of Cauchy's 1814 and 1825 papers,[2] papers supporting this kind of complex substitution, but he did not refer to Cauchy in this connection. Cauchy's 1814 and 1825 papers did not provide the geometrical language of complex integration that would be required for the theory of elliptic functions. It was only in the 1840s, after Cayley and Eisenstein had published their objections to defining elliptic functions as inverses of integrals, that Cauchy began to explicitly employ a geometrical perspective to explain how these objections could be met. We will see that as early as 1811, Gauss had obtained some basic theorems in complex analysis in terms of geometric language; this approach is illustrated in Gauss's letter to Bessel.[3]

After the change of variables, Abel derived a doubly infinite product for $\phi(\alpha)$ analogous to the infinite product for $\sin x$. In fact, since $\phi$ was meromorphic, he expressed $\phi$ as a ratio of two infinite products. He next took $c = e = 1$ in order to consider the case of the lemniscate, for which he proved that for $p$ a prime of the form $2^n + 1$, $\phi\left(\frac{\omega}{p}\right)$,

---

[1] For the justification of Abel's method, see Armitage and Eberlein (2006).
[2] Cauchy (1882–1974) series 1, vol. 1, pp. 329–506 and series 2, vol. 15, pp. 41–89.
[3] Gauss (1863–1927) vol. 8, pp. 90–92.

with $\omega$ as in (2.70), could be solved in terms of square roots. Thus, he was able to verify Gauss's statement in Section 7 of the *Disquisitiones*.

Abel died in 1829 at the age of twenty-six of tuberculosis. In the short period between his first paper and his death, he published seven more papers on elliptic functions, including another paper of one hundred pages, "Prècis d'une theorie des fonctions elliptiques." It is very clear from these papers that, though he set down the foundations of this theory and developed many of its important ideas, he did not have time to work out other aspects of it, including complex multiplication and singular moduli. It remained for Kronecker, Hermite, and their followers to elaborate and expand on these fruitful ideas. Abel's rival in elliptic functions, Jacobi, lived a longer life and had the opportunity to more fully develop the theory of the elliptic function. Jacobi's work and notation were therefore employed by later researchers in this topic, and we present the details of his work, while also mentioning Abel's contributions.

Niels Henrik Abel (1802–1829) became interested in mathematics at the age of sixteen when Bernt Holmboe became a mathematics instructor at his school. Through Holmboe's encouragement, Abel read Euler's books on differential and integral calculus and then studied the works of Lagrange, Lacroix, Gauss, and others. By 1820, Abel was attempting to solve the quintic equation, and in 1821, he thought he had a solution. The solution was sent to the noted Danish mathematician Ferdinand Degen, who could find no error in it. Degen advised Abel to check his method by applying it to particular examples and to study elliptic integrals instead of the sterile subject of algebraic equations. Abel heeded Degen's advice on elliptic integrals, and by 1823, he had rediscovered elliptic functions by taking inverse functions of these integrals. We may be grateful that Abel ignored Degen's counsel to abandon algebraic equations; in fact, he found beautiful connections between the theory of equations and elliptic functions.

By testing particular cases according to Degen's suggestion, Abel discovered an error in his quintic solution; this suggested to him that an algebraic solution, i.e., a solution by radicals, might not be possible. In December, 1823, he proved this impossibility by writing down the general form of a radical expression and determining the properties required for such an expression to be a solution of a quintic. Soon after this, he considered the problem of determining all algebraic equations solvable by radicals. In a January, 1826 letter to Holmboe, he described his work:

> Since my arrival in Berlin, I have also been occupied with the solution of the following general problem: *Find all the equations which are solvable algebraically.* I have not yet completed this, but as far as I can judge, I shall succeed. As long as the degree of the equation is a prime number, the difficulty is not that great, but when that number is composite, the devil is in it. I have worked on equations of the fifth degree, and fortunately I managed to solve the problem for that case. I have found a great number of solvable equations, other than those already known. When I have completed the memoir as I hope to, I flatter myself that it will be good. It will be general, and it will include the method, which seems to me most essential.[4]

Abel employed his work on solvable equations to resolve the problem on the division of the arc of the lemniscate into equal parts; he enthusiastically commented on his result, and others, in another letter to his mentor Holmboe:

---

[4] Abel (1965) vol. 2, p. 256.

You will see how nice it is. I have found that one can divide the lemniscate with ruler and compass into $2^n + 1$ equal parts, as long as the number is prime. The division depends upon the equation of degree $(2^n + 1) - 1$; but I have found the complete solution by using square roots. This has led me to simultaneously investigate the mystery which has hovered over Mr. Gauss's theory on the division of the circumference of the circle. I see clear as day how he reached it. What I just told you about the lemniscate is one of the fruits of my researches on the theory of equations. You cannot imagine how many delightful theorems I have found there, for example this one: If an equation $P = 0$, whose degree is $\mu\nu$, $\mu$ and $\nu$ being prime to each other, is solvable in some manner by radicals, then $P$ is decomposable either into $\mu$ factors of degree $\nu$, whose coefficients depend upon a single equation of degree $\mu$, or else into $\nu$ factors of degree $\mu$, whose coefficients depend upon a single equation of degree $\nu$.[5]

Some of Abel's results on algebraic equations appeared in his 1829 paper, "Mémoire sur une classe particulière d'équations résolubles algébriquement," published in *Crelle's Journal*. Abel's first result supposed that an algebraic equation of degree $n$ had roots that could be expressed in the form

$$x, \theta x, \theta^2 x, \ldots, \theta^{n-1} x, \quad \text{with} \quad \theta^n x = x, \tag{3.1}$$

where $x_1 = \theta x$ was a rational function of $x$, $x_2 = \theta^2 x = \theta(x_1)$, $x_3 = \theta^3 x = \theta(x_2)$, etc. He proved that the equation could be solved algebraically (by radicals).

Secondly, he supposed that all the roots of an algebraic equation could be expressed as rational functions of one of them, meaning that they could be expressed as

$$\theta_1 x = x, \quad \theta_2 x, \quad \ldots, \quad \theta_n x,$$

where the $\theta_i x$, with $i = 1, 2, \ldots, n$, were rational functions of $x$. Also, he assumed that

$$\theta_i \theta_j x = \theta_j \theta_i x. \tag{3.2}$$

Then the equation could be solved by radicals.[6]

As an application of the first theorem, he considered the roots of the equation of degree $n - 1$ with $n$ prime:

$$\frac{x^n - 1}{x - 1} = 0. \tag{3.3}$$

The roots of (3.3) are $e^{\frac{2\pi i k}{n}}$, where $k = 1, 2, \ldots, n - 1$. Since $n$ is prime, there exist primitive roots of the equation

$$x^{n-1} \equiv 1 \pmod{n}; \tag{3.4}$$

this means that there is a number $\alpha$ such that $1, \alpha, \alpha^2, \ldots, \alpha^{n-2}$ are all distinct modulo $n$. So if we set $x = e^{\frac{2\pi i}{n}}$, then all the roots of the equation (3.3) may be given as

$$x, x^\alpha, x^{\alpha^2}, \ldots, x^{\alpha^{n-2}}.$$

---

[5] Abel (1965) vol. 2, pp. 261–262.     [6] Abel (1965) vol. 1, pp. 478–479.

Now set $x^\alpha = \theta x$ and observe that the roots of equation (3.3) take the form

$$x, \theta x, \theta^2 x, \ldots, \theta^{n-2} x. \tag{3.5}$$

Abel encountered a similar situation in his study of the division points of a lemniscate. In this case, the roots of the equation whose roots are the division points would be given by values of the lemniscatic function and these roots can be expressed as rational functions of one of them and be written in the form (3.1).

When $n$ is not prime, we employ Abel's second result. In this case, solutions of (3.3) take the form

$$e^{\frac{2\pi i k}{n}}, \quad k = 1, 2, \ldots, n-1,$$

and we can let $\theta_k(x) = x^k$, where $x = e^{\frac{2\pi i}{n}}$. Thus, we have $\theta_k\big(\theta_j(x)\big) = \theta_j\big(\theta_k(x)\big)$; it follows from (3.2) that the equation is solvable in radicals. In this manner, as Abel wrote to Holmboe,[7] he was able to resolve the mystery surrounding Gauss's theory of the division of the circle.

Note that Abel stated his results in terms of the relations among the roots, that is, with respect to the structure of the roots. Just slightly later, Galois presented his work on algebraic solvability in terms of a certain group of permutations of the roots, later called the Galois group of the equation. Thus, for Galois, the group of equation (3.3) is a cyclic group $G$ of order $n - 1$, generated by the element $g \in G$ that takes $\theta^i x$ to $\theta^{i+1} x$ for $i = 0, 1, \ldots, n - 2$, where $\theta^i x$ is given by (3.5). Since a cyclic group is solvable, the corresponding equation must be solvable by radicals. In a similar way, within Galois theory, condition (3.2) on the roots leads to an Abelian Galois group.

Another paper by Abel on this topic, "Sur la résolution algébrique des équations,"[8] went unpublished until much later because of his very unfortunate early death. In this paper, he considered equations leading to nonabelian solvable groups; in the introduction, Abel stated four theorems; we present the first and third:

(1) If an irreducible equation of prime degree $\mu$ can be solved algebraically, then its roots take the form

$$y = A + \sqrt[\mu]{R_1} + \sqrt[\mu]{R_2} + \cdots + \sqrt[\mu]{R_{\mu-1}},$$

where $A$ is a rational quantity, and $R_1, R_2, R_{\mu-1}$ are roots of an equation of degree $\mu - 1$.

(3) If an irreducible equation of degree $\mu$, where $\mu$ is divisible by distinct primes, is algebraically solvable, then the $\mu$ can be decomposed into two factors $\mu_1$ and $\mu_2$, such that the given equation can be decomposed into $\mu_1$ equations, each of degree $\mu_2$, whose coefficients depend upon the equations of degree $\mu_1$.

Abel's view on algebraic equations contrasts sharply in many ways with that of Galois; in this connection Gårding has written,[9]

The two founders of the theory, Abel and Galois, used different attacks, Abel investigating the structure of the roots and Galois using the Galois group, i.e, permutations of the roots which do not change coefficients. Both wrote unfinished papers. Posterity followed Galois, which has made the theory strongly imbued with group theory, for instance as is done in Weber's *Algebra* (1899,

---

[7] Abel (1965) vol. 2, p. 261.  [8] Abel (1965) vol. 2, pp. 217–243.
[9] Gårding (1997) pp. 27–29. Also see the nice paper by Gårding and Skau (1994).

1912) . . .. Abel's and Galois's manuscripts were enigmas to their contemporaries. That of Galois was deciphered first and therefore he was credited with the discovery of the Galois group for solvable equations of prime degree.

In addition to elliptic functions and the theory of equations and their interconnections, Abel also made important contributions to the theory of infinite series. This work was inspired by Cauchy's *Analyse algébrique* that Abel studied in 1825 while he was in Berlin. In his January 16, 1826 letter[10] to Holmboe, he offered insights on his research:

> Divergent series are generally something quite fatal, and it is a disgrace that one dares to found any demonstration on them. One can show anything one wishes by employing them, and it is these which have caused so many troubles and have given rise to so many paradoxes. Can one imagine anything more horrible than to allege
>
> $$0 = 1 - 2^n + 3^n - 4^n + \text{etc.},$$
>
> *n* being positive integer?[11] My eyes have been opened in a striking manner, since except for the most simple cases, for example geometric series, one finds in all mathematics almost no infinite series whose sum is determined in a rigorous manner, that is to say that the most essential part of mathematics is without foundation. For the most part, the results are accurate, it is true, but this is a very strange thing. I am searching for the explanation, a very interesting problem. I believe that you could present me with only a very small number of theorems containing infinite series, to whose demonstration I cannot make well-founded objections. Do this, and I will respond to you. The binomial theorem itself has not yet been rigorously proved. I have found that one has
>
> $$(1 + x)^m = 1 + mx + \frac{m(m-1)}{2}x^2 + \cdots$$
>
> for all values of *m*, when *x* [|x|] is less than unity.

Abel's contribution to the proof of the binomial theorem provide us with an illustration of the slow development of the concept of complete mathematical rigor. The earliest proofs of the theorem were not incorrect, but rather incomplete. Recall that Newton discovered the binomial theorem for fractional exponents in the winter of 1664–1665, though he had no proof at that time. He later verified the result in a few simple cases. He multiplied the series for $\sqrt{1+x}$ by itself to get $1 + x$; in a similar fashion he multiplied the series for $\sqrt[3]{1+x}$ by itself and then multiplied the resulting series by $\sqrt[3]{1+x}$ again to arrive at $1 + x$. This convinced Newton that his result was correct.[12]

In a paper of 1774,[13] Euler gave a general proof of the binomial theorem for rational exponents by using Newton's idea of multiplication of series. He set

$$[m] = 1 + mx + \frac{m(m-1)}{1 \cdot 2}x^2 + \frac{m(m-1)(m-2)}{1 \cdot 2 \cdot 3}x^3 + \cdots .$$

He showed that the product of the series $[m]$ with the series $[n]$ was the series $[m + n]$; thus,

$$[m + n] = [m][n]. \tag{3.6}$$

[10] Abel (1965) vol. 2, p. 257.    [11] See Eu.I-15, p. 72 or pp. 132–133.

[12] See, for example, Newton's October 1676 letter to Leibniz in Newton (1959–1960) vol. 2, pp. 110–161.

[13] Eu.I-15, pp. 207–216.

Euler also observed that when $m$ was an integer, then

$$[m] = (1 + x)^m; \tag{3.7}$$

He deduced the binomial theorem for rational exponents from (3.6) and (3.7).

Cauchy rendered Euler's proof of the binomial theorem for rational exponents fully rigorous. He defined the concept of an absolutely convergent series and then proved that the product of two absolutely convergent series must be absolutely convergent; he then showed that the binomial series was absolutely convergent for $|x| < 1$.[14] Cauchy also defined the idea of continuity, as was also done by Bolzano in 1817.[15] Cauchy then attempted to extend Euler's result, by using continuity, to the case where $m$ was any real number. He published this work in his 1821 *Analyse algébrique*, though some of the results were obviously completed somewhat earlier. Cauchy presented a proof of a proposition that a convergent series of continuous functions must be continuous; Cauchy then applied this proposition to the case of the binomial series, noting that

$$\binom{m}{k} = \frac{m(m-1)\cdots(m-k+1)}{k!}$$

was continuous in $m$ and hence, by his proposition, the series $[m]$ must be continuous in $m$. Abel pointed out exceptions to Cauchy's proposition; for example, each term in the series

$$\sin\phi - \frac{1}{2}\sin 2\phi + \frac{1}{3}\sin 3\phi - \cdots$$

was a continuous function of $\phi$, and yet when $k$ was an integer, the sum was discontinuous at $\phi = (2k + 1)\pi$. Abel proceeded to state and prove his own continuity theorem and a theorem to replace Cauchy's erroneous proposition.[16] To prove both theorems, Abel employed summation by parts, a method first explicitly stated by Brook Taylor in 1715.[17] Abel may have learned of this method from volume 3 of Lacroix's treatise on differential and integral calculus.[18] Though the method of summation by parts had been generally known for more than a hundred years, Abel was probably the first mathematician to apply it to prove theorems on convergence of series.

We note that the result of Euler referred to by Abel as horrible

$$1 - 2^{2k} + 3^{2k} - 4^{2k} + \cdots = 0 \tag{3.8}$$

is not incorrect. In fact, Euler had proved that

$$\lim_{x \to 1^-} (1 - 2^n + 3^n x^2 - 4^n x^3 + \cdots = 0,$$

---

[14]  See Cauchy (1821) Sections 6.3 and 6.4.    [15]  English translation of Bolzano by Russ (1980).
[16]  Abel (1965) vol. 1, pp. 219–250, especially pp. 221–230.    [17]  Taylor (1715).
[18]  Lacroix (1819) pp. 91–92.

that is, that the Abel mean of the series on the left-hand side of (3.8) was zero. In a paper of 1749,[19] he conjectured that, for $m$ real,

$$\frac{1 - 2^{-m} + 3^{-m} - 4^{-m} + 5^{-m} - 6^{-m} + \text{etc.}}{1 - 2^{m-1} + 3^{m-1} - 4^{m-1} + 5^{m-1} - 6^{m-1} + \text{etc.}}$$

$$= -\frac{1 \cdot 2 \cdot 3 \cdots (m - 1)(2^m - 1)}{(2^{m-1} - 1)\pi^m} \cos \frac{m\pi}{2}. \tag{3.9}$$

By $1 \cdot 2 \cdot 3 \cdots (m - 1)$ Euler meant $\Gamma(m)$ with $m$ an arbitrary real number. He verified this equation for certain classes of values of $m$. For example, he employed (3.8) to verify (3.9) when $m$ was a negative integer. Note that (3.9) is the functional equation for the Riemann zeta function. This follows from Landau's theorem.[20]

The Abel mean of the series $\sum_{n=1}^{\infty} (-1)^{n-1} n^{-s}$ is equal to $(1 - 2^{1-s})\zeta(s)$ for all values of $s$. Thus, Euler's result (3.8), of which Abel was so skeptical, would now be written as $\zeta(-2m) = 0$ when $m$ is a positive integer. Note that $s = -2m$, with $m$ a positive integer, are the trivial zeros of the zeta function.

Abel's work on elliptic functions is intertwined with Jacobi's early research on elliptic functions. Carl Gustav Jacob Jacobi (1804–1851) studied elementary mathematics and ancient languages with a maternal uncle, who gave him a strong enough grounding for Jacobi to enter the Potsdam Gymnasium at age twelve. Initially, he showed little interest in mathematics at his school, but his mathematics teacher, Heinrich Bauer, eventually realized that he had an exceptional student in his class and gave him Euler's *Introductio in Analysin Infinitorum*. Jacobi read Euler and proceeded to Lagrange, gaining sufficient algebra to attempt, at the age of fifteen, the solution of a fifth-degree equation. In a paper of 1834,[21] he mentioned this investigation, in which he reduced the general fifth-degree equation to the form $x^5 = 10q^2x + p$ and then showed that the solution of this equation depended upon the solution of an equation of the tenth degree.

The next year, Jacobi entered the University of Berlin, where he divided his time among philosophy, philology, and mathematics. He attracted the attention of the noted philologist P. A. Böckh, but after two years of study, despite a deep interest in Greek civilization, he decided to concentrate exclusively on mathematics. In a letter to his uncle, he explained the feelings underlying his choice:

> The enormous colossus, created by the work of an Euler, Lagrange, Laplace, requires the most tremendous power and exertion of thought, if one wishes to penetrate its inner essence, and not merely superficially rummage around in it. To become the master of this, so that one will not every instant fear being crushed by it, creates an impulse that cannot stop or rest until one stands at the top and can survey the whole body of work. Only when one has captured its spirit is it possible to truly work in peace at perfecting its individual parts and to further by one's powers the whole, great work,.[22]

Jacobi received his Ph.D. from Berlin with a dissertation on partial fractions, with a new method using infinite series. He applied this method in his later work, including his 1826 paper showing that the effectiveness of Gauss's method of numerical integration

---

[19] Eu.1-15 pp. 70–90.    [20] Landau (1906). See also Titchmarsh (1986) p. 21.
[21] Jacobi (1969) vol. 3, p. 278.    [22] Jacobi (1969) vol. 1, p. 5.

depended upon the orthogonality of the Legendre polynomials. In the same year, Jacobi was appointed lecturer at Königsberg and was promoted to assistant professor a year later. His speedy rise may have been due to his communications with Gauss, including a February 8, 1827 letter[23] Jacobi wrote to Gauss on cubic residues. Gauss was already aware of Jacobi's talents at that time, as a November 20, 1826 letter written to Bessel reflects. Gauss requests Bessel to send him more details about Jacobi, who, he says, "appears to be a very gifted young man."[24] In his letter to Gauss, Jacobi defined Jacobi sums and gave some of their basic properties. It is clear from a paper he wrote ten years later that Jacobi was aware that Gauss and Jacobi sums were finite-field analogs of the gamma and beta functions.[25]

In the spring of 1827, Jacobi started work on elliptic integrals under the influence of Legendre's 1817 work on elliptic integrals.[26] Apparently, Jacobi did not immediately bring forth new ideas upon reading the book. A friend found him looking very unhappy and asked for the cause. Jacobi responded, "You see me just about to return this book (Legendre's *Exercices*) to the library, with which I have had decisive bad luck. Otherwise, when I have studied such a significant work, it has always stimulated my own thought, and it is thus always something from which I profit. This time, I have emerged completely empty, and have been inspired to not the slightest ideas."[27]

Though Jacobi had to wait for original thought to appear in this case, the wait was worth it, leading Jacobi to arguably his greatest accomplishment: the creation, with Abel, of the theory of elliptic functions. He also made very significant contributions to the theory of hypergeometric functions and orthogonal polynomials, functional determinants and the change of variables formula for $n$-dimensional integrals, Hamiltonian dynamics, partial differential equations, number theory, and astronomy.

We have seen that Landen and Gauss published the two forms of a quadratic transformation in 1775 and 1818, respectively. In 1827, Jacobi considered odd-order transformations. Recall Jacobi's definition of a transformation of odd order $n$: For a given modulus $k$, a rational function $y = \frac{U(x)}{V(x)}$, where $U$ and $V$ were relatively prime polynomials of degree $n$ and $n-1$, respectively, such that

$$\frac{dy}{\sqrt{(1-y^2)(1-\lambda^2 y^2)}} = \frac{1}{M}\frac{dx}{\sqrt{(1-x^2)(1-k^2 x^2)}}, \tag{3.10}$$

where $M$ was a constant dependent only on $k$ and $\lambda$, with $\lambda$ the transformed modulus.

In June 1827, Jacobi gave examples of transformations of orders 3 and 5, communicating these results without proof to Heinrich Schumacher, editor of the *Astronomische Nachrichten*. In August of the same year, Jacobi sent Schumacher an example of a transformation of odd order $n$. Since he was not an expert on elliptic integrals, Schumacher did not feel confident in publishing these unproved, though clearly interesting, statements. He was aware that his friend and former teacher, Gauss, had worked extensively in this area, though he had published very few of his results. Schumacher therefore requested Gauss's opinion on the submitted work, and Gauss replied that Jacobi's results were particular cases of Gauss's own theorems; since he intended to publish

---

[23] Jacobi (1969) vol. 7, pp. 393–400.   [24] Auwers (1880) p. 463.   [25] Jacobi (1969) vol. 6, pp. 254–274.
[26] Legendre (1811–1817) vol 3.   [27] Jacobi (1969) vol. 1, p. 7.

these, he asked Schumacher not to contact him further in this connection. Schumacher then published Jacobi's theorems in the September 1827 issue of the *Nachrichten*; both Schumacher and Legendre strongly urged Jacobi to present proofs. In November 1827, reporting that he had worked very hard on this problem, Jacobi was able to mail his proof to Schumacher who promptly published it in December.

However, Jacobi's proof depended upon the fundamental idea of the inverse of the elliptic integral, and this had already appeared in Abel's "Reserches sur les fonctions elliptiques," published in *Crelle's Journal* in September, 1827. This issue of *Crelle's Journal* was received at the University of Königsberg, where Jacobi was a professor, by October 4, 1827. In his November 1827 paper containing his proof of the existence of a transformation of order $n$, Jacobi did not make mention of Abel's work.[28] As far as is known, Jacobi never discussed whether he had independently discovered the idea of the inversion of the elliptic integral, or whether he had gleaned this from Abel's paper. According to the Norwegian scientist Hansteen, who first showed Jacobi's paper to Abel, Abel was quite upset to see that Jacobi, in his *Nachrichten* paper, proved his transformation formula by means of the idea of the inverse of an elliptic integral without mentioning Abel's paper dealing with this concept. In general, Jacobi was in the habit of acknowledging the results of others. On balance, it certainly appears possible that Jacobi independently discovered the inversion and double periodicity and therefore, even if he had read Abel's paper, felt no need to refer to it. Indeed, it was the clear opinion of Dirichlet that the discoveries were independent.[29]

In fact, Abel had inverted the elliptic integral as early as 1823. Abel dated a letter to Holmboe as $\sqrt[3]{6064312219}$ ($\approx 1823.5908$, working out to August 3 or 4, 1823, though there is some doubt as to whether Abel made a computational mistake and the letter was written in late July). Abel wrote, "I requested M. *Degen* to peruse the short paper which, as you recall, treated inverse functions of elliptic transcendentals, and in which I proved something impossible, but he was unable to find any defect in the conclusion, or understand where the mistake might lie. I am hanged if I know how to get out of this."[30]

In Section 41 of the second part of his 1827 elliptic functions paper,[31] Abel addressed the problem of the transformation of elliptic functions and solved the problem for odd-order transformations. As an addition to the paper, he showed how Jacobi's example of an odd-order transformation was a particular case of his transformation theorem. Abel had been thinking about the transformation problem since 1824, and in a more general form than considered by Jacobi. Thus, when Jacobi's *Nachrichten* paper appeared, Abel was in a position to quickly compose "Solution d'un problème général concernant la transformation des fonctions elliptiques"[32] and mail it to Schumacher, editor of the *Nachricten*, on May 27, 1828. The definition of a transformation given in this paper, more general than Jacobi's, was an algebraic function $y$ of $x$ such that (3.10) would hold. In the course of the paper, he showed that this more general situation could be reduced to one in which $y$ was a rational function of $x$, the case considered by Jacobi. Abel formulated the problem in the beginning of his paper:

[28]  Jacobi (1969) vol. 1, pp. 37–48.    [29]  Jacobi (1969) vol. 1, pp. 3–28, especially pp. 10–11.
[30]  Abel (1965) vol. 2, p. 254.    [31]  Abel (1965) vol. 1, pp. 263–388.    [32]  Abel (1965) vol. 1, pp. 403–428.

Find all possible cases in which one can satisfy the differential equation:

$$(1) \qquad \frac{dy}{\sqrt{(1 - c_1^2 y^2)(1 - e_1^2 y^2)}} = \pm a \frac{dx}{\sqrt{(1 - c^2 x^2)(1 - e^2 x^2)}} \qquad (3.11)$$

setting $y$ to be an algebraic function of $x$, rational or irrational.

This problem, given the generality of the function $y$, appears at first glance very difficult, but we can reduce it to the case where $y$ is taken as rational. In effect, we can show that if equation (1) holds for an irrational value of $y$, then we can always arrive at another equation of the same form, in which $y$ is rational, by an appropriate change in the coefficient $a$, the quantities $c_1, e_1, c$, $e$ remaining the same. The method which first presents itself for solving the problem in the case where $y$ is rational is that of indeterminate coefficients; but we would soon be exhausted by the extreme complexity of the equations to be solved. I believe therefore that the following method, which leads in the most simple manner to a complete solution, should perhaps deserve the attention of mathematicians.[33]

Concerning Abel's paper, and supporting the view that Jacobi would not have appropriated Abel's work, Jacobi wrote in a September 9, 1828 letter to Legendre: "[It] contains a rigorous *deduction* of the transformation theorems, the lack of which was felt in my own note on the same subject. It is as far beyond my praise, as it is beyond my own work."[34] Legendre informed Abel of Jacobi's remarks and Abel responded, "I am very much looking forward to understanding the work of Mr. *Jacobi*. One ought to find some marvelous things there. Certainly Mr. *Jacobi* will develop, to a degree unhoped-for, not only the theory of elliptic functions, but also mathematics in general. I could not esteem him more."[35] Thus, it appears that any discord between Jacobi and Abel may have been resolved before Abel's untimely death.

In addition to the two papers in Schumacher's *Nachrichten*, Jacobi published six notes on elliptic functions in *Crelle's Journal* before the appearance of his 1829 book, *Fundamenta Nova Theoriae Functionum Ellipticarum*. These notes contained statements of results almost without proofs. Abel presented proofs for some of these results by applying his own methods and Jacobi in turn mentioned some implications of Abel's methods for his own work.

For example, in Section 10 of his "Recherches sur les fonctions elliptiques," Abel stated two theorems on the differential equation

$$\frac{dy}{\sqrt{(1 - y^2)(1 + \mu y^2)}} = a \frac{dx}{\sqrt{(1 - x^2)(1 + \mu x^2)}}. \qquad (3.12)$$

Theorem I. Suppose $a$ is real and the equation [(3.12)] is algebraically integrable; then $a$ is necessarily a rational number.

Theorem II. Suppose $a$ is a complex number and the equation [(3.12)] is algebraically integrable; then $a$ is necessarily of the form $m + \sqrt{-1}\sqrt{n}$, where $m$ and $n$ are rational numbers. In this case, the quantity $\mu$ is not arbitrary; it must satisfy an equation with an infinite number of real and complex roots. Any such value of $\mu$ fulfills the condition.[36]

Jacobi commented on these theorems in a short paper published in *Crelle's Journal*.[37] In this paper, Jacobi expressed his elliptic functions as ratios of infinite series.

---

[33] Abel (1965) vol. 1, pp. 403–404.  [34] Jacobi (1969) vol. 1, p. 423.
[35] Abel (1965) vol. 2, p. 279.  [36] Ibid. vol. 1, p. 377.  [37] Jacobi (1969) vol. 1, pp. 251–54.

Recall that Abel had represented elliptic functions as ratios of infinite products. Jacobi too had obtained such formulas, but he found a method of converting these products into series; he thus gave his results only in terms of infinite series. Jacobi set up his notation:

$$\int_0^{\frac{\pi}{2}} \frac{d\phi}{\sqrt{1 - k^2 \sin^2 \phi}} = K; \quad \int_0^{\frac{\pi}{2}} \frac{d\phi}{\sqrt{1 - k'^2 \sin^2 \phi}} = K',$$

where $k^2 + k'^2 = 1$ and $k'$ is called the complement of $k$. Using this notation, Jacobi stated a very important result: If $q = e^{-\pi \frac{K'}{K}}$, then

$$\sqrt{\frac{2K}{\pi}} = 1 + 2 \sum_{n=1}^{\infty} q^{n^2}, \tag{3.13}$$

$$\frac{1}{2}\sqrt{k} = \frac{\sum_{n=0}^{\infty} q^{(n+\frac{1}{2})^2}}{1 + 2\sum_{n=1}^{\infty} q^{n^2}}. \tag{3.14}$$

In his *Fundamenta Nova*, Jacobi showed how (3.13) could be applied to find the number of representations of an integer $n$ as a sum of two, four, six, and eight squares, explicitly working out the result only for the case of two and four squares.[38] Another result in this short paper dealt with the algebraic relationship between $k$ and $\lambda$ in (3.10). Suppose $n$ is an odd prime $p$. For a given $p$ and a $k$ taken to lie between 0 and 1, Jacobi noted that there were $p + 1$ possible values of $\lambda$. These values could be obtained when $q$ in (3.14) was replaced by one of the values

$$q^p, q^{\frac{1}{p}}, \alpha q^{\frac{1}{p}}, \ldots, \alpha^{p-1} q^{\frac{1}{p}},$$

where $\alpha = e^{\frac{2\pi i}{p}}$, that is, $\alpha^p = 1$.[39] Two of these values of $\lambda$, those corresponding to $q^p$ and $q^{\frac{1}{p}}$, were real. In the *Fundamenta Nova*, Jacobi showed that when one performed the transformation corresponding to $q^p$ and then the one corresponding to $q^{\frac{1}{p}}$, then the transformation obtained would be

$$\frac{dz}{\sqrt{(1 - z^2)(1 - k^2 z^2)}} = \frac{p\, dx}{\sqrt{(1 - x^2)(1 - k^2 x^2)}}. \tag{3.15}$$

Thus, the multiplication by $p$ of the variable in an elliptic function was obtained in two steps.

In the short paper under discussion, Jacobi also presented the algebraic equations satisfied by $k$ and $\lambda$ in the two cases: $p = 3$ and $p = 5$. These became known as the modular equations of orders 3 and 5. He set $\sqrt[4]{k} = u$, $\sqrt[4]{\lambda} = v$ and presented the modular equations:[40]

$$u^4 - v^4 - 2uv(1 - u^2 v^2) = 0 \quad \text{for } p = 3; \tag{3.16}$$

$$u^6 - v^6 + 5u^2 v^2(u^2 - v^2) - 4uv(1 - u^2 v^2) = 0 \quad \text{for } p = 5. \tag{3.17}$$

At the end of the paper, he commented on Theorems I and II as related to (3.12). He observed that Theorem I was an analog of the result that if $\sin(nx)$ was an algebraic

---

[38] Ibid. pp. 161–163.     [39] Jacobi (1969) vol. 1, p. 252.     [40] Jacobi (1969) vol. 1, p. 253.

function of $\sin x$, then $n$ would be a rational number. Concerning Theorem II, Jacobi commented,[41]

> There exists an infinite number of scales of modules for which $a$ can also have the form $a + b\sqrt{-1}$. They are all those where the module transforms into its complement. One of these modules, for example, is $k = \sqrt{\frac{1}{2}}$. This new method of multiplication is even more remarkable because it holds in the case where the transformation reverts within the multiplication, that is to say, where the transformed module becomes equal to the one from which it originated. For example, for $n = 5$, $k = \sqrt{\frac{1}{2}}, u = \sqrt[8]{\frac{1}{2}}$, one finds that the equation
>
> $$u^6 - v^6 + 5u^2v^2(u^2 - v^2) - 4uv(1 - u^4v^4) = 0$$
>
> has the root $v = (1 + \sqrt{-1})u^5$, from which one obtains $v^8 = \lambda^2 = \frac{1}{2}$, so that one has $\lambda^2 = k^2$. All this follows immediately from the principles established by Mr. Abel.

Abel responded to this paper with his "Note sur quelques formules elliptiques," in which he employed his product formulas for elliptic functions to derive similar representations for Jacobi's functions. He proved some of Jacobi's formulas, but did not derive the series expressions for the elliptic functions. In fact, the conversion of the product to the series can be achieved by using the triple product identity published by Jacobi toward the end of his *Fundamenta Nova* of 1829, though he had found it earlier.

When $a$ is a complex number in (3.12), we have complex multiplication of an elliptic function. In his "Recherches" paper, Abel considered only two simple examples: $a = \sqrt{-3}$ and $a = \sqrt{-5}$; he proved that for the first case, $\sqrt{\mu} = \sqrt{3} \pm 2$ and for the second case, $\sqrt{\mu} = 1, 2 + \sqrt{5} \pm 2\sqrt{2 + \sqrt{5}}$. Note that Theorem II on (3.12) hinted that the modulus $\mu$ would satisfy an algebraic equation if the quantity $a$ in the differential equation (3.12) was of the form $m + \sqrt{-n}$. In the two cases for which he gave a detailed account, $a = \sqrt{-3}$ and $a = \sqrt{-5}$, Abel found that the modulus $\mu$ could be expressed in terms of radicals. He was then uncertain as to whether this would hold in all cases of complex multiplication. With a change of notation, $\mu = e^2$, Abel wrote, "In the two cases we have just considered, it was not difficult to find the value of the quantity $e$, but when the value of $n$ is larger, we will arrive at an algebraic equation that will perhaps not be solvable algebraically [i.e., in terms of radicals]."[42]

A few months later, Abel wrote his "Sur d'un problème général concernant la transformation des fonctions elliptiques" in which he reflected further on these ideas and actually formulated his Theorem II for (3.12) in a more complete form. He had by then understood that $e$ could be expressed in terms of radicals. Among the points he made in this connection, Abel wrote,[43]

> (c) There is a remarkable case of the general problem; it is that in which one seeks all possible solutions of the equation
>
> $$\frac{dy}{\sqrt{(1 - c^2y^2)(1 - e^2y^2)}} = a\frac{dx}{\sqrt{(1 - c^2x^2)(1 - e^2x^2)}}. \tag{3.18}$$
>
> In this connection, we have the following theorem:

---

[41] Jacobi (1965) vol. 1, p. 254.    [42] Abel (1965) vol. 1, p. 383.    [43] Ibid. pp. 425–426.

If the preceding equation admits of an *algebraic* solution in $x$ and $y$, $y$ being rational in $x$ or not, the constant quantity $a$ must necessarily take the form

$$\mu' + \sqrt{-\mu},$$

where $\mu'$ and $\mu$ designate two rational numbers, the latter being essentially positive [i.e. nonnegative]. If we assign to $a$ one such value, we can find an infinity of different values for $e$ and $c$, that render the problem possible. All these values are expressible by *radicals*.

Thus, if we assume that $a$ is a real quantity, it is necessary that it must also be rational. In this case, we also know that we can satisfy the differential equation in question regardless of the values of the quantities $c$ and $e$.

(d) Concerning the preceding theorem, by a simple change of variables, we can deduce this:

If the equation

$$\frac{dy}{\sqrt{(1 - y^2)(1 - b^2 y^2)}} = \frac{a\,dx}{\sqrt{(1 - x^2)(1 - c^2 x^2)}}$$

where $b^2 = 1 - c^2$ [$b$ being the complementary modulus], admits of an algebraic solution between $x$ and $y$, the coefficient $a$ must take the form

$$\sqrt{\mu} + \mu'\sqrt{-1},$$

$\mu'$ and $\mu$ having the same meaning as above. If then we wish $a$ to be real, it is necessary that it be equal to the square root of a rational quantity. If this condition is fulfilled, the problem has an infinity of solutions.

Abel's work on complex multiplication of elliptic functions and the corresponding moduli was continued by Kronecker, starting around 1858. We will touch on this in a later chapter. Here we discuss Jacobi's transformation theory and Jacobi's derivation, from his transformation formulas, of the infinite products for his elliptic functions and for the modulus $k$ and related quantities. We also present Jacobi's theory of theta functions and the triple product identity that converts the mentioned infinite products into infinite series.

## 3.2   Jacobi on Transformations of Orders 3 and 5

Gauss's remark in his *Disquisitiones* on the extent of his theory suggests that he understood something of the theory of complex multiplication of elliptic curves. We have seen that Gauss employed the quadratic transformation (2.66) to express the arithmetic-geometric mean as an elliptic integral. Evidence suggests that he had also discovered examples of higher-order transformations of elliptic integrals. In his first letter to Legendre, dated August 5, 1827, Jacobi wrote that by 1808, Gauss had discovered third-, fifth-, and seventh-order transformations.[44] From his reply to Jacobi, we know that in 1825, Legendre too had found the third-order transformation.[45] Jacobi, the first to publish the fifth-order transformation, communicated the third- and fifth-order transformations without proofs to Heinrich Schumacher in a letter of June 13, 1827. But, as we have seen, Schumacher, a student of Gauss and the founder and longtime editor of the *Astronomische Nachrichten*, passed the letter on to Gauss for verification of the results. Gauss replied that the results were correct, but since he himself was planning to publish his work on elliptic functions and integrals, he requested Schumacher not to

---

[44] Jacobi (1969) vol. 1, p. 394.     [45] Ibid. p. 396.

communicate to him any further papers on the topic. However, Gauss decided not to publish his work in this area, after the publication of Abel's great paper of September, 1827, "Recherches sur les fonctions elliptiques," in *Crelle's Journal*. In fact, Gauss wrote to Crelle that Abel had followed the same path in this study as he himself had taken in 1798.

After hearing from Gauss, Schumacher published Jacobi's notes. For his cubic transformation,[46] set

$$\sin \phi = \frac{\sin \psi \left(ac + \left(\frac{a-c}{2}\right)^2 \sin^2 \psi\right)}{c^2 + \frac{a-c}{2} \cdot \frac{a+3c}{2} \sin^2 \psi}, \tag{3.19}$$

so that we obtain

$$\frac{d\phi}{\sqrt{a^3 c - \frac{a-c}{2} \left(\frac{a+3c}{2}\right)^3 \sin^2 \phi}} = \frac{d\psi}{\sqrt{c^3 a - \left(\frac{a-c}{2}\right)^3 \frac{a+3c}{2} \sin^2 \phi}}. \tag{3.20}$$

In addition, if we have

$$\sin \psi = \frac{\sin \theta \left(-3ac + \left(\frac{a+3c}{2}\right)^2 \sin^2 \theta\right)}{a^2 - 3\frac{a-c}{2} \cdot \frac{a+3c}{2} \sin^2 \theta} \tag{3.21}$$

and

$$\chi = \frac{a-c}{2c} \left(\frac{a+3c}{2a}\right)^3, $$

then we can write

$$\frac{d\phi}{\sqrt{1 - \chi \sin^2 \phi}} = \frac{3d\theta}{\sqrt{1 - \chi \sin^2 \theta}}. \tag{3.22}$$

Note that (3.19) and (3.21) are the two cubic transformations. Jacobi's statements of his quintic transformations is succinct:[47] If we set $a^3 = 2b(1 + a + b)$ and

$$\sin \phi = \frac{\sin \psi (1 + 2a + (a^2 + 2ab + 2b) \sin^2 \psi + b^2 \sin^4 \psi)}{1 + (a^2 + 2a + 2b) \sin^2 \psi + b(b + 2a) \sin^4 \psi}, \tag{3.23}$$

we get

$$\int \frac{d\phi}{\sqrt{(a - 2b)(1 + 2a)^2 - (2 - a)(b + 2a)^2 \sin^2 \phi}}$$
$$= \int \frac{d\psi}{\sqrt{a - 2b - b^2(2 - a) \sin^2 \psi}}. \tag{3.24}$$

Also, if

$$\alpha = \frac{2-a}{1+2a}, \quad \beta = -\frac{b+2a}{1+2a} \cdot \frac{2-a}{a-2b}, \quad \chi = \frac{2-a}{a-2b} \cdot \left(\frac{b+2a}{1+2a}\right)^2,$$

$$\sin \psi = \frac{\sin \theta (1 + 2\alpha + (\alpha^2 + 2\alpha\beta + 2\beta) \sin^2 \theta + \beta^2 \sin^4 \theta)}{1 + (\alpha^2 + 2\alpha + 2\beta) \sin^2 \theta + \beta(\beta + 2\alpha) \sin^4 \theta}, \tag{3.25}$$

---

[46] Ibid. pp. 32–33.     [47] Ibid. p. 33.

then we have

$$\int \frac{d\phi}{\sqrt{1 - \chi \sin^2 \phi}} = 5 \int \frac{d\theta}{\sqrt{1 - \chi \sin^2 \theta}}. \tag{3.26}$$

We can see that here (3.23) and (3.25) are Jacobi's quintic transformations.

In his April 12, 1828 letter[48] to Legendre, Jacobi wrote that he found (3.19) and (3.23) using the algebraic theory of transformation that he developed in March 1827. He also explained that he had found the transformations for trisection and quinsection by the method of undetermined coefficients and then noticed the similarity of (3.19) and (3.23) with these transformations. This led him to discover the second transformations (3.21) and (3.25) by trial and error. In developing his algebraic theory, he began by taking a transformation $y = \frac{U}{V}$, where $U$ and $V$ were polynomials in $x$ differing in degree by at most 1, and such that

$$\frac{dy}{\sqrt{(1 - y^2)(1 - \lambda^2 y^2)}} = \frac{1}{M} \frac{dx}{\sqrt{(1 - x^2)(1 - k^2 x^2)}}, \tag{3.27}$$

where $M$ was a constant depending on $k$ and $\lambda$. Substituting $y = \frac{U}{V}$ in the left-hand side of (3.27), he obtained

$$\frac{dy}{\sqrt{(1 - y^2)(1 - \lambda^2 y^2)}} = \frac{(VU' - UV') \, dx}{\sqrt{(V^2 - U^2)(V^2 - \lambda^2 U^2)}}, \tag{3.28}$$

where $U'$ and $V'$ denoted the derivatives of $U$ and $V$, respectively. If $U$ and $V$ were of degree $p$, then $VU' - UV'$ would be of degree $2p - 2$ because the terms of degree $2p - 1$ in $VU'$ and $UV'$ were identical. The degree of the expression within the square root in the denominator of the right-hand side of (3.28) would be $4p$. Since the numerator could be written as

$$(V - \lambda U)\frac{dU}{dx} - U\frac{d}{dx}(V - \lambda U) = VU' - UV', \tag{3.29}$$

it followed that if any factors of $V \pm U, V \pm \lambda U$ in the denominator had a square factor $(1 - \beta x)^2$, then $1 - \beta x$ must be a factor of $VU' - UV'$. Jacobi concluded that if the denominator were of the form $T^2 X$, where $X$ was a quartic in $x$, then

$$M = \frac{T}{VU' - UV'} \tag{3.30}$$

was a constant, that is, independent of $x$. He therefore looked for polynomials $U$ and $V$ such that there existed polynomials $A, B, C, D$ with

$$V + U = (1 + x)A^2, \quad V - U = (1 - x)B^2, \tag{3.31}$$

$$V + \lambda U = (1 + kx)C^2, \quad V - \lambda U = (1 - kx)D^2. \tag{3.32}$$

He also observed that $y$ was an odd function of $x$, so he could take

$$U = xF(x^2) \quad \text{and} \quad V = \phi(x^2). \tag{3.33}$$

---

[48] Ibid. vol. 1, pp. 409–416, especially pp. 415–416.

In Sections 13–16 of his *Fundamenta Nova*, Jacobi explained how he applied his general transformation theory to derive his cubic and quintic transformations. To derive the cubic transformation (3.19), Jacobi wrote (3.19) and (3.20) in a more convenient way, setting $\frac{a}{c} = 2\alpha + 1$. The transformation then took the form: If

$$y = \frac{x(2\alpha + 1 + \alpha^2 x^2)}{1 + \alpha(\alpha + 2)x^2}, \tag{3.34}$$

then

$$\frac{dy}{\sqrt{(1 - y^2)(1 - \lambda^2 y^2)}} = \frac{(2\alpha + 1)dx}{\sqrt{(1 - x^2)(1 - k^2 x^2)}}, \tag{3.35}$$

where

$$k^2 = \frac{\alpha^3(2 + \alpha)}{2\alpha + 1}, \quad \lambda^2 = \frac{\alpha(2 + \alpha)^3}{(2\alpha + 1)^3}. \tag{3.36}$$

To obtain the third-order transformation (3.34) from the general considerations exemplified by (3.31) through (3.33), observe that since $U = x F(x^2)$ and $V = \phi(x^2)$, Jacobi[49] could assume that

$$V = 1 + b x^2 \quad \text{and} \quad U = x(a + a_1 x^2),$$

where $a$, $a_1$, and $b$ were constants to be determined based on the condition that

$$U + V = (1 + x)A^2.$$

He assumed $A = 1 + \alpha x$ to arrive at

$$U + V = 1 + ax + bx^2 + a_1 x^3 = (1 + x)A^2$$
$$= 1 + (1 + 2\alpha)x + \alpha(2 + \alpha)x^2 + \alpha^2 x^3.$$

Equating the coefficients of the powers of $x$ gave Jacobi

$$a = 1 + 2\alpha, \quad a_1 = \alpha^2, \quad b = \alpha(2 + \alpha),$$

and hence $y = \frac{U}{V}$ would be given by (3.34). To derive (3.36), he observed that (3.35) remained unchanged when $y$ was replaced by $\frac{1}{\lambda y}$ and $x$ by $\frac{1}{kx}$. Implementing this change in (3.34), Jacobi obtained

$$\frac{\lambda x((2\alpha + 1)\alpha^2 + \alpha^4 x^2)}{\alpha^2 + \alpha^3(\alpha + 2)x^2} = \frac{kx(\alpha(\alpha + 2) + k^2 x^2)}{\alpha^2 + (2\alpha + 1)k^2 x^2}.$$

By equating the powers of $x$, he found

$$k^2 = \frac{\alpha^3(2 + \alpha)}{2\alpha + 1} \quad \text{and} \quad \lambda^2 = \frac{k^6}{\alpha^8} = \alpha\left(\frac{2 + \alpha}{2\alpha + 1}\right)^3, \tag{3.37}$$

with the complementary moduli given by

$$k'^2 = 1 - k^2 = \frac{(1 - \alpha)(1 + \alpha)^3}{2\alpha + 1}, \quad \lambda'^2 = 1 - \lambda^2 = \frac{(1 + \alpha)(1 - \alpha)^3}{(2\alpha + 1)^3}. \tag{3.38}$$

---

[49] Ibid. p. 74.

Relations (3.37) and (3.38) produce Legendre's form[50] of the modular equation of third order:

$$\sqrt{k\lambda} + \sqrt{k'\lambda'} = \frac{\alpha(2+\alpha)}{2\alpha+1} + \frac{1-\alpha^2}{2\alpha+1} = 1. \tag{3.39}$$

Jacobi proved (3.35) by observing that, with $D = 1 + \alpha(\alpha+2)x^2$,

$$1 \pm y = \frac{(1 \pm x)(1 \pm \alpha x)^2}{D}$$

$$1 \pm \lambda y = \frac{(1 \pm kx)\left(1 \pm \frac{kx}{\alpha}\right)}{D}$$

and hence

$$\frac{dy}{\sqrt{(1-y^2)(1-\lambda^2 y^2)}} = \frac{(2\alpha+1)\,dx}{\sqrt{(1-x^2)(1-k^2 x^2)}}.$$

We note that Jacobi gave the modular equation in (3.39) a slightly different form. He let $\sqrt[4]{k} = u$ and $\sqrt[4]{\lambda} = v$. From (3.37) he had

$$\sqrt{k\lambda} = \frac{\alpha(\alpha+2)}{2\alpha+1} = u^2 v^2 \tag{3.40}$$

$$\frac{k^3}{\lambda} = \alpha^4 \quad \text{or} \quad \alpha = \frac{u^3}{v}. \tag{3.41}$$

By substituting the value of $\alpha$ from (3.41) into (3.40), Jacobi obtained

$$u^2 v^2 = \frac{u^3(u^3 + 2v)}{v(2u^3 + v)}$$

or

$$u^4 - v^4 + 2uv\left(1 - u^2 v^2\right) = 0 \tag{3.42}$$

and this was Jacobi's form of the modular equation of third order.

To present a summary of Jacobi's derivation of the fifth-order modular equation, let us first work out an implication of Jacobi's observation[51] that (3.27) would remain unchanged when $x$ was replaced by $\frac{1}{kx}$ and $y$ by $\frac{1}{\lambda y}$. Now if $y = \frac{xF(x^2)}{\phi(x^2)}$, where $\phi$ and $F$ are polynomials of degree $\frac{n-1}{2} = m$ ($n$ odd), then, replacing $x$ by $\frac{1}{kx}$, we get

$$\frac{1}{\lambda y} = \frac{\phi(x^2)}{\lambda x F(x^2)} = \frac{F\left(\frac{1}{k^2 x^2}\right)}{kx \phi\left(\frac{1}{k^2 x^2}\right)} = \frac{x^{2m} F\left(\frac{1}{k^2 x^2}\right)}{kx x^{2m} \phi\left(\frac{1}{k^2 x^2}\right)}. \tag{3.43}$$

Since $x^{2m} F\left(\frac{1}{k^2 x^2}\right)$ and $x^{2m} \phi\left(\frac{1}{k^2 x^2}\right)$ are polynomials of degree $2m$, it follows that for a constant $C$

$$\phi(x^2) = C x^{2m} F\left(\frac{1}{k^2 x^2}\right), \qquad \lambda F(x^2) = C k x^{2m} \phi\left(\frac{1}{k^2 x^2}\right). \tag{3.44}$$

---

[50] See Legendre's February 9, 1828 letter to Jacobi in Jacobi (1969) vol.1, pp. 406–408, especially p. 408.
[51] Jacobi (1969) vol. 1, pp. 72–73.

Equation (3.44) implies that

$$\phi\left(\frac{1}{k^2x^2}\right) = \frac{C}{k^{2m}x^{2m}} F(x^2), \qquad \lambda F\left(\frac{1}{k^2x^2}\right) = \frac{Ck}{k^{2m}x^{2m}} \phi(x^2). \qquad (3.45)$$

A comparison of (3.44) with (3.45) shows that

$$\frac{C}{k^{2m}} = \frac{\lambda}{Ck}, \quad \text{or} \quad C = \sqrt{\lambda k^{2m-1}}.$$

Hence

$$\phi(x^2) = \sqrt{\lambda k^{2m-1}}\, x^{2m} F\left(\frac{1}{k^2x^2}\right), \qquad F(x^2) = \sqrt{\frac{k^{2m+1}}{\lambda}}\, x^{2m} \phi\left(\frac{1}{k^2x^2}\right). \qquad (3.46)$$

Note that all relations (3.43) through (3.46) are contained in Jacobi's work; to give Jacobi's derivation[52] of the fifth-order modular equation, we use (3.33) to set

$$V = 1 + b_1 x^2 + b_2 x^4, \quad U = x(a + a_1 x^2 + a_2 x^4), \quad A = 1 + \alpha x + \beta x^2.$$

Applying (3.31) and equating the coefficients of powers of $x$, we get

$$b_1 = 2\alpha + 2\beta + \alpha^2, \quad b_2 = \beta(2\alpha + \beta), \quad a = 1 + 2\alpha,$$
$$a_1 = 2\beta + \alpha^2 + 2\alpha\beta, \quad a_2 = \beta^2. \qquad (3.47)$$

Now (3.46), with $m = 2$, implies that

$$x(a + a_1 x^2 + a_2 x^4) = \sqrt{\frac{k^5}{\lambda}}\left(x^5 + \frac{b_1}{k^2}x^3 + \frac{b_2}{k^4}x\right);$$

thus

$$a_2 = \sqrt{\frac{k^5}{\lambda}}, \quad a_1 = \sqrt{\frac{k}{\lambda}}\, b_1, \quad a = \sqrt{\frac{k}{\lambda}}\frac{b_2}{k^2}. \qquad (3.48)$$

Equation (3.48) yields

$$\frac{a_1^2}{aa_2} = \frac{b_1^2}{b_2}. \qquad (3.49)$$

Substituting the values of $a, a_1, a_2, b_1, b_2$ from (3.47), we obtain the relation

$$\left(\frac{2-\alpha}{\alpha-2\beta}\right)^2 = \frac{2\alpha+\beta}{\beta(1+2\alpha)}. \qquad (3.50)$$

Let $k = u^4$ and $\lambda = v^4$. It follows from (3.47) and (3.48) that

$$\frac{2\alpha+\beta}{\beta(1+2\alpha)} = \frac{b_2}{aa_2} = \frac{b_1^2}{a_1^2} = \frac{\lambda}{k} = \frac{v^4}{u^4}. \qquad (3.51)$$

Next, by (3.47), (3.48), and (3.50) we obtain

$$\frac{2-\alpha}{\alpha-2\beta} = \frac{v^2}{u^2} \quad \text{and} \quad \beta = \sqrt{a_2} = \sqrt[4]{\frac{k^5}{\lambda}} = \frac{u^5}{v}. \qquad (3.52)$$

[52] Ibid. pp. 77–78.

Using $\beta = \frac{u^5}{v}$ and then eliminating $\alpha$ from the equations in (3.51) and (3.52), we obtain Jacobi's fifth-order modular equation:

$$u^6 - v^6 + 5u^2v^2(u^2 - v^2) + 4uv(1 - u^4v^4) = 0.$$

The elimination method used by Jacobi to construct modular equations of orders 3 and 5 is difficult to extend to higher orders. However, it can be applied to show the existence of a modular equation of a general odd-order $n$.[53] In his *Fundamenta*, Jacobi gave the necessary equations to which elimination must be applied in order to obtain the modular equation of order $n$.

Jacobi also stated some general properties of the modular equation of order $n$,

$$F_n(k, \lambda) = 0.$$

He showed that $k$ and $\lambda$ satisfied a third-order differential equation that he wrote in the form[54]

$$2dk\,d\lambda(d\lambda\,d^3k - dk\,d^3\lambda) - 3(d\lambda^2\,d^2k^2 - dk^2\,d^2\lambda^2)$$
$$+ dk^2\,d\lambda^2\left(\left(\frac{1+k^2}{k-k^3}\right)^2 dk^2 - \left(\frac{1+\lambda^2}{\lambda-\lambda^3}\right)^2 d\lambda^2\right) = 0.$$

Note that here $dk^2 = (dk)^2$, $d^2k^2 = (d^2k)^2$. From the symmetry in this equation, Jacobi concluded that $F_n(k, \lambda) = F_n(\lambda, k)$, that is, he saw that $\lambda$ and $k$ could be interchanged in the modular equation. He also noted that

$$F_n(k', \lambda') = 0,$$

where $k'^2 = 1 - k^2$, $\lambda'^2 = 1 - \lambda^2$, and he proved that if $k$ was replaced by $\frac{1}{k}$ and $\lambda$ by $\frac{1}{\lambda}$, then the modular equation remained unchanged.[55]

In 1837, Sohnke gave an alternate method, discussed in our chapter 11, for constructing a general modular equation of an odd prime order $p$. This method was based on a result of Jacobi mentioned in the last paragraph of the present chapter.

### 3.3   The Jacobi Elliptic Functions

In an August 2, 1827 letter to Schumacher,[56] Jacobi stated a particular case of a general odd-order transformation, a case he was later able to verify only after inverting the elliptic integral to obtain the elliptic function. Indeed, as we shall see, the general odd-order transformation is easier to understand in terms of the periods of the elliptic function. Jacobi's proof of this case of the transformation formula, of November 18, 1827,[57] utilized only one period of the elliptic function; in later papers and in his *Fundamenta Nova* of 1829, he considered the double periodicity of such functions. We here follow the treatment of the *Fundamenta*.

---

[53] Jacobi (1969) vol. 1, pp. 72–73.   [54] Ibid. p. 134.   [55] Ibid. pp. 122–128.
[56] Jacobi (1969) vol. 1, pp. 34–36.   [57] Ibid. pp. 37–48.

Jacobi had three basic elliptic functions for which he determined the periods; we present these in Jacobi's own manner and notation.[58] He set

$$u = \int_0^\phi \frac{d\phi}{\sqrt{1 - k^2 \sin^2 \phi}}, \quad 0 < k^2 < 1, \tag{3.53}$$

and called $\phi$ the amplitude of $u$, denoting it by

$$\phi = \operatorname{am} u. \tag{3.54}$$

Substituting $x = \sin \phi$ in the integral (3.53) produced

$$u = \int_0^x \frac{dx}{\sqrt{(1 - x^2)(1 - k^2 x^2)}}, \tag{3.55}$$

$$x = \sin \operatorname{am} u. \tag{3.56}$$

Jacobi followed Legendre in denoting the complete elliptic integral by $K$:

$$\int_0^1 \frac{dx}{\sqrt{(1 - x^2)(1 - k^2 x^2)}} = \int_0^{\frac{\pi}{2}} \frac{d\phi}{\sqrt{1 - k^2 \sin^2 \phi}} = K. \tag{3.57}$$

Jacobi introduced the complementary function, as in trigonometry, denoting it:

$$\operatorname{am}(K - u) = \operatorname{co} \operatorname{am} u.$$

Again following Legendre, he had

$$\Delta \operatorname{am} u \equiv \frac{d}{du} \operatorname{am} u = \sqrt{1 - k^2 \sin^2 \operatorname{am} u}, \tag{3.58}$$

and called $k$ the modulus, defining the complementary modulus $k'$ by

$$k^2 + k'^2 = 1. \tag{3.59}$$

Jacobi denoted the complete integral corresponding to the complementary modulus by $K'$:

$$\int_0^1 \frac{dx}{\sqrt{(1 - x^2)(1 - k'^2 x^2)}} = \int_0^{\frac{\pi}{2}} \frac{d\phi}{\sqrt{1 - k'^2 \sin^2 \phi}} = K'. \tag{3.60}$$

He noted that, in general, any trigonometric function of $\phi$ could be defined; for example, $\cos \phi = \cos \operatorname{am} u$, $\tan \phi = \tan \operatorname{am} u$. These functions satisfied the usual relations among trigonometric functions; for example:

$$\cos^2 \operatorname{am} u + \sin^2 \operatorname{am} u = 1. \tag{3.61}$$

Hereafter, we utilize Gudermann's abbreviated notation for Jacobi's functions:[59]

$$\operatorname{sn} u \equiv \operatorname{sn}(u, k) \equiv \sin \operatorname{am} u,$$

$$\operatorname{cn} u \equiv \operatorname{cn}(u, k) \equiv \cos \operatorname{am} u,$$

$$\operatorname{dn} u \equiv \operatorname{dn}(u, k) \equiv \Delta \operatorname{am} u$$

$$\operatorname{tn} u \equiv \operatorname{tn}(u, k) \equiv \tan \operatorname{am} u.$$

---

[58] For a discussion of how Jacobi's methods can be made rigorous, see Chapter 2 of Armitage and Eberlein (2006).

[59] Gudermann (1838) p. 14.

An application of Euler's addition formula (2.17) allowed Jacobi to then write

$$\text{sn}(u \pm v) = \frac{\text{sn}u\,\text{cn}v\,\text{dn}v \pm \text{sn}v\,\text{cn}u\,\text{dn}u}{1 - k^2\text{sn}^2u\,\text{sn}^2v}, \tag{3.62}$$

$$\text{cn}(u \pm v) = \frac{\text{cn}u\,\text{cn}v \mp \text{sn}u\,\text{sn}v\,\text{dn}u\,\text{dn}v}{1 - k^2\text{sn}^2u\,\text{sn}^2v}, \tag{3.63}$$

$$\text{dn}(u \pm v) = \frac{\text{dn}u\,\text{dn}v \mp k^2\text{sn}u\,\text{sn}v\,\text{cn}u\,\text{cn}v}{1 - k^2\text{sn}^2u\,\text{sn}^2v}. \tag{3.64}$$

Now it follows from (3.57) that

$$\text{sn}K = 1, \quad \text{sn}(\pm 2K) = 0;$$
$$\text{cn}K = 0, \quad \text{cn}(\pm 2K) = -1;$$
$$\text{dn}K = k', \quad \text{dn}(\pm 2K) = 0.$$

The addition formulas imply that

$$\text{sn}(K - u) = \frac{\text{cn}u}{\text{dn}u}, \quad \text{cn}(K - u) = \frac{k'\text{sn}u}{\text{dn}u}, \quad \text{dn}(K - u) = \frac{k'}{\text{dn}u}. \tag{3.65}$$

Moreover, (3.62) and (3.63) show that

$$\text{sn}(u + 2K) = -\text{sn}u, \quad \text{cn}(u + 2K) = -\text{cn}u$$

and this means that $4K$ is a period of sn as well as cn, since

$$\text{sn}(u + 4K) = \text{sn}u, \quad \text{cn}(u + 4K) = \text{cn}u. \tag{3.66}$$

To determine the imaginary periods, Jacobi applied his imaginary transformation[60]

$$\sin\phi = i\tan\psi, \quad d\phi = \frac{id\psi}{\cos\psi},$$

so that

$$\frac{d\phi}{\sqrt{1 - k^2\sin^2\phi}} = \frac{id\psi}{\sqrt{\cos^2\psi + k^2\sin^2\psi}} = \frac{id\psi}{\sqrt{1 - k'^2\sin^2\psi}}. \tag{3.67}$$

This yielded the equations

$$\text{sn}(iu, k) = i\tan\text{am}(u, k') = i\frac{\text{sn}(u, k')}{\text{cn}(u, k')}, \tag{3.68}$$

$$\text{cn}(iu, k) = \sec\text{am}(u, k') = \frac{1}{\text{cn}(u, k')}, \tag{3.69}$$

$$\text{dn}(iu, k) = \frac{\text{dn}(u, k')}{\text{cn}(u, k')}. \tag{3.70}$$

From (3.68) through (3.70) and the addition formula, Jacobi could derive the relations

$$\text{sn}2iK' = 0,$$
$$\text{sn}iK' = \infty, \tag{3.71}$$

[60] Jacobi (1969) vol. 1, pp. 85–87.

and

$$\text{sn}(u + 2iK') = \text{sn}u,$$
$$\text{cn}(u + 2iK') = -\text{cn}u,$$
$$\text{dn}(u + 2iK') = -\text{dn}u,$$
$$\text{sn}(u + iK') = \frac{1}{k\,\text{sn}u}, \tag{3.72}$$
$$\text{cn}(u + iK') = \frac{-i\text{dn}u}{k\,\text{sn}u},$$
$$\text{dn}(u + iK') = \frac{-i\text{cn}u}{\text{sn}u}.$$

Thus, snu has periods $4K$, $2iK'$; cnu has periods $4K$ and $2K + 2iK'$; dnu has periods $2K$ and $4iK'$. The zeros of sn are at $2mK + 2niK'$, those of cn are at $(2m + 1)K + 2niK'$, and those of dn are at $(2m + 1)K + (2n + 1)iK'$, where $m$ and $n$ are arbitrary integers. Also observe that snu has simple poles at $u \equiv 2mK + (2n + 1)iK'$; sn has residue $\pm\frac{1}{k}$, with the $+$ sign when $m$ is even and $-$ when $m$ is odd. Similar results can be stated for the cn and dn functions.

## 3.4 Transformations of Order $n$ and Infinite Products

In his *Fundamenta Nova*, Jacobi gave a presentation of transformation theory, including as an application his derivation of the infinite products of his elliptic functions. Since Jacobi's arguments are computationally complex, we here give an outline of these derivations, with details contained in the next sections. To state his basic theorem, let $K$ and $K'$ denote the integrals (3.57) and (3.60), respectively, and let

$$\omega = \frac{mK + m'iK'}{n}, \tag{3.73}$$

where $n$ is an odd positive integer and $m$ and $m'$ are integers having no common divisor (other than 1) that also divides $n$. Also let

$$\lambda = k^n \prod_{s=1}^{\frac{n-1}{2}} \text{sn}^4(K - 4s\,\omega), \tag{3.74}$$

$$M = (-1)^{\frac{n-1}{2}} \prod_{s=1}^{\frac{n-1}{2}} \frac{\text{sn}^2(K - 4s\,\omega)}{\text{sn}^2(4s\,\omega)}, \tag{3.75}$$

$$y = \frac{x}{M} \prod_{s=1}^{\frac{n-1}{2}} \frac{1 - \frac{x^2}{\text{sn}^2 4s\,\omega}}{1 - (k^2\,\text{sn}^2\,4s\,\omega)(x^2)}. \tag{3.76}$$

Then

$$\frac{dy}{\sqrt{(1 - y^2)(1 - \lambda^2 y^2)}} = \frac{1}{M} \frac{dx}{\sqrt{(1 - x^2)(1 - k^2 x^2)}}. \tag{3.77}$$

Consequently, if we write $x = \text{sn}(u, k)$, then we have $y = \text{sn}\left(\frac{u}{M}, \lambda\right)$. When $n$ is prime, the algebraic equation connecting the modules $k$ and $\lambda$, called the modular equation of order $n$, is of degree $n + 1$. For each value of $k$, there are $n + 1$ values of $\lambda$. Therefore, there are $n + 1$ transformations corresponding to a given $k$. These transformations can be described in terms of the periods by choosing specific values of $m$ and $m'$ in (3.73). Jacobi[61] presented four lists of values for $m$ and $m'$ in (3.73) that would produce the $n + 1$ transformations:

$$\frac{K}{n}, \frac{iK'}{n}, \frac{K + im'K'}{n}, \quad m' = 1, 2, \ldots, n - 1; \tag{3.78}$$

$$\frac{K}{n}, \frac{iK'}{n}, \frac{mK + iK'}{n}, \quad m = 1, 2, \ldots, n - 1; \tag{3.79}$$

$$\frac{K}{n}, \frac{iK'}{n}, \frac{K \pm m'iK'}{n}, \quad m' = 1, 2, \ldots, \frac{n - 1}{2}; \tag{3.80}$$

$$\frac{K}{n}, \frac{iK'}{n}, \frac{mK \pm iK'}{n}, \quad m = 1, 2, \ldots, \frac{n - 1}{2}. \tag{3.81}$$

Jacobi denoted the case in which $\omega = \frac{K}{n}$ as the first transformation. It can be seen that $\text{sn}\left(K - \frac{4sK}{n}\right)$ is real and less than 1 in absolute value. Hence, the new modulus $\lambda^2$ is real and $\lambda^2 < k^2$ when $0 < k^2 < 1$. Thus, the first transformation takes a modulus $k^2 < 1$ to a smaller modulus $\lambda^2$. When $n = 3$ and $n = 5$, (3.76) is equal, respectively, to (3.19) and (3.23).

The case $\omega = \frac{iK'}{n}$ was named by Jacobi the second transformation. He denoted the corresponding values of $\lambda$ and $M$ by $\lambda_1$ and $M_1$; he showed that these values were again real. Moreover, Jacobi showed that $0 < k^2 < 1$ implied that $k^2 < \lambda_1^2$; the second transformation must take a modulus $k^2 < 1$ to a larger modulus $\lambda_1^2$. When $n = 3$ and $n = 5$, (3.76) is equivalent, respectively, to (3.21) and (3.25).

Denote the integrals (3.57) and (3.60) by $\Lambda$ and $\Lambda'$ when $k$ and $k'$ are replaced by $\lambda$ and $\lambda'$ respectively. Similarly define the integrals $\Lambda_1$ and $\Lambda_1'$. Jacobi employed the relation between $\text{sn}(u, k)$ and $\text{sn}\left(\frac{u}{M}, \lambda\right)$, given by (3.76), to show that[62]

$$\frac{K}{nM} = \Lambda \quad \text{and} \quad \frac{K}{M_1} = \Lambda_1. \tag{3.82}$$

To obtain similar relations involving the complementary moduli $k'$, $\lambda'$, and $\lambda_1'$, Jacobi applied the imaginary transformation, (3.68), $\text{sn}(iu, k) = i\,\text{tn}(u, k')$ to (3.76). He called this the complementary transformation; he thus had a complementary transformation corresponding to both the first and the second transformations. The first and second complementary transformations gave him[63]

$$\frac{K'}{M} = \Lambda' \quad \text{and} \quad \frac{K'}{nM_1} = \Lambda_1'. \tag{3.83}$$

[61] Jacobi (1969) vol. 1, pp. 100–101.     [62] Jacobi (1969) vol. 1, p. 108.
[63] Jacobi (1969) vol. 1, p. 110.

Combining (3.82) and (3.83) yielded

$$\frac{\Lambda'}{\Lambda} = n\frac{K'}{K} \quad \text{and} \quad \frac{K'}{K} = n\frac{\Lambda'_1}{\Lambda_1}. \tag{3.84}$$

The first equation in (3.84) connects $\lambda$ and $k$; the second equation connects $k$ and $\lambda_1$. Thus, if $\lambda = \phi(k)$ for some function $\phi$, then $k = \phi(\lambda_1)$. With this in mind, consider the second transformation, relating $\mathrm{sn}\left(\frac{u}{M_1}, \lambda_1\right)$ to $\mathrm{sn}(u, k)$: When Jacobi changed $k$ into $\lambda$, this changed $\lambda_1$ into $k$. Jacobi denoted the corresponding value of $M_1$ by $M'$ so that $z = \mathrm{sn}\left(\frac{v}{M'}, k\right)$ and $y = \mathrm{sn}(v, \lambda)$ satisfied the differential equation[64]

$$\frac{dz}{\sqrt{(1 - z^2)(1 - k^2 z^2)}} = \frac{dy}{M'\sqrt{(1 - y^2)(1 - \lambda^2 y^2)}}. \tag{3.85}$$

Combining (3.85) with the first transformation

$$\frac{dy}{\sqrt{(1 - y^2)(1 - \lambda^2 y^2)}} = \frac{dx}{M\sqrt{(1 - x^2)(1 - k^2 x^2)}} \tag{3.86}$$

gave him

$$\frac{dz}{\sqrt{(1 - z^2)(1 - k^2 z^2)}} = \frac{dx}{MM'\sqrt{(1 - x^2)(1 - k^2 x^2)}}, \tag{3.87}$$

where $z = \mathrm{sn}\left(\frac{u}{MM'}, k\right)$ and $x = \mathrm{sn}(u, k)$. Since Jacobi had made the changes $k \to \lambda$ and $\lambda_1 \to k$, (3.82) was converted into[65]

$$\frac{K}{nM} = \Lambda \quad \text{and} \quad \frac{\Lambda}{M'} = K. \tag{3.88}$$

Thus

$$\frac{1}{MM'} = n. \tag{3.89}$$

A little more work gave him a relation between $z = \mathrm{sn}(nu, k)$ and $y = \mathrm{sn}\left(\frac{u}{M}, \lambda\right)$:[66]

$$\mathrm{sn}(nu, k) = nM \,\mathrm{sn}\left(\frac{u}{M}, \lambda\right) \prod_{s=1}^{\frac{n-1}{2}} \frac{\left(1 - \frac{\mathrm{sn}^2\left(\frac{u}{M}, \lambda\right)}{\mathrm{sn}^2\left(\frac{2si\Lambda'}{n}, \lambda\right)}\right)}{\left(1 - \frac{\mathrm{sn}^2\left(\frac{u}{M}, \lambda\right)}{\mathrm{sn}^2\left(\frac{(2s-1)i\Lambda'}{n}, \lambda\right)}\right)}. \tag{3.90}$$

Jacobi denoted (3.90) the first supplementary transformation, from which he derived his infinite product formulas. For this purpose, he let $n \to \infty$ so that $\lambda \to 0$, since $k^2 < 1$. In that case, $\mathrm{sn}(\theta, \lambda) = \sin\theta$, $\Lambda = \frac{\pi}{2}$; by (3.88) and (3.84), he had[67]

$$nM = \frac{2K}{\pi}, \quad \frac{\Lambda'}{n} = \frac{\pi K'}{2K}. \tag{3.91}$$

---

[64] Ibid. p. 111.    [65] Ibid. p. 112.    [66] Ibid. p. 113.    [67] Ibid. p. 141.

He changed *nu* to *u* in (3.90) and let *n* tend to infinity. Then, using (3.91), he obtained the infinite product

$$\operatorname{sn} u = \frac{2K}{\pi} \sin \frac{\pi u}{2K} \prod_{s=1}^{\infty} \left(1 - \frac{\sin^2 \frac{\pi u}{2K}}{\sin^2 \frac{s i \pi K'}{K}}\right) \div \prod_{s=1}^{\infty} \left(1 - \frac{\sin^2 \frac{\pi u}{2K}}{\sin^2 \frac{(2s-1) i \pi K'}{2K}}\right). \tag{3.92}$$

Note that in his first paper on elliptic functions,[68] Abel used another method to derive (3.92).

Before turning to the proofs of the transformation formulas, we note that if the periods $K$, $iK'$, and the periods $\Lambda$, $i\Lambda'$ are related by (3.84), then the algebraic relation between the corresponding moduli $k$ and $\lambda$ is called the modular equation of order $n$. After a study of Jacobi's theta functions, a more general definition of the modular equation will emerge.

### 3.5   Jacobi's Transformation Formulas

We have seen that Jacobi proved that if $y = \frac{U(x)}{V(x)}$, where $U$ and $V$ were polynomials of odd degree $n$ and even degree $n - 1$, respectively, then the differential equation

$$\frac{dy}{\sqrt{(1 - y^2)(1 - \lambda^2 y^2)}} = \frac{dx}{M\sqrt{(1 - x^2)(1 - k^2 x^2)}} \tag{3.93}$$

would be satisfied if there existed polynomials $A, B, C, D$ such that

$$V + U = (1 + x)A^2, \; V - U = (1 - x)B^2, \tag{3.94}$$

$$V + \lambda U = (1 + kx)C^2, \; V - \lambda U = (1 - kx)D^2. \tag{3.95}$$

Jacobi applied elliptic functions to solve this problem.[69] He showed that with no factor of gcd $(m, m')$ dividing $n$ and with

$$\omega = \frac{mK + m'iK'}{n}, \tag{3.96}$$

he could take

$$U = \frac{x}{M} \prod_{s=1}^{\frac{n-1}{2}} \left(1 - \frac{x^2}{\operatorname{sn}^2 4s\, \omega}\right), \tag{3.97}$$

$$V = \prod_{s=1}^{\frac{n-1}{2}} (1 - k^2 x^2 \operatorname{sn}^2 4s\, \omega), \tag{3.98}$$

$$A = \prod_{s=1}^{\frac{n-1}{2}} \left(1 + \frac{x}{\operatorname{sn}(K - 4s\, \omega)}\right), \tag{3.99}$$

$$B = \prod_{s=1}^{\frac{n-1}{2}} \left(1 - \frac{x}{\operatorname{sn}(K - 4s\, \omega)}\right), \tag{3.100}$$

---

[68] Abel (1965) vol. 1, p. 345.   [69] Jacobi (1969) vol. 1, pp. 87–93.

$$C = \prod_{s=1}^{\frac{n-1}{2}} (1 + kx\,\mathrm{sn}(K - 4s\,\omega)), \tag{3.101}$$

$$D = \prod_{s=1}^{\frac{n-1}{2}} (1 - kx\,\mathrm{sn}(K - 4s\,\omega)), \tag{3.102}$$

$$\lambda = k^n \prod_{s=1}^{\frac{n-1}{2}} \mathrm{sn}^4(K - 4s\,\omega), \tag{3.103}$$

$$M = (-1)^{\frac{n-1}{2}} \prod_{s=1}^{\frac{n-1}{2}} \frac{\mathrm{sn}^2(K - 4s\,\omega)}{\mathrm{sn}^2(4s\,\omega)}. \tag{3.104}$$

We need only check that (3.97) through (3.104) satisfy (3.94) and (3.95). The discussion given with (3.27) through (3.32) then shows that the differential equation (3.93) is automatically satisfied. Observe that in (3.94), when the second equation is divided by the first, we get

$$\frac{1 - \frac{U}{V}}{1 + \frac{U}{V}} = \frac{1 - x}{1 + x} \left(\frac{A}{C}\right)^2, \tag{3.105}$$

where $\frac{A}{C}$ is given by (3.99) and (3.100). We have to show that if we write $y = \frac{U}{V}$, then the equation

$$\frac{1 - y}{1 + y} = \frac{1 - x}{1 + x} \prod_{s=1}^{\frac{n-1}{2}} \left(\frac{1 - \frac{x}{\mathrm{sn}(K - 4s\,\omega)}}{1 + \frac{x}{\mathrm{sn}(K - 4s\,\omega)}}\right)^2, \tag{3.106}$$

where $x = \mathrm{sn}(u, k)$, implies that $y$ can be taken to be the ratio of (3.97) and (3.98). To see this, apply the formulas[70]

$$\frac{(1 \pm \mathrm{sn}(u + \alpha))(1 \pm \mathrm{sn}(u - \alpha))}{\mathrm{cn}^2 \alpha} = \frac{\left(1 \pm \frac{\mathrm{sn}\,u}{\mathrm{sn}(K - \alpha)}\right)^2}{1 - k^2\,\mathrm{sn}^2 u\,\mathrm{sn}^2 \alpha} \tag{3.107}$$

$$\mathrm{sn}(u - 4s\,\omega) = \mathrm{sn}(u + 4(n - s)\,\omega)$$

so that (3.106) may be written as

$$\frac{1 - y}{1 + y} = \prod_{s=0}^{n-1} \frac{1 - \mathrm{sn}(u + 4s\,\omega)}{1 + \mathrm{sn}(u + 4s\,\omega)}. \tag{3.108}$$

Again from (3.106), we deduce that $y$ has the form

$$y = \frac{xf(x^2)}{g(x^2)}, \tag{3.109}$$

where $f$ and $g$ are polynomials of degree $\frac{1}{2}(n - 1)$. Now observe that the right-hand side of (3.108) remains unchanged when $u$ is changed to $u + 4\,\omega$; this is a consequence

---

[70] See Jacobi (1969) vol. 1, p. 88, where a hint for a proof is given.

of the fact that

$$\text{sn}(u + 4\omega + 4(n-1)\omega) = \text{sn}(u + 4n\omega) = \text{sn}(u + 4mK + 4m'iK') = \text{sn}\,u.$$
(3.110)

Also note that for $u = 0$, $x = \text{sn}\,u = 0$ and by (3.106), $y = 0$. From (3.108) we see that $y = 0$ for $u = 4s\omega$, $s = 0, 1, \ldots, n-1$, that is, when $x = \text{sn}\,4s\omega$ for $s = 0, 1, \ldots, n-1$. Moreover, since

$$\text{sn}\,4(n-k)\omega = \text{sn}(4n\omega - 4k\omega) = -\text{sn}\,4k\omega,$$

we obtain $y = 0$ for $x = \pm\text{sn}\,4s\omega$, $s = 0, 1, \ldots, \frac{n-1}{2}$. This means that the polynomial $xf(x^2)$ in (3.109) is of the form

$$xf(x^2) = \frac{x}{C} \prod_{s=1}^{\frac{n-1}{2}} \left(1 - \frac{x^2}{\text{sn}^2\,4s\omega}\right).$$
(3.111)

On the other hand, by (3.71), for $u = iK'$ we have $x = \infty$; therefore, by (3.109), $y = \infty$. From (3.108) we also know that $y = \infty$ for $u = iK' + 4s\omega$, $s = 1, 2, \ldots, n-1$, that is, when $x = \text{sn}(iK' + 4s\omega)$. Now note that $\text{sn}(iK' + 4s\omega)$, $s = 1, 2, \ldots, n-1$, comprises the same set of numbers as $\text{sn}(iK' \pm 4s\omega)$, $s = 1, 2, \ldots, \frac{n-1}{2}$. Since, by (3.72),

$$\text{sn}(iK' \pm 4s\omega) = \frac{\pm 1}{k\,\text{sn}\,4s\omega},$$

we see that $y = \infty$ when $x = \frac{\pm 1}{k\,\text{sn}\,4s\omega}$. Thus, $g(x^2)$ in (3.109) is given by

$$g(x^2) = \prod_{s=1}^{\frac{n-1}{2}} \left(1 - k^2\,\text{sn}^2\,4s\omega \cdot x^2\right).$$
(3.112)

It is possible that a constant may come in here, but this could be absorbed in the constant $C$ in (3.111). To find the value for $C$, we observe that by (3.108), $y = 1$ when $x = 1$. So by taking $x = 1$ in (3.109), (3.111), and (3.112), we arrive at

$$1 = \frac{1}{C} \prod_{s=1}^{\frac{n-1}{2}} \left(1 - \frac{1}{\text{sn}^2\,4s\omega}\right) \div \prod_{s=1}^{\frac{n-1}{2}} (1 - k^2\,\text{sn}^2\,4s\omega)$$

or, by using the first formula in exercise (2) of this chapter and (3.65), we can write

$$C = (-1)^{\frac{n-1}{2}} \prod_{s=1}^{\frac{n-1}{2}} \frac{1 - \text{sn}^2\,4s\omega}{1 - k^2\,\text{sn}^2\,4s\omega} \cdot \frac{1}{\text{sn}^2\,4s\omega}$$
(3.113)

$$= (-1)^{\frac{n-1}{2}} \prod_{s=1}^{\frac{n-1}{2}} \frac{\text{sn}^2(K - 4s\omega)}{\text{sn}^2\,4s\omega}$$
(3.114)

$$= M.$$
(3.115)

Note that (3.115) follows from the definition of $M$ in (3.104). This proves that $y = \frac{U}{V}$ where $U$ and $V$ are given by (3.97) and (3.98).

We now show that (3.94) implies (3.95) where $C$ and $D$ are given by (3.101) and (3.102). We first determine that if $x$ is changed to $\frac{1}{kx}$, then $y$ becomes $\frac{1}{\lambda y}$. By (3.103) and (3.111) through (3.115), it follows that

$$
\begin{aligned}
\frac{1}{Mkx} \frac{f\left(\frac{1}{k^2x^2}\right)}{g\left(\frac{1}{k^2x^2}\right)} &= \frac{1}{Mkx} \prod_{s=1}^{\frac{n-1}{2}} \left(1 - \frac{1}{k^2x^2 \operatorname{sn}^2 4s\,\omega}\right) \div \prod_{s=1}^{\frac{n-1}{2}} \left(1 - \frac{\operatorname{sn}^2 4s\,\omega}{x^2}\right) \\
&= \frac{1}{Mkx} \prod_{s=1}^{\frac{n-1}{2}} \frac{1 - k^2x^2 \operatorname{sn}^2 4s\,\omega}{k^2x^2 \operatorname{sn}^2 4s\,\omega} \cdot \frac{x^2}{\operatorname{sn}^2 4s\,\omega - x^2} \\
&= \frac{1}{Mkx} \prod_{s=1}^{\frac{n-1}{2}} \frac{1 - k^2x^2 \operatorname{sn}^2 4s\,\omega}{k^2 \operatorname{sn}^4 4s\,\omega} \cdot \frac{1}{1 - \frac{x^2}{\operatorname{sn}^2 4s\,\omega}} \\
&= \frac{1}{M^2 k^n \prod \operatorname{sn}^4 4s\,\omega} \prod_{s=1}^{\frac{n-1}{2}} (1 - k^2 \operatorname{sn}^2 4s\,\omega \cdot x^2) \\
&\qquad \div \frac{x}{M} \prod_{s=1}^{\frac{n-1}{2}} \left(1 - \frac{x^2}{\operatorname{sn}^2 4s\,\omega}\right) \\
&= \frac{1}{\lambda y},
\end{aligned}
$$

since $\lambda = M^2 k^n \prod \operatorname{sn}^4 4s\,\omega$ by (3.103) and (3.104). One may then check that (3.94) implies (3.95). We have thus verified relations (3.94) through (3.104), as well as the equation $y = \frac{U}{V}$.

We can likewise find analogous formulas for $\operatorname{cn}\left(\frac{u}{M}, \lambda\right)$ and $\operatorname{dn}\left(\frac{u}{M}, \lambda\right)$ by noting that[71]

$$
\begin{aligned}
\operatorname{cn}\left(\frac{u}{M}, \lambda\right) &= \sqrt{1 - y^2} \\
&= \operatorname{cn}(u, k) \prod_{s=1}^{\frac{n-1}{2}} \left(1 - \frac{x^2}{\operatorname{sn}^2(K - 4s\,\omega)}\right) \div \prod_{s=1}^{\frac{n-1}{2}} (1 - k^2x^2 \operatorname{sn}^2 4s\,\omega)
\end{aligned}
$$
(3.116)

$$
\begin{aligned}
\operatorname{dn}\left(\frac{u}{M}, \lambda\right) &= \sqrt{1 - \lambda^2 y^2} \\
&= \operatorname{dn}(u, k) \prod_{s=1}^{\frac{n-1}{2}} \frac{1 - k^2x^2 \operatorname{sn}^2(K - 4s\,\omega)}{1 - k^2x^2 \operatorname{sn}^2 4s\,\omega},
\end{aligned}
$$
(3.117)

where $x = \operatorname{sn}(u, k)$.

---

[71] See Jacobi (1969) vol. 1, p. 102, where Jacobi has given these formulas for $\omega = \frac{K}{n}$; or see Cayley (1895), p. 256 for the general case.

### 3.6   Equivalent Forms of the Transformation Formulas

Jacobi's basic $n$th-order transformation is given by $y = \frac{U}{V}$, where $U$ and $V$ are defined by (3.97) and (3.98). He also gave some equivalent forms that were useful for specific purposes. For example, take the formula[72]

$$\operatorname{sn}(u + \alpha)\operatorname{sn}(u - \alpha) = \frac{\operatorname{sn}^2 u - \operatorname{sn}^2 \alpha}{1 - k^2 \operatorname{sn}^2 u \operatorname{sn}^2 \alpha} \tag{3.118}$$

along with the values of $M$ and $\lambda$, gleaned from (3.74) and (3.75), and the fact that $\operatorname{sn}(u - \alpha) = \operatorname{sn}(u + 4n\,\omega - \alpha)$. Apply these to equation (3.76) to obtain

$$y = \operatorname{sn}\left(\frac{u}{M}, \lambda\right) = \prod_{s=0}^{n-1} \operatorname{sn}(u + 4s\,\omega) \div \prod_{s=1}^{\frac{n-1}{2}} \operatorname{sn}^2(K - 4s\,\omega)$$

$$= \sqrt{\frac{k^n}{\lambda}} \prod_{s=0}^{n-1} \operatorname{sn}(u + 4s\,\omega). \tag{3.119}$$

In a similar way, by the use of (3.116) one may show that[73]

$$\sqrt{1 - y^2} = \operatorname{cn}\left(\frac{u}{M}, \lambda\right) = \sqrt{\frac{\lambda' k^n}{\lambda k'^n}} \prod_{s=0}^{n-1} \operatorname{cn}(u + 4s\,\omega). \tag{3.120}$$

Dividing (3.119) by (3.120) produces

$$\operatorname{tn}\left(\frac{u}{M}, \lambda\right) = \sqrt{\frac{k'^n}{\lambda'}} \prod_{s=0}^{n-1} \operatorname{tn}(u + 4s\,\omega). \tag{3.121}$$

For another equivalent form of the transformation, note that the sequence

$$\pm \operatorname{sn} 4\,\omega, \ \pm \operatorname{sn} 8\,\omega, \ \ldots, \ \pm \operatorname{sn} 2(n - 1)\,\omega$$

may be reordered as

$$\pm \operatorname{sn} 2\,\omega, \ \pm \operatorname{sn} 4\,\omega, \ \ldots, \ \pm \operatorname{sn}(n - 1)\,\omega.$$

This allows us to write formulas (3.97), (3.98), (3.103), and (3.104) as[74]

$$U = \frac{x}{M} \prod_{s=1}^{\frac{n-1}{2}} \left(1 - \frac{x^2}{\operatorname{sn}^2 2s\,\omega}\right), \tag{3.122}$$

$$V = \prod_{s=1}^{\frac{n-1}{2}} \left(1 - k^2 x^2 \operatorname{sn}^2 2s\,\omega\right), \tag{3.123}$$

---

[72] Jacobi (1969) vol. 1, p. 84.    [73] Ibid. p. 98.    [74] Jacobi (1969) vol. 1, pp. 91–93.

$$\lambda = k^n \prod_{s=1}^{\frac{n-1}{2}} \text{sn}^4(K - 2s\,\omega), \qquad (3.124)$$

$$M = (-1)^{\frac{n-1}{2}} \prod_{s=1}^{\frac{n-1}{2}} \frac{\text{sn}^2(K - 2s\,\omega)}{\text{sn}^2(2s\,\omega)}. \qquad (3.125)$$

## 3.7 The First and Second Transformations

Of the several transformations of order $n$ contained in (3.97) and (3.98), Jacobi singled out two transformations that gave real values of $\lambda$ in (3.103) when $0 < k^2 < 1$. He used these to develop relations between $K$ and $\Lambda$. He denoted as "first" the transformation for the case $\omega = \frac{K}{n}$, that is when $m = 1$ and $m' = 0$ in (3.96). Note that for this value of $\omega$, the values of $\lambda$ and $M$, given in (3.103) and (3.104), respectively, are necessarily real when $0 < k^2 < 1$. He obtained his second transformation by taking $\omega = \frac{iK'}{n}$, that is with $m = 0$ and $m' = 1$ in (3.96). He designated the corresponding values of $\lambda$ and $M$ as $\lambda_1$ and $M_1$. By use of Jacobi's imaginary transformation, it is possible to write the second transformation in a real form. Recall (3.68) through (3.70), showing that the imaginary transformation may be given as

$$\text{sn}(iu, k) = i\frac{\text{sn}(u, k')}{\text{cn}(u, k')}, \quad \text{cn}(iu, k) = \frac{1}{\text{cn}(u, k')}, \quad \text{dn}(iu, k) = \frac{\text{dn}(u, k')}{\text{cn}(u, k')}. \quad (3.126)$$

By (3.65) and the last two equations in (3.126), we may write

$$\text{sn}(K - iu, k) = \frac{\text{cn}(iu, k)}{\text{dn}(iu, k)} = \frac{1}{\text{dn}(u, k')}.$$

Hence, by the use of (3.122) through (3.125), the second transformation may be represented as[75]

$$\text{sn}\left(\frac{u}{M_1}, \lambda_1\right) = \frac{\text{sn}\,u}{M_1} \prod_{s=1}^{\frac{n-1}{2}} \left(1 + \frac{\text{sn}^2 u}{\text{tn}^2\left(\frac{2sK'}{n}, k'\right)}\right) \div \prod_{s=1}^{\frac{n-1}{2}} \left(1 + k^2 x^2 \, \text{tn}^2\left(\frac{2sK'}{n}, k'\right)\right),$$

$$(3.127)$$

where

$$x = \text{sn}\,u,$$

$$\lambda_1 = k^n \div \prod_{s=1}^{\frac{n-1}{2}} \text{dn}^4\left(\frac{2sK'}{n}, k'\right),$$

$$M_1 = \prod_{s=1}^{\frac{n-1}{2}} \frac{\text{sn}^2\left(K - \frac{2sK'}{n}, k'\right)}{\text{sn}^2\left(\frac{2sK'}{n}, k'\right)}.$$

[75] Ibid. p. 106.

Clearly, $\lambda_1$ and $M_1$, like $\lambda$ and $M$ in (3.74) and (3.75), are real when $0 < k^2 < 1$. Note also that $k'^2 = 1 - k^2$ lies between 0 and 1.

Jacobi established[76] a relation between $K$ and $\Lambda$ by using the first transformation in the form (3.122) through (3.125) with $\omega = \frac{K}{n}$:

$$\operatorname{sn}\left(\frac{u}{M}, \lambda\right) = \frac{\operatorname{sn} u}{M} \prod_{s=1}^{\frac{n-1}{2}} \left(1 - \frac{\operatorname{sn}^2 u}{\operatorname{sn}^2 \frac{2sK}{n}}\right) \div \prod_{s=1}^{\frac{n-1}{2}} \left(1 - k^2 \operatorname{sn}^2 \frac{2sK}{n} \operatorname{sn}^2 u\right). \quad (3.128)$$

Observe that the least real and positive value for which $\operatorname{sn}\left(\frac{u}{M}, \lambda\right)$, the left-hand side of (3.128), vanishes is $\frac{u}{M} = 2\Lambda$ and the least real, positive value for which the right-hand side vanishes must arise from the factor

$$1 - \frac{\operatorname{sn}^2 u}{\operatorname{sn}^2 \frac{2K}{n}}.$$

This gives us $u = \frac{2K}{n}$. Therefore,

$$2M\Lambda = \frac{2K}{n} \quad \text{or} \quad \frac{K}{nM} = \Lambda. \quad (3.129)$$

A similar analysis of the second transformation (3.127) shows that the smallest positive zero of the left-hand side is $u = 2M_1 \Lambda_1$. For the right-hand side, this value occurs when $\operatorname{sn} u = 0$ since the other factors do not vanish for real $u$. So $u = 2K$ and we conclude

$$2M_1 \Lambda_1 = 2K \quad \text{or} \quad \frac{K}{M_1} = \Lambda_1. \quad (3.130)$$

### 3.8   Complementary Transformations

Jacobi defined complementary transformations, that is, transformations of elliptic functions with respect to the complementary moduli $k'$ and $\lambda'$, for the purpose of finding relationships among the periods $iK'$, $i\Lambda'$, and $i\Lambda'_1$.[77] The first complementary transformation gives the equation connecting $\operatorname{sn}\left(\frac{u}{M}, \lambda'\right)$ with $\operatorname{sn}(u, k')$. Begin by taking (3.121), with $\omega = \frac{K}{n}$, as an expression of the first transformation:

$$\operatorname{tn}\left(\frac{u}{M}, \lambda\right) = \sqrt{\frac{k'^n}{\lambda'}} \prod_{s=0}^{n-1} \operatorname{tn}\left(u + \frac{4sK}{n}, k\right). \quad (3.131)$$

Change $u$ to $iu$ and recall that Jacobi's imaginary transformation (3.68) gives

$$\operatorname{tn}(iu, k) = i \operatorname{sn}(u, k'),$$

---

[76]  Ibid. p. 108.      [77]  Ibid. pp. 108–111.

so that (3.131) takes the form

$$
\operatorname{sn}\left(\frac{u}{M},\lambda'\right) = (-1)^{\frac{n-1}{2}}\sqrt{\frac{k'^n}{\lambda'}}\prod_{s=0}^{n-1}\operatorname{sn}\left(u-\frac{4si\,K}{n},k'\right)
$$

$$
=\sqrt{\frac{k'^n}{\lambda'}}\operatorname{sn}(u,k')\prod_{s=1}^{\frac{n-1}{2}}\operatorname{sn}\left(u+\frac{2si\,K}{n},k'\right)\operatorname{sn}\left(u-\frac{2si\,K}{n},k'\right)
$$

$$
=(-1)^{\frac{n-1}{2}}\sqrt{\frac{k'^n}{\lambda'}}\prod_{s=1}^{\frac{n-1}{2}}\operatorname{sn}^2\left(\frac{2si\,K}{n},k'\right)\operatorname{sn}(u,k')\prod_{s=1}^{\frac{n-1}{2}}\left(1-\frac{\operatorname{sn}^2(u,k')}{\operatorname{sn}^2\left(\frac{2si\,K}{n},k'\right)}\right)
$$

$$
\div\prod_{s=1}^{\frac{n-1}{2}}\left(1-k'^2\operatorname{sn}^2\left(\frac{2siK}{n},k'\right)\operatorname{sn}^2(u,k')\right), \tag{3.132}
$$

where the last step is a result of an application of (3.118). Next, by using (3.125), it can be shown that

$$
M=(-1)^{\frac{n-1}{2}}\sqrt{\frac{k'^n}{\lambda'}}\prod_{s=1}^{\frac{n-1}{2}}\operatorname{sn}^2\left(\frac{2si\,K}{n},k'\right).
$$

This allows us to write relation (3.132) as[78]

$$
\operatorname{sn}\left(\frac{u}{M},\lambda'\right)=\frac{\operatorname{sn}(u,k')}{M}\prod_{s=1}^{\frac{n-1}{2}}\left(1+\frac{\operatorname{sn}^2(u,k')}{\operatorname{tn}^2\left(\frac{2sK}{n},k\right)}\right)
$$

$$
\div\prod_{s=1}^{\frac{n-1}{2}}\left(1+k'^2\operatorname{tn}^2\left(\frac{2sK}{n},k\right)\operatorname{sn}^2(u,k')\right) \tag{3.133}
$$

and this is the first complementary transformation. When we consider the least positive zero on the two sides of (3.133), we obtain the relation

$$
\Lambda'=\frac{K'}{M}. \tag{3.134}
$$

In a similar manner, if we start with the second transformation written in the form

$$
\operatorname{tn}\left(\frac{u}{M_1},\lambda_1\right)=\sqrt{\frac{k'^n}{\lambda_1'}}\prod_{s=0}^{n-1}\operatorname{tn}\left(u+\frac{4si\,K'}{n}\right), \tag{3.135}
$$

---

[78] Ibid. p. 117.

then we can derive the second complementary transformation[79]

$$
\text{sn}\left(\frac{u}{M_1}, \lambda_1'\right) = \frac{\text{sn}(u, k')}{M_1} \prod_{s=0}^{\frac{n-1}{2}} \left(1 - \frac{\text{sn}^2(u, k')}{\text{sn}^2\left(\frac{2sK'}{n}, k'\right)}\right)
$$

$$
\div \prod_{s=0}^{\frac{n-1}{2}} \left(1 - k'^2 \, \text{sn}^2\left(\frac{2sK'}{n}, k'\right) \text{sn}^2(u, k')\right). \tag{3.136}
$$

Calculating the least positive zero of each side yields $\frac{u}{M_1} = 2\Lambda_1'$ and $u = \frac{2K'}{n}$; thus

$$
M_1 \Lambda_1' = \frac{K'}{n} \quad \text{or} \quad \Lambda_1' = \frac{K'}{nM_1}. \tag{3.137}
$$

By combining (3.129), (3.130), (3.134), and (3.137), we arrive at

$$
\frac{\Lambda'}{\Lambda} = n\frac{K'}{K}, \quad \frac{K'}{K} = n\frac{\Lambda_1'}{\Lambda_1}. \tag{3.138}
$$

We may thus conclude that $\lambda$ is the same function of $k$ as $k$ is of $\lambda_1$; when we change $k$ to $\lambda$, then $\lambda_1$ is changed to $k$.[80]

### 3.9   Jacobi's First Supplementary Transformation

Recall that, in order to obtain the infinite product of sn $u$ given by (3.92), Jacobi needed to write the transformation formula in the form of (3.90). Now rewrite the second transformation (3.127) as

$$
\text{sn}\left(\frac{u}{M_1}, \lambda_1\right) = \frac{\text{sn } u}{M_1} \prod_{s=1}^{\frac{n-1}{2}} \left(1 - \frac{\text{sn}^2 u}{\text{sn}^2 \frac{2siK'}{n}}\right) \div \prod_{s=1}^{\frac{n-1}{2}} \left(1 - \frac{\text{sn}^2 u}{\text{sn}^2 \frac{(2s-1)iK'}{n}}\right). \tag{3.139}
$$

Next, we change $k$ to $\lambda$ so that $\lambda_1$ is changed to $k$. Write the new value of $M_1$ as $M'$ so that (3.139) becomes

$$
\text{sn}\left(\frac{u}{M'}, k\right) = \frac{\text{sn}(u, \lambda)}{M'} \prod_{s=1}^{\frac{n-1}{2}} \left(1 - \frac{\text{sn}^2(u, \lambda)}{\text{sn}^2\left(\frac{2si\Lambda'}{n}, \lambda\right)}\right) \div \prod_{s=1}^{\frac{n-1}{2}} \left(1 - \frac{\text{sn}^2(u, \lambda)}{\text{sn}^2\left(\frac{(2s-1)i\Lambda'}{n}, \lambda\right)}\right).
$$
$$\tag{3.140}$$

Recall (3.130), showing that $M_1 = \frac{K}{\Lambda_1}$. Since we have changed $K$ to $\Lambda$, $\Lambda_1$ to $K$, and $M_1$ to $M'$, we may write

$$
M' = \frac{\Lambda}{K} = \frac{1}{nM}, \tag{3.141}
$$

---

[79] Ibid. p. 119.     [80] Ibid. p. 111.

where the last equation follows from (3.129). Thus, we have $n = \frac{1}{M'M}$, so that if we replace $u$ with $\frac{u}{M}$, then (3.140) becomes the first supplementary transformation (3.90):[81]

$$
\mathrm{sn}(nu, k) = nM\,\mathrm{sn}\left(\frac{u}{M}, \lambda\right) \prod_{s=1}^{\frac{n-1}{2}} \left(1 - \frac{\mathrm{sn}^2\left(\frac{u}{M}, \lambda\right)}{\mathrm{sn}^2\left(\frac{2si\,\Lambda'}{n}, \lambda\right)}\right) \div \prod_{s=1}^{\frac{n-1}{2}} \left(1 - \frac{\mathrm{sn}^2\left(\frac{u}{M}, \lambda\right)}{\mathrm{sn}^2\left(\frac{(2s-1)i\Lambda'}{n}, \lambda\right)}\right).
$$
(3.142)

It can also be shown that

$$
\mathrm{cn}(nu, k) = \sqrt{1-x^2} \prod_{s=1}^{\frac{n-1}{2}} \left(1 - \frac{x^2}{\mathrm{sn}^2\left(\Lambda - \frac{2si\Lambda'}{n}\right)}\right) \div \prod_{s=1}^{\frac{n-1}{2}} \left(1 - \frac{x^2}{\mathrm{sn}^2 \frac{(2s-1)i\Lambda'}{n}}\right),
$$
(3.143)

where $x = \mathrm{sn}\left(\frac{u}{M}, \lambda\right)$. We also have the useful formula for $\mathrm{dn}(nu, k)$, with $x = \mathrm{sn}\left(\frac{u}{M}, \lambda\right)$:

$$
\mathrm{dn}(nu, k) = \sqrt{1-\lambda^2 x^2} \prod_{s=1}^{\frac{n-1}{2}} \left(1 - \frac{x^2}{\mathrm{sn}^2\left(\Lambda - \frac{(2s-1)i\Lambda'}{n}\right)}\right) \div \prod_{s=1}^{\frac{n-1}{2}} \left(1 - \frac{x^2}{\mathrm{sn}^2 \frac{(2s-1)i\Lambda'}{n}}\right).
$$
(3.144)

### 3.10 Jacobi's Infinite Products for Elliptic Functions

We have seen how Jacobi derived (3.92) from (3.90); he derived similar infinite products for $\mathrm{cn}\,u$ and $\mathrm{dn}\,u$ from (3.143) and (3.144), respectively. He presented these infinite products in Section 35 of his *Fundamenta Nova*:

$$
\mathrm{sn}\,u = \frac{2Ky}{\pi} \frac{\left(1 - \frac{y^2}{\sin^2 \frac{i\pi K'}{K}}\right)\left(1 - \frac{y^2}{\sin^2 \frac{2i\pi K'}{K}}\right)\left(1 - \frac{y^2}{\sin^2 \frac{3i\pi K'}{K}}\right)\cdots}{\left(1 - \frac{y^2}{\sin^2 \frac{i\pi K'}{2K}}\right)\left(1 - \frac{y^2}{\sin^2 \frac{3i\pi K'}{2K}}\right)\left(1 - \frac{y^2}{\sin^2 \frac{5i\pi K'}{2K}}\right)\cdots};
$$
(3.145)

$$
\mathrm{cn}\,u = \sqrt{1-y^2}\frac{\left(1 - \frac{y^2}{\cos^2 \frac{i\pi K'}{K}}\right)\left(1 - \frac{y^2}{\cos^2 \frac{2i\pi K'}{K}}\right)\left(1 - \frac{y^2}{\cos^2 \frac{3i\pi K'}{K}}\right)\cdots}{\left(1 - \frac{y^2}{\sin^2 \frac{i\pi K'}{2K}}\right)\left(1 - \frac{y^2}{\sin^2 \frac{3i\pi K'}{2K}}\right)\left(1 - \frac{y^2}{\sin^2 \frac{5i\pi K'}{2K}}\right)\cdots};
$$
(3.146)

$$
\mathrm{dn}\,u = \frac{\left(1 - \frac{y^2}{\cos^2 \frac{i\pi K'}{2K}}\right)\left(1 - \frac{y^2}{\cos^2 \frac{3i\pi K'}{2K}}\right)\left(1 - \frac{y^2}{\cos^2 \frac{5i\pi K'}{2K}}\right)\cdots}{\left(1 - \frac{y^2}{\sin^2 \frac{i\pi K'}{2K}}\right)\left(1 - \frac{y^2}{\sin^2 \frac{3i\pi K'}{2K}}\right)\left(1 - \frac{y^2}{\sin^2 \frac{5i\pi K'}{2K}}\right)\cdots},
$$
(3.147)

[81] Ibid. p. 113.

where $y = \sin \frac{\pi u}{2K}$. Jacobi then set $e^{-\pi \frac{K'}{K}} = q$, $u = \frac{2Kx}{\pi}$, $y = \sin x$ to obtain[82]

$$\sin \frac{m \, i\pi \, K'}{K} = \frac{q^m - q^{-m}}{2i} = \frac{i(1 - q^{2m})}{2q^m},$$

$$\cos \frac{m \, i\pi \, K'}{K} = \frac{q^m + q^{-m}}{2} = \frac{1 + q^{2m}}{2q^m},$$

$$1 - \frac{y^2}{\sin^2 \frac{m \, i\pi \, K'}{K}} = 1 + \frac{4q^{2m} \sin^2 x}{(1 - q^{2m})^2} = \frac{1 - 2q^{2m} \cos 2x + q^{4m}}{(1 - q^{2m})^2},$$

$$1 - \frac{y^2}{\cos^2 \frac{m \, i\pi \, K'}{K}} = 1 - \frac{4q^{2m} \sin^2 x}{(1 + q^{2m})^2} = \frac{1 + 2q^{2m} \cos 2x + q^{4m}}{(1 + q^{2m})^2}.$$

He then defined the quantities $A, B, C$ as the products

$$A = \prod_{n=1}^{\infty} \frac{\left(1 - q^{2n-1}\right)^2}{\left(1 - q^{2n}\right)^2},$$

$$B = \prod_{n=1}^{\infty} \frac{\left(1 - q^{2n-1}\right)^2}{\left(1 + q^{2n}\right)^2},$$

$$C = \prod_{n=1}^{\infty} \frac{\left(1 - q^{2n-1}\right)^2}{\left(1 + q^{2n+1}\right)^2}.$$

Jacobi was thus able to rewrite the elliptic functions as

$$\operatorname{sn} \frac{2Kx}{\pi} = \frac{2AK}{\pi} \sin x \prod_{n=1}^{\infty} \frac{1 - 2q^{2n} \cos 2x + q^{4n}}{1 - 2q^{2n-1} \cos 2x + q^{4n-2}}, \tag{3.148}$$

$$\operatorname{cn} \frac{2Kx}{\pi} = B \cos x \prod_{n=1}^{\infty} \frac{1 + 2q^{2n} \cos 2x + q^{4n}}{1 - 2q^{2n-1} \cos 2x + q^{4n-2}}, \tag{3.149}$$

$$\operatorname{dn} \frac{2Kx}{\pi} = C \prod_{n=1}^{\infty} \frac{1 + 2q^{2n-1} \cos 2x + q^{4n-2}}{1 - 2q^{2n-1} \cos 2x + q^{4n-2}}. \tag{3.150}$$

He next set $x = \frac{\pi}{2}$ and observed that, since

$$\operatorname{sn} K = 1 \quad \text{and} \quad \operatorname{dn} K = \sqrt{1 - k^2 \operatorname{sn}^2 K} = \sqrt{1 - k^2} = k',$$

it followed from the expression for $\operatorname{dn} \frac{2Kx}{\pi}$ that[83]

$$k' = C \cdot C = C^2 \quad \text{or} \quad C = \sqrt{k'}. \tag{3.151}$$

---

[82] Ibid. pp. 142–143.    [83] Ibid. pp. 144–145.

From the product, given in (3.148), for sn $\frac{2Kx}{\pi}$ and upon setting $x = \frac{\pi}{2}$, he could obtain

$$1 = \frac{2AK}{\pi} \prod_{n=1}^{\infty} \frac{(1+q^{2n})^2}{(1+q^{2n-1})^2} = \frac{2AK}{\pi} \cdot \frac{C}{B} = \frac{2\sqrt{k'}\,AK}{\pi B},$$

$$\text{or} \qquad B = \frac{2\sqrt{k'}\,AK}{\pi}. \qquad (3.152)$$

He noted that the formula for sn $\frac{2Kx}{\pi}$ could be rewritten:

$$\text{sn}\,\frac{2Kx}{\pi} = \frac{2AK}{\pi}\left(\frac{e^{ix}-e^{-ix}}{2i}\right)\prod_{n=1}^{\infty} \frac{\left(1-q^{2n}e^{2ix}\right)\left(1-q^{2n}e^{-2ix}\right)}{\left(1-q^{2n-1}e^{2ix}\right)\left(1-q^{2n-1}e^{-2ix}\right)}. \qquad (3.153)$$

Jacobi observed that by (3.72), the change $x \to x + \frac{i\pi K'}{2K}$ would yield

$$\text{sn}\left(\frac{2Kx}{\pi}+iK'\right) = \frac{1}{k\sin\frac{2Kx}{\pi}};$$

$$\cos 2\left(x+\frac{i\pi K'}{2K}\right) = \frac{1}{2}\left(e^{2ix-\frac{\pi K'}{K}}+e^{-2ix+\frac{\pi K'}{K}}\right) = \frac{1}{2}\left(qe^{2ix}+\frac{1}{q}e^{-2ix}\right);$$

$$e^{i\left(x+\frac{i\pi K'}{2K}\right)} = \sqrt{q}\,e^{ix}.$$

Also, taking $x \to x + \frac{i\pi K'}{2K}$ in (3.153) would produce

$$\frac{1}{k\,\text{sn}\,\frac{2Kx}{\pi}} = \frac{2AK}{\pi}\left(\frac{\sqrt{q}e^{ix}-\frac{1}{\sqrt{q}}e^{-ix}}{2i}\right)\prod_{n=1}^{\infty}\frac{\left(1-q^{2n+1}e^{2ix}\right)\left(1-q^{2n-1}e^{-2ix}\right)}{\left(1-q^{2n}e^{2ix}\right)\left(1-q^{2n-2}e^{-2ix}\right)}. \qquad (3.154)$$

Observe that by multiplying (3.153) and (3.154) Jacobi got the value of $\sqrt{k}\,K$:

$$\frac{1}{k} = \frac{1}{\sqrt{q}}\left(\frac{AK}{\pi}\right)^2 \quad \text{or} \quad A = \frac{\pi\sqrt[4]{q}}{\sqrt{k}\,K}. \qquad (3.155)$$

To find an expression for $B$ in terms of $k'$ and $k$, Jacobi inserted the value of $A$ from (3.155) into (3.152), obtaining

$$B = 2\sqrt{\frac{k'}{k}}\sqrt[4]{q}. \qquad (3.156)$$

Substituting the values of $A$, $B$, and $C$ in the formulas for sn, cn, and dn, Jacobi could finally state the infinite products:

$$\text{sn}\,\frac{2Kx}{\pi} = \frac{2q^{\frac{1}{4}}}{\sqrt{k}}\sin x\prod_{n=1}^{\infty}\frac{1-2q^{2n}\cos 2x + q^{4n}}{1-2q^{2n-1}\cos 2x + q^{4n-2}}, \qquad (3.157)$$

$$\operatorname{cn} \frac{2Kx}{\pi} = \sqrt{\frac{k'}{k}} \cdot 2q^{\frac{1}{4}} \cos x \prod_{n=1}^{\infty} \frac{1 + 2q^{2n} \cos 2x + q^{4n}}{1 - 2q^{2n-1} \cos 2x + q^{4n-2}}, \tag{3.158}$$

$$\operatorname{dn} \frac{2Kx}{\pi} = \sqrt{k'} \prod_{n=1}^{\infty} \frac{1 + 2q^{2n-1} \cos 2x + q^{4n-2}}{1 - 2q^{2n-1} \cos 2x + q^{4n-2}}. \tag{3.159}$$

Jacobi published (3.157) without proof in a short paper of 1828. He provided detailed proofs in his *Fundamenta* of the next year. Abel published these product formulas with proofs[84] after Jacobi's 1828 paper.

From (3.151), (3.155), and (3.156), one may obtain the product expressions for $k, k'$, $K, kK$, and so on. Thus, (3.151) gives us[85]

$$k' = C^2 = \prod_{n=1}^{\infty} \left( \frac{1 - q^{2n-1}}{1 + q^{2n-1}} \right)^4. \tag{3.160}$$

Again, from (3.156) and (3.160), we have

$$k = 4\sqrt{q} \frac{k'}{B^2} = 4\sqrt{q} \prod_{n=1}^{\infty} \left( \frac{1 + q^{2n}}{1 + q^{2n-1}} \right)^4; \tag{3.161}$$

and (3.160) combined with (3.152) allows us to write

$$\frac{2K}{\pi} = \prod_{n=1}^{\infty} \left( \frac{1 - q^{2n}}{1 - q^{2n-1}} \cdot \frac{1 + q^{2n-1}}{1 + q^{2n}} \right)^2. \tag{3.162}$$

Jacobi also stated formulas obtainable from (3.160), (3.161), and (3.162):

$$\frac{2kK}{\pi} = 4\sqrt{q} \prod_{n=1}^{\infty} \left( \frac{1 - q^{4n}}{1 - q^{4n-2}} \right)^2, \tag{3.163}$$

$$\frac{2k'K}{\pi} = \prod_{n=1}^{\infty} \left( \frac{1 - q^n}{1 + q^n} \right)^2, \tag{3.164}$$

$$\frac{2\sqrt{k}K}{\pi} = 2\sqrt[4]{q} \prod_{n=1}^{\infty} \left( \frac{1 - q^{2n}}{1 - q^{2n-1}} \right)^2, \tag{3.165}$$

$$\frac{2\sqrt{k'}K}{\pi} = \prod_{n=1}^{\infty} \left( \frac{1 - q^{2n}}{1 + q^{2n}} \right)^2. \tag{3.166}$$

As a more detailed example, to derive (3.163), we may multiply (3.161) and (3.162) to obtain

$$\frac{2kK}{\pi} = 4\sqrt{q} \prod_{n=1}^{\infty} \left( \frac{1 + q^{2n}}{1 + q^{2n-1}} \right)^2 \cdot \left( \frac{1 - q^{2n}}{1 - q^{2n-1}} \right)^2 = 4\sqrt{q} \prod_{n=1}^{\infty} \left( \frac{1 - q^{4n}}{1 - q^{4n-2}} \right)^2.$$

---

[84] Abel (1965) vol. 1, pp. 467–477, especially p. 472.　　[85] Jacobi (1969) vol. 1, p. 146.

Jacobi then employed the triple product identity (2.116) to write $\sqrt{\frac{2K}{\pi}}$ as a theta series. To see how this can be done, first observe that

$$\sqrt{\frac{2K}{\pi}} = \prod_{n=1}^{\infty} \left( \frac{1 - q^{2n}}{1 - q^{2n-1}} \cdot \frac{1 + q^{2n-1}}{1 + q^{2n}} \right). \tag{3.167}$$

Now write the triple product identity:

$$1 + 2 \sum_{n=1}^{\infty} q^{n^2} x^n = \prod_{n=1}^{\infty} (1 - q^{2n})(1 + q^{2n-1}x) \left( 1 + \frac{q^{2n-1}}{x} \right),$$

and take $x = 1$ to get

$$\prod_{n=1}^{\infty} (1 - q^{2n})(1 + q^{2n-1})^2 = 1 + 2 \sum_{n=1}^{\infty} q^{n^2}.$$

Thus, from (3.167), we may prove that

$$\sqrt{\frac{2K}{\pi}} = 1 + 2 \sum_{n=1}^{\infty} q^{n^2} \tag{3.168}$$

if we can prove that

$$\prod_{n=1}^{\infty} \frac{1}{\left(1 - q^{2n-1}\right) \left(1 + q^{2n}\right)} = \prod_{n=1}^{\infty} (1 + q^{2n-1}). \tag{3.169}$$

To complete the proof of Jacobi's famous formula[86] (3.168), write the left-hand side of (3.169) as

$$\prod_{n=1}^{\infty} \frac{1 - q^{2n}}{(1 - q^n)(1 + q^{2n})} = \prod_{n=1}^{\infty} \frac{1 + q^n}{1 + q^{2n}} = \prod_{n=1}^{\infty} (1 + q^{2n-1}).$$

In his July 1828 paper in *Crelle's Journal*,[87] Jacobi gave a remarkable application of (3.168) to obtain a one-line derivation of the transformation formula

$$\sqrt{\frac{1}{x}} = \frac{1 + 2 \sum_{n=1}^{\infty} e^{-n^2 \pi x}}{1 + 2 \sum_{n=1}^{\infty} e^{-\frac{n^2 \pi}{x}}}. \tag{3.170}$$

Jacobi noted that by interchanging the moduli $k$ and $k'$, he could effect an interchange of the periods $K$ and $K'$. Since $q = e^{-\frac{\pi K'}{K}}$, (3.168) would imply that with $x = \frac{K'}{K}$

$$\sqrt{\frac{2K'}{\pi}} = 1 + 2 \sum_{n=1}^{\infty} e^{-\frac{n^2 \pi}{x}}. \tag{3.171}$$

He then derived the transformation by dividing (3.168) by (3.171). We note that Abel had obtained formulas analogous to (3.145) through (3.147) in his first paper on elliptic functions, "Recherches sur les fonctions elliptiques." Abel, unlike Jacobi, justified

---

[86] Ibid. p. 235.    [87] Ibid. p. 260.

his limiting processes. In an 1846 letter to von Humboldt, Jacobi commented on the lack of rigor in his own work and that of other scientists, as compared with Dirichlet's approach: "He [Dirichlet] alone, not myself, not Cauchy, not Gauss knows what a perfectly rigorous mathematical proof is ...."[88]

### 3.11 Jacobi's Theory of Theta Functions

In a paper of July 1828,[89] Jacobi defined two theta functions, $\Theta(x)$ and $H(x)$, by means of infinite series; he then wrote his elliptic functions as ratios of these theta functions. He also expressed the theta series as infinite products, but he did not give the triple product identity needed to convert the series into the product. In his *Fundamenta*, he initially defined $\Theta(u)$ by the infinite product that occurred in the denominators of (3.148) through (3.150). He set

$$\frac{\Theta\left(\frac{2Kx}{\pi}\right)}{\Theta(0)} = \frac{\prod_{n=1}^{\infty}(1 - 2q^{2n-1}\cos 2x + q^{4n-2})}{\prod_{n=1}^{\infty}(1 - q^{2n-1})^2}, \quad q = e^{-\pi\frac{K'}{K}}. \tag{3.172}$$

In Section 61 of his *Fundamenta*, Jacobi introduced another important theta function:

$$\frac{H\left(\frac{2Kx}{\pi}\right)}{\Theta(0)} = \frac{2\sqrt[4]{q}\,\sin x\prod_{n=1}^{\infty}(1 - 2q^{2n}\cos 2x + q^{4n})}{\prod_{n=1}^{\infty}(1 - q^{2n-1})^2}, \quad q = e^{-\pi\frac{K'}{K}}. \tag{3.173}$$

He was then in a position to express his elliptic functions, given by (3.148) through (3.150), as ratios of theta functions:

$$\mathrm{sn}u = \frac{1}{\sqrt{k}}\frac{H(u)}{\Theta(u)}; \quad \mathrm{cn}u = \sqrt{\frac{k'}{k}}\frac{H(u+K)}{\Theta(u)}; \quad \mathrm{dn}u = \sqrt{k'}\frac{\Theta(u+K)}{\Theta(u)}. \tag{3.174}$$

Jacobi showed handily that the theta functions $H$ and $\Theta$ had the properties

$$\Theta(u + 2K) = \Theta(-u) = \Theta(u),$$

$$H(u + 2K) = H(-u) = -H(u), \quad H(u + 4K) = H(u),$$

$$\Theta(u + iK') = ie^{\pi\frac{(K'-2iu)}{4K}}H(u), \quad \Theta(u + 2iK') = -e^{\pi\frac{(K'-iu)}{K}}\Theta(u),$$

$$H(u + iK') = ie^{\pi\frac{(K'-2iu)}{4K}}\Theta(u), \quad H(u + 2iK') = -e^{\pi\frac{(K'-iu)}{K}}H(u).$$

These properties of $\Theta$ and $H$ have a number of implications, including, for example, that $\Theta$ has period $2K$ and "quasi-period" $2iK'$. In fact, in Section 62, Jacobi applied his properties to derive the series expansions:

$$\frac{\Theta\left(\frac{2Kx}{\pi}\right)}{\Theta(0)} = A\left(1 + 2\sum_{n=1}^{\infty}(-1)^n q^{n^2}\cos 2nx\right) \tag{3.175}$$

$$\frac{H\left(\frac{2Kx}{\pi}\right)}{\Theta(0)} = A\left(2\sum_{n=1}^{\infty}(-1)^{n-1}\sqrt[4]{q^{(2n-1)^2}}\sin(2n-1)x\right). \tag{3.176}$$

---

[88] See Pieper (1987) p. 99.   [89] Jacobi (1969) vol. 1, p. 256.

Equation (3.175) was proved by assuming that the left-hand side could be expanded as a cosine series

$$A + 2 \sum_{n=1}^{\infty} (-1)^n A_n \cos 2nx$$

and then applying the quasi-periodic property of $\Theta(u)$ to show that

$$A_n = q^{2n-1} A_{n-1},$$

which in turn implied that

$$A_n = q^{n^2} A.$$

Equation (3.176) was proved in the same manner; he then employed the equation

$$\Theta(u + iK') = ie^{\pi \frac{(K'-2iu)}{4K}} H(u)$$

to show that the constant $A$ was the same in both equations, (3.175) and (3.176).

Jacobi applied a very interesting method in Section 63 to determine $A$. Using (3.172), (3.173), (3.175), and (3.176), he wrote

$$\prod_{n=1}^{\infty} \left(1 - 2q^{2n-1} \cos 2x + q^{4n-2}\right) = P(q) \left(1 + \sum_{n=1}^{\infty} (-1)^n q^{n^2} \cos 2nx\right), \qquad (3.177)$$

$$\sin x \prod_{n=1}^{\infty} \left(1 - 2q^{2n} \cos 2x + q^{4n}\right) = P(q) \sum_{n=1}^{\infty} (-1)^{n-1} q^{(n-1)n} \sin(2n-1), \qquad (3.178)$$

so that, once $P(q)$ was determined, he had

$$A = \frac{P(q)}{\prod_{n=1}^{\infty} (1 - q^{2n-1})^2}. \qquad (3.179)$$

To find $P(q)$, Jacobi observed that when (3.177) was multiplied by (3.178), with $q$ changed to $q^2$ in both equations, the result was

$$P(q^2)P(q^2)(\sin x - q^4 \sin 3x + q^{12} \sin 5x - \cdots)(1 - 2q^2 \cos 2x + 2q^8 \cos 4x - \cdots)$$
$$= P(q)(\sin x - q^2 \sin 3x + q^6 \sin 5x - \cdots). \qquad (3.180)$$

He noted that since $2 \sin mx \cos nx = \sin(m+n)x + \sin(m-n)x$, the coefficient of $\sin x$ in the product on the left-hand side of (3.180) would be

$$P^2(q^2)(1 + q^2 + q^6 + q^{12} + q^{20} + \cdots),$$

and hence

$$\frac{P(q)}{P(q^2)P(q^2)} = 1 + q^2 + q^6 + q^{12} + q^{20} + \cdots. \qquad (3.181)$$

Setting $x = \frac{\pi}{2}$ in (3.178) then gave him

$$\left((1 + q^2)(1 + q^4)(1 + q^6) \cdots\right)^2 = P(q)(1 + q^2 + q^6 + q^{12} + q^{20} + \cdots), \qquad (3.182)$$

so that he could combine (3.181) with (3.182) to arrive at

$$\frac{P(q)P(q)}{P(q^2)P(q^2)} = \left((1+q^2)(1+q^4)(1+q^6)\cdots\right)^2$$

or

$$P(q) = P(q^2)\left((1+q^2)(1+q^4)(1+q^6)\cdots\right)$$
$$= P(q^2)\frac{(1-q^4)(1-q^8)(1-q^{12})\cdots}{(1-q^2)(1-q^4)(1-q^6)\cdots}.$$

Repeated iteration then yielded

$$P(q) = \frac{1}{(1-q^2)(1-q^4)(1-q^6)\cdots}. \tag{3.183}$$

When the value of $P(q)$ in (3.183) is substituted in (3.177), we obtain what is usually called the triple product identity. Thus, Jacobi obtained another proof of the triple product identity, in addition to the one contained in Section 64 of the *Fundamenta*. In this manner, he found the value of $A$ in (3.179) to be

$$A = \prod_{n=1}^{\infty}\left(\frac{1+q^n}{1-q^n}\right),$$

or, by using (3.164),

$$\frac{1}{A} = \sqrt{\frac{2k'K}{\pi}}.$$

He then set

$$\Theta(0) = \frac{1}{A} = \sqrt{\frac{2k'K}{\pi}}$$

so that he could define the two theta functions:

$$\Theta\left(\frac{2Kx}{\pi}\right) = 1 + 2\sum_{n=1}^{\infty}(-1)^n q^{n^2}\cos 2nx, \tag{3.184}$$

$$H\left(\frac{2Kx}{\pi}\right) = 2\sum_{n=1}^{\infty}(-1)^{n-1}\sqrt[4]{q^{(2n-1)^2}}\sin(2n-1)x. \tag{3.185}$$

Note that since $|q| < 1$, $\Theta$ and $H$ were defined as entire functions of $x$. By showing that the elliptic functions sn, cn, and dn were quotients of $\Theta$ and $H$, Jacobi was able to unfetter the elliptic function from its origin as the inverse of an elliptic integral. Inverting an elliptic integral to obtain a doubly periodic function cannot be accomplished in a rigorous manner without a fairly well-developed theory of functions of a complex variable. Jacobi most probably sensed this lack of rigor; after publishing his *Fundamenta* in 1829, he began to approach an elliptic function as a doubly periodic function, constructed as the ratio of two theta functions. He was also able to raise his famous two questions:

- Is there an analytic function having two independent periods whose ratio is real?
- Is there an analytic function of one variable with three or more periods?

Jacobi correctly answered both questions in the negative, though there is a gap in his discussion of the second question.[90]

In the winter of 1835–1836, Jacobi gave a series of lectures on theta functions defined as infinite series; he gave lectures on this topic at a later date as well. Carl Borchardt took notes of the later lectures and published them in 1881 in Jacobi's collected papers.[91] Jacobi began by defining four related theta series:

$$\theta_0(x, q) = \sum_{n=-\infty}^{\infty} (-1)^n q^{n^2} e^{2nxi} = 1 + 2 \sum_{n=1}^{\infty} (-1)^n q^{n^2} \cos 2nx, \tag{3.186}$$

$$\theta_1(x, q) = -\sum_{n=-\infty}^{\infty} i^{2n+1} q^{(n+\frac{1}{2})^2} e^{(2n+1)xi}$$

$$= 2 \sum_{n=1}^{\infty} (-1)^{n-1} \sqrt[4]{q^{(2n-1)^2}} \sin(2n-1)x, \tag{3.187}$$

$$\theta_2(x, q) = \sum_{n=-\infty}^{\infty} q^{(n+\frac{1}{2})^2} e^{(2n+1)xi} = 2 \sum_{n=1}^{\infty} \sqrt[4]{q^{(2n-1)^2}} \cos(2n-1)x, \tag{3.188}$$

$$\theta_3(x, q) = \sum_{n=-\infty}^{\infty} q^{n^2} e^{2nxi} = 1 + 2 \sum_{n=1}^{\infty} q^{n^2} \cos 2nx. \tag{3.189}$$

In his lectures, Jacobi took $0 < q < 1$, though we can take $q = e^{\omega \pi i}$, where Im $\omega > 0$. Jacobi himself defined $q$ in this manner, as did Eisenstein, Hermite, and others. We also note that Jacobi wrote no subscript in $\theta_0$, and that he omitted the $q$ in $\theta(x, q)$, writing it simply as $\theta(x)$. In order to define his elliptic functions, Jacobi derived a number of formulas, including:

$$\theta_3^3(0)\,\theta_3(2x) = \theta_3^4(x) + \theta_1^4(x) = \theta_0^4(x) + \theta_2^4(x), \tag{3.190}$$

$$\theta_3^2(0)\,\theta_0(0)\,\theta_0(2x) = \theta_0^2(x)\theta_3^2(x) + \theta_1^2(x)\theta_2^2(x), \tag{3.191}$$

$$\theta_2^3(0)\,\theta_2(2x) = \theta_2^4(x) - \theta_1^4(x) = \theta_3^4(x) - \theta_0^4(x), \tag{3.192}$$

$$\theta_3^2(0)\,\theta_3^2(x) = \theta_0^2(0)\,\theta_0^2(x) + \theta_2^2(0)\,\theta_2^2(x), \tag{3.193}$$

$$\theta_3^2(0)\,\theta_0^2(x) = \theta_0^2(x)\theta_3^2(x) + \theta_2^2(0)\,\theta_1^2(x), \tag{3.194}$$

$$\theta_3^2(0)\,\theta_2^2(x) = \theta_2^2(0)\,\theta_3^2(x) - \theta_0^2(0)\,\theta_1^2(x), \tag{3.195}$$

$$\theta_3^2(0)\,\theta_1^2(x) = \theta_2^2(0)\,\theta_0^2(x) - \theta_0^2(0)\,\theta_2^2(x). \tag{3.196}$$

Jacobi derived these identities from much more general identities involving four variables, having obtained the general identities from properties of lattices in the

---

[90] Jacobi (1969) vol. 2, pp. 23–32. See Markushevich (1992) pp. 14–23.
[91] See Jacobi (1969) vol. 1, pp. 499–538.

four-dimensional space $\mathbb{R}^4$. The reader might wish to read Jacobi's lectures on this material.[92] For a modern treatment of the connection between lattices and theta identities, see Stalker's book.[93] We present a very succinct derivation of the basic identity, due to H. J. S. Smith, in our Chapter 5. Smith employed Hermite's notation for the theta function, a notation allowing one to deal simultaneously with the four functions by varying two subscripts.

To define the modular functions $k^2$ and $k'^2$, Jacobi took $x = 0$ in (3.190) or in (3.191) to get

$$\theta_3^4(0) = \theta_0^4(0) + \theta_2^4(0). \tag{3.197}$$

Note that Gauss gave the result (3.197) in an unpublished and undated paper[94] on the arithmetic-geometric mean. According to the editor of Gauss's works, Schering, this paper appears in a notebook immediately following an astronomical calculation that was written by Gauss on April 28, 1809.

Jacobi next set

$$\sqrt{k} = \frac{\theta_2(0)}{\theta_3(0)}, \quad \sqrt{k'} = \frac{\theta_0(0)}{\theta_3(0)}, \tag{3.198}$$

so that $k^2 + k'^2 = 1$. Then, dividing (3.196) by $\theta_2^2(0)\,\theta_0^2(x)$, Jacobi obtained

$$\frac{\theta_0(0)}{\theta_2(0)} \cdot \frac{\theta_2(x)}{\theta_0(x)} = \sqrt{1 - \left(\frac{\theta_3(0)}{\theta_2(0)} \cdot \frac{\theta_1(x)}{\theta_0(x)}\right)^2}. \tag{3.199}$$

Similarly,

$$\frac{\theta_0(0)}{\theta_3(0)} \cdot \frac{\theta_3(x)}{\theta_0(x)} = \sqrt{1 - \left(\frac{\theta_2(0)}{\theta_3(0)} \cdot \frac{\theta_1(x)}{\theta_0(x)}\right)^2}. \tag{3.200}$$

He could then define his elliptic functions

$$\frac{\theta_1(x)}{\theta_0(x)} = \sqrt{k}\,\sin\phi, \quad \frac{\theta_2(x)}{\theta_0(x)} = \sqrt{\frac{k}{k'}}\,\cos\phi, \quad \frac{\theta_3(x)}{\theta_0(x)} = \frac{1}{\sqrt{k'}}\sqrt{1 - k^2\sin^2\phi}.$$

Jacobi's lectures contain another important and useful result:[95]

$$\theta_1'(0, q) = \theta_0(0, q)\,\theta_2(0, q)\,\theta_3(0, q). \tag{3.201}$$

To obtain (3.201), he needed (3.192) and (3.197); he first noted the directly verifiable relations

$$\theta_3(x, q) = \theta_3(2x, q^4) + \theta_2(2x, q^4), \tag{3.202}$$

$$\theta_0(x, q) = \theta_3(2x, q^4) - \theta_2(2x, q^4). \tag{3.203}$$

[92] Jacobi (1969) vol. 1, pp. 501–511.     [93] Stalker (1998) pp. 164–175.
[94] Gauss (1863–1927) vol. 3, p. 447.     [95] Jacobi (1969) vol. 1, pp. 516–517.

Then, by (3.192), (3.202), and (3.203),

$$\theta_2^3(0, q)\,\theta_2(2x, q) = \theta_3^4(x, q) - \theta_0^4(x, q)$$
$$= 8\theta_2(2x, q^4)\,\theta_3(2x, q^4)\,\big(\theta_3^2(2x, q^4) + \theta_2^2(2x, q^4)\big). \qquad (3.204)$$

Next, observe that

$$\theta_3\Big(x + \frac{\pi}{2}, q\Big) = \theta_0(x), \qquad \theta_2\Big(x + \frac{\pi}{2}\Big) = -\theta_1(x);$$

so that, by replacing $2x$ by $x$ and then changing $x$ to $x + \frac{\pi}{2}$, we get

$$\theta_2^3(0, q)\,\theta_1(x, q) = 8\theta_1(x, q^4)\,\theta_0(x, q^4)\,\big(\theta_0^2(x, q^4) + \theta_1^2(x, q^4)\big). \qquad (3.205)$$

Differentiating (3.205) with respect to $x$ and then setting $x = 0$, while noting that $\theta_1(0) = 0$, produces

$$\theta_2^3(0, q)\,\theta_1'(0, q) = 8\theta_0^3(0, q^4)\,\theta_1'(0, q^4). \qquad (3.206)$$

Upon setting $x = 0$ in (3.202), (3.203), and (3.204), we obtain

$$\theta_3(0, q) = \theta_3(0, q^4) + \theta_2(0, q^4), \qquad (3.207)$$
$$\theta_0(0, q) = \theta_3(0, q^4) - \theta_2(0, q^4), \qquad (3.208)$$
$$\theta_2^4(0, q) = 8\theta_2(0, q^4)\,\theta_3(0, q^4)\,\big(\theta_3^2(0, q^4) + \theta_2^2(0, q^4)\big). \qquad (3.209)$$

Set $x = 0$ in (3.190) and replace $q$ by $q^4$ to obtain

$$\theta_3^4(0, q^4) - \theta_2^4(0, q^4) = \theta_0^4(0, q^4);$$

hence, the product of the three relations (3.207) through (3.209) gives us

$$\theta_2^4(0, q)\,\theta_3(0, q)\,\theta_0(0, q) = 8\theta_0^4(0, q^4)\,\theta_2(0, q^4)\,\theta_3(0, q^4). \qquad (3.210)$$

Dividing (3.206) by (3.210), we finally arrive at

$$\xi(q) \equiv \frac{\theta_1'(0, q)}{\theta_0(0, q)\,\theta_2(0, q)\,\theta_3(0, q)} = \frac{\theta_1'(0, q^4)}{\theta_0(0, q^4)\,\theta_2(0, q^4)\,\theta_3(0, q^4)} \equiv \xi(q^4). \qquad (3.211)$$

Iteration then produces

$$\xi(q) = \xi(q^4) = \xi\big(q^{4^2}\big) = \cdots = \xi(0) = 1.$$

This concludes Jacobi's very ingenious proof of (3.201). This proof requires a high degree of familiarity with theta identities. Had Jacobi also employed the product representation in (3.177), he could have given a simpler proof. However, it is clear that Jacobi wished to illustrate that the important relations among the theta functions could be obtained by using only the theta series. Recognizing Jacobi's mastery of this subject, Dirichlet, in his 1852 obituary notice of Jacobi, mentioned that $\theta$-series should be named Jacobi series.[96] And Kronecker, in his lectures during the 1880s on definite integrals, wrote concerning Jacobi's work in 1836 on theta series, "Thence began a new epoch for the theory of theta functions."[97]

---

[96] Jacobi (1969) vol. 1, p. 15.  [97] Kronecker (1894) p. 198.

### 3.12  Jacobi's Triple Product Identity

In Section 64 of his *Fundamenta Nova*, Jacobi gave a more direct proof of the triple product identity than given in our Section 3.11. This proof is of intrinsic interest, but is also significant because Jacobi himself set a high value on the triple product identity. In 1848, he wrote to P. H. Fuss, Secretary of the Petersburg Academy, that this was probably his "wichtigste und fruchtbarste" [most important and fruitful] discovery in pure mathematics. In that letter,[98] he stated the triple product formula in the form

$$(z - z^{-1}) \prod_{n=1}^{\infty} (1 - q^n)(1 - q^n z)(1 - q^n z^{-1}) = \sum_{n=1}^{\infty} (-1)^{n-1} q^{\frac{n(n-1)}{2}} (z^{2n-1} - z^{-2n+1}).$$
$$(3.212)$$

He also noted in the letter that by taking the derivative with respect to $z$ and then setting $z = 0$, one obtains the formula

$$\prod_{n=1}^{\infty} (1 - q^n)^3 = \sum_{n=0}^{\infty} (-1)^n (2n + 1) q^{\frac{n(n+1)}{2}}; \qquad (3.213)$$

Jacobi gave an alternative proof of (3.213) in his *Fundamenta*.[99]

When $q$ is changed to $q^2$ and we set $z = e^{xi}$ on the right-hand side of (3.212), we arrive at the series for $\theta_1(x, q)$, except for a factor $iq^{\frac{1}{4}}$. Thus, the triple product identity in the form (3.212) implies that

$$\theta_1(x, q) = 2q^{\frac{1}{4}} \sin x \prod_{n=1}^{\infty} (1 - q^{2n})(1 - q^{2n} e^{2nxi})(1 - q^{2n} e^{-2nxi}). \qquad (3.214)$$

From this it is easy to obtain

$$\theta_0(x, q) = \prod_{n=1}^{\infty} (1 - q^{2n})(1 - q^{2n-1} e^{2nxi})(1 - q^{2n-1} e^{-2nxi}), \qquad (3.215)$$

$$\theta_2(x, q) = 2q^{\frac{1}{4}} \cos x \prod_{n=1}^{\infty} (1 - q^{2n})(1 + q^{2n} e^{2nxi})(1 + q^{2n} e^{-2nxi}), \qquad (3.216)$$

$$\theta_3(x, q) = \prod_{n=1}^{\infty} (1 - q^{2n})(1 + q^{2n-1} e^{2nxi})(1 + q^{2n-1} e^{-2nxi}). \qquad (3.217)$$

The formulas (3.214) through (3.217) follow from (3.212).

It remains to give a second proof of the triple product identity. In Section 64 of his *Fundamenta*, Jacobi stated this identity in the form:

$$\prod_{n=1}^{\infty} (1 - q^{2n})(1 + q^{2n-1} z)(1 + q^{2n-1} z^{-1}) = \sum_{n=-\infty}^{\infty} q^{n^2} z^n. \qquad (3.218)$$

---

[98] Stäckel and Ahrens (1908) p. 60.     [99] Jacobi (1969) vol. 1, p. 237.

He applied the following two identities to reprove this result, the first due to Euler who found it in 1740,[100] the second due to Jacobi himself.

$$\prod_{n=1}^{\infty} \left(1 + q^{2n-1}z\right) = 1 + \sum_{n=1}^{\infty} \frac{q^{n^2} z^n}{(1 - q^2)(1 - q^4) \cdots (1 - q^{2n})}, \qquad (3.219)$$

$$\prod_{n=1}^{\infty} (1 - q^n z)^{-1} = 1 + \sum_{n=1}^{\infty} \frac{q^{n^2} z^n}{(1 - q)(1 - q^2) \cdots (1 - q^n)(1 - qz)(1 - q^2 z) \cdots (1 - q^n z)}. \qquad (3.220)$$

To prove (3.219), let $f(z)$ denote the left-hand side of (3.219) and suppose that $f(z)$ has the series expansion

$$f(z) = 1 + \sum_{n=1}^{\infty} A_n z^n.$$

Then

$$f(q^2 z) = 1 + \sum_{n=1}^{\infty} A_n q^{2n} z^n.$$

On the other hand, it is clear that $f(z) = (1 + qz)f(q^2 z)$, so that

$$1 + \sum_{n=1}^{\infty} A_n z^n = (1 + qz) \left( 1 + \sum_{n=1}^{\infty} A_n q^{2n} z^n \right).$$

Equating coefficients of $z^n$, we have

$$A_n = A_n q^{2n} + A_{n-1} q^{2n-1}.$$

Therefore

$$A_n = \frac{q^{2n-1}}{1 - q^{2n}} A_{n-1} = \frac{q^{2n-1+2n-3+\cdots+3+1}}{(1 - q^{2n})(1 - q^{2n-2}) \cdots (1 - q^2)}$$

$$= \frac{q^{n^2}}{(1 - q^2)(1 - q^4) \cdots (1 - q^{2n})}.$$

This proves (3.219). This argument of Jacobi is identical to the one given earlier by Euler. To prove (3.220), let $g(z)$ denote the infinite product on the left-hand side of (3.220) and suppose that

$$g(z) = 1 + \sum_{n=1}^{\infty} \frac{B_n z^n}{(1 - qz) \cdots (1 - q^n z)}.$$

---

[100] Euler (1988) Chapter 16, and Eu.I-2 pp. 163–193.

The relation $g(z) = \frac{g(qz)}{1-qz}$ then produces, with $B_0 \equiv 1$,

$$1 + \sum_{n=1}^{\infty} \frac{B_n z^n}{(1-qz)(1-q^2 z)\cdots(1-q^n z)} = \sum_{n=1}^{\infty} \frac{B_{n-1} q^{n-1} z^{n-1}}{(1-qz)(1-q^2 z)\cdots(1-q^n z)}$$

$$= 1 + \sum_{n=1}^{\infty} \frac{(q^{2n-1} B_{n-1} + q^n B_n) z^n}{(1-qz)(1-q^2 z)\cdots(1-q^n z)}.$$

This leads us to

$$B_n = q^{2n-1} B_{n-1} + q^n B_n,$$

and hence

$$B_n = \frac{q^{n^2}}{(1-q)(1-q^2)\cdots(1-q^n)}.$$

It is now clear from (3.219) that

$$\prod_{n=1}^{\infty} (1 + q^{2n-1} z)(1 + q^{2n-1} z^{-1}) = \left( 1 + \sum_{n=1}^{\infty} \frac{q^{n^2}}{(1-q^2)(1-q^4)\cdots(1-q^{2n})} \cdot z^n \right)$$

$$\times \left( 1 + \sum_{n=1}^{\infty} \frac{q^{n^2}}{(1-q^2)(1-q^4)\cdots(1-q^{2n})} \cdot \frac{1}{z^n} \right). \tag{3.221}$$

If $C_n$ denotes the coefficient of $z^n$, or of $\frac{1}{z^n}$, since they are identical, then it is not difficult to see that

$$C_n = \frac{q^{n^2}}{(1-q^2)\cdots(1-q^{2n})}$$

$$\times \left( 1 + \sum_{m=1}^{\infty} \frac{q^{2m^2}}{(1-q^2)\cdots(1-q^{2m})} \cdot \frac{q^{2mn}}{(1-q^{2n+2})(1-q^{2n+4})\cdots(1-q^{2n+2m})} \right). \tag{3.222}$$

This series can be summed by (3.220), when we change $q$ to $q^2$ and set $z = q^{2n}$; the result is then

$$C_n = \frac{q^{n^2}}{(1-q^2)(1-q^4)(1-q^6)\cdots}.$$

This completes Jacobi's second proof of the triple product identity. Another proof by Jacobi is given as an exercise at the end of this chapter. In his lectures during the 1880s on definite integrals, Kronecker included a chapter on the theta series in which he derived the triple product identity, starting with the series. After giving the derivation, he remarked,[101] "Herein lies the tremendous discovery of Jacobi; the transformation

---

[101] Kronecker (1894) p. 186.

of the series into the product was very difficult. Abel also had the product, but not the series. Therefore, Dirichlet used to also call them Jacobi series."

## 3.13 Modular Equations and Transformation Theory

In view of Jacobi's aims, first to define doubly periodic functions and then to find their connection with elliptic integrals, we will define an $n$th-order transformation in terms of periods of elliptic functions. Relations (3.73) through (3.76) also suggest this definition.

Suppose $\phi$ is an elliptic function with two fundamental periods, $\omega_1$ and $\omega_2$. This means that every period of $\phi$ can be written as $m\omega_1 + n\omega_2$, where $m$ and $n$ are integers. Let Im $\frac{\omega_1}{\omega_2} > 0$. Note that this is not a restriction because either $\frac{\omega_1}{\omega_2} > 0$ or $\frac{\omega_2}{\omega_1} > 0$; if the latter case holds, then we can simply renumber the subscripts. Let $n$ be a positive integer. Suppose $\omega_1'$ and $\omega_2'$ are related to $\omega_1$ and $\omega_2$ by

$$M\omega_1' = a\omega_1 + b\omega_2 \qquad (3.223)$$

$$M\omega_2' = c\omega_1 + d\omega_2, \qquad (3.224)$$

where $M$ is as in (3.75) and (3.77), $a, b, c, d$ are integers, and $ad - bc = n$. An $n$th-order transformation of the elliptic function $\phi$ produces an elliptic function $\psi$ with fundamental periods $M\omega_1'$ and $M\omega_2'$; it can be shown that $\psi$ is a rational function of $\phi$. A fairly simple argument shows that if there exists an algebraic relation between two elliptic functions $x = f(u)$ and $y = g(u)$, then the periods $\omega_1, \omega_2$ of $f(u)$ and the periods of $\omega_1', \omega_2'$ of $g(u)$ are related by the equations

$$m_1\omega_1 = a\omega_1' + b\omega_2' \qquad (3.225)$$

$$m_2\omega_2 = c\omega_1' + d\omega_2', \qquad (3.226)$$

where $m_1, m_2, a, b, c, d$ are all integers. To verify (3.225), suppose $F(x, y) = 0$ is the algebraic relation between $x$ and $y$, of degree $n$ in $y$; then the identity $F\big(f(u), g(u)\big) \equiv 0$ holds for $u \in \mathbb{C}$, except for the poles of $f$ and $g$. If $u$ is changed to $u + m\omega_1$, $m$ an integer, then

$$F\big(f(u + m\omega_1), g(u + m\omega_1)\big) \equiv 0$$

or

$$F\big(f(u), g(u + m\omega_1)\big) \equiv 0.$$

Since $g$ satisfies an $n$th-degree equation for any given $u$, there are at most $n$ different values of $g(u + m\omega_1)$ at the infinite number of different values $u + m\omega_1$. Thus, there must exist a relation of the form

$$u + l\omega_1 = u + l'\omega_1 + a\omega_1' + b\omega_2',$$

or

$$m_1\omega_1 = a\omega_1' + b\omega_2',$$

where $l, l', m_1, a, b$ are integers. This proves (3.225); the proof of (3.226) may be accomplished in a similar manner.

We have seen that the modulus $k$ is a function of $q = e^{-\pi \frac{K'}{K}} = q^{\pi i \frac{iK'}{K}}$, and that $k$ can be expressed in terms of theta functions. Thus, we may define an $n$th-order modular equation more generally in terms of theta functions. We first combine the two equations (3.223) and (3.224) into one equation

$$\omega' = \frac{\omega_1'}{\omega_2'} = \frac{a\omega + b}{c\omega + d}, \quad \omega = \frac{\omega_1}{\omega_2}, \quad ad - bc = n. \tag{3.227}$$

Let $q = e^{i\pi\omega}$ and $q' = e^{i\pi\omega'}$. An $n$th-order transformation of $\theta(x, q)$, with $\theta$ as given by (3.186) through (3.189), is the theta function $\theta(x, q')$. An $n$th-order modular equation is a relation between $\theta(0, q) \equiv \theta(q)$ and $\theta(q')$. It is customary to take $a = n, d = 1$, $b = c = 0$, and refer to an $n$th-order modular equation as a relation between $\theta(q)$ and $\theta(q^n)$.

Now we know that a modular equation of order 3 is a polynomial of degree 4 in the moduli $\sqrt[4]{k}$ and $\sqrt[4]{\lambda}$. This implies that for each value of $k$, there are four values of $\sqrt[4]{\lambda}$ and that each one of the latter yields a cubic transformation of the elliptic function or the elliptic integral. Similarly, the modular equation of order 5 gives an equation of degree 6. More generally, a modular equation of order $p$, where $p$ is an odd prime, is an equation of degree $p + 1$. For each value of $k$, there are $p + 1$ values of $\lambda$ and each of these values of $\lambda$ yields a $p$th-order transformation of the elliptic integral or function. In terms of (3.223) and (3.224), a $p$th-order transformation is given by the equation $ad - bc = p$. The $p + 1$ basic transformations of this order are given by

$$a = p, \quad d = 1, \quad b = c = 0, \quad a = 1, \quad d = p, \quad c = 0, \quad 0 \le b < p.$$

Any other transformation of order $p$ can be obtained from these by an application of an appropriate first-order transformation to one of these $p + 1$ transformations. Note that a first-order transformation is one in which $ad - bc = 1$. In the nineteenth century such a transformation was called a linear transformation.

Observe that the elliptic functions $\mathrm{sn}^2(u, k)$ and $\mathrm{cn}^2(u, k)$ have periods $2K$ and $2iK'$. If we let $\omega = \frac{iK'}{K}$ and $q = e^{i\pi\omega} = e^{-\frac{\pi K'}{K}}$, then it can be shown that the $p + 1$ transformations of $k(q)$ will be given by

$$q \to q^p, \quad q^{\frac{1}{p}}, \quad \alpha q^{\frac{1}{p}}, \quad \alpha^2 q^{\frac{1}{p}}, \quad \ldots \alpha^{p-1} q^{\frac{1}{p}}, \quad \text{where } \alpha^p = 1. \tag{3.228}$$

Jacobi mentioned these $p + 1$ transformations in his 1828 paper[102] that appeared in *Crelle's Journal*.

## 3.14 Exercises

1. Fill in the details of Abel's method of proving (3.62): Let $r$ denote the right-hand side of (3.62). Show that $\frac{\partial r}{\partial u} = \frac{\partial r}{\partial v}$. Then deduce that $r = \mathrm{sn}(u + v)$. See Abel (1965) vol. 1, pp. 268–269.

[102] Jacobi (1969) vol. 1, pp. 251–254.

2. Prove the following identities of Jacobi:

$$\operatorname{sn}(u+\alpha)\operatorname{sn}(u-\alpha) = \frac{\operatorname{sn}^2 u - \operatorname{sn}^2\alpha}{1 - k^2\operatorname{sn}^2 u\operatorname{sn}^2\alpha}$$

$$\operatorname{sn}(u+\alpha) + \operatorname{sn}(u-\alpha) = \frac{2\operatorname{sn}u\operatorname{cn}\alpha\operatorname{dn}\alpha}{1 - k^2\operatorname{sn}^2 u\operatorname{sn}^2\alpha}$$

$$\frac{\bigl(1\pm\operatorname{sn}(u+\alpha)\bigr)\bigl(1\pm\operatorname{sn}(u-\alpha)\bigr)}{\operatorname{cn}^2\alpha} = \frac{\bigl(1\pm\frac{\operatorname{sn}u}{\operatorname{sn}(K-\alpha)}\bigr)^2}{1 - k^2\operatorname{sn}^2 u\operatorname{sn}^2\alpha}$$

$$\frac{\bigl(1\pm k\operatorname{sn}(u+\alpha)\bigr)\bigl(1\pm k\operatorname{sn}(u-\alpha)\bigr)}{\operatorname{dn}^2\alpha} = \frac{(1\pm k\operatorname{sn}u\operatorname{sn}(K-\alpha))^2}{1 - k^2\operatorname{sn}^2 u\operatorname{sn}^2\alpha}$$

$$\frac{\operatorname{cn}(u+\alpha)\operatorname{cn}(u-\alpha)}{\operatorname{cn}^2\alpha} = \frac{1 - \frac{\operatorname{sn}^2 u}{\operatorname{sn}^2(K-\alpha)}}{1 - k^2\operatorname{sn}u\operatorname{sn}^2\alpha}$$

$$\frac{\operatorname{dn}(u+\alpha)\operatorname{dn}(u-\alpha)}{\operatorname{dn}^2\alpha} = \frac{1 - k^2\operatorname{sn}^2 u\operatorname{sn}^2(K-\alpha)}{1 - k^2\operatorname{sn}^2 u\operatorname{sn}^2\alpha}.$$

See Jacobi (1969) vol. 1, pp. 83, 84, 88.

3. Prove that

$$\left(\frac{1-\operatorname{sn}\frac{2Kx}{\pi}}{1+\operatorname{sn}\frac{2Kx}{\pi}}\right)^{\frac12} = \left(\frac{1-\sin x}{1+\sin x}\right)^{\frac12}\prod_{n=1}^{\infty}\frac{1-2q^n\sin x+q^{2n}}{1+2q^n\sin x+q^{2n}} \tag{3.229}$$

$$\left(\frac{1-k\operatorname{sn}\frac{2Kx}{\pi}}{1+k\operatorname{sn}\frac{2Kx}{\pi}}\right)^{\frac12} = \prod_{n=1}^{\infty}\frac{1-2q^{\frac{2n-1}{2}}\sin x+q^{2n-1}}{1+2q^{\frac{2n-1}{2}}\sin x+q^{2n-1}}. \tag{3.230}$$

See Jacobi (1969) vol. 1, p. 155.

4. Take the logarithmic derivatives of (3.157) through (3.159) and of (3.229) and (3.230) to obtain, respectively,

$$\frac{2k'K}{\pi}\cdot\frac{\operatorname{cn}\frac{2Kx}{\pi}}{\operatorname{cn}\bigl(K-\frac{2Kx}{\pi}\bigr)} = \cot x - \sum_{n=1}^{\infty}\frac{4q^n\sin 2nx}{1+q^n} \tag{3.231}$$

$$\frac{2K}{\pi}\cdot\frac{\operatorname{sn}\frac{2Kx}{\pi}}{\operatorname{sn}\bigl(K-\frac{2Kx}{\pi}\bigr)} = \tan x + \sum_{n=1}^{\infty}\frac{4q^n\sin 2nx}{1+(-q)^n} \tag{3.232}$$

$$\frac{2k^2K}{\pi}\operatorname{sn}\frac{2Kx}{\pi}\operatorname{sn}\bigl(K-\frac{2Kx}{\pi}\bigr) = 8\sum_{n=1}^{\infty}\frac{q^{2n-1}\sin(4n-2)x}{1-q^{4n-2}} \tag{3.233}$$

$$\frac{2K}{\pi\operatorname{sn}\bigl(K-\frac{2Kx}{\pi}\bigr)} = \frac{1}{\cos x} + 4\sum_{n=1}^{\infty}(-1)^{n-1}\frac{q^{2n-1}\cos(2n-1)x}{1-q^{2n-1}} \tag{3.234}$$

$$\frac{2kK}{\pi}\cdot\operatorname{sn}\bigl(K-\frac{2Kx}{\pi}\bigr) = 4\sum_{n=1}^{\infty}(-1)^{n-1}\frac{q^{\frac{2n-1}{2}}\cos(2n-1)x}{1-q^{2n-1}}. \tag{3.235}$$

See Jacobi (1969) vol. 1, pp. 156–157.

5. From (3.232), derive the formulas

$$\frac{2K}{\pi} = 1 + 4 \sum_{n=1}^{\infty} \frac{(-1)^{n-1} q^{2n-1}}{1 - q^{2n-1}}$$

$$\left(\frac{2K}{\pi}\right)^2 = 1 + 8 \sum_{n=1}^{\infty} \frac{nq^n}{1 + (-q)^n}.$$

Apply (3.168) to interpret these formulas as results on sums of squares of integers. See Jacobi (1969) vol. 1, pp. 159, 160, 163.

6. Let $K^{(m)}, k^{(m)}$ be quantities that depend on $q^m$ in the same manner as $K, k$ depend on $q$. Show that

$$\prod_{j=0}^{m-1} \Theta\left(x + \frac{2j\pi}{n}\right) = C\Theta(mx, q^m)$$

$$(-1)^{\frac{m-1}{2}} \prod_{j=0}^{m-1} H\left(x + \frac{2j\pi}{n}\right) = CH(mx, q^m)$$

$$\mathrm{sn}\left(\frac{2mK^{(m)}}{\pi}, k^{(m)}\right) = (-1)^{\frac{m-1}{2}} \sqrt{\frac{k^m}{k^{(m)}}} \prod_{j=0}^{m-1} \mathrm{sn}\frac{2K}{\pi}\left(x + \frac{2j\pi}{m}\right).$$

See Jacobi (1969) vol. 1, p. 258.

7. Reproduce Jacobi's third proof of the triple product identity along these lines:

Let

$$S = 1 + \sum_{n=1}^{\infty} (-1)^n q^{n^2} (z^n + z^{-n})$$

$$\Pi = \prod_{n=1}^{\infty} (1 - q^{2n})(1 - q^{2n-1}z)(1 - q^{2n-1}z^{-1}).$$

Set

$$P = -q\frac{\partial \log \Pi}{\partial q}$$

$$R = -z\frac{\partial \log \Pi}{\partial z}.$$

Show that

$$P = \sum_{m=1}^{\infty} \left(2\sigma_1(m)q^{2m} + \frac{q^m(1 + q^{2m})}{(1 - q^{2m})^2}(z^m + z^{-m})\right),$$

where $\sigma_1(m) = \sum_{d|m} d,$ and show that

$$R = \sum_{m=1}^{\infty} \frac{q^m}{1 - q^{2m}} (z^m - z^{-m}).$$

Verify that

$$-q\frac{\partial S}{\partial q} = SP \quad \text{and} \quad -z\frac{\partial S}{\partial z} = SR;$$

finally, deduce that $S = \Pi$. See Jacobi (1969) vol. 2, pp. 153–160.

# 4

## Eisenstein and Hurwitz

### 4.1 Preliminary Remarks

In April, 1846, Alexander von Humboldt wrote to Gauss,[1] explaining that Eisenstein (1823–1852) was somewhat isolated, lonely, and, despite von Humboldt's past assistance, still financially needy; von Humboldt requested Gauss to support him in his efforts to secure a stipend for Eisenstein. Gauss responded to his old friend, first dwelling on the memories of their great friendship and then continuing, "These reminiscences lead me naturally to the request, that you bestow the protection of your great name on a young man whom I hold in great esteem, namely the young mathematician, Eisenstein.... I consider his talent to be such as nature bestows upon only a few in a century."[2] We are fortunate to have Eisenstein's own moving account of his intellectual and spiritual development, included in his application to take the Berlin entrance examination.[3] We learn that he entered the Cauersche Anstalt in Charlottenburg at the age of ten or eleven; suffering from weak health throughout his short life, Eisenstein found the military discipline there taxing; nevertheless, it was there that he became aware of his mathematical talent. The director, von der Lage, utilized a teaching method similar to the one more recently championed by R. L. Moore. There were no lectures; each student was assigned a theorem to prove and when he had succeeded, he received another theorem to tackle. There was no discussion among the students; each student worked independently to provide complete proofs of the theorems to be proved. Eisenstein very much enjoyed learning in this challenging manner, but he had insightful reservations on the exclusive use of this pedagogical method:[4]

> The method was very good, firstly because the intellectual powers of deduction were strengthened through independent thinking and then again, a sense of rivalry among the students was stimulated by the general effort. Still, this method should probably not be widely adopted. For as much as I recognize all that can be said of its benefits, one must nevertheless admit that it isolates one's abilities too much, and that it allows for no overview of the whole subject, which can be provided only by a good lecture. Now that such an enormous wealth of varied and beautiful recent

---

[1] Bierman (1977) pp. 89–91.   [2] Ibid. p. 93.   [3] See Eisenstein (1975) vol. 2, pp. 879–904.

[4] For the original, see Ibid. p. 889; for the translation by M. Schmitz, see Res. Lett. Inf. Math. Sci., 2004 Vol. 6, pp. 1–13.

discoveries has accumulated, one is truly compelled to work as part of a group, if one wishes to grasp that which is primary and essential and even, to some extent, make it one's own. The greatest mathematical genius cannot, after all, discover alone all that has emerged through the collaboration of many outstanding minds, and which has arisen out of their collective accomplishments. This method of independent discovery is appropriate for students only if it involves the consideration of a small area of simple, especially geometrical, propositions, where everyone will, after all, use the same method of approach and no new or ingenious ideas will be required.

After leaving Charlottenburg, Eisenstein studied at the Gymnasium, where he devoted himself to studying the works of Euler and Lagrange and then Jacobi and Dirichlet. The works of Dirichlet revealed to him the close relationship between number theory and analysis. In fact, during 1840–1842, Eisenstein attended Dirichlet's Berlin lectures and those of Martin Ohm, brother of Georg Ohm. In 1842, Eisenstein left Germany with his mother to join his father in Britain; during his travels there, Eisenstein made a close study of Gauss's *Disquisitiones*. These studies and a fortuitous meeting with The Irish mathematician William Rowan Hamilton (1805–1866) motivated Eisenstein to delve into unsolved problems in number theory and elliptic functions. He presented to the Berlin Academy Hamilton's paper on Abel's work in algebraic equations of degree greater than four, also submitting a paper of his own on cubic forms. When the famous journal editor A. L. Crelle saw the latter paper, he requested Eisenstein to publish in his journal and during 1844–45, twenty-five of the young mathematician's papers appeared in *Crelle's Journal*. These papers treated topics of abiding interest to Gauss: elliptic functions; quadratic, cubic, and quartic reciprocity. Thus, Gauss invited Eisenstein to visit Göttingen.

In fact, Eisenstein's 1847 *Mathematische Abhundlung*, a collection of twelve of his *Crelle's Journal* papers, was given a foreword by Gauss himself. He wrote, "The present papers contain so much to be admired and of such dignity that through them a place of honor will be secured for their author among his predecessors, whose work these are worthy of joining."[5] The *Math. Abh.* included the long paper on elliptic functions: "Genaue Untersuchung der unendlichen Doppelproducte, aus welchen die elliptischen Functionen als Quotienten zusammengesetzt sind, und der mit ihnen zusammenhangenden Doppelreichen."[6]

In this paper, Eisenstein considered double sums of partial fractions that converged to doubly periodic functions. For example, let $\omega_1$ and $\omega_2$ be complex numbers such that the imaginary part of $\omega = \frac{\omega_1}{\omega_2}$ is positive. He defined the series

$$(s, x) = \lim_{N \to \infty} \lim_{M \to \infty} \sum_{n=-N}^{N} \sum_{m=-M}^{M} \frac{1}{(x + m\omega_1 + n\omega_2)^s}, \qquad (4.1)$$

where $s = 1, 2, 3, \ldots$, and

$$(s^*, 0) = \lim_{N \to \infty} \lim_{M \to \infty} \sum_{n=-N}^{N} {\sum_{m=-M}^{M}}' \frac{1}{(m\omega_1 + n\omega_2)^s}, \qquad (4.2)$$

---

[5] See Eisenstein (1847) p. IV. Also see Eisenstein (1975) vol. 2, p. 918, and Schmitz's translation: Res. Lett. Inf. Math. Sci., 2004, Vol. 6, p. 2.

[6] Eisenstein (1847) pp. 213–334 and Eisenstein (1975) pp. 357–478.

where the prime symbol on the sum in (4.2) indicates that $(m, n) \neq (0, 0)$. Eisenstein's notation $(s, x)$, unfortunately, did not indicate dependence on $\omega_1$ and $\omega_2$. The sums (4.1) and (4.2) converge absolutely for $s \geq 3$ and conditionally for $s = 1$ and $s = 2$. It was from Dirichlet's 1837–1839 papers on primes in an arithmetic progression[7] that Eisenstein learned of absolute and conditional convergence. We remark that Dirichlet himself gleaned the idea of absolute convergence from Cauchy's *Analyse Algébrique* of 1821. Although he was not aware of it, Eisenstein needed the concept of uniform convergence in order to differentiate and integrate the series $(s, x)$. In fact, Cauchy too had made a mistake in this connection, when he stated that the sum of a convergent series of continuous functions was continuous. As we have noted earlier, Abel gave a counterexample[8] to this claim and proved that it was correct in the case of convergent power series. We also mention that it appears that in 1838 Gauss's student Christoph Gudermann (1798–1853) became the first mathematician to conceive of the general concept of uniform convergence.[9] His student Weierstrass applied it extensively in his own work, although uniform convergence was not generally known until Weierstrass began his Berlin lectures on analysis around 1860.[10]

The double product mentioned in the title of Eisenstein's long paper was defined as

$$\phi(x) = \lim_{N \to \infty} \lim_{M \to \infty} \prod_{n=-N}^{N} \prod_{m=-M}^{M} \left(1 - \frac{x}{m\omega_1 + n\omega_2}\right). \tag{4.3}$$

He proved that

$$(2, x) - (2, y) = \frac{\phi(x + y)\phi(y - x)}{\phi^2(x)\phi^2(y)}. \tag{4.4}$$

Note that since $(s, x)$ is absolutely convergent for $s \geq 3$, $(s, x + \omega_i) = (s, x)$ for $i = 1, 2$ and thus $(s, x)$ is doubly periodic. Moreover, because he had employed a special method for summing the series, he was able to show that $(2, x)$ was also doubly periodic. Therefore, $(2, x)$ must be an elliptic function. Eisenstein had thus given a new foundation and mode of development for the theory of elliptic and modular functions, an approach remarkably well suited for applications to number theory; he himself gave some of these applications.

As early as 1844, Eisenstein pointed out in his paper "Bemerkungen zu den elliptischen und Abelschen Transcendenten,"[11] that there was a difficulty in inverting an elliptic integral, so that it could not be used to define an elliptic function:

> The ordinary definition which one gives of elliptic functions, quite contrary to the analogy with the exponential functions, is that they are inverse functions of the given integrals, in which the polynomial function under the square root goes up to the fourth power. But since even the clear meaning of such an integral, whose differential abruptly moves from the reals to the imaginaries, is not that simple, it appears to the student nearly impossible to construct *a priori* a clear conception of the inverse of such a function, especially since here the geometric conception is missing, which

---

[7] Eisenstein (1975) vol. 1, p. 370. See Dirichlet (1969) vol. 1. pp. 313–342 and pp. 411–496.

[8] Abel cites Cauchy's book *Cours d'analyse* and the counterexample in Abel (1965) pp. 224–225.

[9] Gudermann (1838) pp. 251–252.

[10] For a brief history of uniform convergence, with relevant references, see R. Remmert (1991) pp. 96–98.

[11] Eisenstein (1975) vol. 1, pp. 30–32.

one can call upon for help at least for the circular function. A special difficulty is presented by the periodicity. Let us return for a moment to the circular function in order to talk of a simple period. We wish to define, for example, the sine as that function $y$ of $x$ which is given by the equation

$$\int_0 \frac{dy}{\sqrt{1-y^2}} = x$$

so that one must, in accordance with the known properties of the sine, claim that the integral for every given value of $y$ takes on infinitely many values, although this stands in contradiction with the ordinary meaning which one attributes to such an integral (e.g. for $y < 1$). In the same manner, one must state, on account of the double periodicity of sine am $x$, that the elliptic integral

$$\int_0 \frac{dy}{\sqrt{(1-y^2)(1-k^2y^2)}} = x,$$

for every value of $y$, actually takes on an infinite number of different values an infinite number of times.

This paper raised the question of how an integral, by definition single-valued, could take on infinitely many values. Eisenstein went on to observe that the difficulties with the Abelian integrals were even more formidable since the inverses of such integrals were required to have three fold periodicity or greater. He noted that according to Jacobi, this implied that an Abelian integral with fixed lower limits for any given value of the variables would take on all possible real and imaginary values; if he assumed this, then the Abelian integral completely ceased to be a function of its variables. Thus, either such an integral had no meaning or the given definition of such an integral had been insufficient.

The answers to Eisenstein's puzzles were provided by Cauchy's theory of integration in the complex domain, developed by Cauchy beginning in 1814.[12] However, Eisenstein himself avoided the use of integrals and resorted to infinite products and sums instead.

In his lectures on elliptic functions dating from 1862, Weierstrass defined the function known as the Weierstrass $\wp$ function:

$$\wp(x) = \frac{1}{x^2} + \sum_{m,n}' \left( \frac{1}{(x+m\omega_1+n\omega_2)^2} - \frac{1}{(m\omega_1+n\omega_2)^2} \right), \qquad (4.5)$$

where the double sum is over all integers $m, n$ except $(m, n) = (0, 0)$. It is clear that

$$\wp(x) = (2, x) - (2^*, 0), \qquad (4.6)$$

and that $\wp$ is absolutely convergent. Weierstrass also had an absolutely convergent product $\sigma(x)$ corresponding to Eisenstein's $\phi(x)$ defined in (4.3):

$$\sigma(x) = x \prod_{m,n}' \left( 1 - \frac{x}{m\omega_1+n\omega_2} \right) e^{\frac{x}{m\omega_1+n\omega_2} + \frac{1}{2}\left(\frac{x}{m\omega_1+n\omega_2}\right)^2}. \qquad (4.7)$$

Now it can be shown that

$$\phi(x) = e^{-\frac{1}{2}(2^*,0)x^2} \sigma(x). \qquad (4.8)$$

---

[12] Cauchy (1882–1974), vol. 10, ser. 1, pp. 133–138 and pp. 158–167.

Weierstrass also defined the Weierstrass $\zeta$ function

$$\zeta(x) = \frac{d}{dx} \log \sigma(x),$$

so that by (4.8),

$$\zeta(x) = (1, x) + (2^*, 0)x. \tag{4.9}$$

It is evident that Eisenstein had defined all the basic functions used by Weierstrass in his theory of elliptic functions. As we shall see, Eisenstein also verified all the important properties of his function $(2, x) - (2^*, 0)$, including the first- and second-order differential equations satisfied by this function, an addition formula, and a formula for $(2, x) - (2, q)$, $q = \frac{\omega_1}{2}$, $\frac{\omega_2}{2}$, or $\frac{\omega_1 + \omega_2}{2}$ in terms of $\phi(x)$.

Weierstrass did not mention Eisenstein's work on elliptic functions. André Weil has conjectured that this may be attributed to the fact that Eisenstein assumed and did not explicitly prove that $\frac{d}{dx}(s, x)$ was equal to the term-by-term derivative of the series $(s, x)$. But Weil also showed that Eisenstein could have proved this result on the derivative of his series by the very methods he employed in his work.[13]

Eisenstein established a connection between his product function $\phi(x)$ and the theta function

$$\theta_1(z, \omega) = 2q^{\frac{1}{4}} \sin \pi z \prod_{m=1}^{\infty} (1 - q^{2m})(1 - q^{2m}e^{2\pi i z})(1 - q^{2m}e^{-2\pi i z}), \quad q = e^{\pi i \omega},$$

where $\operatorname{Im} \omega > 0$. He also proved a result that implies, taking $\frac{x}{\omega_1} = z$ and $\omega = \frac{\omega_2}{\omega_1}$

$$\phi(x) = \frac{\omega_1}{\pi} \frac{\theta_1(z, \omega)}{\eta^3(\omega)},$$

where $\eta(\omega)$ denoted what we now call Dedekind's function,

$$\eta(\omega) = q^{\frac{1}{12}} \prod_{n=1}^{\infty} (1 - q^{2m}), \quad q = e^{\pi i \omega}.$$

He also obtained a $k$th-order transformation for $\phi(x)$. Let

$$\omega_1' = \lambda \omega_1 + \nu \omega_2,$$
$$\omega_2' = \mu \omega_1 + \rho \omega_2,$$

where $\lambda, \nu, \mu, \rho$ are integers with $\lambda \rho - \mu \nu = k > 0$. Eisenstein set

$$\phi'(x) = \lim_{N \to \infty} \lim_{M \to \infty} \prod_{n=-N}^{N} \prod_{m=-M}^{M} \left(1 - \frac{x}{m \omega_1' + n \omega_2'}\right).$$

Note here that Eisenstein's $\phi'$ did not denote the derivative. He then succeeded in proving the important formula[14]

$$\phi'(x) = e^{-\frac{\partial \nu \pi i x^2}{\omega_1 \omega_1'}} x^2 \phi(x), \tag{4.10}$$

where $\partial = +1$ if $\operatorname{Im} \frac{\omega_2}{\omega_1} < 0$, and $\partial = -1$ if $\operatorname{Im} \frac{\omega_2}{\omega_1} > 0$.

---

[13] Weil (1976), p. 5.    [14] Eisenstein (1975) vol. 1, p. 424.

In his 1891 paper "Die Legendre'sche Relation," Kronecker wrote about the results of Eisenstein, such as (4.10):

> In his fundamental but seldom cited paper, based on completely original ideas, "Beiträgen zur Theorie der elliptischen Functionen," which appeared in 1847 in the volume 35 of *Crelle's* Journal, *Eisenstein* has established essentially new points of view for the theory of the $\theta$-function in general, and also in particular for their linear transformations, as well as by implication for the *Legendre* relation.[15]

Note that by a linear transformation, Kronecker meant a transformation $\omega \to \frac{\lambda\omega+\nu}{\mu\omega+\rho}$ with $\lambda\rho - \mu\nu = 1$. In 1845 in Berlin, Kronecker had met with Eisenstein frequently; Kronecker soon left Berlin, leaving Eisenstein somewhat lonely. It seems that Kronecker did not carefully study Eisenstein's 1847 paper until much later, perhaps in connection with the study of Hurwitz's 1882 paper on modular forms, wherein Hurwitz gave high praise to this work of Eisenstein. Kronecker planned to present a lecture on Eisenstein's work at the first meeting of the Deutsche Mathematiker-Vereinigung, held in 1890, but could not do so because of the death of his wife. Unfortunately, he himself died before he could write up the complete lecture.

Despite his correspondence with Kronecker, friendship with Gauss, support of von Humboldt, 1845 doctorate degree from Breslau, and 1847 *habilitation* at Berlin, Eisenstein was still financially insecure and apparently exceedingly lonely, suffering from ill health. He was also distressed by a priority dispute with Jacobi and the political turmoil of the time. He was elected to the Göttingen Academy in 1851 and, with the support of Dirichlet and Jacobi, to the Berlin Academy in 1852. The elderly von Humboldt succeeded in raising funds to send him to a warmer climate for his health, but before he could travel, Eisenstein died of tuberculosis in 1852, leaving one to wonder what he might have accomplished within a full lifetime.

Adolf Hurwitz (1859–1919) was one of the very small number of nineteenth-century mathematicians who studied Eisenstein's 1847 paper on elliptic functions. He drew upon the ideas in this paper in preparing his 1881 Leipzig doctoral dissertation, written under the supervision of Felix Klein. Using a result of Eisenstein, he proved without the use of elliptic or theta functions that the function defined by the infinite product,

$$\Delta(\omega_1, \omega_2) = \left(\frac{2\pi}{\omega_2}\right)^{12} q^2 \prod_{n=1}^{\infty} (1 - q^{2n})^{24}, \quad q = e^{i\pi\omega}, \tag{4.11}$$

where $\omega = \frac{\omega_1}{\omega_2}$ and Im $\omega > 0$, was in fact the discriminant function:

$$\Delta(\omega_1, \omega_2) = 4\big(15(4^*, 0)\big)^3 - 27\big(35(6^*, 0)\big)^2. \tag{4.12}$$

At the time Hurwitz wrote his Ph.D. thesis, formula (4.11) was already known from Weierstrass's 1862–1863 lectures on elliptic functions.[16] Then, in 1878, Klein derived (4.11)[17] from some formulas due to Jacobi. But Hurwitz wished to derive it within the theory of modular forms, independent of the theory of elliptic functions. He succeeded in this by employing Eisenstein's series $(s^*, 0)$; note that these series are modular forms

---

[15] Kronecker (1968) vol. 5, pp. 149–50.  [16] See Weierstrass (1894–1927) vol. 5, pp. 161–164.
[17] See Klein (1921–23) vol. 3, p. 21.

for $s \geq 3$. For $s = 2$, the series is conditionally convergent, leading Hurwitz to work out a new notation showing the dependence on the order of summation. He set

$$G_2(\omega_1, \omega_2) = \lim_{k \to \infty} \lim_{j \to \infty} \sum_{m=-k}^{k} \sideset{}{'}\sum_{n=-j}^{j} \frac{1}{(m\,\omega_1 + n\,\omega_2)^2}, \quad (m, n) \neq (0, 0), \qquad (4.13)$$

so that

$$G_2(\omega_1, \omega_2) - G_2(-\omega_2, \omega_1) = \frac{2i\pi}{\omega_1 \omega_2}. \qquad (4.14)$$

Note that the transformation of $\omega_1$ and $\omega_2$ is given by

$$\begin{pmatrix} \omega_1' \\ \omega_2' \end{pmatrix} = \begin{pmatrix} 0 & -1 \\ 1 & 0 \end{pmatrix} \begin{pmatrix} \omega_1 \\ \omega_2 \end{pmatrix}. \qquad (4.15)$$

As we noted, Eisenstein had applied transformations more general than (4.15) to his sums; instead of pursuing Eisenstein's general results, Hurwitz gave a proof of the particular case in (4.14) by applying Eisenstein's overall approach. To obtain (4.11) from (4.14), Hurwitz required the formula

$$3G_2(\omega_1, \omega_2) = \left(\frac{\pi}{\omega_2}\right)^2 \left[ 1 - 24 \sum_{k=1}^{\infty} \frac{kq^k}{1 - q^k} \right] \qquad (4.16)$$

and he was able to derive it from the Fourier expansion of the $m$th derivative of the partial fraction expansion for $\cot \omega$:

$$\sum_{n=-\infty}^{\infty} \frac{1}{(\omega + n)^m} = (-1)^m \frac{(2i\pi)^m}{(m-1)!} \sum_{k=1}^{\infty} k^{m-1} q^{2k}. \qquad (4.17)$$

Hurwitz wrote[18] that (4.17) was a particular case of a formula given by his teacher, Wilhelm Scheibner (1826–1908), who was Jacobi's student. In 1860, Scheibner published this formula:[19]

$$\sum_{n=-\infty}^{\infty} \frac{\Gamma(x)}{(\omega + 2n\,\pi i)^x} = \sum_{n=1}^{\infty} n^{x-1} e^{-n\omega}, \quad \mathrm{Re}\,\omega > 0, \ \mathrm{Re}\,x > 1. \qquad (4.18)$$

In fact, Hurwitz employed (4.16) to prove that the infinite product on the right-hand side of (4.11) was a modular form of weight 12. He then used the Dedekind-Klein modular function $J$ to derive (4.12). In this chapter, we will give the first portion of Hurwitz's proof; we will discuss his later portion after defining the function $J$. Note that some modern texts retain Hurwitz's first step while replacing the second step with the observation that the space of cusp forms of weight 12 has dimension 1.

---

[18] Hurwitz (1962) vol. 1, p. 21.      [19] Scheibner (1860) p. 48.

## 4.2 Eisenstein's Theory of Trigonometric Functions

Eisenstein defined periodic functions by means of infinite series. For positive integers $s$, consider the series

$$\sum_{m=-\infty}^{\infty} \frac{1}{(x+m)^s}.$$

For $x$ not an integer, the series converges absolutely for $s \geq 2$ by the integral test. When $s = 1$, the series is divergent. However the series converges absolutely for the special method of summation

$$\sum_{m=-\infty}^{\infty} \frac{1}{x+m} = \lim_{M \to \infty} \sum_{m=-M}^{M} \frac{1}{x+m}$$

$$= \frac{1}{x} + \sum_{m=1}^{\infty} \left( \frac{1}{x+m} + \frac{1}{x-m} \right)$$

$$= \frac{1}{x} + \sum_{m=1}^{\infty} \frac{2x}{x^2 - m^2}. \tag{4.19}$$

Eisenstein adopted this method of summation and for integers $s \geq 1$, set

$$(s, x) = \sum_{m=-\infty}^{\infty} \frac{1}{(x+m)^s} := \lim_{M \to \infty} \sum_{m=-M}^{M} \frac{1}{(x+m)^s}. \tag{4.20}$$

The limit of the symmetric sum given in (4.20) has been called by André Weil the Eisenstein sum of the series. Note here that (4.19) defines a periodic function with period 1, since

$$\frac{1}{x+1} + \lim_{M \to \infty} \sum_{m=1}^{m} \left( \frac{1}{x+1+m} + \frac{1}{x+1-m} \right)$$

$$= \frac{1}{x+1} + \lim_{M \to \infty} \left( \frac{1}{x} - \frac{1}{x+1} + \frac{1}{x+M+1} + \sum_{m=1}^{M-1} \left( \frac{1}{x+m} + \frac{1}{x-m} \right) \right)$$

$$= \frac{1}{x} + \lim_{M \to \infty} \sum_{m=1}^{M} \left( \frac{1}{x+m} + \frac{1}{x-m} \right).$$

It is clear that

$$(s, x+1) = (s, x); \tag{4.21}$$

hence $(s, x)$ is a periodic function of $x$ with period 1 for all integers $s \geq 1$. Eisenstein derived relations among the functions $(s, x)$ by applying the elementary partial fractions formula

$$\frac{1}{pq} = \frac{1}{(p+q)} \left( \frac{1}{p} + \frac{1}{q} \right) \tag{4.22}$$

and its derivatives with respect to $p$ and $q$. For example, taking the derivative with respect to $p$ and then with respect to $q$ produces[20]

$$\frac{1}{p^2 q^2} = \frac{1}{(p+q)^2}\left(\frac{1}{p^2} + \frac{1}{q^2}\right) + \frac{2}{(p+q)^3}\left(\frac{1}{p} + \frac{1}{q}\right). \tag{4.23}$$

Choosing $p$ and $q$ appropriately, he obtained relations among $(s, x)$ for different values of $s$. Apply term-by-term differentiation, so that

$$\frac{d}{dx}(s, x) = -s(s+1, x). \tag{4.24}$$

As we noted in the Introductory Remarks to this chapter, Eisenstein was justified in implementing term-by-term differentiation, although he did not explicitly verify it. By applying (4.24) to the relations among the $(s, x)$, Eisenstein then obtained differential equations satisfied by these functions and in particular, by the simplest function $(1, x)$. Again by a suitable choice of $p$ and $q$ in (4.22), he obtained, for example, the addition formula for $(1, x)$. Thus, to show that $y = (1, x)$ satisfies

$$\frac{dy}{dx} = -(y^2 + \pi^2), \tag{4.25}$$

he set out by taking $p = x + m$, $q = -x - n$, $p + q = m - n$ in (4.23) to get

$$\frac{1}{(x+m)^2} \cdot \frac{1}{(x+n)^2}$$
$$= \frac{1}{(m-n)^2}\left(\frac{1}{(x+m)^2} + \frac{1}{(x+n)^2}\right) + \frac{2}{(m-n)^3}\left(\frac{1}{(x+m)} - \frac{1}{(x+n)}\right). \tag{4.26}$$

Note that, while the left-hand side can be summed over all $m$ and $n$, the right-hand side requires that $m \neq n$. Now summing the left-hand side over all $n$ except for $n = m$ yields

$$\frac{1}{(x+m)^2}\left(\sum_{n=-\infty}^{\infty} \frac{1}{(x+n)^2} - \frac{1}{(x+m)^2}\right) = \frac{1}{(x+m)^2}(2, x) - \frac{1}{(x+m)^4}$$

and summing over all $m$ gives

$$(2, x)^2 - (4, x). \tag{4.27}$$

For the right-hand side, set $m - n = m'$, $m = m' + n$ with $m' \neq 0$. Summing over all $n$ gives

$$\frac{1}{m'^2}\left((2, x+m') + (2, x)\right) + \frac{2}{m'^3}\left((1, x+m') - (1, x)\right). \tag{4.28}$$

From the periodicity given by (4.21), the sum (4.28) reduces to $\frac{2}{m'^3}(2, x)$; summing over all $m' \neq 0$ then produces

$$2(2^*, 0)(2, x), \tag{4.29}$$

---

[20]  See Eisenstein (1975) vol. 1, pp. 395–409.

where

$$(s^*, 0) = \sum_{n=-\infty}^{\infty}{}' \frac{1}{n^s}, \quad n \neq 0. \tag{4.30}$$

Note that

$$(s^*, 0) = 0, \quad s \text{ an odd integer.} \tag{4.31}$$

Combining (4.27) and (4.29) gives the result

$$(4, x) = (2, x)^2 - 2(2^*, 0)(2, x). \tag{4.32}$$

Next, in (4.23) take $p = x + m$, $q = n$, $p + q = x + m + n$, $m + n = m'$, $m = m' - n$:

$$\frac{1}{(x+m)^2} \cdot \frac{1}{n^2}$$
$$= \frac{1}{(x+m')^2} \left( \frac{1}{(x+m'-n)^2} + \frac{1}{n^2} \right) + \frac{2}{(x+m')^3} \left( \frac{1}{x+m'-n} + \frac{1}{n} \right). \tag{4.33}$$

Summing the left-hand side of (4.33), first with respect to $m$ and then with respect to $n \neq 0$, we get

$$(2, x)(2^*, 0). \tag{4.34}$$

Similarly, summing the right-hand side over all $n$ except $n = 0$ gives

$$\frac{1}{(x+m')^2} \left( (2, x) - \frac{1}{(x+m')^2} + (2^*, 0) \right) + \frac{2}{(x+m')^3} \left( (1, x) - \frac{1}{(x+m')} + (1^*, 0) \right). \tag{4.35}$$

Observing that $(1^*, 0) = 0$ and summing (4.35) over $m'$, we obtain

$$(2, x)^2 - (4, x) + (2, x)(2^*, 0) + 2(3, x)(1, x) - 2(4, x)$$
$$= (2, x)^2 + (2, x)(2^*, 0) + 2(3, x)(1, x) - 3(4, x). \tag{4.36}$$

By equating (4.34) and the right-hand side of (4.36) we arrive at

$$3(4, x) = (2, x)^2 + 2(3, x)(1, x). \tag{4.37}$$

Now set $y = (1, x)$ and apply (4.24) to write (4.32) and (4.37) as differential equations:

$$-y''' = 6y'^2 + 12cy', \quad c = (2^*, 0), \tag{4.38}$$
$$-y''' = 2y'^2 + 2yy''. \tag{4.39}$$

Eliminating $y'''$ from (4.38) and (4.39) leads to

$$yy'' = 2y'^2 + 6cy'. \tag{4.40}$$

Differentiating (4.40) and using (4.38), we obtain

$$4y'y'' + 6cy'' = y'y'' + yy''' = y'y'' - 6yy'^2 - 12cyy'$$

or

$$3y'y'' + 6cy'' = -6yy'^2 - 12cyy';$$

dividing across by $3y' + 6c$, we find that

$$y'' = -2yy'. \tag{4.41}$$

Eliminating $y''$ from (4.40) and (4.41) produces

$$-2y^2y' = 2y'^2 + 6cy'$$

or

$$-y^2 = y' + 3c$$
$$= y' + \pi^2, \tag{4.42}$$

since by Euler's formula

$$3c = 3\sum_{-\infty}^{\infty}{}' \frac{1}{n^2} = 6\sum_{n=1}^{\infty} \frac{1}{n^2} = \pi^2. \tag{4.43}$$

Equation (4.42) can be solved by rewriting it as

$$-\int \frac{dy}{y^2 + \pi^2} = x + c_1$$

or

$$\frac{1}{\pi}\cot^{-1}\frac{y}{\pi} = x + c_1, \quad y = \pi\cot(\pi x + d). \tag{4.44}$$

To show that $d = 0$, observe that $y = (1, x)$ is an odd function; thus, by periodicity

$$\left(1, \frac{1}{2}\right) = -\left(1, -\frac{1}{2}\right) = -\left(1, -\frac{1}{2} + 1\right) = -\left(1, \frac{1}{2}\right)$$
$$\left(1, \frac{1}{2}\right) = 0.$$

Take $x = \frac{1}{2}$ and $y = (1, x) = 0$ in (4.44) to see that $d = 0$. Thus

$$(1, x) = \pi\cot\pi x. \tag{4.45}$$

Eisenstein integrated (4.45) to get

$$\lim_{M\to\infty}\sum_{m=-M}^{M} \log(x + m) = \log\sin\pi x + c$$

or

$$\lim_{M\to\infty}\sum_{m=-M}^{M}{}' \log(x + m) = \log\frac{\sin\pi x}{x} + c.$$

Eisenstein used * instead of a prime over the summation sign to indicate that the term corresponding to $m = 0$ was missing from the summation. He let $x \to 0$, so that $\frac{\sin \pi x}{x} \to \pi$ to arrive at

$$\lim_{M \to \infty} \sum_{m=-M}^{M} \left( \log(x+m) - \log m \right) = \log \frac{\sin \pi x}{x} - \log \pi. \tag{4.46}$$

We note parenthetically that Eisenstein wrote $\frac{\sin \pi x}{x} = \pi$ for $x = 0$. Equation (4.46) gave him product formulas for $\sin \pi x$ and $\sin x$:

$$\sin \pi x = \pi x (1+x)(1-x)\left(1 + \frac{x}{2}\right)\left(1 - \frac{x}{2}\right) \cdots$$

$$\sin x = x \prod_{m=-\infty}^{\infty}{}' \left(1 + \frac{x}{m\pi}\right) \equiv \lim_{M \to \infty} x \prod_{m=-M}^{M}{}' \left(1 + \frac{x}{m\pi}\right). \tag{4.47}$$

## 4.3 Eisenstein's Derivation of the Addition Formula

Using (4.22), Eisenstein worked out the addition formula for the cotangent function.[21] He began by setting $p = x + m$, $q = -y - m$ and obtained

$$\frac{1}{x+m} - \frac{1}{y+m} = \frac{y-x}{(x+m)(y+m)}.$$

He next considered the product

$$\left(\frac{1}{x+m} - \frac{1}{y+m}\right)\left(\frac{1}{x+n} - \frac{1}{y+n}\right) \tag{4.48}$$

and summed over all $m$ and $n$ to arrive at

$$\big((1, x) - (1, y)\big)^2. \tag{4.49}$$

Now when $m \neq n$, the product (4.48) may be rewritten as

$$\frac{1}{m-n}\left(-\frac{1}{x+m} + \frac{1}{x+n} - \frac{1}{y+m} + \frac{1}{y+n}\right)$$

$$+ \frac{1}{x-y+m-n}\left(\frac{1}{x+m} - \frac{1}{y+n}\right) + \frac{1}{y-x+m-n}\left(\frac{1}{y+m} - \frac{1}{x+n}\right) \tag{4.50}$$

and when $m = n$, the last two terms in (4.50) retain the same form, while the first term must be replaced by

$$\frac{1}{(x+m)^2} + \frac{1}{(y+m)^2}. \tag{4.51}$$

Summing (4.50) and (4.51) over all $m$ and $n$, we obtain

$$(1, x-y)\big((1,x) - (1,y)\big) + (1, y-x)\big((1,y) - (1,x)\big) + (2, x) + (2, y)$$
$$= (2, x) + (2, y) + 2(1, x-y)\big((1,x) - (1,y)\big). \tag{4.52}$$

[21] Eisenstein (1975) vol. 1, pp. 408–409.

Equating (4.49) and (4.52), and then changing $y$ to $-y$ gives

$$\big((1, x) + (1, y)\big)^2 - (2, x) - (2, y) = 2(1, x+y)\big((1, x) + (1, y)\big). \qquad (4.53)$$

Put (4.42) into the form

$$(2, x) = (1, x)^2 + \pi^2 \qquad (4.54)$$

and then apply it to (4.53) to get the addition formula:

$$(1, x+y) = \frac{(1, x)(1, y) - \pi^2}{(1, x) + (1, y)} \qquad (4.55)$$

or

$$\cot(x+y) = \frac{\cot x \cot y - 1}{\cot x + \cot y}. \qquad (4.56)$$

In his work, Eisenstein used Euler's formula

$$(1, x) = \pi \cot \pi x = \pi i \, \frac{e^{\pi i x} + e^{-\pi i x}}{e^{\pi i x} - e^{-\pi i x}}; \qquad (4.57)$$

André Weil has pointed out that, in fact, from (4.55) Eisenstein could have derived the formulas of Euler expressing the trigonometric functions in terms of the exponential function. This derivation could be accomplished by using the function

$$e(x) = \frac{(1, x) + \pi i}{(1, x) - \pi i} \qquad (4.58)$$

and then (4.55) to see that

$$e(x+y) = e(x)e(y). \qquad (4.59)$$

From this we have

$$e(x) = e^{2\pi i x}, \qquad (4.60)$$

since the power series expansion of $e(x)$ in (4.58) starts with $1 + 2\pi i x$. We can give another proof of (4.60) using (4.54) by noting that

$$\frac{de(x)}{dx} = \frac{2\pi i(2, x)}{\big((1, x) - \pi i\big)^2} = 2\pi i \, \frac{(1, x)^2 + \pi^2}{\big((1, x) - \pi i\big)^2}$$

$$= 2\pi i \, \frac{(1, x) + \pi i}{(1, x) - \pi i} = 2\pi i e(x);$$

(4.60) follows upon integration. Also note that (4.57) follows from (4.58) and (4.60).

## 4.4 Eisenstein's Theory of Elliptic Functions

Eisenstein constructed elliptic functions as double infinite sums of partial fractions, in a manner similar to his method of constructing trigonometric functions. Let $\omega_1$ and $\omega_2$ denote the fundamental periods of an elliptic function so that the set of all periods of the elliptic function is given by $m\,\omega_1 + n\,\omega_2$, where $m$ and $n$ are integers. For integers $s \geq 1$, Eisenstein defined

$$(s, x) = \lim_{N \to \infty} \lim_{M \to \infty} \sum_{n=-N}^{N} \sum_{m=-M}^{M} \frac{1}{(x + m\,\omega_1 + n\,\omega_2)^s}. \tag{4.61}$$

Eisenstein's notation for denoting the sum in (4.61) is the same as that used for the sum in (4.20). Unfortunately, this notation does not reflect the dependence of the sum on the periods $\omega_1$, $\omega_2$. Thus, though Eisenstein did not write $(s, x; \omega_1, \omega_2)$, at times we do so for clarity. Note that the series

$$\sum_{m,n} \frac{1}{(x + m\,\omega_1 + n\,\omega_2)^s} \tag{4.62}$$

converges absolutely by the double integral test when $x \neq m\,\omega_1 + n\,\omega_2$, with $m, n \in \mathbb{Z}$ and $s \geq 3$. Therefore, for $s \geq 3$, $(s, x)$ has periods $\omega_1$ and $\omega_2$: For integers $a$ and $b$

$$(s, x + a\,\omega_1 + b\,\omega_2) = (s, x) \quad \text{for} \quad s \geq 3. \tag{4.63}$$

The Eisenstein method of summation is needed only when $s = 1$ or $s = 2$. We can show that $(1, x)$, defined by (4.61), is absolutely convergent. Using (4.45) we can write

$$\lim_{M \to \infty} \sum_{m=-M}^{M} \frac{1}{x + m\,\omega_1 + n\,\omega_2} = \frac{1}{\omega_1} \lim_{M \to \infty} \sum_{m=-M}^{M} \frac{1}{\frac{x + n\,\omega_2}{\omega_1} + m} \tag{4.64}$$

$$= \frac{\pi}{\omega_1} \cot \frac{(x + n\,\omega_2)\pi}{\omega_1}. \tag{4.65}$$

Thus, by (4.61)

$$(1, x) = \frac{\pi}{\omega_1} \cot \frac{x\pi}{\omega_1} + \frac{\pi}{\omega_1} \sum_{n=1}^{\infty} \left( \cot \frac{(x + n\,\omega_2)\pi}{\omega_1} + \cot \frac{(x - n\,\omega_2)\pi}{\omega_1} \right). \tag{4.66}$$

Let $\frac{\omega_2}{\omega_1} = \pm\omega$, where the $+$ sign is chosen if Im $\omega > 0$; otherwise, the negative sign holds. Put $z = e^{\frac{2\pi i x}{\omega_1}}$ and $q = e^{i\pi\,\omega}$ so that by (4.57) one obtains

$$\cot \frac{(x + n\,\omega_2)\pi}{\omega_1} + \cot \frac{(x - n\,\omega_2)\pi}{\omega_1} = i \left( \frac{q^{2n}z + 1}{q^{2n}z - 1} + \frac{q^{-2n}z + 1}{q^{-2n}z - 1} \right) \tag{4.67}$$

$$= -2i \left( \frac{1}{1 - q^{2n}z} - \frac{1}{1 - q^{2n}z^{-1}} \right). \tag{4.68}$$

Since the absolute value of the final expression in (4.68) is $\leq C(z)|q|^{2n}$ (where $C(z)$ is some function depending only on $z$) and since $|q| < 1$, it must follow that the series (4.66) is absolutely convergent. Arguing along the same lines, it can be shown that the corresponding series for $s = 2$ is absolutely convergent. Observe that (4.66) implies that for an integer $a$

$$(1, x + a\,\omega_1) = (1, x). \tag{4.69}$$

Similarly,

$$(2, x + a\,\omega_1) = (2, x).\tag{4.70}$$

Now observe that after summing with respect to $n$, from (4.61) we get

$$(s, x + \omega_2) - (s, x) = \lim_{N \to \infty} \lim_{M \to \infty} \left( \sum_{m=-M}^{M} \frac{1}{(x + m\,\omega_1 + (N+1)\,\omega_2)^s} \right.$$
$$\left. - \sum_{m=-M}^{M} \frac{1}{(x + m\,\omega_1 - N\,\omega_2)^s} \right).\tag{4.71}$$

Following Eisenstein,[22] we have by (4.57) and (4.65)

$$\lim_{M \to \infty} \sum_{m=-M}^{M} \frac{1}{x + m\,\omega_1 + l\,\omega_2} = \frac{\pi i}{\omega_1} \frac{e^{iy} + e^{-iy}}{e^{iy} - e^{-iy}}, \quad \text{where } y = \frac{x + l\,\omega_2}{\omega_1}.\tag{4.72}$$

Taking $l = N + 1$ in (4.72) and letting $N \to \infty$, we can see that

$$\lim_{N \to \infty} \lim_{M \to \infty} \sum_{m=-M}^{M} \frac{1}{x + m\,\omega_1 + (N+1)\omega_2} = \frac{\pi i}{\omega_1} \lim_{y \to \infty} \frac{e^{iy} + e^{-iy}}{e^{iy} - e^{-iy}} = -\frac{\delta \pi i}{\omega_1},$$

where $\delta = 1$ if $\operatorname{Im} \frac{\omega_2}{\omega_1} > 0$ and $\delta = -1$ if $\operatorname{Im} \frac{\omega_2}{\omega_1} < 0$. After a similar calculation taking $l = -N$, it follows from (4.71) that

$$(1, x + \omega_2) - (1, x) = -\frac{\delta 2\pi i}{\omega_1}.\tag{4.73}$$

Now from (4.24) we have

$$\lim_{M \to \infty} \sum_{m=-M}^{M} \frac{1}{(x + m\,\omega_1 + l\,\omega_2)^2} = -\frac{\pi}{\omega_1} \frac{d}{dx} \cot \pi \left( \frac{x + l\,\omega_2}{\omega_1} \right)$$
$$= \frac{\pi^2}{\omega_1^{\,2}} \csc^2 \pi \left( \frac{x + l\,\omega_2}{\omega_1} \right).$$

Since

$$\lim_{l \to \infty} \csc^2 \pi \left( \frac{x + l\,\omega_2}{\omega_1} \right) = 0,$$

we obtain from (4.71) the result

$$(2, x + \omega_2) - (2, x) = 0.\tag{4.74}$$

Thus, $(2, x)$ is a doubly periodic function. However, by (4.70) and (4.73), we have for integers $a$ and $b$

$$(1, x + a\,\omega_1 + b\,\omega_2) - (1, x) = -\frac{2\delta\, b\pi i}{\omega_1}.\tag{4.75}$$

---

[22] Eisenstein (1975) vol. 1, pp. 381–387.

Just as in the case of trigonometric functions, Eisenstein defined sums

$$(s^*, 0) = \lim_{N \to \infty} \lim_{M \to \infty} \sum_{n=-N}^{N} \sum_{m=-M}^{M} {}' \frac{1}{(m\,\omega_1 + n\,\omega_2)^s}, \tag{4.76}$$

where the term corresponding to $m = 0$ and $n = 0$ would be omitted from the sum. As before,

$$(s^*, 0) = 0 \tag{4.77}$$

when $s$ is an odd integer. Note here that $(s^*, 0)$ is a modular form of weight $s$. In fact, Eisenstein considered the effect of the $n$th-order transformation of the periods $\omega_1, \omega_2$ on $(s, x)$ and $(s^*, 0)$. We consider only the effect of a "linear transformation," that is, the case in which $p, q, r, s$ are integers with $pq - rs = 1$ and

$$\omega_1' = p\,\omega_1 + q\,\omega_2$$
$$\omega_2' = r\,\omega_1 + s\,\omega_2.$$

Eisenstein gave the values of

$$(s, x; \; \omega_1', \omega_2') - (s, x; \; \omega_1, \omega_2) \tag{4.78}$$

for $s = 1, 2$. When $s \geq 3$, the value of (4.78) is clearly zero because of the absolute convergence of the series (4.62).

## 4.5   Differential Equations for Elliptic Functions

Using the same method he employed for trigonometric functions, Eisenstein found differential equations satisfied by his elliptic functions $(s, x)$. For example, he took[23]

$$p = x + m\omega_1 + n\omega_2, \; q = -x - m_1\,\omega_1 - n_1\,\omega_2, \; p + q = (m - m_1)\,\omega_1 + (n - n_1)\,\omega_2$$

in (4.23), given again here:

$$\frac{1}{p^2 q^2} = \frac{1}{(p+q)^2} \left( \frac{1}{p^2} + \frac{1}{q^2} \right) + \frac{2}{(p+q)^3} \left( \frac{1}{p} + \frac{1}{q} \right).$$

Note that $p + q = 0$ when $m = m_1$ and $n = n_1$. Summing over all $m, n$ except $m = m_1$ and $n = n_1$; then summing over all $m_1, n_1$, gives the left-hand side as

$$(2, x)^2 - (4, x). \tag{4.79}$$

For the right-hand side, set $m' = m - m_1, n' = n - n_1$, so that $m = m' + m_1, n = n' + n_1$. Summing over all $m_1$ and $n_1$ produces

$$\frac{1}{(m'\,\omega_1 + n'\,\omega_2)^2} \left( (2, x + m'\,\omega_1 + n'\,\omega_2) + (2, x) \right)$$

$$+ \frac{2}{(m'\,\omega_1 + n'\,\omega_2)^3} \left( (1, x + m'\,\omega_1 + n'\,\omega_2) - (1, x) \right). \tag{4.80}$$

[23] Eisenstein (1975) vol. 1, pp. 467–468.

By (4.74) and (4.75), the sum (4.80) reduces to

$$\frac{2}{(m'\,\omega_1 + n'\,\omega_2)^2}\,(2, x) - \frac{4\delta n'\,\pi i}{\omega_1(m'\,\omega_1 + n'\,\omega_2)^3}.$$

Sum over all $m'$ and $n'$ except $m' = 0$, $n' = 0$ to arrive at

$$2(2^*, 0)(2, x) - \frac{4\delta\pi i}{\omega_1}\sum_{m'\,n'}{}'\frac{n'}{(m'\,\omega_1 + n'\,\omega_2)^3}$$

$$= 2(2^*, 0)(2, x) + \frac{2\delta\pi i}{\omega_1}\frac{\partial}{\partial\,\omega_2}(2^*, 0). \tag{4.81}$$

Equating (4.79) and (4.81) yields

$$(4, x) = (2, x)^2 - 2(2^*, 0)(2, x) - 2\frac{\delta\pi i}{\omega_1}\frac{\partial}{\partial\,\omega_2}(2^*, 0)$$

$$= \big((2, x) - (2^*, 0)\big)^2 - (2^*, 0)^2 - \frac{2\delta\pi i}{\omega_1}\frac{\partial}{\partial\,\omega_2}(2^*, 0). \tag{4.82}$$

As we show, Eisenstein could have easily derived (4.84) from (4.82). Such a derivation would use the Taylor expansion of $(s, x)$. Note that (4.84) is equivalent to the second-order differential equation satisfied by the Weierstrass $\wp$ function. However, Eisenstein had already derived (4.84) some forty pages earlier in his long paper by a somewhat more complicated procedure. In fact, the Taylor expansion of $(s, x)$ is given as the final formula in his paper, although Eisenstein had implicitly employed it in an earlier part of the paper. Thus, we see that Eisenstein, perhaps because of ill-health, composed his papers in "fits and starts," as Weil observed. Eisenstein gave the Taylor expansion of $(s, x) - \frac{1}{x^s}$ as:

$$(s, x) = \frac{1}{x^s} + (s^*, 0) - s(s + 1^*, 0)x + \frac{s(s + 1)}{2}(s + 2^*, 0)x^2 - \cdots. \tag{4.83}$$

When this is applied to (4.82), and (4.77) is used, we see that

$$(4, x) = \frac{1}{x^4} + (4^*, 0) + 10(6^*, 0)x^2 + \cdots$$

$$= \left(\frac{1}{x^2} + 3(4^*, 0)x^2 + \cdots\right)^2 - (2^*, 0)^2 - \frac{2\delta\pi i}{\omega_1}\frac{\partial}{\partial\,\omega_2}(2^*, 0).$$

Equating the terms independent of $x$ yields

$$(2^*, 0)^2 + \frac{2\delta\pi i}{\omega_1}\frac{\partial}{\partial\,\omega_2}(2^*, 0) = 5(4^*, 0).$$

Thus, (4.82) may be rewritten as

$$(4, x) = \big((2, x) - (2^*, 0)\big)^2 - 5(4^*, 0). \tag{4.84}$$

Recall that in Weierstrass's notation

$$\wp(x) = (2, x) - (2^*, 0); \tag{4.85}$$

note also that

$$\frac{d}{dx}(s, x) = -s(s + 1, x) \quad \text{for } s \geq 2. \tag{4.86}$$

We can now see that (4.84) is in fact identical with the equation well-known in its Weierstrassian form:

$$\frac{1}{6}\wp''(x) = \wp^2(x) - 5(4^*, 0) \quad \text{or} \quad \wp''(x) = 6\wp^2(x) - \frac{1}{2}g_2, \tag{4.87}$$

where $g_2 = 60(4^*, 0)$. Now Eisenstein could have multiplied (4.84) by $(3, x)$ and then integrated to obtain the first-order differential equation satisfied by $(2, x) - (2^*, 0)$. However, he had already discovered this first-order differential equation by starting out with a $p, q$ identity other than (4.23).[24] He obtained this identity by differentiating (4.23) with respect to $p$ and then $q$. The result was

$$\frac{1}{p^3} \cdot \frac{1}{q^3} = \frac{1}{(p+q)^3}\left(\frac{1}{p^3} + \frac{1}{q^3}\right) + \frac{3}{(p+q)^4}\left(\frac{1}{p^2} + \frac{1}{q^2}\right) + \frac{6}{(p+q)^5}\left(\frac{1}{p} + \frac{1}{q}\right). \tag{4.88}$$

Briefly summarizing his next steps, set

$$p = x + m\omega_1 + n\omega_2, \quad q = -x - m_1\omega_1 - n_1\omega_2, \quad p + q = (m - m_1)\omega_1 + (n - n_1)\omega_2.$$

The left-hand side, summing over all $m, n$, and then over all $m_1, n_1$, except $m = m_1$ and $n = n_1$, becomes

$$(6, x) - (3, x)^2. \tag{4.89}$$

Set $m' = m - m_1$, $n' = n - n_1$, and $n = n' + n_1$, $m = m' + m_1$. For fixed $m', n'$, sum all $m_1$ and $n_1$ to get, for $\omega = m'\omega_1 + n'\omega_2$,

$$-\frac{1}{\omega^3}\big((3, x + \omega) - (3, x)\big) + \frac{3}{\omega^4}\big((2, x + \omega) + (2, x)\big) - \frac{6}{\omega^5}\big((1, x + \omega) - (1, x)\big). \tag{4.90}$$

By periodicity and (4.75), this reduces to

$$\frac{6}{\omega^4}(2, x) - \frac{12\delta\pi i}{\omega_1} \cdot \frac{n'}{\omega^5}. \tag{4.91}$$

Sum over all $m'$ and $n'$ except $m' = 0$ and $n' = 0$ and equate with (4.89) to get

$$(6, x) = (3, x)^2 + 6(4^*, 0)(2, x) - 6c, \tag{4.92}$$

where

$$c = \frac{\delta 2\pi i}{\omega_1} {\sum}' \frac{n}{(m\omega_1 + n\omega_2)^5}. \tag{4.93}$$

Eisenstein referred to the constant $c$ as "eigenthümlichen" or peculiar. Differentiating (4.92) with respect to $x$ yields

$$(7, x) = (3, x)\big((4, x) + 2(4^*, 0)\big). \tag{4.94}$$

---

[24] Eisenstein (1975) vol. 1, pp. 427–429.

To find another equation involving $(7, x)$, consider

$$\frac{1}{p^4 q^3} - \frac{1}{p^3 q^4} = \frac{1}{r^3}\left(\frac{1}{p^4} - \frac{1}{q^4}\right) + \frac{2}{r^4}\left(\frac{1}{p^3} - \frac{1}{q^3}\right) + \frac{2}{r^5}\left(\frac{1}{p^2} - \frac{1}{q^2}\right), \quad (4.95)$$

where $r = p + q$, obtainable by subtracting the derivative with respect to $p$ of (4.88) from its derivative with respect to $q$. Relation (4.95) leads to

$$5(7, x) = 2(5, x)\big((2, x) - (2^*, 0)\big) + 3(3, x)(4, x). \quad (4.96)$$

Differentiating both (4.94) and (4.96) twice with respect to $x$ results in six equations from which the five functions $(5, x)$ through $(9, x)$ can be eliminated to obtain (4.84):

$$\big((2, x) - (2^*, x)\big)^2 = (4, x) + 5(4^*, 0).$$

To next obtain the first-order differential equation, he twice differentiated (4.84) with respect to $x$ to get

$$5(6, x) = 3(4, x)\big((2, x) - (2^*, 0)\big) + 2(3, x)^2 \quad (4.97)$$

and then eliminated $(4, x)$ and $(6, x)$ from (4.84), (4.92), and (4.97) to arrive at the requisite first-order differential equation:

$$(3, x)^2 = \big((2, x) - (2^*, 0)\big)^3 - 15(4^*, 0)\big((2, x) - (2^*, 0)\big) + 10\big(c - (2^*, 0)(4^*, 0)\big). \quad (4.98)$$

In his 1891 paper on Eisenstein's work, Kronecker mentioned that an application of (4.83) gives the constant $c$ in (4.98) as

$$10\big(c - (2^*, 0)(4^*, 0)\big) = -35(6^*, 0) \quad (4.99)$$

so that in Weierstrass's notation, (4.98) may be given, as Hurwitz also observed,[25] in the form

$$\wp'^2(x) = 4\wp^3(x) - g_2\wp(x) - g_3, \quad g_2 = 60(4^*, 0), \quad g_3 = 140(6^*, 0); \quad (4.100)$$

We mention that Kronecker referred to $\wp(x)$ as Schwarz's notation for Eisenstein's elliptic function $(2, x) - (2^*, 0)$.[26] Continuing his analysis of equation (4.98), Eisenstein observed that $(3, x) = 0$ when $x = \frac{\omega_1}{2}, \frac{\omega_2}{2}, \frac{\omega_1 + \omega_2}{2}$. To prove this, he argued that $(3, x)$ was an odd periodic function and so

$$\left(3, \frac{\omega_1}{2}\right) = -\left(3, -\frac{\omega_1}{2}\right) = -\left(3, -\frac{\omega_1}{2} + \omega_1\right) = -\left(3, \frac{\omega_1}{2}\right) = 0.$$

The other two cases may be handled in the same manner. Therefore, the third-degree polynomial on the right-hand side of (4.98) could be factored, so that (4.98) would become

$$(3, x)^2 = \left((2, x) - \left(2, \frac{\omega_1}{2}\right)\right)\left((2, x) - \left(2, \frac{\omega_2}{2}\right)\right)\left((2, x) - \left(2, \frac{\omega_1 + \omega_2}{2}\right)\right), \quad (4.101)$$

a well-known equation when written in Weierstrassian form.

---

[25]  Hurwitz (1932) vol. 1, p. 31.     [26]  See Kronecker (1968) vol. 5, p. 181.

## 4.6 The Addition Theorem for the Elliptic Function

Eisenstein stated the addition formula for the elliptic function $(2, x)$ in the form[27]

$$(2, x + y) + (2, x) + (2, y) - 3(2^*, 0) = \left( \frac{(3, x) - (3, y)}{(2, x) - (2, y)} \right)^2, \qquad (4.102)$$

written in Weierstrass's notation as

$$\wp(x + y) + \wp(x) + \wp(y) = \frac{1}{4} \left( \frac{\wp'(x) - \wp'(y)}{\wp(x) - \wp(y)} \right)^2. \qquad (4.103)$$

First set $u_1 = m\omega_1 + n\omega_2$, $u_2 = m_1\omega_1 + n_1\omega_2$, $u_1 - u_2 = u$. In a manner very similar to that employed in deriving the addition formula for the cotangent function using (4.48) and (4.50), Eisenstein started with the two-variables formula[28]

$$\left( \frac{1}{x + u_1} - \frac{1}{y + u_1} \right) \left( \frac{1}{x + u_2} - \frac{1}{y + u_2} \right)$$
$$= \frac{1}{u} \left( -\frac{1}{x + u_1} + \frac{1}{x + u_2} - \frac{1}{y + u_1} + \frac{1}{y + u_2} \right)$$
$$+ \frac{1}{x - y + u} \left( \frac{1}{x + u_1} - \frac{1}{y + u_2} \right) + \frac{1}{y - x + u} \left( \frac{1}{y + u_1} - \frac{1}{x + u_2} \right). \qquad (4.104)$$

We proceed with the understanding that summing over all $u_1$ will mean summing over all $m$ and $n$, and summing over all $u_2$ will mean over all $m_1, n_1$, etc. Note that the right-hand side of (4.104) requires that $u = (m - m_1)\omega_1 + (n - n_1)\omega_2$ be nonzero, that is, $m \neq m_1$ and $n \neq n_1$. Similar to the derivation of (4.53), the left-hand side of (4.104), when summed over all $u_1$ and $u_2$ except $u_1 = u_2$, sums to

$$\big((1, x) - (1, y)\big)^2 - (2, x) - (2, y). \qquad (4.105)$$

Keeping $u$ fixed and summing the right-hand side over $u_2$ with $u_1 = u + u_2$, the right-hand side becomes

$$\frac{1}{u} \big( -(1, x + u) + (1, x) - (1, y + u) + (1, y) \big)$$
$$+ \frac{1}{x - y + u} \big( (1, x + u) - (1, y) \big) + \frac{1}{y - x + u} \big( (1, y + u) - (1, x) \big). \qquad (4.106)$$

An application of (4.75) then gives

$$-(1, x + u) + (1, x) = \frac{2\delta\pi i}{\omega_1} \frac{\partial u}{\partial \omega_2}.$$

---

[27] Ibid. p. 471.  [28] Ibid. p. 469.

Using this in (4.106) and summing over all $u \neq 0$ and also applying (4.118), we arrive at

$$\big((1, x) - (1, y)\big)\big((1, x - y) - (1, y - x)\big)$$
$$+ \frac{2\delta\pi i}{\omega_1} \sum_u{}' \left( \frac{1}{x - y + u} + \frac{1}{y - x + u} - \frac{2}{u} \right) \frac{\partial u}{\partial \omega_2}$$
$$= \big((1, x) - (1, y)\big)\big((1, x - y) - (1, y - x)\big) + \frac{4\delta\pi i}{\omega_1} \frac{\partial}{\partial \omega_2} \log \phi(x - y). \quad (4.107)$$

Equating (4.105) and (4.107) and changing $y$ to $-y$ produces

$$\big((1, x) + (1, y)\big)^2 = 2(1, x + y)\big((1, x) + (1, y)\big)$$
$$+ (2, x) + (2, y) + \frac{4\delta\pi i}{\omega_1} \frac{\partial}{\partial \omega_2} \log \phi(x + y). \quad (4.108)$$

An application of the expansion (4.83) to $(1, y)$, $(1, x + y)$, and $(2, y)$ in powers of $y$ gives

$$(1, y) = \frac{1}{y} - (2^*, 0)y + (3^*, 0)y^2 - \cdots$$
$$(1, x + y) = (1, x) - (2, x)y + (3, x)y^2 - \cdots$$
$$(2, y) = \frac{1}{y^2} + (2^*, 0) - \cdots .$$

Substituting these in (4.108) and letting $y \to 0$, we obtain

$$(2, x) - \frac{4\delta\pi i}{\omega_1} \frac{\partial}{\partial \omega_2} \log \phi(x) = (1, x)^2 + 3(2^*, 0). \quad (4.109)$$

Changing $x$ to $x + y$ in (4.109) and combining this result with (4.108), we have

$$\big((1, x) + (1, y) - (1, x + y)\big)^2 = (2, x) + (2, y) + (2, x + y) - 3(2^*, 0). \quad (4.110)$$

Now set $y = c - x$, where $c$ is a constant; next take the derivative of (4.110) with respect to $x$ and use (4.86) to see that

$$2\big((1, x) + (1, y) - (1, x + y)\big)\big((2, x) - (2, y)\big) = 2\big((3, x) - (3, y)\big). \quad (4.111)$$

Finally, equations (4.110) and (4.111) give (4.102), Eisenstein's addition formula:

$$(2, x) + (2, y) + (2, x + y) - 3(2^*, 0) = \left( \frac{(3, x) - (3, y)}{(2, x) - (2, y)} \right)^2 .$$

Note that by using the Weierstrass $\zeta$ function as defined in (4.9), we may write (4.111) as the well-known result:

$$2\zeta(x) + 2\zeta(y) - 2\zeta(x + y) = -\frac{\wp'(x) - \wp'(y)}{\wp(x) - \wp(y)}.$$

### 4.7   Eisenstein's Double Product

Analogous to the product for the sine function is Eisenstein's double product, in which the prime symbol indicates that the term corresponding to $m = n = 0$ is to be excluded:

$$\phi(x) = x \prod{}' \left( 1 - \frac{x}{m\omega_1 + n\omega_2} \right)$$

$$= \lim_{N\to\infty} \lim_{M\to\infty} x \prod_{n=-N}^{N}{}' \prod_{m=-M}^{M} \left( 1 - \frac{x}{m\omega_1 + n\omega_2} \right). \qquad (4.112)$$

We shall continue to denote a product with a finite number of excluded terms as $\prod'$. Let $u$ denote the general term $m\omega_1 + n\omega_2$. To prove that the product converged,[29] Eisenstein removed the finite number of terms for which $|u| \le |x|$ so that he could consider the product

$$\prod{}' \left( 1 - \frac{x}{u} \right), \quad \text{where } |x| < |u|.$$

Now he noted that since

$$\log \left( 1 - \frac{x}{u} \right) = -\frac{x}{u} - \frac{x^2}{2u^2} - \frac{x^3}{3u^3} - \frac{x^4}{4u^4} - \cdots \quad \text{for } |x| < |u|,$$

he could conclude that

$$\log \prod{}' \left( 1 - \frac{x}{u} \right) = -x \sum{}' \frac{1}{u} - \frac{x^2}{2} \sum{}' \frac{1}{u^2} - \frac{x^3}{3} \sum{}' \frac{1}{u^3} - \frac{x^4}{4} \sum{}' \frac{1}{u^4} - \cdots .$$

$$(4.113)$$

Note that here the summation $\sum'$ was taken in the same order as the product. Eisenstein observed that since the series $\sum' \frac{1}{u^\mu}$ converged absolutely for $\mu \ge 3$, the order of summation did not matter for these values of $\mu$. He set $p = \sum' \frac{1}{u}$ and $q = \sum' \frac{1}{u^2}$, where these sums were defined as in (4.76) and saw that in (4.113) $e^{-px - \frac{1}{2}qx^2}$ was a part of the product $\prod' \left( 1 - \frac{x}{u} \right)$. The remaining part would be independent of the order of summation; in fact, it turned out to be

$$\prod{}' \left( 1 - \frac{x}{u} \right) e^{\frac{x}{u} + \frac{x^2}{2u^2}}. \qquad (4.114)$$

We can now relate Eisenstein's function $\phi(x)$ to Weierstrass's $\sigma$ function, defined by the absolutely convergent product

$$\sigma(x) = x \prod_{m,n}{}' \left( 1 - \frac{x}{u} \right) e^{\frac{x}{u} + \frac{x^2}{2u^2}}, \quad u = m\omega_1 + n\omega_2. \qquad (4.115)$$

Since $\sum'_{m,n} \frac{1}{u} = (1^*, 0) = 0$ when the sum is over all $m$ and $n$ except $m = n = 0$, it follows from (4.114) that

$$\phi(x) = \sigma(x) e^{-\frac{x^2(2^*,0)}{2}}, \quad \text{where } (2^*, 0) = \lim_{N\to\infty} \lim_{M\to\infty} \sum{}' \frac{1}{(m\omega_1 + n\omega_2)^2}, \qquad (4.116)$$

---

[29] Eisenstein (1975) vol. 1, pp. 374–376.

and the summation is over all $m$, $n$ except $m = n = 0$. Eisenstein also noted[30] and used the formulas

$$\frac{d^s}{dx^s} \log \phi(x) = (-1)^{s-1}(s-1)!\,(s, x) \tag{4.117}$$

$$\frac{\partial}{\partial \omega_2} \log \phi(x) = \sum_{m,n}' \left( \frac{1}{x+u} - \frac{1}{u} \right) \frac{\partial u}{\partial \omega_2}, \tag{4.118}$$

where $u = m\omega_1 + n\omega_2$.

## 4.8   Elliptic Functions in Terms of the $\phi$ Function

One of the aims of Eisenstein's project, as indicated by the title of his paper, was to express an elliptic function as a quotient of infinite products. In fact, he proved[31] that

$$(2, x) - (2, y) = \frac{\phi(y+x)\phi(y-x)}{\phi^2(y)\phi^2(x)}. \tag{4.119}$$

It is clear from (4.85) and (4.112) that (4.116) is identical with

$$\wp(x) - \wp(y) = \frac{\sigma(x+y)\sigma(y-x)}{\sigma^2(x)\sigma^2(y)}, \tag{4.120}$$

the Weierstrass $\wp$-function in terms of the $\sigma$-function. To prove (4.119), Eisenstein started with the formula obtained from (4.111):

$$(1, x+y) = (1, x) + (1, y) - \frac{(3, x) - (3, y)}{(2, x) - (2, y)}. \tag{4.121}$$

Change $y$ to $-y$ to obtain

$$(1, x-y) = (1, x) - (1, y) - \frac{(3, x) + (3, y)}{(2, x) - (2, y)}. \tag{4.122}$$

Add (4.121) and (4.122) to get

$$(1, x+y) + (1, x-y) = 2(1, x) - \frac{2(3, x)}{(2, x) - (2, y)}$$

and then use (4.117) to rewrite as

$$\frac{d}{dx} \log \phi(x+y) + \frac{d}{dx} \log \phi(x-y) = 2\frac{d}{dx} \log \phi(x) + \frac{d}{dx} \log \big((2, x) - (2, y)\big);$$

integration and then exponentiation give us

$$\phi(x+y)\,\phi(x-y) = c\phi^2(x)\big((2, x) - (2, y)\big). \tag{4.123}$$

Next note that

$$\phi(x+y)\,\phi(x-y) = \phi(y)\phi(-y) + \cdots + \text{higher powers of } x$$

[30] Eisenstein (1975) vol. 1, pp. 427, 470–474.     [31] Eisenstein (1975) vol. 1, pp. 473–474.

and

$$\phi^2(x)\big((2, x) - (2, y)\big) = (x^2 + \cdots)\left(\frac{1}{x^2} + \cdots\right) = 1 + \text{higher powers of } x.$$

Finally, let $x \to 0$ in (4.123) to get $c = \phi(y)\phi(-y)$ from which (4.119) follows.

Eisenstein also discovered the well-known formula for $\wp(z) - \wp(\frac{\omega_i}{2})$ (with $i = 1, 2$) in the form: If $q$ is a zero of $(3, x) = 0$ and in particular, if $q = \frac{\omega_1}{2}, \frac{\omega_2}{2}, \frac{\omega_1 + \omega_2}{2}$, then

$$(2, x) - (2, q) = e^{-2(1,q)x} \frac{\phi^2(x + q)}{\phi^2(x)\phi^2(q)}. \tag{4.124}$$

Note that the formula (4.124) cannot be obtained from (4.119) merely by setting $y = q$. Rather, we have to show that

$$\phi(x - q) = -e^{-2(1,q)x} \phi(x + q), \tag{4.125}$$

where $q$ is one of the half-periods $\frac{\omega_1}{2}, \frac{\omega_2}{2}$, or $\frac{\omega_1 + \omega_2}{2}$. Note that (4.125) is equivalent to

$$\sigma(x - q) = -e^{-2\zeta(q)x} \sigma(x + q),$$

where $\zeta(q)$ is the Weierstrass $\zeta$-function. To then prove (4.124), Eisenstein set $y = q$ in (4.121) so that $(3, q) = 0$ and then integrated to arrive at

$$\log \phi(x + q) - \log \phi(x) - (1, q)x = \frac{1}{2} \log \big((2, x) - (2, q)\big) + C_1 \tag{4.126}$$

where $C_1$ is independent of $x$. Exponentiating (4.124) gives

$$\big((2, x) - (2, q)\big)C = e^{-2(1,q)x} \frac{\phi^2(x + q)}{\phi^2(x)} \tag{4.127}$$

where $C$ is independent of $x$. Expand in powers of $x$ as in (4.123), multiply by $x^2$ and let $x \to 0$ to see that

$$C = \phi^2(q).$$

## 4.9 Connection of $\phi$ with Theta Functions

Yet another accomplishment of Eisenstein's 1847 elliptic functions paper was to reveal a critical connection between his product function $\phi(x)$ and the theta function $\theta_1(z, \omega)$.[32] Kronecker pointed out in his 1891 paper "Die Legendre'sche Relation," that Eisenstein's results actually allow easy access to the transformation of theta functions. Indeed, Eisenstein's discussion of this topic began with the more general product

$$f(x, y) = \prod_{n=-\infty}^{\infty} \prod_{m=-\infty}^{\infty} \left(1 - \frac{x}{m\omega_1 + n\omega_2 + y}\right) \tag{4.128}$$

$$:= \lim_{N \to \infty} \lim_{M \to \infty} \prod_{n=-N}^{N} \prod_{m=-M}^{M} \left(1 - \frac{x}{m\omega_1 + n\omega_2 + y}\right). \tag{4.129}$$

[32] Eisenstein (1975) pp. 419–425 or Eisenstein (1847) pp. 275–281.

We have already seen that all Eisenstein's infinite products were defined in a similar manner. He observed that, with $\eta = \frac{\pi \omega_2}{\omega_1} = \pi \omega, \xi = \frac{\pi y}{\omega_1}$, and $t = \frac{\pi x}{\omega_1}$ (4.47) would imply that

$$\prod_{m=-\infty}^{\infty} \left(1 - \frac{x}{m\omega_1 + n\omega_2 + y}\right) = \frac{\sin(n\eta + \xi - t)}{\sin(n\eta + \xi)}$$

$$= \prod_{m=-\infty}^{\infty} \frac{e^{(n\eta+\xi-t)i} - e^{-(n\eta+\xi-t)i}}{e^{(n\eta+\xi)i} - e^{-(n\eta+\xi)i}}.$$

Thus, he could write (4.128) as

$$f(x,y) = \prod_{n=-\infty}^{\infty} \frac{\sin(n\eta + \xi - t)}{\sin(n\eta + \xi)} = \prod_{n=-\infty}^{\infty} \frac{e^{(n\eta+\xi-t)i} - e^{-(n\eta+\xi-t)i}}{e^{(n\eta+\xi)i} - e^{-(n\eta+\xi)i}}. \quad (4.130)$$

Next, Eisenstein set $e^{\eta i} = p, e^{\xi i} = \zeta, e^{ti} = z$, so that the $n$th term in the product (4.130) could be written as

$$\frac{p^n \zeta z^{-1} - p^{-n} \zeta^{-1} z}{p^n \zeta - p^{-n} \zeta^{-1}}; \quad (4.131)$$

and (4.131) could in turn be expressed as

$$z^{-1} \frac{1 - p^{-2n} \zeta^{-2} z^2}{1 - p^{-2n} \zeta^{-2}} \quad \text{or} \quad z \frac{1 - p^{2n} \zeta^2 z^{-2}}{1 - p^{2n} \zeta^2},$$

depending on whether $|p| > 1$ or $|p| < 1$, respectively. Setting $q = p^{-\delta}$, while choosing $\delta$ to be $+1$ or $-1$ so that $|p|^{-\delta} < 1$, Eisenstein combined the $n$th term with the $-n$th term to obtain

$$(1 - q^{2n} \zeta^2 z^{-2})(1 - q^{2n} \zeta^{-2} z^2) \div (1 - q^{2n} \zeta^2)(1 - q^{2n} \zeta^{-2}) \quad (4.132)$$

for positive $n$, and

$$\frac{\zeta z^{-1} - \zeta^{-1} z}{\zeta - \zeta^{-1}} \quad (4.133)$$

for $n = 0$. Now expressions (4.132) and (4.133) led him to consider the theta product

$$\mathcal{X}(\zeta) = (\zeta - \zeta^{-1}) \prod_{n=1}^{\infty} (1 - q^{2n} \zeta^2)(1 - q^{2n} \zeta^{-2}), \quad (4.134)$$

so that he could write

$$f(x,y) = \prod_{m,n} \left(1 - \frac{x}{m\omega_1 + n\omega_2 + y}\right) = \frac{\mathcal{X}(\zeta z^{-1})}{\mathcal{X}(\zeta)}$$

$$= \frac{\mathcal{X}(\zeta^{-1} z)}{\mathcal{X}(\zeta^{-1})} = \frac{\mathcal{X}(e^{\frac{x-y}{\omega_1} \pi i})}{\mathcal{X}(e^{-\frac{y}{\omega_1} \pi i})}. \quad (4.135)$$

He then demonstrated for positive integer $h$ that

$$\mathcal{X}(q^h z) = (-1)^h q^{-h^2} z^{-2h} \mathcal{X}(z); \tag{4.136}$$

whose proof boils down to verifying that

$$(q^h z - q^{-h} z^{-1}) \prod_{n=1}^{h} \left(1 - q^{2(n-h)} z^{-2}\right) = (-1)^h z^{-2h} q^{-h^2} (z - z^{-1}) \prod_{n=1}^{h} \left(1 - q^{2n} z^2\right).$$

He also observed that since

$$\phi(x) = x \prod_{m,n}' \left(1 - \frac{x}{m\omega_1 + n\omega_2}\right),$$

he could obtain

$$f(x, y) = \frac{\phi(x - y)}{\phi(-y)} = \frac{\phi(y - x)}{\phi(y)} \tag{4.137}$$

and

$$\phi(x) = -\lim_{y \to 0} \left(y f(x, y)\right). \tag{4.138}$$

Eisenstein himself wrote (4.138) not as a limit, but as

$$\phi(x) = \left\{-y f(x, y)\right\}_{y=0}.$$

Eisenstein thus showed that $\phi(x)$ may be written in terms of the $\theta$-function:

$$\phi(x) = \frac{\omega_1}{2\pi i} \mathcal{X}(z) \prod_{n=1}^{\infty} (1 - q^{2n})^{-2} \tag{4.139}$$

$$= \frac{\omega_1}{\pi} q^{-\frac{1}{4}} \theta_1(z, \omega) \prod_{n=1}^{\infty} (1 - q^{2n})^{-3} \tag{4.140}$$

with $q = e^{\omega \pi i}, z = e^{\frac{x}{\omega_1}\pi i}$, and where Im $\frac{\omega_2}{\omega_1} > 0$. Now if we change $x$ to $x + m\omega_1 + n\omega_2$ in (4.139), then $z$ changes to $(-1)^m q^n z$, so that by (4.136) we obtain another important formula of Eisenstein:

$$\phi(x + m\omega_1 + n\omega_2) = (-1)^{m+n} \phi(x) e^{\delta \frac{n^2 \omega_2 + 2nx}{\omega_1} \pi i}, \tag{4.141}$$

where $\delta = -1$ when Im $\frac{\omega_2}{\omega_1} > 0$.

Eisenstein may also be credited with discovering the general transformation formula for the theta function (4.134), though he did not give it in its entirety for lack of space.[33] Suppose

$$\omega_1' = \lambda \omega_1 + \nu \omega_2, \quad \omega_2' = \mu \omega_1 + \rho \omega_2, \quad \lambda \rho - \mu \nu = k \neq 0, \tag{4.142}$$

---

[33] Eisenstein (1847) pp. 278–279 or Eisenstein (1975) pp. 422–423.

where $\lambda$, $\nu$, $\mu$, $\rho$ are integers. Let

$$\omega = \frac{\omega_2}{\omega_1}, \quad \omega' = \frac{\omega_2'}{\omega_1'} = \frac{\mu + \rho\,\omega}{\lambda + \nu\,\omega}, \quad x' = \frac{x}{\lambda + \nu\,\omega} \qquad (4.143)$$

and write $\mathcal{X}(\zeta)$ in (4.134) as $\mathcal{X}(\zeta, q)$. Eisenstein made a calculation similar to that used in the derivation of (4.141) to show that

$$\mathcal{X}(e^{\frac{\pi x'}{\omega_1'}i}, q') = C \cdot e^{-\frac{\delta\nu i\pi}{\lambda + \nu\,\omega}\left(\frac{x}{\omega_1}\right)^2} \mathcal{X}(e^{\frac{\pi x}{\omega_1}i}, q), \qquad (4.144)$$

with $k$ as in (4.142), $\omega'$ as in (4.143), and where $q = e^{-\delta\pi i\omega}$, $q' = e^{-k\delta\pi i\omega'}$. Also, $\delta = \pm 1$, depending on whether Im $\omega < 0$ or Im $\omega > 0$. Thus, if Im $\omega > 0$, and if $\omega \to \omega'$ in (4.143) is a modular transformation, then $\delta = -1$, $k = 1$, $q = e^{\pi i\omega}$, $q' = e^{\pi i\omega'}$. Eisenstein noted that $C$ in (4.144) was independent of $x$ but dependent upon $\omega$, writing that since his paper was already too long, he would omit his description of his method for calculating $C$. He also observed that a particular case of the formula (4.144), $\lambda = \rho = 0$ and $\nu = -\mu = 1$, would yield a changed order of the product in $f(x, y)$ in (4.135).

It is natural to ask the question: How might Eisenstein have calculated, in terms of the theta constants, the values of

$$(2, q_i) - (2, q_j), \quad i \neq j, \qquad (4.145)$$

where

$$q_1 = \frac{\omega_1}{2}, \quad q_2 = \frac{\omega_2}{2}, \quad q_3 = \frac{\omega_1 + \omega_2}{2}\,?$$

Observe that the product of the squares of the values given in (4.145) will produce the discriminant of the cubic on the right-hand side of (4.101). We note that these values were derived by Weierstrass in his lectures during the 1860s.[34] For this derivation, we could apply (4.124). However, it would be more convenient to use (4.119) in combination with (4.141); in that case, we see that

$$(2, x) - (2, q_i) = C_i \frac{\phi^2(x - q_i)}{\phi^2(x)\phi^2(q_i)} e^{\frac{2\delta n_i x}{\omega_1}\pi i}, \qquad (4.146)$$

where $C_i$ is a constant and $n_1 = 0, n_2 = 1, n_3 = 1$. Thus

$$(2, x) - (2, q_i) = K_i \frac{\phi^2(x - q_i)}{\phi^2(x)} e^{\frac{2\delta n_i x}{\omega_1}\pi i},$$

where $K_i$ is independent of $x$. Expanding the two sides around $x = 0$ shows that $K_i$ is $\phi^{-2}(q_i)$ and we finally obtain

$$(2, x) - (2, q_i) = \frac{\phi^2(x - q_i)}{\phi^2(x)\phi^2(q_i)} e^{\frac{2\delta n_i x}{\omega_1}\pi i}. \qquad (4.147)$$

Setting $x = q_j$, we get

$$(2, q_j) - (2, q_i) = \frac{\phi^2(q_j - q_i)}{\phi^2(q_j)\phi^2(q_i)} e^{\frac{2\delta n_i q_j}{\omega_1}\pi i}. \qquad (4.148)$$

---

[34] Weierstrass (1894–1927) vol. 5, pp. 160–174.

Setting $j = 3$ and $i = 2$, we have $n_2 = 1$ and this gives

$$\left(2, \frac{\omega_1 + \omega_2}{2}\right) - \left(2, \frac{\omega_2}{2}\right) = \frac{\phi^2\left(\frac{\omega_1}{2}\right) e^{\delta \frac{(\omega_1+\omega_2)}{\omega_1}\pi i}}{\phi^2\left(\frac{\omega_1+\omega_2}{2}\right)\phi^2\left(\frac{\omega_2}{2}\right)}. \tag{4.149}$$

We take Im $\frac{\omega_2}{\omega_1} > 0$ so that $\delta = -1$ and the exponential factor is $e^{-(1+\omega)\pi i} = -q^{-1}$. We compute one of the factors, say $\phi^2\left(\frac{\omega_1+\omega_2}{2}\right)$, by using (4.139). The other factors are calculated in a similar manner. Now since $x = \frac{\omega_1+\omega_2}{2}$, we have

$$z = e^{\frac{\omega_1+\omega_2}{2\omega_1}\pi i} = iq^{\frac{1}{2}}.$$

Thus by (4.134)

$$\mathcal{X}(iq^{\frac{1}{2}}) = i(q^{\frac{1}{2}} + q^{-\frac{1}{2}}) \prod_{n=1}^{\infty}(1 + q^{2n+1})(1 + q^{2n-1})$$

$$= \frac{i}{q^{\frac{1}{2}}} \prod_{n=1}^{\infty}(1 + q^{2n-1})^2$$

and

$$\phi^2\left(\frac{\omega_1 + \omega_2}{2}\right) = \left(\frac{\omega_1}{2\pi}\right)^2 \frac{1}{q} \prod_{n=1}^{\infty}(1 + q^{2n-1})^4 (1 - q^{2n})^{-4}. \tag{4.150}$$

Similarly, we can show that

$$\phi^2\left(\frac{\omega_1}{2}\right) = \left(\frac{\omega_1}{\pi}\right)^2 \prod_{n=1}^{\infty}(1 + q^{2n})^4 (1 - q^{2n})^{-4}, \tag{4.151}$$

$$\phi^2\left(\frac{\omega_2}{2}\right) = -\left(\frac{\omega_1}{2\pi}\right)^2 \frac{1}{q} \prod_{n=1}^{\infty}(1 - q^{2n-1})^2 (1 - q^{2n})^{-4}. \tag{4.152}$$

Setting $e_i = (2, q_i) - (2^*, 0)$, we see that upon applying (4.150) through (4.152), we arrive at the formula given by Weierstrass in his lectures:[35]

$$e_3 - e_2 = \frac{16\pi^2}{\omega_1^2} q \prod_{n=1}^{\infty}(1 - q^{2n})^8 (1 + q^{2n})^4$$

$$= \frac{16\pi^2}{\omega_1^2} \theta_0^4 (0, \omega). \tag{4.153}$$

It is also possible to verify in a similar manner that

$$e_1 - e_3 = \frac{\pi^2}{\omega_1^2} \theta_2^4 (0, \omega) \quad \text{and} \quad e_1 - e_2 = \frac{\pi^2}{\omega_1^2} \theta_3^4 (0, \omega). \tag{4.154}$$

---

[35] Weierstrass (1894–1927) vol. 5, p. 164.

Now applying (4.101) to the discriminant of the third-degree polynomial on the right-hand side of (4.100) gives us

$$\Delta(\omega_1, \omega_2) \equiv g_2^3 - 27 g_3^2 = 16(e_3 - e_2)^2 (e_1 - e_3)^2 (e_1 - e_2)^2$$

$$= \left(\frac{2\pi}{\omega_1}\right)^{12} q^2 \prod_{n=1}^{\infty} (1 - q^{2n})^{24}. \qquad (4.155)$$

Observe that from its definition, since $g_2(\omega_1, \omega_2)$ and $g_3(\omega_1, \omega_2)$ are invariant under the action of the modular group, $\Delta(\omega_1, \omega_2)$ must be invariant under the action of the modular group: Thus, if

$$\omega_1' = a\omega_1 + b\omega_2, \quad \omega_2' = c\omega_1 + d\omega_2, \quad \text{where} \quad ad - bc = 1,$$

then

$$\Delta(\omega_1', \omega_2') = \Delta(\omega_1, \omega_2). \qquad (4.156)$$

If we take $\omega = \frac{\omega_2}{\omega_1}$, so that its imaginary part is positive, then $\omega' = \frac{\omega_2'}{\omega_1'}$ also has its imaginary part positive, and

$$\omega' = \frac{c + d\omega}{a + b\omega}, \quad ad - bc = 1. \qquad (4.157)$$

Hence, (4.155) through (4.157) imply that

$$q'^2 \prod_{n=1}^{\infty} (1 - q'^{2n})^{24} = (a + b\omega)^{12} q^2 \prod_{n=1}^{\infty} (1 - q^{2n})^{24}, \qquad (4.158)$$

where $q = e^{i\omega\pi}, q' = e^{i\omega'\pi}$. If we set $\Delta(1, \omega) \equiv \Delta(\omega)$, then (4.158) may be written as

$$\Delta\left(\frac{d\omega + c}{b\omega + a}\right) (b\omega + a)^{-12} = \Delta(\omega) \qquad (4.159)$$

or

$$\Delta(\omega') \left(\frac{d\omega'}{d\omega}\right)^6 = \Delta(\omega), \qquad (4.160)$$

where $\omega'$ is given by (4.157). Note that Hurwitz gave a direct proof of (4.158) using Eisenstein's approach. Also recall that Hurwitz employed the $J$ function to show that the infinite product in (4.155) was the discriminant function.

In his 1881 paper, "Grundlagen einer independenten Theorie der elliptischen Modulfunktionen," Hurwitz mentioned Eisenstein's work at several points; in fact, he made extensive use of the methods and results of Eisenstein on modular forms. However, he wrote a puzzling footnote on Eisenstein's paper: "This content-rich work, already cited many times, is also historically very interesting. It even contains, for example, the function introduced by Weierstrass $\wp(u)$ (for Eisenstein: $(2, x) - (2^*, 0)$), but without Eisenstein having recognized its fundamental significance,"[36] We can see clearly that Eisenstein understood the fundamental significance of his function $(2, x) - (2^*, 0)$. He

---

[36] Hurwitz (1962) vol. 1, p. 31.

gave the first- and second-order differential equations satisfied by this function, gave its addition formula, expressed it as a quotient of the double product $\phi$, and proved other related results. Perhaps Hurwitz was led to his view because, as we have seen, Eisenstein's writings were presented in a somewhat disorganized way, making it difficult to perceive their full import. For example, in his paper on elliptic functions, we find that Eisenstein might at one point derive a certain formula and then move on to a new topic. But then later on he would return to the earlier topic, presenting a different proof of that original formula, this time expressed in a different but equivalent form. Then, later in the paper, he would come back to this topic, stating a result showing that the two forms of the formula were equivalent. In this connection, André Weil has written in the introduction to his famous work, *Elliptic Functions According to Eisenstein and Kronecker*:[37]

> As any reader of Eisenstein must realize, he felt hard pressed for time during the whole of his short mathematical career. As a young man he complains of nervous ailments which often compel him to interrupt his work; later, he developed tuberculosis and died of it in 1852 at the age of 29. His papers, although brilliantly conceived, must have been written by fits and starts, with the details worked out only as occasion arose; sometimes a development is cut short, only to be taken up again at a later stage.

### 4.10 Hurwitz's Fourier Series for Modular Forms

Hurwitz made the important contribution of deriving the Fourier expansions of Eisenstein modular forms $(s^*, 0)$. He also introduced the now commonly used notation $G_s(\omega_1, \omega_2)$ instead of $(s^*, 0)$; he wrote $G_{\frac{s}{2}}$ in his 1882 paper but later changed it to $G_s$. To understand Hurwitz's work in this connection, first recall Euler's formula, given by(4.45) and (4.57), for which Eisenstein gave a new proof:

$$\frac{1}{z} + \sum_{n=1}^{\infty} \left( \frac{1}{z+n} + \frac{1}{z-n} \right) = \pi \cot \pi z = \pi i \frac{e^{2\pi i z} + 1}{e^{2\pi i z} - 1}$$

$$= -\pi i - 2\pi i \sum_{k=1}^{\infty} q^{2k}, \quad q = e^{\pi i z}, \ \text{Im } z \geq 0. \quad (4.161)$$

Hurwitz differentiated this formula $s$ times with respect to $z$ to get a particular case of Scheibner's (4.18):

$$\sum_{n=-\infty}^{\infty} \frac{1}{(z+n)^s} = (-1)^s \frac{(2\pi i)^s}{(s-1)!} \sum_{k=1}^{\infty} k^{s-1} q^{2k}. \quad (4.162)$$

Eisenstein noted[38] that for $s = 2, 3, 4$ respectively, the sum on the left-hand side of (4.162) would be given by

$$\frac{\pi^2}{\sin^2 \pi z}, \quad \frac{\pi^3 \cos \pi z}{\sin^3 \pi z}, \quad \pi^4 \left( -\frac{2}{3} \cdot \frac{1}{\sin^2 \pi z} + \frac{1}{\sin^4 \pi z} \right)$$

---

[37] Weil (1976) p. 4.  [38] Eisenstein (1975) vol. 1, pp. 402–403.

and, in general, the left-hand side for even $s = 2g$ would be

$$\pi^{2g} \left( \frac{a}{\sin^2 \pi z} + \frac{a'}{\sin^4 \pi z} + \frac{a''}{\sin^6 \pi z} + \text{etc} \right);$$

for $s$ odd, $s = 2g + 1$, Eisenstein had the left-hand side of (4.162) as:

$$\pi^{2g+1} \left\{ \frac{b \cos \pi z}{\sin^3 \pi z} + \frac{b' \cos \pi z}{\sin^5 \pi z} + \frac{b'' \cos \pi z}{\sin^7 \pi z} + \text{etc.} \right\}.$$

Concerning the coefficients $a, a', a'', b, b', b''$, Eisenstein remarked that they were dependent upon the Bernoulli numbers.

Hurwitz remarked[39] that the formula (4.162) holds for $z$ in the upper half-plane, since $|q| < 1$ for such $z$. To obtain the Fourier series for modular forms given by the Eisenstein series, set $z = m\omega$ so that $\text{Im}\,\omega > 0$, and then sum over all positive integer values of $m$ to arrive at

$$\sum_{m=1}^{\infty} \sum_{n=-\infty}^{\infty} \frac{1}{(m\omega + n)^s} = (-1)^s \frac{(2\pi i)^s}{(s-1)!} \sum_{k=1}^{\infty} k^{s-1} \frac{q^{2k}}{1 - q^{2k}}, \quad q = e^{\pi i \omega}. \quad (4.163)$$

Recall that for $s \geq 3$ and $s$ odd

$$G_s(\omega_1, \omega_2) = \sum_{m,n}' \frac{1}{(m\omega_1 + n\omega_2)^s} = 0.$$

By (4.163) we know that if $\omega = \frac{\omega_1}{\omega_2}$ has $\text{Im}\,\omega > 0$, then for an even integer $2s$

$$G_{2s}(\omega_1, \omega_2) = \frac{1}{\omega_2^{2s}} \left( \sum_{n=-\infty}^{\infty}{}' \frac{1}{n^{2s}} + 2(-1)^s \frac{(2\pi)^{2s}}{(2s-1)!} \sum_{k=1}^{\infty} k^{2s-1} \frac{q^{2k}}{1 - q^{2k}} \right). \quad (4.164)$$

Hurwitz next employed Euler's well-known formula[40]

$$\sum_{n=-\infty}^{\infty}{}' \frac{1}{n^{2s}} = 2 \sum_{n=1}^{\infty} \frac{1}{n^{2s}} = (-1)^{s-1} (2\pi)^{2s} \frac{B_{2s}}{(2s)!} \quad \text{for } s \geq 1. \quad (4.165)$$

With the use of (4.165), (4.164) can be rewritten for $s \geq 1$ as

$$G_{2s}(\omega_1, \omega_2) = \left( \frac{2\pi}{\omega_2} \right)^{2s} \frac{(-1)^{s-1}}{(2s)!} \left( B_{2s} - 4s \sum_{k=1}^{\infty} k^{2s-1} \frac{q^{2k}}{1 - q^{2k}} \right). \quad (4.166)$$

Hurwitz made use of an arithmetic function he denoted by $\psi$, now written as $\sigma$:

$$\sigma_l(k) = \sum_{d|k} d^l.$$

Then he could write (in modern notation)

$$\sum_{k=1}^{\infty} k^{2s-1} \frac{q^{2k}}{1 - q^{2k}} = \sum_{k=1}^{\infty} k^{2s-1} q^{2k} (1 + q^{2k} + q^{4k} + \cdots) = \sum_{k=1}^{\infty} \sigma_{2s-1}(k) q^{2k}.$$

---

[39] Hurwitz (1962) vol. 1, p. 21.    [40] Eu.I-14, pp. 434–439.

Thus, (4.166) can be rewritten as the Fourier series expansion of the Eisenstein series[41]

$$G_{2s}(\omega_1, \omega_2) = \frac{2\zeta(2s)}{\omega_2{}^{2s}} \left( 1 - \frac{4s}{B_{2s}} \sum_{k=1}^{\infty} \sigma_{2s-1}(k) q^{2k} \right), \quad s \geq 1, \qquad (4.167)$$

where

$$\zeta(l) = \sum_{n=1}^{\infty} \frac{1}{n^l}, \quad l > 1.$$

To show that $G_{2s}$ can be used to define a modular form of one variable, first note that if $a, b, c, d$ are integers with $ad - bc = 1$, and if

$$\omega_1' = a\omega_1 + b\omega_2, \quad \text{and} \quad \omega_2' = c\omega_1 + d\omega_2,$$

then for $s > 2$

$$G_s(\omega_1', \omega_2') = G_s(\omega_1, \omega_2). \qquad (4.168)$$

To prove this requires only a change in the order of summation, since

$$m\omega_1' + n\omega_2' = (ma + nc)\omega_1 + (mb + nd)\omega_2 =: m'\,\omega_1 + n'\,\omega_2$$

and hence the pair $m, n$ is related to the pair $m', n'$ by means of the invertible matrix of determinant 1: $\left( \begin{smallmatrix} a & b \\ c & d \end{smallmatrix} \right)$. Next, if we let $\omega = \frac{\omega_1}{\omega_2}$ with $\operatorname{Im} \omega > 0$, and let $\omega' = \frac{\omega_1'}{\omega_2'}$, then $\operatorname{Im} \omega' > 0$ and

$$\omega' = \frac{a\omega + b}{c\omega + d}. \qquad (4.169)$$

Therefore, (4.168) may be written as

$$\sideset{}{'}\sum_{m,n} \frac{1}{(m\omega' + n)^s} = \left( \frac{\omega_2'}{\omega_2} \right)^s \sideset{}{'}\sum_{m,n} \frac{1}{(m\omega + n)^s}$$

$$= (c\omega + d)^s \sideset{}{'}\sum_{m,n} \frac{1}{(m\omega + n)^s}, \quad s \geq 3. \qquad (4.170)$$

Thus, for $k \geq 3$, the function

$$G_k(\omega) := \sideset{}{'}\sum_{m,n} \frac{1}{(m\omega + n)^k} \qquad (4.171)$$

satisfies

$$G_k \left( \frac{a\omega + b}{c\omega + d} \right) = (c\omega + d)^k\, G(\omega). \qquad (4.172)$$

Note that $G_s(\omega)$ is an analytic function of $\omega$ in the upper half-plane. The real line forms the boundary of this function because every rational number $\omega = -\frac{n}{m}$ is a singularity

---

[41] Hurwitz (1962) vol. 1, p. 30.

of the function. Note also that (4.167) can be written as

$$G_{2k}(\omega) = 2\zeta(2k)\left(1 - \frac{4k}{B_{2k}}\sum_{n=1}^{\infty}\sigma_{2k-1}(n)q^{2n}\right), \quad q = e^{\pi i\omega}. \qquad (4.173)$$

Equations (4.172) and (4.173) show that $G_{2k}(\omega)$ is a modular form of weight $k$. Recall that $G_{2k+1}(\omega) \equiv 0$.

An important contribution of Hurwitz to the theory of modular forms was his use of a formula of Eisenstein to prove that the function defined by the formula

$$\Delta(\omega) = (2\pi)^{12}q^2\prod_{n=1}^{\infty}(1 - q^{2n})^{24}, \quad q = e^{\pi i\omega} \qquad (4.174)$$

was a cusp form of weight 12. Note that Klein[42] employed formulas due to Jacobi to show that $g_2^3 - 27g_3^2$ equals the right-hand side of (4.174).

### 4.11 Hurwitz's Proof That $\Delta(\omega)$ Is a Modular Form

Note that we can write the formula $f\left(\frac{a\omega+b}{c\omega+d}\right) = (c\omega + d)^k f(\omega)$ in the form

$$f(\gamma\omega)(\gamma'\omega)^{\frac{k}{2}} = f(\omega) \quad \text{for all } \gamma \in \Gamma(1). \qquad (4.175)$$

By the chain rule, if $\gamma_1, \gamma_2 \in \Gamma$, then

$$\begin{aligned}
f(\gamma_1 \circ \gamma_2\,\omega)\big((\gamma_1 \circ \gamma_2)'\,\omega\big)^{\frac{k}{2}} &= f\big(\gamma_1(\gamma_2\,\omega)\big)\big(\gamma_1'(\gamma_2\,\omega)\big)^{\frac{k}{2}}(\gamma_2'\,\omega)^{\frac{k}{2}} \\
&= f(\gamma_2\,\omega)(\gamma_2'\,\omega)^{\frac{k}{2}} \\
&= f(\omega).
\end{aligned}$$

To verify (4.175) it is therefore sufficient to check it for the generators of $\Gamma$. From Lagrange's work of 1773 on the reduction of binary quadratic forms, it was known to Hurwitz that

$$S = \begin{pmatrix} 1 & 1 \\ 0 & 1 \end{pmatrix} \quad \text{and} \quad T = \begin{pmatrix} 0 & -1 \\ 1 & 0 \end{pmatrix} \qquad (4.176)$$

generate $\Gamma = SL_2(\mathbb{Z})$. This means that to confirm (4.175), it is sufficient to check it for the generators $S$ and $T$. Thus, to prove that $\Delta(\omega)$ as defined in (4.174) is a modular form of weight 12, it is enough for us to show that

$$\Delta(S\,\omega) = \Delta(\omega + 1) = \Delta(\omega), \qquad (4.177)$$

$$\Delta(T\,\omega) = \Delta\left(-\frac{1}{\omega}\right) = \omega^{12}\Delta(\omega). \qquad (4.178)$$

[42] Klein (1921–1923) vol. 3, p. 21.

Since $q^2 = e^{2\pi i(\omega+1)} = e^{2\pi i\omega}$, (4.174) implies that (4.177) must be true. To prove (4.178), Hurwitz showed that if $\omega = \frac{\omega_1}{\omega_2}$, and

$$\Delta(\omega_1, \omega_2) = \frac{1}{\omega_2^{12}} \Delta(\omega) \qquad (4.179)$$

then

$$\Delta(-\omega_2, \omega_1) = \Delta(\omega_1, \omega_2). \qquad (4.180)$$

Observe that (4.180) is equivalent to (4.178). Hurwitz first proved a formula derived in a more general form[43] by Eisenstein. Hurwitz showed that

$$G_2(\omega_1, \omega_2) - G_2(-\omega_2, \omega_1) = \frac{2\pi i}{\omega_1 \omega_2}, \qquad (4.181)$$

to be proved in the next section; on this basis we reproduce the remainder of Hurwitz's derivation of (4.180).[44] Note that if we set

$$q = e^{i\pi \frac{\omega_1}{\omega_2}}, \quad q' = e^{-i\pi \frac{\omega_2}{\omega_1}},$$

then using (4.167) and the fact that $B_2 = \frac{1}{6}$, we can write (4.181) in the form

$$\log q \left(1 - 24 \sum_{k=1}^{\infty} k \frac{q^{2k}}{1 - q^{2k}}\right) + \log q' \left(1 - 24 \sum_{k=1}^{\infty} k \frac{q'^{2k}}{1 - q'^{2k}}\right) = -6. \quad (4.182)$$

Next, since $\log q \cdot \log q' = \pi^2$, upon differentiation we obtain

$$\frac{d \log q}{\log q} = \frac{dq}{q \log q} = \frac{-dq'}{q' \log q'}.$$

Multiplying (4.182) by $\frac{d \log q}{\log q}$ yields

$$\frac{dq}{q} \left(1 - 24 \sum_{k=1}^{\infty} k \frac{q^{2k}}{1 - q^{2q}}\right) - \frac{dq'}{q'} \left(1 - 24 \sum_{k=1}^{\infty} k \frac{q'^{2k}}{1 - q'^{2k}}\right) = -6 \frac{d \log q}{\log q}. \quad (4.183)$$

After term-by-term integration and then exponentiation, we arrive at

$$(\log q)^6 q \prod_{k=1}^{\infty} (1 - q^{2k})^{12} = Cq' \prod_{k=1}^{\infty} (1 - q'^{2k})^{12},$$

where $C$ denotes the constant of integration. The value of $C$ can be found by taking $\frac{\omega_1}{\omega_2} = i$, since in that case $q = q'$ so that we have

$$\left(\frac{2\pi}{\omega_2}\right)^6 q \prod_{k=1}^{\infty} (1 - q^{2k})^{12} = -\left(\frac{2\pi}{\omega_1}\right)^6 q' \prod_{k=1}^{\infty} (1 - q'^{2k})^{12}. \qquad (4.184)$$

We finally obtain (4.180) by squaring (4.184).

---

[43] Eisenstein (1975) vol. 1, pp. 416–418.    [44] Hurwitz (1962) vol. 1, p. 26.

## 4.12   Hurwitz's Proof of Eisenstein's Result

Hurwitz wanted to derive Eisenstein's result (4.181)

$$G_2(\omega_1, \omega_2) - G_2(-\omega_2, \omega_1) = \frac{2\pi i}{\omega_1 \omega_2}$$

in a new way, while still retaining Eisenstein's method for deriving elliptic function formulas. He set[45]

$$p = m\omega_1 + n\omega_2, \quad q = -m_1\omega_1 - n_1\omega_2, \quad p + q = (m - m_1)\omega_1 + (n - n_1)\omega_2$$

in (4.88), given again here:

$$\frac{1}{p^3} \cdot \frac{1}{q^3} = \frac{1}{(p+q)^3}\left(\frac{1}{p^3} + \frac{1}{q^3}\right) + \frac{3}{(p+q)^4}\left(\frac{1}{p^2} + \frac{1}{q^2}\right) + \frac{6}{(p+q)^5}\left(\frac{1}{p} + \frac{1}{q}\right),$$

and summed over all $m, n, m_1, n_1$ except for $m = m_1$, $n = n_1$. Keeping in mind that $G_3(\omega_1, \omega_2) = 0$, it is not difficult to see that the left-hand side will sum to $3G_6 \equiv 3G_6(\omega_1, \omega_2)$. Applying the same idea, after summation the right-hand side becomes

$$6G_2(\omega_1, \omega_2) \sum_\omega {}' \frac{1}{\omega^4} - 18 \sum_\omega {}' \frac{1}{\omega^6} - \frac{12\pi i}{\omega_2} \sum_\omega \frac{m}{\omega^5};$$

thus,

$$3G_6 = 6G_2(\omega_1, \omega_2)G_4 - 18G_6 + \frac{3\pi i}{\omega_2}\frac{\partial G_4}{\partial \omega_1}$$

or

$$21G_6 = 6G_2(\omega_1, \omega_2)G_4 + \frac{3\pi i}{\omega_2}\frac{\partial G_4}{\partial \omega_1}. \tag{4.185}$$

Replace $\omega_1$ by $-\omega_2$ and $\omega_2$ by $\omega_1$, so that (4.185) becomes

$$21G_6 = 6G_2(-\omega_2, \omega_1)G_4 - \frac{3\pi i}{\omega_1}\frac{\partial G_4}{\partial \omega_2}. \tag{4.186}$$

Subtracting (4.186) from (4.185), we get

$$6G_4(G_2(\omega_1, \omega_2) - G_2(-\omega_2, \omega_1)) + \frac{3\pi i}{\omega_1\omega_2}\left(\omega_1\frac{\partial G_4}{\partial \omega_1} + \omega_2\frac{\partial G_4}{\partial \omega_2}\right) = 0. \tag{4.187}$$

Since $G_4$ is a homogeneous function of degree $-4$, we have, from Euler's well-known formula for homogeneous functions,[46] the result

$$\omega_1\frac{\partial G_4}{\partial \omega_1} + \omega_2\frac{\partial G_4}{\partial \omega_2} = -4G_4. \tag{4.188}$$

Substituting (4.188) in (4.187), Hurwitz had (4.181).

[45] Hurwitz (1962) vol. 1, pp. 24–25.

[46] The earliest known reference to this result of Euler is in his manuscript, *De differentiatione*, dating from about 1730, first published as an appendix in Engelsman (1984). Euler also mentioned this result in his *Mechanica* of 1736.

## 4.13 Kronecker's Proof of Eisenstein's Result

In his 1892 paper, "Die Legendre'sche Relation," Kronecker wrote that though Eisenstein had proved his result, (4.181), by a highly ingenious method, its presentation was somewhat lengthy. Kronecker wished to present Eisenstein's deduction in a somewhat simpler form.[47] First, Kronecker introduced a new notation that revealed the dependence of the sums $(s, x)$ on the periods $\omega_1$ and $\omega_2$. For example:

$$f_1(x, \omega_1, \omega_2) = \lim_{N \to \infty} \lim_{M \to \infty} \sum_{n=-N}^{N} \sum_{m=-M}^{M} \frac{1}{x + m\omega_1 + n\omega_2}. \tag{4.189}$$

Let $a, b, c, d$ be integers with $ad - bc = 1$ and let

$$\omega_1' = a\omega_1 + b\omega_2, \quad \omega_2' = c\omega_1 + d\omega_2. \tag{4.190}$$

Kronecker showed that if $\delta$ had the same sign as the imaginary part of $\frac{\omega_1}{\omega_2}$, then

$$\lim_{N \to \infty} \lim_{M \to \infty} \sum_{n=-N}^{N} {\sum_{m=-M}^{M}}' \left( \frac{1}{(m\omega_1' + n\omega_2')^2} - \frac{1}{(m\omega_1 + n\omega_2)^2} \right) = -\frac{2\delta b\pi i}{\omega_1 \omega_1'}. \tag{4.191}$$

We observe that (4.191) contains (4.181) as particular case when $a = 0, b = -1, c = 1$, $d = 0$. Interestingly, Eisenstein had found an even more general result from which (4.191) was a consequence.

Kronecker started his proof with Eisenstein's formula (4.75) and wrote it in the form

$$f_1(x + p\omega_1 + q\omega_2, \omega_1, \omega_2) - f_1(x, \omega_1, \omega_2) = -\frac{2\delta q\pi i}{\omega_1}. \tag{4.192}$$

With $a, b, c, d$ as in (4.190), set

$$p' = dp - cq \quad \text{and} \quad q' = -bp + aq$$

so that

$$p'\omega_1' + q'\omega_2' = p\omega_1 + q\omega_2 =: -t.$$

By (4.192),

$$f_1(x + p'\omega_1' + q'\omega_2', \omega_1', \omega_2') - f_1(x, \omega_1', \omega_2') = -\frac{2\delta q'\pi i}{\omega_1'}. \tag{4.193}$$

Subtracting (4.192) from (4.193) we get

$$f_1(x - t, \omega_1', \omega_2') - f_1(x - t, \omega_1, \omega_2) - f_1(x, \omega_1', \omega_2') + f_1(x, \omega_1, \omega_2)$$
$$= \frac{2\delta\pi i(q\omega_1' - q'\omega_1)}{\omega_1 \omega_1'} = -\frac{2\delta bt\pi i}{\omega_1 \omega_1'}. \tag{4.194}$$

[47] Kronecker (1968) vol. 5, pp. 152–154.

Note that since

$$f_1(u, v, w) = \frac{1}{u} + \lim_{N \to \infty} \lim_{M \to \infty} \sum_{n=-N}^{N} \sum_{m=-M}^{M}{}' \frac{1}{u + (mv + nw)}$$

and

$$\frac{1}{u + (mv + nw)} = \frac{1}{mv + nw} - \frac{u}{(mv + nw)^2} + \frac{u^2}{(mv + nw)^3} - \cdots$$

it follows that the expression

$$-\frac{1}{u} + f_1(u, v, w) + \lim_{N \to \infty} \lim_{M \to \infty} \sum_{n=-N}^{N} \sum_{m=-M}^{M}{}' \frac{u}{(mv + nw)^2} \qquad (4.195)$$

contains in its denominator only terms of third and higher powers of $mv + nw$; thus, the series is absolutely convergent. When this fact is combined with (4.194), we obtain

$$\lim_{N \to \infty} \lim_{M \to \infty} \sum_{n=-N}^{N} \sum_{m=-M}^{M}{}' \left( \frac{t}{(m\omega_1' + n\omega_2')^2} - \frac{t}{(m\omega_1 + n\omega_2)^2} \right) = -\frac{2\delta b t \pi i}{\omega_1 \omega_1'}.$$

After canceling $t$ on each side, Kronecker obtained (4.191).

## 4.14 Exercises

1. Use the following identity, with $r = p + q$,

$$\frac{1}{p^4 q^3} - \frac{1}{p^3 q^4} = \frac{1}{r^3} \left( \frac{1}{p^4} - \frac{1}{q^4} \right) + \frac{2}{r^4} \left( \frac{1}{p^3} - \frac{1}{q^3} \right) + \frac{2}{r^5} \left( \frac{1}{p^2} - \frac{1}{q^2} \right)$$

to prove that

$$5(7, x) = 2(5, x)\big((2, x) - (2^*, 0)\big) + 3(3, x)(4, x).$$

See Eisenstein (1847) p. 285 or Eisenstein (1975) p. 429.

2. Let $(s, x)$ be defined by (4.61). Prove that

$$(5, x) = (3, x)\big((2, x) - (2^*, 0)\big).$$

Also write the equation as a differential equation involving the Weierstrass $\wp$-function. See Eisenstein (1975) vol. 1, p. 468.

3. Let $(s, x)$ for $s \geq 1$ be defined by (4.61). Show that

$$(1, x) + (1, y) - (1, x + y)$$
$$= \frac{1}{2(3, x + y)}$$
$$\times \big((2, x)(2, y) - 3(4, x + y) - (2, x + y)((2, x) + (2, y) - (2, x + y))\big).$$

See Eisenstein (1975) vol. 1, p. 469.

4. Let $\omega_1, \omega_2$ be complex numbers such that $\operatorname{Im} \frac{\omega_2}{\omega_1} > 0$. Let $\lambda, \nu, \mu, \rho$ be integers such that $\lambda\rho - \mu\nu = \epsilon > 0$ and $\omega_1' = \lambda\omega_1 + \nu\omega_2$, $\omega_2' = \mu\omega_1 + \rho\omega_2$. Let $f(x, y)$ be defined by (4.129) and let $f'(x, y)$ denote the function obtained from $f(x, y)$ when $\omega_1, \omega_2$ are replaced by $\omega_1', \omega_2'$. Show that

$$f'(x, y) = e^{-\frac{2\nu\pi i}{\omega_1 \omega_1'}(yx - \frac{1}{2}x^2)} f(x, y).$$

5. With the same notation as in exercise 4, let $\omega = \frac{\omega_2}{\omega_1}$, $q = e^{i\pi\omega}$, $q' = e^{\epsilon\pi i\frac{\mu+\rho\omega}{\lambda+\nu\omega}}$, $y' = \frac{y}{\lambda+\nu\omega}$. Show that

$$\mathcal{X}(e^{y'i}, q') = C \frac{\nu i y^2}{e^{\pi(\lambda+\nu\omega)}} y^2 \mathcal{X}(e^{yi}, q),$$

where $C$ depends only on $\omega$. For this and the previous problem, see Eisenstein (1847) p. 278 or Eisenstein (1975) p. 422.

6. Let

$$E_{2s}(\omega) = 1 - \frac{4s}{B_{2s}} \sum_{k=1}^{\infty} \sigma_{2s-1}(k) e^{2\pi ik\omega}.$$

Prove that

$$E_2\left(\frac{a\omega + b}{c\omega + d}\right) = (c\omega + d)^2 E_2(\omega) + \frac{6c(c\omega + d)}{\pi i},$$

and that

$$E_{2s}\left(\frac{a\omega + b}{c\omega + d}\right) = (c\omega + d)^{2s} E_{2s}(\omega),$$

when $s > 1$.

# 5

# *Hermite's Transformation of Theta Functions*

### 5.1 Preliminary Remarks

Charles Hermite (1822–1901) is most noted for being the creator, along with Joseph Liouville, of the theory of transcendental numbers. In 1844,[1] Liouville showed that there was an infinite number of transcendental numbers and in 1873,[2] Hermite proved that a specific number, *e*, was transcendental. Hermite also made important contributions to the theory of quadratic forms in two or more variables, the theory of elliptic and Abelian functions, and several topics in analysis. In this chapter, we concentrate on Hermite's 1858 papers on the transformation of theta functions and its application to the solution of a quintic by modular functions.

In 1840, Hermite entered Louis-le-Grand college to prepare for the entrance examination for the École Polytechnique. Note that Évariste Galois had done the same thing a decade earlier, with disastrous results. Hermite fared somewhat better, admitted to the Polytechnique ranked sixty-eighth in order of merit. This mediocre performance was the result of his having spent his time studying Lagrange's *Traité de la résolution des équations numériques* and Gauss's *Disquisitiones Arithmeticae* in French translation. That Hermite had mastered algebra at a high level can be gathered from his 1842 paper, "Considérations sur la résolution algébrique de l'équation du cinquième degré." He described his purpose in the first paragraph, here given in translation:[3]

> We know that Lagrange showed the algebraic solution of a general fifth degree equation to be dependent upon the determination of a root of a *particular* sixth-degree equation, which he called *reduced* [i.e., resolvent] (*Résolution des équations numérieques, Note XIII*). Thus, if this reduced [equation] were able to be decomposed into rational factors of the second or of the third degrees, we would have the solution of the fifth-degree equation. I will try to demonstrate that such a decomposition is impossible.

Thus, Hermite took up the fifth-degree algebraic equation at the very outset of his mathematical career. Then in 1858, he derived a transformation formula for theta

---

[1] Liouville (1844).    [2] Hermite (1905–1917) vol. 3. pp. 150–181.    [3] Hermite, (1905–1917) vol. 1, p. 3.

functions. In the paper containing this formula,[4] he wrote that in 1848 Jacobi had obtained transformation formulas for theta constants $\theta_0(\omega)$, $\theta_2(\omega)$, and $\theta_3(\omega)$.[5] Jacobi defined these series in such a way that, for example,

$$\theta_3(-i\rho) = 1 + 2e^{\pi\rho} + 2e^{4\pi\rho} + 2e^{9\pi\rho} + \cdots \tag{5.1}$$

converged when Re $\rho < 0$. In his 1848 paper, written in 1847,[6] Jacobi stated a transformation formula: Let $a, b, a', b'$ be integers and let

$$r = \frac{a\rho + \sqrt{-1}b'}{a' + \sqrt{-1}b\rho}, \qquad aa' + bb' = 1. \tag{5.2}$$

Of $a$ and $b$, if one is even and the other odd, then

$$\delta \cdot \frac{1 + 2e^{\pi r} + 2e^{4\pi r} + 2e^{9\pi r} + \cdots}{\sqrt{a' + \sqrt{-1}b\rho}} = 1 \pm 2e^{\pi\rho} + 2e^{4\pi\rho} \pm 2e^{9\pi\rho} + \cdots, \tag{5.3}$$

where $\delta$ is an eighth root of unity and the upper sign holds if, of $a'$ and $b'$, one is even and one is odd; the lower sign holds if both are odd. If both $a$ and $b$ are odd, then the series on the right-hand side of (5.3) is replaced by

$$2e^{\frac{\pi\rho}{4}} + 2e^{\frac{9\pi\rho}{4}} + 2e^{\frac{25\pi\rho}{4}} + \cdots. \tag{5.4}$$

Hermite generalized this result to theta functions instead of theta constants. Now from (5.3) and (5.4), it is clear that under a modular transformation, a theta constant may transform into another theta constant. To deal with this situation in a unified manner, Hermite defined the theta series

$$\theta_{\mu,\nu}(x) = \sum_{m=-\infty}^{\infty} (-1)^{m\nu} e^{i\pi((2m+\mu)x + \frac{\omega}{4}(2m+\mu)^2)} = \theta_{\mu\nu}(x, \omega), \tag{5.5}$$

where Im $\omega > 0$, so that the series is absolutely convergent. We can see that Jacobi's theta functions are related to (5.5):

$$\theta_0(\pi x, \omega) = \theta_{0,1}(x, \omega), \quad \theta_1(\pi x, \omega) = -i\theta_{1,1}(x, \omega),$$
$$\theta_2(\pi x, \omega) = \theta_{1,0}(x, \omega), \quad \theta_3(\pi x, \omega) = \theta_{0,0}(x, \omega). \tag{5.6}$$

Hermite showed that for integers $\mu$, $\nu$, the function $\theta_{\mu,\nu}(x)$ was essentially one of the four functions in (5.6). He stated the transformation theorem: If $a, b, c, d$ are integers with $ad - bc = 1$, $b > 0$, and if $M = a\mu + b\nu + ab$, $N = c\mu + d\nu + cd$, with $\mu$, $\nu$ integers, then

$$\theta_{\mu,\nu}\big((a + b\omega)x, \omega\big)e^{i\pi b(a+b\omega)x^2} = T\theta_{M,N}\left(x, \frac{c + d\omega}{a + b\omega}\right), \tag{5.7}$$

---

[4] Hermite Ibid. pp. 487–496.

[5] Note that in referring to Jacobi's paper, Hermite mistakenly gave vol. 34 of *Crelle's Journal* instead of vol. 36, pp. 97–112.

[6] Jacobi (1969) vol. 2, pp. 171–190, especially pp. 186–190.

where

$$T = \frac{\delta \sum_{\rho=0}^{b-1} e^{-i\pi \frac{a(\rho - \frac{1}{2}b)^2}{b}}}{\sqrt{-ib(a+b\omega)}}, \tag{5.8}$$

$$\delta = e^{-\frac{1}{4}i\pi(ac\mu^2 + 2bc\mu\nu + bd\nu^2 + 2abc\mu + 2abd\nu + ab^2c)}. \tag{5.9}$$

If $b = 0$, then

$$\theta_{\mu,\nu}(x, \omega) = e^{-\frac{i\pi}{4}\alpha\mu^2} \theta_{M,N}(x, \omega + \alpha), \tag{5.10}$$

$\alpha$ being an arbitrary integer, $M = \mu$, and $N = \alpha(\mu + 1) + \nu$.

In particular, Hermite's theorem implies that

$$\theta_1\left(x, \frac{c + d\omega}{a + b\omega}\right) = \delta\sqrt{a + b\omega}\, e^{i\pi b(a+b\omega)x^2}\, \theta_1\big((a + b\omega)x, \omega\big), \tag{5.11}$$

where $\delta$ is an eighth root of unity, depending upon $a, b, c, d$.

Although he did not state it completely, we know that Hermite had in his possession the more general odd-order transformation formula for theta functions, that is, a formula for the case in which $ad - bc = n$, with $n$ odd; he noted in his paper that for this case, $T$ in (5.7) would be replaced by $T_{\frac{n-1}{2}}$, a homogeneous function of degree $\frac{n-1}{2}$ in two of the functions $\theta_{\mu,\nu}\big(x, \frac{c+d\omega}{a+b\omega}\big)$.

Hermite applied his transformation formula (5.7) to solve the quintic equation. His 1858 paper[7] on this topic began by mentioning a general result on modular equations of order $p$, where $p$ is an odd prime. He set

$$\phi^8(\omega) = k^2(\omega) \quad \text{and} \quad \psi^8(\omega) = k'^2(\omega) = 1 - k^2(\omega) \tag{5.12}$$

and let $u = \phi(\omega)$ and $v = \phi(p\omega)$. The result stated that $u$ and $v$ were connected by an equation of degree $p + 1$ and that for a given value of $u$, the $n + 1$ values of $v$ would be

$$\epsilon\phi(p\omega) \quad \text{and} \quad \phi\left(\frac{\omega + 16m}{p}\right), \quad m = 0, 1, \ldots, p - 1, \tag{5.13}$$

where $\epsilon = \pm 1$ according as 2 is or is not a quadratic residue modulo $p$.[8] Hermite next brought up a theorem given by Galois in his famous May 29, 1832 letter to his friend Auguste Chevalier:[9] When $p = 5, 7, 11$, the modular equation is reducible to an equation of degree $p$; this reduction is not possible for higher values of $p$. Hermite explained that, while it was possible to recover the proof of this theorem from the sketch given in Galois's letter, a gap remained with respect to the solution of an algebraic equation of the fifth degree. Filling this gap required not just the verification of the possibility of the reduction of the modular equation to an equation of a lesser degree, but an explicit method for the reduction. Hermite wrote that after several attempts, dating back a long time (perhaps to his college days), he found that the Jacobi modular equation of degree

---

[7] Hermite (1905–1917) vol. 2, pp. 5–12.     [8] Hermite (1905–1917) vol. 2, pp. 9–10.
[9] See Neumann (2011) pp. 84–97, especially p. 91.

six

$$u^6 - v^6 + 5u^2v^2(u^2 - v^2) + 4uv(1 - u^4v^4) = 0 \tag{5.14}$$

could easily be reduced by means of the function

$$\Phi(\omega) = \left(\phi(5\omega) + \phi\left(\frac{\omega}{5}\right)\right)\left(\phi\left(\frac{\omega + 16}{5}\right) - \phi\left(\frac{\omega + 4 \cdot 16}{5}\right)\right)$$
$$\cdot\left(\phi\left(\frac{\omega + 2 \cdot 16}{5}\right) - \phi\left(\frac{\omega + 3 \cdot 16}{5}\right)\right). \tag{5.15}$$

Note that the $+$ sign in the first factor on the right-hand side of (5.15) occurred because 2 was not a quadratic residue modulo 5, so that $\epsilon\phi(5\,\omega) = -\phi(5\,\omega)$. Thus,

$$\Phi(\omega), \quad \Phi(\omega + 16), \quad \Phi(\omega + 2 \cdot 16), \quad \Phi(\omega + 3 \cdot 16), \quad \Phi(\omega + 4 \cdot 16) \tag{5.16}$$

were roots of the fifth-degree equation

$$\Phi^5 - 2^4 \cdot 5^3\phi^4(\omega)\,\psi^{16}(\omega)\Phi - 2^6\sqrt{5^5}\,\phi^3(\omega)\,\psi^{16}(\omega)(1 + \phi^8(\omega)) = 0. \tag{5.17}$$

Hermite set

$$\Phi = \sqrt[4]{2^4\,5^3}\,\phi(\omega)\psi^4(\omega)x$$

to obtain the simpler equation

$$x^5 - x - \frac{2}{\sqrt[4]{5^5}}\frac{1 + \phi^8(\omega)}{\phi^2(\omega)\psi^4(\omega)} = 0. \tag{5.18}$$

Here Hermite referred to a result of George Jerrard, dating from 1834, that a general fifth-degree equation could be reduced by means of algebraic transformations to an equation of the form

$$x^5 - x - a = 0. \tag{5.19}$$

As Felix Klein[10] pointed out, however, Erland Bring had already performed this reduction in 1786. It was then sufficient for Hermite to solve (5.19). He set

$$\frac{1 + \phi^8(\omega)}{\phi^2(\omega)\psi^4(\omega)} = \frac{2}{\sqrt[4]{5^5}}a \equiv A$$

and used (5.12) to obtain the fourth degree equation

$$k^4 + A^2k^3 + 2k^2 - A^2k + 1 = 0, \quad k(\omega) = \phi^4(\omega). \tag{5.20}$$

He noted that by setting $\frac{4}{A^2} = \sin\alpha$, the four solutions of (5.20) would be

$$k = \tan\frac{\alpha}{4}, \quad \tan\frac{\alpha + 2\pi}{4}, \quad \tan\frac{\pi - \alpha}{4}, \quad \tan\frac{3\pi - \alpha}{4}.$$

By choosing one of these solutions as the value of the module $k$, the corresponding value of $\omega$ gave Hermite the solution of the Bring-Jerrard equation in terms of modular

[10] Klein (1956) p. 157.

functions. The values of $x$ that solved (5.18) were thus the values in (5.16), each divided by

$$\sqrt[4]{2^4 \, 5^3} \, \phi(\omega) \, \psi^4(\omega).$$

In this manner, Hermite could solve a fifth-degree equation in terms of modular functions.[11]

Hermite required the transformation of $\theta$ functions in this work because $\phi(\omega)$ and $\psi(\omega)$ can be expressed in terms of ratios of $\theta$ functions, and he needed the values of $\phi\left(\frac{c+d\omega}{a+b\omega}\right)$. With $ad - bc = 1$, he considered six cases of the matrices $\begin{pmatrix} a & b \\ c & d \end{pmatrix}$(mod 2):

$$\begin{pmatrix} 1 & 0 \\ 0 & 1 \end{pmatrix}, \; \begin{pmatrix} 0 & 1 \\ 1 & 0 \end{pmatrix}, \; \begin{pmatrix} 1 & 1 \\ 0 & 1 \end{pmatrix}, \; \begin{pmatrix} 1 & 1 \\ 1 & 0 \end{pmatrix}, \; \begin{pmatrix} 1 & 0 \\ 1 & 1 \end{pmatrix}, \; \begin{pmatrix} 0 & 1 \\ 1 & 1 \end{pmatrix}; \qquad (5.21)$$

he found

$$\phi\left(\frac{c+d\omega}{a+b\omega}\right) = \phi(\omega)e^{\frac{i\pi}{8}(d(c+d)-1)} : \qquad \text{case 1} \qquad (5.22)$$

$$= \psi(\omega)e^{\frac{i\pi}{8}(c(c-d)-1)} : \qquad \text{case 2} \qquad (5.23)$$

$$= \frac{1}{\phi(\omega)}e^{\frac{i\pi}{8}(d(d-c)-1)} : \qquad \text{case 3} \qquad (5.24)$$

$$= \frac{1}{\psi(\omega)}e^{\frac{i\pi}{8}(c(c+d)-1)} : \qquad \text{case 4} \qquad (5.25)$$

$$= \frac{\phi(\omega)}{\psi(\omega)}e^{\frac{i\pi}{8}cd} : \qquad \text{case 5} \qquad (5.26)$$

$$= \frac{\psi(\omega)}{\phi(\omega)}e^{-\frac{i\pi}{8}cd} : \qquad \text{case 6.} \qquad (5.27)$$

Hermite did not present a derivation of these results in his paper of 1858. Early in the 1890s, when Jules Tannery and Jules Molk were writing their treatise on elliptic functions, Tannery asked Hermite about this derivation. Hermite wrote Tannery a letter[12] with details of his proof, employing mainly his transformation theorem. Tannery and Molk included this letter and also incorporated these details with commentary in their comprehensive work.[13]

Observe that case 1 consists of all modular transformations such that

$$\begin{pmatrix} a & b \\ c & d \end{pmatrix} \equiv \begin{pmatrix} 1 & 0 \\ 0 & 1 \end{pmatrix} \; (\text{mod } 2),$$

that is, the group $\Gamma(2)$. In this case

$$d(c+d) - 1 \equiv 0(\text{mod } 2) \qquad (5.28)$$

---

[11] Two recently published books present details of Hermite's solution of the fifth-degree equation: Armitage and Eberlein (2006), McKean and Moll (1997).

[12] Hermite (1905–1917) vol. 2, pp. 12–21.    [13] See Tannery and Molk (1972) vol. 4, pp. 282–303.

and this implies that

$$\phi^8 \left( \frac{c + d\omega}{a + b\omega} \right) = \phi^8(\omega) = k^2(\omega); \tag{5.29}$$

hence, $k^2$ is invariant under the action of $\Gamma(2)$. In 1828, the result given in (5.29) was stated without proof by Jacobi; Hermite may have missed this among the myriad results of Jacobi, since he did not refer to it in his paper. However, Hermite's respect for and indebtedness to Jacobi are reflected in his letter to Tannery concerning the material in formulas (5.2) through (5.4): "...elle repose en entier sur le hasard d'une formule de Jacobi, oubliée et comme perdue parmi tant de découvertes due à son génie."[14] ["...it rests entirely on a chance formula of Jacobi, overlooked and as good as lost among so many discoveries due to his genius."]

After Dedekind read this paper of Hermite, he verified the invariance of $k^2(\omega)$ under the action of $\Gamma(2)$ and this led him to create the theory of modular functions independent of the theory of elliptic functions.

Returning to the six cases discussed by Hermite, observe that $\Gamma(2)$ is of index six in $\Gamma = \Gamma(1)$. By taking the eighth power of each of the formulas (5.22) through (5.27), one obtains the six different values taken by $k^2$ under the action of $\Gamma(1)$:

$$k^2, \quad 1 - k^2, \quad \frac{1}{k^2}, \quad \frac{1}{1 - k^2}, \quad \frac{k^2}{k^2 - 1}, \quad \frac{k^2 - 1}{k^2}. \tag{5.30}$$

We may check the first four cases in the same way as (5.29). For the fifth and sixth cases, note that $cd \equiv 1 \pmod{2}$ and we arrive at

$$-\frac{\phi^8}{\psi^8} = -\frac{k^2}{1 - k^2} = \frac{k^2}{k^2 - 1} \quad \text{and} \quad -\frac{\psi^8}{\phi^8} = \frac{k^2 - 1}{k^2}. \tag{5.31}$$

Hermite discussed these six cases in his paper, "Sur la théorie des équations modulaires,"[15] where he considered the function

$$y = \frac{(k^4 - k^2 + 1)^3}{(k^4 - k^2)^2}, \tag{5.32}$$

observing that $y$ would remain unchanged when $k^2$ was replaced by any of the six functions in (5.30). This implies that $y$ is invariant under the action of $\Gamma(1)$. In fact, $y$ is related to Klein's $J$ invariant by the equation

$$y = \frac{27}{4} J = \frac{1}{2^8} j.$$

In this paper on modular equations, Hermite observed that by applying Jacobi's formula

$$\sqrt[4]{kk'} = \sqrt{2} q^{\frac{1}{8}} \frac{\sum_{n=-\infty}^{\infty} (-1)^n q^{2n^2 + n}}{\sum_{n=-\infty}^{\infty} q^{n^2}} \tag{5.33}$$

---

[14] Hermite (1905–1917) vol. 2, pp. 12–21, especially p. 21. Also see Tannery and Molk (1972) vol. 4, pp. 282–303.

[15] Hermite (1905–1917) vol. 2, pp. 38–82, especially pp. 58–61.

in (5.32), one obtained the result

$$j(\omega) = \frac{1}{q^2} \frac{(1 + 2^4 \cdot 3 \cdot 5q^2 + 2^4 \cdot 3^3 \cdot 5q^4 + 2^6 \cdot 3 \cdot 5 \cdot 7q^6 + \cdots)^3}{(1 - 3q^2 + 5q^6 - 6q^{10} + \cdots)^8}, \qquad (5.34)$$

where $q = e^{i\pi\omega}$. Note that Hermite did not employ the symbol $j$; he wrote it instead as $-2^8\alpha$, but used the symbol $\alpha$ to denote other similar quantities.

Hermite then wrote down the result of expanding the expression (5.34) as a series, to obtain his famous expansion for $j(\omega)$:

$$-j\left(\frac{1 + i\sqrt{\Delta}}{2}\right) = e^{\pi\sqrt{\Delta}} - 744 + \frac{196884}{e^{\pi\sqrt{\Delta}}} - \cdots . \qquad (5.35)$$

He calculated

$$e^{\pi\sqrt{43}} = 884736743.999777\ldots, \qquad (5.36)$$

also mentioning that $e^{\pi\sqrt{163}}$ had twelve 9s after the decimal point. In fact,

$$e^{\pi\sqrt{163}} = 262537412640768743.9999999999992\ldots. \qquad (5.37)$$

Hermite was able to deduce that

$$-j\left(\frac{1 + \sqrt{-43}}{2}\right) + 744 \quad \text{and} \quad -j\left(\frac{1 + \sqrt{-163}}{2}\right) + 744$$

were the integers closest to the values given in (5.36) and (5.37), respectively. Thus

$$-j\left(\frac{1 + \sqrt{-43}}{2}\right) = 884736000$$

$$= (960)^3$$

and

$$-j\left(\frac{1 + \sqrt{-163}}{2}\right) = 262537412640768000$$

$$= (640320)^3.$$

For a proof of (5.33), see exercises 5, 6, and 7 at the end of this chapter.

## 5.2   Hermite's Proof of the Transformation Formula

To begin our discussion of Hermite's proof, recall the definition of the $\theta$-function as we also gave in (5.5):

$$\theta_{\mu,\nu}(x) = \sum_{m=-\infty}^{\infty} (-1)^{m\nu} e^{i\pi((2m+\mu)x + \frac{\omega}{4}(2m+\mu)^2)} = \theta_{\mu\nu}(x, \omega), \quad \text{Im } \omega > 0.$$

Next, it is easy to check that[16]

$$\theta_{\mu,\nu}(x+1) = (-1)^{\mu}\,\theta_{\mu,\nu}(x), \tag{5.38}$$

$$\theta_{\mu,\nu}(x+\omega) = (-1)^{\nu}\,\theta_{\mu,\nu}(x)e^{-i\pi(2x+\omega)}, \tag{5.39}$$

$$\theta_{\mu,\nu}(-x) = (-1)^{\mu\nu}\,\theta_{\mu,\nu}(x), \tag{5.40}$$

$$\theta_{\mu+2,\nu}(x) = (-1)^{\nu}\,\theta_{\mu,\nu}(x), \tag{5.41}$$

$$\theta_{\mu,\nu+2}(x) = \theta_{\mu,\nu}(x), \tag{5.42}$$

$$\theta_{\mu+\mu',\nu+\nu'}(x) = \theta_{\mu',\nu'}\left(x + \frac{\mu\omega+\nu}{2}\right)e^{i\pi(\mu x + \frac{\mu^2\omega}{4} - \frac{\nu\mu'}{2})}. \tag{5.43}$$

Relations (5.41) and (5.42) show that $\theta_{\mu\nu}(x)$ produce essentially four functions and (5.40) shows that the case $\mu = 1$, $\nu = 1$ produces the only odd functions among those four. Now set

$$\Omega = \frac{c+d\omega}{a+b\omega}, \quad ad - bc = 1, \tag{5.44}$$

$$M = a\mu + b\nu + ab, \quad N = c\mu + d\nu + cd, \tag{5.45}$$

$$\Pi(x) = \theta_{\mu,\nu}\big((a+b\omega)x\big)e^{i\pi b(a+b\omega)x^2}. \tag{5.46}$$

Turning to the proof of (5.7), note that the case $b = 0$, contained in (5.10), is not difficult; consider that when $b \neq 0$, we can take $b > 0$, for if $b < 0$ we can simply replace $a, b, c, d$ in (5.44) by $-a, -b, -c, -d$. We check the validity of the two equations

$$\Pi(x+1) = (-1)^{M}\,\Pi(x), \tag{5.47}$$

$$\Pi(x+\Omega) = (-1)^{N}\,\Pi(x)e^{-i\pi(2x+\Omega)}. \tag{5.48}$$

Following Hermite, we can write[17]

$$\Pi(x) = \sum_{m=-\infty}^{\infty} e^{i\pi\phi(x,m)}, \tag{5.49}$$

where

$$\phi(x,m) = b(a+b\omega)x^2 + (2m+\mu)(a+b\omega)x + \frac{\omega}{4}(2m+\mu)^2 - m\nu. \tag{5.50}$$

It is easy to check that

$$\phi(x+1,m) - \phi(x,m+b) = 2am + M \equiv M \bmod 2; \tag{5.51}$$

this implies that

$$\Pi(x+1) = \sum_{m=-\infty}^{\infty} e^{i\pi\phi(x+1,m)} = (-1)^{M}\sum_{m=\infty}^{\infty} e^{i\pi\phi(x,m+b)}$$

$$= (-1)^{M}\sum_{m=-\infty}^{\infty} e^{i\pi\phi(x,m)},$$

---

[16] Hermite (1905–1917) vol. 1, p. 487.     [17] Hermite (1905–1917) vol. 1, pp. 491–495.

confirming (5.47). To check equation (5.48), first set

$$\Pi_0(x) = e^{i\pi \Omega x^2} \, \Pi(\Omega x) = \sum_{m=-\infty}^{\infty} e^{i\pi \phi_0(x,m)}, \qquad (5.52)$$

so that

$$\phi_0(x, m) = \Omega x^2 + \phi(\Omega x, m) \qquad (5.53)$$

or

$$\phi_0(x, m) = d(c + d\,\omega)x^2 + (2m + \mu)(c + d\,\omega)x + \frac{\omega}{4}(2m + \mu)^2 - m\nu. \qquad (5.54)$$

Comparing (5.54) with (5.50) shows that it is possible to move from $\phi(x, m)$ to $\phi_0(x, m)$ by changing $a$ and $b$ to $c$ and $d$, respectively. It follows that

$$\phi_0(x + 1, m) - \phi_0(x, m + d) = 2cm + M \equiv N \bmod 2,$$

and thus by (5.52)

$$\Pi_0(x + 1) = (-1)^N \, \Pi_0(x)$$

and also

$$e^{i\pi \Omega(x+1)^2} \Pi(\Omega x + \Omega) = (-1)^N \, e^{i\pi \Omega x^2} \Pi(x).$$

Canceling $e^{i\pi \Omega x^2}$ from each side and then replacing $\Omega x$ by $x$, we obtain (5.48). Now (5.47) and (5.48) lead to

$$\Pi(x) = T\theta_{M,N}(x, \Omega), \qquad (5.55)$$

where $T$ is a constant. Using (5.49), we may write (5.55) as

$$\sum_{m=-\infty}^{\infty} e^{i\pi \phi(x,m)} = T \sum_{m=-\infty}^{\infty} (-1)^{mN} e^{i\pi \left((2m+M)x + \frac{\Omega}{4}(2m+M)^2\right)}$$

or as

$$\sum_{m=-\infty}^{\infty} e^{i\pi \psi(x,m)} = T \sum_{m=-\infty}^{\infty} (-1)^{mN} e^{i\pi \left(2mx + \frac{\Omega}{4}(2m+M)^2\right)}, \qquad (5.56)$$

where $\psi(x, m) = \phi(x, m) - Mx$. By integrating each side of (5.56) with respect to $x$ from 0 to 1, we obtain

$$\int_0^1 \sum_{m=-\infty}^{\infty} e^{i\pi \psi(x,m)} \, dx = Te^{M^2 \frac{i\pi \Omega}{4}}. \qquad (5.57)$$

Now observe that

$$\psi(x + 1, m) \equiv \psi(x, m + b) \quad (\bmod 2),$$

or, more generally,

$$\psi(x + n, m) \equiv \psi(x, m + nb) \quad (\bmod 2), \qquad (5.58)$$

where $n$ is an arbitrary integer. From (5.58) it is clear that

$$\sum_{m=-\infty}^{\infty} e^{i\pi\,\psi(x,m)} = \sum_{\rho=0}^{b-1} \sum_{n=-\infty}^{\infty} e^{i\pi\,\psi(x,nb+\rho)}$$

$$= \sum_{\rho=0}^{b-1} \sum_{n=-\infty}^{\infty} e^{i\pi\,\psi(x+n,\rho)}. \tag{5.59}$$

Using term-by-term integration and making a change of variables, we may write

$$\int_0^1 \sum_{n=-\infty}^{\infty} e^{i\pi\,\psi(x+n,\rho)}\,dx = \sum_{n=-\infty}^{\infty} \int_n^{n+1} e^{i\pi\,\psi(x,\rho)}\,dx$$

$$= \int_{-\infty}^{\infty} e^{i\pi\,\psi(x,\rho)}\,dx. \tag{5.60}$$

At this point in his proof, Hermite invoked a result of Cauchy that can be proved by Cauchy's integral theorem in complex analysis:

If $\mathrm{Im}\,p > 0$, then

$$\int_{-\infty}^{\infty} e^{i\pi\,(px^2+qx+r)}\,dx = \frac{1}{\sqrt{-ip}}\,e^{i\pi\,\frac{4pr-q^2}{4p}}. \tag{5.61}$$

From (5.57), (5.60), and (5.61) it follows that

$$T = \frac{\delta \sum_{\rho=0}^{b-1} e^{i\pi\,\frac{a(\rho-\frac{1}{2}b)^2}{b}}}{\sqrt{-ib(a+b\omega)}}$$

where

$$\delta = e^{-\frac{i\pi}{4b}(a\mu^2-2\mu M+dM^2-ab^2)}. \tag{5.62}$$

By substituting the expression for $M$ from (5.45) into (5.62) and using the condition $ad - bc = 1$, we can see that this value of $\delta$ reduces to the one given in (5.9) and (5.7) is verified.

### 5.3 Smith on Jacobi's Formula for the Product of Four Theta Functions

The reader might wish to read the treatment of Hermite's work on transformation of theta functions by Henry Smith (1826–1883). Smith followed Hermite very closely, but also presented details, omitted by Hermite, for the general $n$th-order transformation. Unfortunately, Smith died in 1883, before his long monograph, "Memoir on theta and omega functions," could be fully completed or published, but this work is contained in his collected papers.[18] Smith began this memoir in 1874 at the request of Glaisher, who wanted it to serve as an introduction to his tables of theta functions.[19] Glaisher added missing diagrams to Smith's memoir, but the project of publishing the tables of theta functions accompanied by Smith's paper was apparently never realized.

---

[18] Smith, H. J. S. (1965) vol. 2, pp. 415–621; especially see pp. 460–481.  [19] Ibid. vol. 1, p. lxiv.

Smith also composed two long notes, amounting to about fifty pages each, on the theory of elliptic functions, published in Glaisher's *Messenger of Mathematics*. Smith's memoir and these notes constitute an insightful and original summary by a great mathematical mind of the fundamental ideas of elliptic and modular functions up to about 1882. Benjamin Jowett, the noted judge Lord Bowen, and others offered moving biographical notes and reminiscences to accompany Smith's collected papers, describing him as an extraordinarily gracious and modest person; about Smith, Glaisher himself wrote:[20]

> I have no words to express the admiration and affection with which I regarded him myself. As regards his qualities and abilities, if I had not known him as I did it would have seemed to me incredible that such varied gifts and powers could be combined in the same person. All the assistance that I have ever received with respect to the direction of my own work, or the manner of conducting research, came from him, and I have never ceased to miss his advice and help: and more and more with each succeeding year. It will be long indeed before his place in Mathematics can be held by another, but in the lives of those who were personally indebted to him the void can never be filled.

Smith employed Hermite's notation for the theta functions, to derive Jacobi's formula for the product of four theta functions. Jacobi presented this formula in his 1835–36 Königsberg lectures. Recall, however, that this formula was not published during Jacobi's lifetime, but appeared in print only in 1881 when Borchardt published Jacobi's lectures in the first volume of Jacobi's collected mathematical works. Much before its publication, Smith became aware of the existence of this product formula, because Jacobi mentioned it in an August 6, 1845 letter to Hermite, and then included this letter in the 1846 first volume of his *Opuscula Mathematica*.[21] Concerning his theory of theta functions, Jacobi wrote to Hermite:[22]

> In seeking to extract the transformation from the properties of the functions $\Theta$, without making use of their decomposition into infinite factors, you have wisely considered the most general case where one must probably resign oneself to the impossibility of factorization.
>
> In my university lectures at Koenigsberg, I too was accustomed to begin with $\Theta$ functions. In those lectures, in multiplying the four series
>
> $$\sum_{-\infty}^{\infty}{}_{i} e^{-(ax+bi)^2}$$
>
> by various values of $x$, and in transforming the exponents by the formula
>
> $$i^2 + i'^2 + i''^2 + i'''^2 = \left(\frac{i + i' + i'' + i'''}{2}\right)^2 + \left(\frac{i + i' - i'' - i'''}{2}\right)^2 + \text{etc.},$$
>
> I immediately obtained a formula from which follow, as particular cases and without any calculation, the fractional expressions of elliptic functions, three kinds of addition theorems, and hundreds more interesting formulas which one may otherwise obtain only by tedious algebraic calculations.

[20] Ibid. vol. 1, pp. lxi–xcv, especially p. lxxxvii.

[21] Jacobi (1846) vol. 1, pp. 357–362. Also Jacobi (1969) vol. 2, pp. 115–120.

[22] Jacobi (1846) vol. 1, p. 358 or Jacobi (1969) vol. 2, pp. 116–117.

On this account of Jacobi, Smith remarked in his 1866 paper entitled "On a formula for the multiplication of four Theta functions":[23]

> The formula to which this passage refers has not (it would seem) been given by any writer on elliptic functions. The object of the present paper is to enunciate and demonstrate it; and to justify what Jacobi says of it by showing that many of the fundamental formulae of the theory of elliptic functions are either particular cases of it, or corollaries from it. For the sake of symmetry, however, it is convenient to employ the arithmetic equality

$$i^2 + i'^2 + i''^2 + i'''^2 = \left(\frac{-i + i' + i'' + i'''}{2}\right)^2 + \left(\frac{i - i' + i'' + i'''}{2}\right)^2$$
$$+ \left(\frac{i + i' - i'' + i'''}{2}\right)^2 + \left(\frac{i + i' + i'' - i'''}{2}\right)^2,$$

instead of that indicated by Jacobi.

Smith wrote Hermite's $\theta_{\mu,\mu'}(x)$ in (5.5) in terms of $q = e^{i\pi\omega}$ and $v = e^{i\frac{\pi}{a}}$, with $a$ some constant:

$$\theta_{\mu,\mu'}(x) = \sum_{n=-\infty}^{\infty} (-1)^{n\mu'} q^{\frac{1}{4}(2n+\mu)^2} v^{(2n+\mu)x}; \tag{5.63}$$

he stated the formula for the product of four theta functions as:

$$2\theta_{\mu_1,\mu'_1}(x_1)\,\theta_{\mu_2,\mu'_2}(x_2)\,\theta_{\mu_3,\mu'_3}(x_3)\,\theta_{\mu_4,\mu'_4}(x_4)$$
$$= \prod_{i=1}^{4} \theta_{\sigma-\mu_i,\sigma'-\mu'_i}(s - x_i) + \prod_{i=1}^{4} \theta_{\sigma-\mu_i,\sigma'-\mu'_i+1}(s - x_i)$$
$$+ (-1)^{\sigma'} \prod_{i=1}^{4} \theta_{\sigma-\mu_i+1,\sigma'-\mu'_i}(s - x_i) + (-1)^{\sigma'+1} \prod_{i=1}^{4} \theta_{\sigma-\mu_i+1,\sigma'-\mu'_i+1}(s - x_i), \tag{5.64}$$

where

$$2s = \sum_{i=1}^{4} x_i, \quad 2\sigma = \sum_{i=1}^{4} \mu_i, \quad 2\sigma' = \sum_{i=1}^{4} \mu'_i, \tag{5.65}$$

with $\sigma$ and $\sigma'$ assumed to be integers. Smith started with the observation that, by using (5.65), he could easily verify the identities:

$$\sum_{i=1}^{4} \mu_i^2 = \sum_{i=1}^{4} (\sigma - \mu_i)^2, \tag{5.66}$$

$$\sum_{i=1}^{4} \mu_i\mu'_i = \sum_{i=1}^{4} (\sigma - \mu_i)(\sigma - \mu'_i). \tag{5.67}$$

---

[23] Smith (1965) vol. 1, pp. 443–454, especially pp. 443–444.

Denote the product of the four theta functions on the left-hand side of the equation (5.64) by $S$. Smith wrote the product in the form:

$$2S = 2 \sum (-1)^{\sum \mu_i' n_i} \, q^{\frac{1}{4} \sum (2n_i + \mu_i)^2} \, v^{\sum (2n_i + \mu_i) x_i}, \tag{5.68}$$

where the sum was over all integer values of $n_1, n_2, n_3, n_4$ from $-\infty$ to $\infty$ and the index $i$ in the exponents ran from 1 to 4. He then set

$$N_1 = -n_1 + n_2 + n_3 + n_4,$$
$$N_2 = n_1 - n_2 + n_3 + n_4,$$
$$N_3 = n_1 + n_2 - n_3 + n_4,$$
$$N_4 = n_1 + n_2 + n_3 - n_4,$$

so that

$$4n_1 = -N_1 + N_2 + N_3 + N_4$$
$$4n_2 = N_1 - N_2 + N_3 + N_4$$
$$4n_3 = N_1 + N_2 - N_3 + N_4$$
$$4n_4 = N_1 + N_2 + N_3 - N_4.$$

Smith next transformed the exponent of $q$ in (5.68) by (5.66) and the exponents of $-1$ and $v$ by (5.67) to obtain

$$2S = 2 \sum (-1)^{\frac{1}{2} \sum N_i (\sigma' - \mu_i')} \, q^{\frac{1}{4} \sum (N_i + \sigma - \mu_i)^2} \, v^{\sum (N_i + \sigma - \mu_i)(s - x_i)}, \tag{5.69}$$

where the summation extended over all integer values of $N_1, N_2, N_3, N_4$ from $-\infty$ to $\infty$ such that substituting these values of $N_i$ in the equations for $4n_i$, $i = 1, 2, 3, 4$, would produce integer values of $n_i$. To determine appropriate values of $N_i$, Smith noted that the difference between any two $N_i$ was always even; hence either all the $N_i$, $i = 1, 2, 3, 4$, must be even or all must be be odd. He considered the two cases:

- All $N_i$ were even, implying that $N_i = 2v_i$, $i = 1, \ldots, 4$, and that the $n_i$, $i = 1, \ldots, 4$, obtained from the equations for $4n_i$, would all be integers if and only if $v_1 + v_2 + v_3 + v_4$ was even;
- All $N_i$ were odd, implying that $N_i = 2v_i + 1$, $i = 1, \ldots, 4$, and that the $n_i$ would all be integers if and only if $v_1 + v_2 + v_3 + v_4$ was odd.

Thus, Smith separated the terms in (5.69) in which all the $N_i$ were even from the terms in which all the $N_i$ were odd. In this way, he obtained

$$2S = 2 \sum{}' (-1)^{\sum v_i(\sigma' - \mu_i')} \, q^{\frac{1}{4} \sum (2v_i + \sigma - \mu_i)^2} \, v^{\sum (2v_i + \sigma - \mu_i)(s - x_i)}$$
$$+ 2(-1)^{\sigma'} \sum{}'' (-1)^{\sum v_i(\sigma' - \mu_i')} \, q^{\frac{1}{4} \sum (2v_i + \sigma - \mu_i + 1)^2} \, v^{\sum (2v_i + \sigma - \mu_i + 1)(s - x_i)}, \tag{5.70}$$

where $\sum'$ and $\sum''$ refer to all values of $v_1, v_2, v_3, v_4$ from $-\infty$ to $\infty$ for which the sum $v_1 + v_2 + v_3 + v_4$ was, respectively, even and odd. Representing the general terms of

$\sum'$ and $\sum''$ by $P$ and $Q$, respectively, Smith next had

$$2\sum{}' P = \sum P + \sum (-1)^{\sum v_i} P, \quad 2\sum{}'' Q = \sum Q - \sum (-1)^{\sum v_i} Q, \quad (5.71)$$

where the sum $\sum$ extended over all integer values of $v_1, v_2, v_3, v_4$ from $-\infty$ to $\infty$. Substituting (5.71) in (5.70) finally granted Smith the required formula:

$$2S = \sum P + \sum (-1)^{\sum v_i} P + (-1)^{\sigma'} \sum Q + (-1)^{\sigma'+1} \sum (-1)^{\sum v_i} Q, \quad (5.72)$$

completing the proof of (5.64). Important special cases of formula (5.64) can be obtained by carefully choosing the four variables $x_1, x_2, x_3, x_4$ and the eight parameters $\left(\begin{smallmatrix} \mu_1, & \mu_2, & \mu_3, & \mu_4 \\ \mu'_1, & \mu'_3, & \mu'_3, & \mu'_4 \end{smallmatrix}\right)$. If we let $x_1 = x$, $x_2 = x$, $x_3 = 0$, $x_4 = 0$ and set the eight parameters successively as

$$\begin{pmatrix} 0, & 0, & 0, & 0 \\ 0, & 0, & 0, & 0 \end{pmatrix}, \quad \begin{pmatrix} 0, & 0, & 0, & 0 \\ 1, & 1, & 0, & 0 \end{pmatrix}, \quad \begin{pmatrix} 1, & 1, & 0, & 0 \\ 0, & 0, & 0, & 0 \end{pmatrix}, \quad \begin{pmatrix} 1, & 1, & 0, & 0 \\ 1, & 1, & 0, & 0 \end{pmatrix},$$

then, as Smith noted, we obtain the four equations

$$\theta_{0,0}^2(x)\,\theta_{0,0}^2(0) = \theta_{0,1}^2(x)\,\theta_{0,1}^2(0) + \theta_{1,0}^2(x)\,\theta_{1,0}^2(0) \quad (5.73)$$

$$\theta_{0,1}^2(x)\,\theta_{0,0}^2(0) = \theta_{0,0}^2(x)\,\theta_{0,1}^2(0) - \theta_{1,1}^2(x)\,\theta_{1,0}^2(0) \quad (5.74)$$

$$\theta_{1,0}^2(x)\,\theta_{0,0}^2(0) = \theta_{1,1}^2(x)\,\theta_{0,1}^2(0) + \theta_{0,0}^2(x)\,\theta_{1,0}^2(0) \quad (5.75)$$

$$\theta_{1,1}^2(x)\,\theta_{0,0}^2(0) = \theta_{1,0}^2(x)\,\theta_{0,1}^2(0) - \theta_{0,1}^2(x)\,\theta_{1,0}^2(0). \quad (5.76)$$

We have noted that a consequence of (5.41) and (5.42) is that there are essentially four different theta functions $\theta_{\mu,v}$, given by $\theta_{0,0}, \theta_{1,0}, \theta_{0,1}, \theta_{1,1}$. These can be identified with the Jacobi theta functions by means of the relations

$$\theta_{0,1}(x) = \theta_0(x), \quad \theta_{1,1}(x) = i\theta_1(x), \quad \theta_{1,0}(x) = \theta_2(x), \quad \theta_{0,0}(x) = \theta_3(x). \quad (5.77)$$

Now we take the variables $x_1 = x - y$, $x_2 = x + y$, $x_3 = 0$, $x_4 = 0$. Since there are four functions, there will be twelve products of different theta functions; six of these may be given by

$$\theta_{0,0}(x-y)\,\theta_{0,1}(x+y), \quad \theta_{0,0}(x-y)\,\theta_{1,0}(x+y), \quad \theta_{0,0}(x-y)\,\theta_{1,1}(x+y),$$
$$\theta_{0,1}(x-y)\,\theta_{1,0}(x+y), \quad \theta_{0,1}(x-y)\,\theta_{1,1}(x+y), \quad \theta_{1,0}(x-y)\,\theta_{1,1}(x+y). \quad (5.78)$$

The other six products can be obtained by changing $y$ to $-y$ in the products in (5.78). The formulas for the products in (5.78) are given in the exercises.

There are four products in which the two theta functions are the same:

$$\theta_{0,0}(x-y)\,\theta_{0,0}(x+y), \quad \theta_{0,1}(x-y)\,\theta_{0,1}(x+y),$$
$$\theta_{1,0}(x-y)\,\theta_{1,0}(x+y), \quad \theta_{1,1}(x-y)\,\theta_{1,1}(x+y). \quad (5.79)$$

Smith observed that each of the four products in (5.79) could be written in terms of squares of theta functions of $x$ and $y$ in six different ways. He wrote:[24]

$$\theta_{\mu,\mu'}(x-y)\,\theta_{\mu,\mu'}(x+y)\,\theta_{0,0}^2(0)$$
$$= \theta_{0,1}^2(y)\,\theta_{\mu,\mu'+1}^2(x) + (-1)^{\mu'}\,\theta_{1,0}^2(y)\,\theta_{\mu+1,\mu'}^2(x)$$
$$= \theta_{0,0}^2(y)\,\theta_{\mu,\mu'}^2(x) + (-1)^{\mu'}\,\theta_{1,1}^2(y)\,\theta_{\mu+1,\mu'+1}^2(x), \qquad (5.80)$$

$$\theta_{\mu,\mu'}(x-y)\,\theta_{\mu,\mu'}(x+y)\,\theta_{0,1}^2(0)$$
$$= \theta_{0,0}^2(y)\,\theta_{\mu,\mu'+1}^2(x) + (-1)^{\mu'}\,\theta_{1,0}^2(y)\,\theta_{\mu+1,\mu'+1}^2(x)$$
$$= \theta_{0,1}^2(y)\,\theta_{\mu,\mu'}^2(x) + (-1)^{\mu'}\,\theta_{1,1}^2(y)\,\theta_{\mu+1,\mu'}^2(x), \qquad (5.81)$$

$$\theta_{\mu,\mu'}(x-y)\,\theta_{\mu,\mu'}(x+y)\,\theta_{1,0}^2(0)$$
$$= (-1)^{\mu'}\theta_{0,0}^2(y)\,\theta_{\mu+1,\mu'}^2(x) - (-1)^{\mu'}\,\theta_{0,1}^2(y)\,\theta_{\mu+1,\mu'+1}^2(x)$$
$$= \theta_{1,0}^2(y)\,\theta_{\mu,\mu'}^2(x) - \theta_{1,1}^2(y)\,\theta_{\mu,\mu'+1}^2(x). \qquad (5.82)$$

To form these equations, we represent by $(A)$, $(A')$, $(B)$, $(B')$, $(C)$, $(C')$, the formulae obtained by attributing to the indices in (5.64) the values

$$\begin{pmatrix} \mu, & \mu, & 0, & 0 \\ \mu', & \mu', & 0 & 0 \end{pmatrix}, \ (A) \qquad \begin{pmatrix} \mu+1, & \mu+1, & 1, & 1 \\ \mu'+1, & \mu'+1, & 1 & 1 \end{pmatrix}, \ (A')$$

$$\begin{pmatrix} \mu, & \mu, & 0, & 0 \\ \mu', & \mu', & 1 & 1 \end{pmatrix}, \ (B) \qquad \begin{pmatrix} \mu+1, & \mu+1, & 1, & 1 \\ \mu', & \mu', & 1 & 1 \end{pmatrix}, \ (B')$$

$$\begin{pmatrix} \mu, & \mu, & 1, & 1 \\ \mu', & \mu', & 0 & 0 \end{pmatrix}, \ (C) \qquad \begin{pmatrix} \mu, & \mu, & 1, & 1 \\ \mu'+1, & \mu'+1, & 1 & 1 \end{pmatrix}; \ (C')$$

the indeterminates $x_1$, $x_2$, $x_3$, $x_4$ receiving the values $x-y$, $x+y$, $0$, $0$: the left-hand members of the three formulae $(A')$, $(B')$, $(C')$ contain the factor $\theta_{1,1}^2(0)$, and are therefore zero; and the equations $(A) \pm (A')$, $(B) \pm (B')$, $(C) \pm (C')$ will be found to coincide, respectively, with (5.80), (5.81), (5.82).

Jacobi's formulas on theta functions, mentioned without proofs in our Chapter 3, can now be derived as special cases of formulas (5.72) through (5.82). We now derive the formulas used in Chapter 3. To obtain (3.190), take $\mu = 0$, $\mu' = 0$, and $x = y$ in (5.80) and then use (5.77); to obtain (3.192), take $\mu = 1$, $\mu' = 0$ in (5.82); (3.194) and (3.196) are identical, respectively, with (5.74) and (5.76) by an application of (5.77).

As mentioned by Smith in a footnote, Hermite gave a particular case of (5.64) in his paper on the transformation of theta functions;[25] his formula, with $\alpha = \mu - \mu'$, $\beta = v' - v$ was stated as:

$$2\theta_{\mu,v}(x+y)\theta_{\mu',v'}(x-y)\,\theta_{\alpha,0}(0)\,\theta_{0,\beta}(0)$$
$$= \theta_{\mu,v}(x)\,\theta_{\mu',v'}(x)\,\theta_{\alpha,0}(y)\,\theta_{0,\beta}(y)$$
$$+ (-1)^{v}\theta_{\mu+1,v}(x)\,\theta_{\mu'+1,v'}(x)\,\theta_{\alpha+1,0}(y)\,\theta_{1,\beta}(y)$$
$$+ (-1)^{v}\theta_{\mu+1,v+1}(x)\,\theta_{\mu'+1,v'+1}(x)\,\theta_{\alpha+1,1}(y)\,\theta_{1,\beta+1}(y)$$
$$+ \theta_{\mu,v+1}(x)\,\theta_{\mu',v'+1}(x)\,\theta_{\alpha,1}(y)\,\theta_{0,\beta+1}(y). \qquad (5.83)$$

---

[24] Smith (1965) vol. 2, p. 449. Of course, Smith's equation numbers are different from those given here.
[25] Hermite (1905–1917) vol. 1, p. 488.

## 5.4 Exercises

1. Show that the third-order modular equation (3.42)

$$v^4 - u^4 + 2uv(u^2v^2 - 1) = 0$$

has solutions

$$v = -\phi(3\omega), \quad \phi(\frac{\omega}{3}), \quad \phi\left(\frac{\omega + 16}{3}\right), \quad \phi\left(\frac{\omega + 2 \cdot 16}{3}\right)$$

   when $u = \phi(\omega)$ is given by (5.12). See Hermite (1905–1917) vol. 2, p. 24.

2. Prove Hermite's formula (5.83).

3. Prove the following formulas for theta functions:

$$\theta_{0,0}(x - y)\,\theta_{0,1}(x + y)\,\theta_{0,0}(0)\,\theta_{0,1}(0)$$
$$= \theta_{0,0}(x)\,\theta_{0,1}(x)\,\theta_{0,0}(y)\,\theta_{0,1}(y) - \theta_{1,0}(x)\,\theta_{1,1}(x)\,\theta_{1,0}(y)\,\theta_{1,1}(y),$$
$$\theta_{0,0}(x - y)\,\theta_{1,0}(x + y)\,\theta_{0,0}(0)\,\theta_{1,0}(0)$$
$$= \theta_{0,0}(x)\,\theta_{1,0}(x)\,\theta_{0,0}(y)\,\theta_{1,0}(y) + \theta_{0,1}(x)\,\theta_{1,1}(x)\,\theta_{0,1}(y)\,\theta_{1,1}(y),$$
$$\theta_{0,0}(x - y)\,\theta_{1,1}(x + y)\,\theta_{0,1}(0)\,\theta_{1,0}(0)$$
$$= \theta_{1,0}(x)\,\theta_{0,1}(x)\,\theta_{0,0}(y)\,\theta_{1,1}(y) + \theta_{0,0}(x)\,\theta_{1,1}(x)\,\theta_{0,1}(y)\,\theta_{1,0}(y),$$
$$\theta_{0,1}(x - y)\,\theta_{1,0}(x + y)\,\theta_{0,1}(0)\,\theta_{1,0}(0)$$
$$= \theta_{0,1}(x)\,\theta_{1,0}(x)\,\theta_{0,1}(y)\,\theta_{1,0}(y) + \theta_{0,0}(x)\,\theta_{1,1}(x)\,\theta_{0,0}(y)\,\theta_{1,1}(y),$$
$$\theta_{0,1}(x - y)\,\theta_{1,1}(x + y)\,\theta_{0,0}(0)\,\theta_{1,0}(0)$$
$$= \theta_{0,1}(x)\,\theta_{1,1}(x)\,\theta_{0,0}(y)\,\theta_{1,0}(y) + \theta_{0,0}(x)\,\theta_{1,0}(x)\,\theta_{0,1}(y)\,\theta_{1,1}(y),$$
$$\theta_{1,0}(x - y)\,\theta_{1,1}(x + y)\,\theta_{0,0}(0)\,\theta_{0,1}(0)$$
$$= \theta_{1,0}(x)\,\theta_{1,1}(x)\,\theta_{0,0}(y)\,\theta_{0,1}(y) + \theta_{0,0}(x)\,\theta_{0,1}(y)\,\theta_{1,0}(y)\,\theta_{1,1}(y).$$

   According to Smith, Enrico Betti derived these formulas from Hermite's (5.83). See Smith (1965) vol. 1, pp. 448–449.

4. Differentiate the six formulas in the previous exercise with respect to $y$, and then set $y = 0$. For example, the result for the first formula is:

$$\theta_{0,0}'(x)\,\theta_{0,1}(x) - \theta_{0,0}(x)\,\theta_{0,1}'(x) = \frac{\theta_{1,1}'(0)\,\theta_{1,0}(0)}{\theta_{0,0}(0)\,\theta_{0,1}(0)}\,\theta_{1,0}(x)\,\theta_{1,1}(x).$$

   This exercise illustrates the fact that the derivative of the quotient of two theta functions, say $\frac{\theta_{0,0}(x)}{\theta_{0,1}(x)}$, can be expressed in terms of the theta functions themselves. Thus, this exercise yields the derivatives of sn, cn, and dn in terms of these elliptic functions. See Smith (1965) vol. 1, pp. 448 and 451.

5. Note that (3.160) gives us

$$\sqrt[4]{k'} = \prod_{n=1}^{\infty} \frac{1 - q^{2n-1}}{1 + q^{2n-1}}.$$

Now prove Jacobi's formulas

$$\sqrt[4]{k'} = \frac{\prod_{n=1}^{\infty}(1-q^n)}{\prod_{n=1}^{\infty}(1+q^{2n-1})(1-q^{2n})} = \frac{\sum_{m=-\infty}^{\infty}(-1)^m q^{\frac{m(3m+1)}{2}}}{\sum_{m=-\infty}^{\infty}(-1)^{\frac{m(m+1)}{2}} q^{\frac{m(3m+1)}{2}}},$$

$$= \frac{\prod_{n=1}^{\infty}(1-q^{2n-1})(1-q^{4n})}{\prod_{n=1}^{\infty}(1+q^{2n-1})(1-q^{4n})} = \frac{\sum_{m=-\infty}^{\infty}(-1)^m q^{m(2m+1)}}{\sum_{m=-\infty}^{\infty} q^{m(2m+1)}},$$

$$= \frac{\prod_{n=1}^{\infty}(1-q^{2n-1})^2(1-q^{2n})}{\prod_{n=1}^{\infty}(1-q^{4n-2})^2(1-q^{4n})} = \frac{\sum_{m=-\infty}^{\infty}(-1)^m q^{m^2}}{\sum_{m=-\infty}^{\infty}(-1)^m q^{2m^2}},$$

$$= \frac{\prod_{n=1}^{\infty}(1-q^{4n-2})^2(1-q^{4n})}{\prod_{n=1}^{\infty}(1+q^{2n-1})^2(1-q^{2n})} = \frac{\sum_{m=-\infty}^{\infty}(-1)^m q^{2m^2}}{\sum_{m=-\infty}^{\infty} q^{m^2}}.$$

The proofs of the series formulas on the right-hand side may require the triple product identity. See Jacobi (1969) vol. 2, pp. 234–235.

6. Observe that (3.161) produces

$$\sqrt[4]{k} = \sqrt{2}\, q^{\frac{1}{8}} \prod_{n=1}^{\infty} \frac{1+q^{2n}}{1+q^{2n-1}}.$$

Now prove Jacobi's formulas

$$2^{-\frac{1}{2}} q^{-\frac{1}{8}} \sqrt[4]{k} = \frac{\prod_{n=1}^{\infty}(1-q^{4n})}{\prod_{n=1}^{\infty}(1-q^{2n-1})(1-q^{2n})} = \frac{\sum_{m=-\infty}^{\infty}(-1)^m q^{2m(3m+1)}}{\sum_{m=-\infty}^{\infty}(-1)^{\frac{m(m+1)}{2}} q^{\frac{m(3m+1)}{2}}},$$

$$= \frac{\prod_{n=1}^{\infty}(1-q^{2n-1})(1-q^{8n})}{\prod_{n=1}^{\infty}(1-q^{4n-2})(1-q^{4n})} = \frac{\sum_{m=-\infty}^{\infty} q^{2m(2m+1)}}{\sum_{m=-\infty}^{\infty} q^{m(2m+1)}},$$

$$= \frac{\prod_{n=1}^{\infty}(1-q^{2n-1})(1-q^{8n})}{\prod_{n=1}^{\infty}(1-q^{4n-2})^2(1-q^{4n})} = \frac{\sum_{m=-\infty}^{\infty}(-1)^m q^{2m(2m+1)}}{\sum_{m=-\infty}^{\infty}(-1)^m q^{2m^2}},$$

$$= \frac{\prod_{n=1}^{\infty}(1+q^{2n-1})(1-q^{4n})}{\prod_{n=1}^{\infty}(1+q^{2n-1})^2(1-q^{2n})} = \frac{\sum_{m=-\infty}^{\infty} q^{m(2m+1)}}{\sum_{m=-\infty}^{\infty} q^{m^2}}.$$

See Jacobi (1969) vol. 2, pp. 235–236.

7. Deduce from the formulas in exercises 5 and 6 Jacobi's formula

$$\sqrt{kk'} = \frac{\sqrt{2}\, q^{\frac{1}{8}} \sum_{n=-\infty}^{\infty}(-1)^n q^{n(2n+1)}}{\sum_{n=-\infty}^{\infty} q^{n^2}}.$$

See Jacobi (1969) vol. 2, p. 283. Note that all the formulas mentioned in this exercise and exercises 5 and 6 were used by Hermite in his 1858 papers on the quintic and on the modular equation.

# 6

---

## *Complex Variables and Elliptic Functions*

### 6.1 Historical Remarks on the Roots of Unity

For its full development, the theory of modular functions requires the theory of functions of a complex variable; the necessary work in this area was worked out in the 1840s and 1850s by Cauchy, Weierstrass, and Riemann. In fact, Cauchy began developing his complex function theory in 1814. But even before that, British mathematicians Simpson (1710–1761) and Waring (1736–1796) had independently shown[1] that if $f(x) = \sum_{n=0}^{\infty} a_n x^n$, then one has the Simpson dissection of the series for $f(x)$:

$$\frac{1}{p} \sum_{k=0}^{p-1} f(e^{\frac{2\pi i k}{p}} x) = \sum_{n=0}^{\infty} a_{pn} x^{pn}, \tag{6.1}$$

and Euler had obtained the factorization

$$1 - x^p = \prod_{k=0}^{p-1} (1 - e^{\frac{2\pi i k}{p}} x). \tag{6.2}$$

It is instructive and interesting to view these results in the context of their gradual development. Newton (1642–1727) was the first person to consider the factorization of $1 \pm x^m$. He raised this question in connection with the integration of rational functions. In notes dating from 1676,[2] he presented a method for finding quadratic factors of $1 \pm x^m$. His method was to write

$$(1 + nx + x^2)(1 - nx + px^2 - qx^3 + rx^4 - \cdots \pm x^{m-2}) = 1 \pm x^m$$

and to equate the coefficients of the powers of $x$ to discover relations among $n, p, q, r \ldots$ and then, using elimination, find an algebraic equation satisfied by $n$. For example, when $m = 4$, Newton obtained the equations $p + 1 - n^2 = 0$ and $pn - n = 0$ by equating the coefficients of $x^2$ and $x^3$. Eliminating $p$ from these two equations gave him $2n - n^3 = 0$ so that $n = 0$ or $n = \pm\sqrt{2}$. The value $n = 0$ produced the factorization

---

[1] Waring reported that he submitted his results to the Royal Society, but the paper appears to have gotten lost. Simpson's work is contained in Simpson (1759).

[2] Newton (1967–1981) vol. 4, pp. 205–213.

149

$(1 + x^2)(1 - x^2) = 1 - x^4$; $n = \pm\sqrt{2}$ resulted in the factorization

$$(1 + \sqrt{2}x + x^2)(1 - \sqrt{2}x + x^2) = 1 + x^4. \tag{6.3}$$

By the same method, taking $m = 5$, he found the equation

$$n^4 - 3n^2 + 1 = 0 \text{ so that } n = \pm\frac{1 \pm \sqrt{5}}{2}.$$

The case $n = \frac{1\pm\sqrt{5}}{2}$ yielded the factorization

$$1 - x^5 = \left(1 + \frac{1 + \sqrt{5}}{2}x + x^2\right)\left(1 - \frac{1 + \sqrt{5}}{2}x + \frac{1 + \sqrt{5}}{2}x^2 - x^3\right).$$

Factoring this cubic, we obtain

$$(1 - x)\left(1 + \frac{1 - \sqrt{5}}{2}x + x^2\right),$$

although Newton's notes do not contain this explicitly. We know that Newton was aware of the existence of complex numbers, but he did not factorize the quadratic factors in terms of linear factors that involved complex numbers. Probably around 1666, he gave an extension of Descartes' rule of signs, from which he got a good lower bound on the number of complex roots of algebraic equations.[3] He called these impossible roots.

In fact, the Italian mathematician Gerolamo Cardano (1501–1576) had earlier encountered complex numbers in the course of solving quadratic and cubic algebraic equations.[4] He solved the quadratic $x^2 - 10x + 40 = 0$ to obtain $x = 5 \pm \sqrt{-15}$, and then verified that the product of $5 + \sqrt{-15}$ and $5 - \sqrt{-15}$ was 40. However, Cardano regarded the use of such numbers as meaningless sophistication. For the cubic $x^3 = px + q$, his solution, here given in more modern notation, was:

$$x = \sqrt[3]{\frac{q + \sqrt{q^2 - (\frac{4p^3}{27})}}{2}} + \sqrt[3]{\frac{q - \sqrt{q^2 - (\frac{4p^3}{27})}}{2}}. \tag{6.4}$$

Cardano did not discuss complex numbers in connection with the cubic, but his successor Rafael Bombelli (1526–1572) was much more willing to work with complex numbers. He introduced them in the first part of his 1572 book, *L'Algebra*, in which he outlined rules for addition, subtraction, and multiplication of complex numbers. He also gave many examples of computations involving complex numbers. In the second part of his book, he presented applications of his rules; for example, he solved the equation $x^3 = 15x + 4$. Clearly, Cardano's formula (6.4) gave the solution as

$$x = \sqrt[3]{2 + \sqrt{-121}} + \sqrt[3]{2 - \sqrt{-121}}; \tag{6.5}$$

---

[3] Newton (1967–1981) vol. 1, pp. 520–531. Also see Whiteside's footnotes on these pages.
[4] Cardano (1993). See also Smith (1959) vol. 1, pp. 201–202.

Bombelli wrote this value (in modern notation) as $x = 2 + i + 2 - i = 4$. Bombelli's rules gave the calculation:

$$(2 + \sqrt{-1})^3 = 2 + 11\sqrt{-1} = 2 + \sqrt{-121} \qquad (6.6)$$

and

$$(2 - \sqrt{-1})^3 = 2 - 11\sqrt{-1} = 2 - \sqrt{-121}, \qquad (6.7)$$

concluding that (6.5) reduced to

$$x = 2 + \sqrt{-1} + 2 - \sqrt{-1} = 4.$$

Complex numbers were also treated by some mathematicians of the seventeenth century. John Wallis (1616–1703) devoted a chapter of his 1675 work on algebra to complex numbers, where he defined $\sqrt{-bc}$ as signifying a mean proportional between $+b$ and $-c$ or between $-b$ and $+c$. However, as late as 1702, Leibniz (1646–1716) was apparently unable to deal with $\sqrt{\sqrt{-1}}$.[5] In that year, he published a paper on the integration of rational functions, in which he factorized $x^2 + a^2 = (x + a\sqrt{-1})(x - a\sqrt{-1})$ and

$$x^4 + a^4 = \left(x + a\sqrt{\sqrt{-1}}\right)\left(x - a\sqrt{\sqrt{-1}}\right)\left(x + a\sqrt{-\sqrt{-1}}\right)\left(x - a\sqrt{-\sqrt{-1}}\right).$$
$$(6.8)$$

He then wondered whether $\int \frac{1}{x^4 + a^4}$ and $\int \frac{1}{x^8 + a^8}$ could be calculated in terms of the logarithmic and arctangent functions. Had he realized that $\sqrt{\sqrt{-1}}$ was a complex number of the form $c + id$, he might have tried

$$(c \pm d\sqrt{-1})^2 = \pm\sqrt{-1}$$

to get

$$c^2 - d^2 = 0 \ \text{ and } \ 2cd = 1 \quad \text{or} \quad c = \frac{\sqrt{2}}{2} \ ; \ d = \frac{\sqrt{2}}{2},$$

and this would have given Leibniz the factorization (6.3) obtained by Newton. We note in passing that in 1676 Newton applied (6.3) to obtain the formula

$$\int_0^x \frac{1 + t^2}{1 + t^4}\, dt = \frac{1}{\sqrt{2}} \arctan \frac{x\sqrt{2}}{1 - x^2}. \qquad (6.9)$$

By expressing the integral in (6.9) as a series, he then found that

$$\frac{\pi}{2\sqrt{2}} = 1 + \frac{1}{3} - \frac{1}{5} - \frac{1}{7} + \frac{1}{9} + \frac{1}{11} - \cdots ,[6]$$

---

[5] Leibniz (1971) vol. 5, pp. 350–361.     [6] Newton (1959–60) vol. 2, pp. 138.

and he offered this to Leibniz, as a contrast to Leibniz's own formula of 1673 (obtainable by integrating the simpler function $\frac{1}{1+x^2}$):[7]

$$\frac{\pi}{4} = 1 - \frac{1}{3} + \frac{1}{5} - \frac{1}{7} + \frac{1}{9} - \frac{1}{11} + \cdots .$$

In 1702, Johann Bernoulli (1667–1748) published a paper in the journal of the French Academy in which he made use of complex numbers to draw a connection between the arctangent and the logarithmic function.[8] This paper also treated the integration of rational functions and Bernoulli observed at the end of the paper that

$$\int \frac{dx}{a^2 + x^2} = \frac{1}{2a} \left( \int \frac{dx}{a + ix} + \int \frac{dx}{a - ix} \right). \tag{6.10}$$

Clearly, (6.10) implies that

$$\frac{1}{a} \arctan \frac{x}{a} = \frac{1}{2ia} \log \frac{a + ix}{a - ix},$$

though Bernoulli did give this explicitly. If we then set $x = a \tan \theta$, we have

$$2i\theta = \log \frac{\cos \theta + i \sin \theta}{\cos \theta - i \sin \theta}$$

$$= \log(\cos \theta + i \sin \theta)^2$$

$$= 2 \log(\cos \theta + i \sin \theta)$$

or

$$\log(\cos \theta + i \sin \theta) = i\theta. \tag{6.11}$$

Roger Cotes (1682–1716) discovered (6.11),[9] perhaps in a manner similar to that presented here. Bernoulli, on the other hand, apparently made no further use of his discovery of (6.10). In fact, when Euler mentioned to Bernoulli that his formula (6.10) implied that the logarithm of $-1$ was imaginary, Bernoulli could not accept it and maintained that it must be 0, because $2 \log(-1) = \log(-1)^2 = \log 1 = 0$. Euler fully clarified this point in a 1740 paper[10] in which he formulated the idea that the logarithm was an infinitely many-valued function.

The foundation for solving the problem of finding roots of unity was finally laid by Abraham de Moivre (1667–1754); in 1707 he stated a formula[11] implying that

$$\sin \theta = \pm \frac{1}{2} \left( \sqrt[n]{\sin n\theta + i \cos n\theta} + \sqrt[n]{\sin n\theta - i \cos n\theta} \right), \tag{6.12}$$

where $n$ is an odd integer. Perhaps on the basis of this formula, Roger Cotes succeeded in factorizing $x^n \pm 1$. In 1716, Cotes wrote to the noted mathematical correspondent and preserver of books and manuscripts, William Jones, that he had solved the problem of

[7] Leibniz gave an account of this work in his 1712 paper: "Historia et origo calculi differentialis." Leibniz (1920) provides an English translation of this paper by J. Child on pp. 22–57; especially see p. 45.
[8] Johann Bernoulli (1968) vol. 1, pp. 393–400.     [9] Cotes (1722) p. 28.
[10] Eu.I-19, pp. 417–438.     [11] de Moivre (1707) vol. 25, pp. 2368–2371.

integrating functions of the form $\frac{1}{x^n \pm 1}$, a problem raised by Leibniz in his 1702 paper. Recall that this took place during the bitter calculus priority dispute and, as a supporter of Newton, Cotes took the opportunity to sneer at Leibniz:[12]

> M. Leibniz, in the Leipsic Acts of 1702 p. 218 and 219, has very rashly undertaken to demonstrate that the fluent of $1/(x^4 + a^4)$ cannot be expressed by measures of ratios and angles [in terms of the logarithm and the arctangent]; and he swaggers upon the occasion (according to his usual vanity), as having by this demonstration determined a question of the greatest moment.

Tragically for mathematics, two months later, Cotes was dead. However, Robert Smith, editor of Cotes's mathematical papers, found the factorization of $x^n \pm 1$ among the unpublished manuscripts; in 1722, Smith published this result in the collected papers of Cotes, titled *Harmonia Mensurarum*. As stated by Smith,[13] the result is a property of the circle corresponding to the factorizations:

$$x^n - 1 = (x^2 - 1) \prod_{k=1}^{m-1} \left( x^2 - 2x \cos \frac{2k\pi}{n} + 1 \right), \quad n = 2m,$$

$$= (x - 1) \prod_{k=1}^{m} \left( x^2 - 2x \cos \frac{2k\pi}{n} + 1 \right), \quad n = 2m + 1, \tag{6.13}$$

and

$$x^n + 1 = \prod_{k=1}^{m} \left( x^2 - 2x \cos \frac{(2k-1)\pi}{n} + 1 \right), \quad n = 2m,$$

$$= (x + 1) \prod_{k=1}^{m} \left( x^2 - 2x \cos \frac{(2k-1)\pi}{n} + 1 \right), \quad n = 2m + 1. \tag{6.14}$$

In 1730, de Moivre presented the method for applying (6.12) to the derivation of a generalization of (6.13) and (6.14). He began with a formula stated without proof:[14] If $l$ and $x$ are cosines of arcs $A$ and $B$ of the unit circle where $A$ is to $B$ as the integer $n$ to one, then

$$x = \frac{1}{2} \left( \sqrt[n]{l + \sqrt{l^2 - 1}} + \frac{1}{\sqrt[n]{l + \sqrt{l^2 - 1}}} \right). \tag{6.15}$$

As a corollary, he noted that if he set $z = \sqrt[n]{l + \sqrt{l^2 - 1}}$, then

$$z^n - l = \sqrt{l^2 - 1} \quad \text{or} \quad z^{2n} - 2lz^n + l^2 = l^2 - 1 \quad \text{or} \quad z^{2n} - 2lz^n + 1 = 0.$$

Moreover, using (6.15), de Moivre concluded that

$$x = \frac{1}{2} \left( z + \frac{1}{z} \right) \quad \text{or} \quad z^2 - 2xz + 1 = 0.$$

---

[12] See Rigaud (1841) vol. 1, p. 271.    [13] Gowing (1983) pp. 68–78.
[14] de Moivre (1730) pp. 1–2. For a translation into English, see Smith (1959) vol. 2, p. 446.

This meant that if $z^2 - 2xz + 1 = 0$, then $z^{2n} - 2lz^n + 1 = 0$; in other words, $z^2 - 2xz + 1$ was a factor of $z^{2n} - 2lz^n + 1$. To obtain the remaining $n - 1$ factors of $z^{2n} - 2lz^n + 1$, de Moivre noted another corollary:

> And in particular if $z$ be eliminated between the equations $1 \mp 2lz^n + z^{2n} = 0$, $1 - 2xz + z^2 = 0$ there will arise a new equation expressing a relation between the cosine of the arc $A$ (less or greater than the quadrant according as $l$ has the negative or positive sign) and all the cosines of the arcs $\frac{A}{n}$, $\frac{C-A}{n}$, $\frac{C+A}{n}$, $\frac{2C-A}{n}$, $\frac{2C+A}{n}$, $\frac{3C-A}{n}$, $\frac{3C+A}{n}$, etc. in which series of arcs $C$ denotes the entire circumference.

Thus we see that solving for $z$ in

$$1 - 2lz^n + z^{2n} = 0$$

gives

$$z = \sqrt[n]{l \pm \sqrt{l^2 - 1}}; \tag{6.16}$$

solving for $z$ in

$$1 - 2xz + z^2 = 0$$

gives

$$z = x \pm \sqrt{x^2 - 1}, \tag{6.17}$$

where $x = \cos\frac{A}{n}$, $\cos\frac{2\pi + A}{n}$, $\cdots$. Therefore, by eliminating $z$ from the two equations (6.16) and (6.17), we arrive at de Moivre's relation

$$(\cos A \pm i \sin A)^{\frac{1}{n}} = \cos\frac{2k\pi \pm A}{n} + i \sin\frac{2k\pi \pm A}{n}. \tag{6.18}$$

This result produced de Moivre's generalization of Cotes's factorizations:

$$z^{2n} - (2\cos A)z^n + 1 = \prod_{k=0}^{n-1}\left(z^2 - 2\cos\frac{2k\pi + A}{n} + 1\right). \tag{6.19}$$

In a paper of 1728, Daniel Bernoulli provided a proof of (6.15) by solving a second-order difference equation.[15] This paper contained the first use of the method of substituting $a_x = \lambda^x$ in the difference equation (with constant coefficient)

$$a_{n+k} = \alpha_1 a_{n+k-1} + \alpha_2 a_{n+k-2} + \cdots + \alpha_k a_n,$$

then solving the resulting algebraic equation for the values of $\lambda$, and finally taking a linear combination of the solutions $\lambda_i{}^x$, $i = 1, 2, \ldots, n$. Note that a modification was required if the equation for $\lambda$ contained repeated roots. To prove (6.15), Bernoulli first

---

[15] Daniel Bernoulli (1982–1996) vol. 2, pp. 49–64. See also Bottazzini's discussion of Bernoulli's early mathematical work in vol. 1, pp. 133–189.

considered the sequence $1, \cos A, \cos 2A, \cos 3A, \ldots$ . Letting $a_n = \cos nA$, the addition formula for the cosine function gave him the difference equation

$$a_{n+1} + a_{n-1} = (2\cos A)a_n,$$

and this led him to the algebraic equation

$$\lambda^2 - (2\cos A)\lambda + 1 = 0$$

with solutions $\lambda_1 = \cos A + \sqrt{\cos^2 A - 1}$ and $\lambda_2 = \cos A - \sqrt{\cos^2 A - 1}$. Next, by using the initial values $1$ and $\cos A$, he obtained the required result in a slightly different form:

$$a_n = \cos nA = \frac{1}{2}(\lambda_1^n + \lambda_2^n) = \frac{1}{2}\left((\cos A + i\sin A)^n + (\cos A - i\sin A)^n\right).$$

The work of Cotes and de Moivre gives the $n$th roots of unity as

$$\cos\frac{2k\pi}{n} + i\sin\frac{2k\pi}{n}, \quad \text{for } k = 0, 1, 2, \ldots, n-1.$$

Euler later expressed this as $e^{\frac{2k\pi i}{n}}$. Now in his *Harmonia Mensurarum* Cotes had the relation[16] $\log(\cos\theta + i\sin\theta) = i\theta$ and this implies

$$\log\left(\cos\frac{2k\pi}{n} + i\sin\frac{2k\pi}{n}\right) = \frac{i2k\pi}{n}. \tag{6.20}$$

Observe the contrast between (6.20) and Euler's expression

$$e^{\frac{2k\pi i}{n}} = \cos\frac{2k\pi}{n} + i\sin\frac{2k\pi}{n}.$$

To understand this difference of approach, we note that in the seventeenth century, the objects of interest to mathematicians were curves, as opposed to functions, and these curves were defined in terms of the relation between the variables. Inverting the relation between the variables would result in the identical curve. The transition from the study of curves to the modern study of functions was initiated by Johann Bernoulli and executed by his illustrious student, Euler. Euler emphasized the efficacy of taking the functional point of view in his famous book of 1748, *Introductio in Analysin Infinitorum*. He there defined the functions $\log x$ and $e^x$ analytically; although they defined the same essential curve as inverses of one another, he regarded them as distinct functions. It appears that Euler was the first to employ the symbol $e$ and define it by the series[17]

$$e = 1 + 1 + \frac{1}{2!} + \frac{1}{3!} + \cdots .$$

Thus, we may conclude that the factorization (6.2) is in essence due to Cotes, though its form is attributable to Euler. Formula (6.1) is also due to British mathematicians, Simpson and Waring. However, as Waring pointed out, the germ of the idea for a proof of (6.1) originated with de Moivre, who incidentally emigrated to England from France due to religious persecution. In 1717, de Moivre became the first to publish the method

---

[16] The original of Cotes appears to have a typographical error, as others have also pointed out.

[17] Euler (1988) p. 97. See also Smith (1959) vol. 1, pp. 95–98.

of using a generating function to solve a difference equation with constant coefficient. He named the generating function a recurrent series and showed that such a series could be represented by a rational function.

In fact, the series studied by de Moivre were all recurrent, that is, series whose coefficients $a_0, a_1, a_2, \ldots$ satisfied a recurrence relation, or difference equation

$$a_n + \alpha_1 a_{n-1} + \alpha_2 a_{n-2} + \cdots + \alpha_k a_{n-k} = 0, \tag{6.21}$$

for $n = k, k+1, k+2, \ldots$. He showed[18] that the recurrent series whose coefficients satisfied (6.21) would have to be a rational function with denominator

$$1 + \alpha_1 x + \alpha_2 x^2 + \cdots + \alpha_k x^k.$$

Let us work out the particular case $k = 2$ since the general case can be tackled in an analogous manner. Let $S$ denote the generating function (i.e., recurrent series) for $a_0, a_1, a_2, \ldots$. Clearly,

$$S = a_0 + a_1 x + a_2 x^2 + a_3 x^3 + \cdots, \tag{6.22}$$

$$\alpha_1 x S = \alpha_1 a_0 x + \alpha_1 a_1 x^2 + \alpha_1 a_2 x^3 + \cdots,$$

and

$$\alpha_2 x^2 S = \alpha_2 a_0 x^2 + \alpha_2 a_1 x^3 + \alpha_2 a_2 x^4 + \cdots.$$

Upon adding these three equations, we arrive at

$$(1 + \alpha_1 x + \alpha_2 x^2) S$$
$$= a_0 + (a_1 + \alpha_1 a_0) x + (a_2 + \alpha a_1 + a_0) x^2 + (a_3 + \alpha_1 a_2 + \alpha_2 a_1) x^3 + \cdots. \tag{6.23}$$

Now note that the recurrence (6.21) with $k = 2$ gives

$$a_n + \alpha_1 a_{n-1} + \alpha_2 a_{n-2} = 0 \tag{6.24}$$

for $n = 2, 3, 4, \ldots$. This implies that the coefficients of $x^2, x^3, x^4, \ldots$ in the series on the right-hand side of (6.23) are all zero. Therefore, (6.23) is reduced to

$$(1 + \alpha_1 x + \alpha_2 x^2) S = a_0 + (a_1 + \alpha_1 a_0) x$$

and $S$ may be expressed by the rational function

$$S = \frac{a_0 + (a_1 + \alpha_1 a_0) x}{1 + \alpha_1 x + \alpha_2 x^2}. \tag{6.25}$$

De Moivre next showed how to sum the two series

$$a_0 + a_2 x^2 + a_4 x^4 + \cdots \quad \text{and} \quad a_1 x + a_3 x^3 + a_5 x^5 + \cdots \tag{6.26}$$

when the sum of the recurrent series (6.22) was known. He gave details for the cases of the recurrence relation of second order and of third order. Suppose the recurrence

---

[18] de Moivre (1967) pp. 220–229.

relation is of the third order, given by

$$a_n - \alpha_1 a_{n-1} + \alpha_2 a_{n-2} - \alpha_3 a_{n-3} = 0.$$

In this case, the rational function representing $S$ in (6.22) would have denominator given by $1 - \alpha_1 x + \alpha_2 x^2 - \alpha_3 x^3$. De Moivre noted that the denominators of the two series (6.26) could be obtained by eliminating $x$ from the two equations

$$1 - \alpha_1 x + \alpha_2 x^2 - \alpha_3 x^3 = 0 \text{ and } x^2 = z. \tag{6.27}$$

The second equation yielded $x = \sqrt{z}$ and, substituting in the first equation, he obtained

$$1 + \alpha_2 z = \sqrt{z}(\alpha_1 + \alpha_3 z)$$

or

$$(1 + \alpha_2 z)^2 = z(\alpha_1 + \alpha_3 z)^2$$

or

$$1 + (2\alpha_2 - \alpha_1^2)x^2 + (\alpha_2^2 - 2\alpha_1\alpha_3)x^4 - \alpha_3^2 x^6 = 0. \tag{6.28}$$

The polynomial on the left-hand side of (6.28) was thus shown to be the denominator of the sum of the two series in (6.26). To determine the numerator, note that the recurrence relation satisfied by the coefficients of the two recurrent series in (6.26) would be given by

$$b_n + (2\alpha_2 - \alpha_1^2)b_{n-1} + (\alpha_2^2 - 2\alpha_1\alpha_3)b_{n-2} - \alpha_3^2 b_{n-3} = 0. \tag{6.29}$$

By applying the recurrence relation (6.29), one may finally determine the numerators of the series in (6.26). For the general case, where the recurrence relation was of order $k$ and where he wished to sum the $m$ series

$$a_j x^j + a_{m+j} x^{m+j} + a_{2m+j} x^{2m+j} + \cdots, \quad j = 0, 1, \ldots, m - 1, \tag{6.30}$$

de Moivre noted that the denominator for the sum of subseries in (6.30) could be obtained by eliminating $x$ from the two equations

$$1 + \alpha_1 x + \alpha_2 x^2 + \cdots + \alpha_k x^k = 0 \quad \text{and} \quad x^m = z. \tag{6.31}$$

We observe that de Moivre made implicit use of the roots of unity in the equation $x^m = z$.

In 1758, Simpson published a paper with the detailed title, "The invention of a general method for determining the sum of every 2d, 3d,4th, or 5th, &c. term of a series, taken in order; the sum of the whole series being known."[19] He considered the problem of representing the partitioned series (6.1) in terms of the function $f(x)$; recall that $f(x)$ was given by

$$f(x) = a_0 + a_1 x + a_2 x^2 + a_3 x^3 + \cdots. \tag{6.32}$$

[19] Simpson (1759).

Simpson illustrated his method by partitioning a series (6.32) into three parts where one of these parts was the trisection of $f(x)$

$$a_0 + a_3 x^3 + a_6 x^6 + a_9 x^9 + \cdots . \tag{6.33}$$

From (6.32), he had for some quantities $p, q, r$:

$$\frac{1}{3} f(px) = \frac{1}{3}(a_0 + a_1 px + a_2\, p^2 x^2 + a_3\, p^3 x^3 + \cdots)$$

$$\frac{1}{3} f(qx) = \frac{1}{3}(a_0 + a_1 qx + a_2\, q^2 x^2 + a_3\, q^3 x^3 + \cdots)$$

$$\frac{1}{3} f(rx) = \frac{1}{3}(a_0 + a_1 rx + a_2\, r^2 x^2 + a_3\, r^3 x^3 + \cdots).$$

Simpson wrote that

$$\frac{1}{3}(f(px) + f(qx) + f(rx)) = a_0 + a_3 x^3 + a_6 x^6 + a_9 x^9 + \cdots \tag{6.34}$$

if

$$p + q + r = 0$$
$$p^2 + q^2 + r^2 = 0$$
$$p^3 + q^3 + r^3 = 3$$
$$p^4 + q^4 + r^4 = 0, \text{ etc.}$$

by which he meant

$$p^k + q^k + r^k = 0 \text{ when } 3 \nmid k \text{ and } = 3 \text{ when } 3 \mid k \tag{6.35}$$

He then noted that the infinite number of relations in (6.35) would hold when $p, q, r$ were cube roots of unity, that is when $p, q, r$ were the three roots of the equation $z^3 - 1 = 0$. He reasoned that $-(p + q + r)$ would be the coefficient of $z^2$ in the polynomial $z^3 - 1$ and would therefore be zero. The coefficient of $z$ would be $pq + qr + rp$, also zero. Thus

$$p^2 + q^2 + r^2 = (p + q + r)^2 - 2(pq + qr + rp) = 0.$$

Next, Simpson argued that, since $p^3 = 1$, $q^3 = 1$, $r^3 = 1$, it must follow that $p^4 = p$, $q^4 = q$, and $r^4 = r$; thus, all the relations (6.35) were verified, and formula (6.34) followed. Simpson also considered the case for which the given series was partitioned into three parts, and then wrote about the general case:

> And, by the very same reasoning, and the process above laid down, it is evident, that, if every $n$th term (instead of every third term) of the given series be taken, the values $p, q, r, s,$ &c. will be the roots of the equation $z^n - 1 = 0$; and that, the sum of all the terms so taken, will be truly obtained by substituting $px, qx, rx, sx,$ &c. successively for $x$, in the given value of $S$ [the sum of the series], and then dividing the sum of all the quantities thence arising from the given number $n$.

Simpson also observed that the series

$$a_1 x + a_4\, x^4 + a_7\, x^7 + \cdots$$

could be expressed in terms of $f(x)$ by applying the method for determining (6.34) to the series

$$\frac{S - a_0}{x} = a_1 + a_2 x + a_3 x^2 + a_4 x^3 + \cdots .$$

Waring also made contributions to this problem; he reported that he communicated a statement of his result to the Royal Society in 1757. His result was first published in the first edition of his *Miscellanea Analytica* of 1762. Later editions were divided into two parts with various extensions and additions and, unfortunately, a good many confusing typographical errors; like Waring's other works, this was plagued by a very obscure mode of expression. The portion of this work devoted to the partition of series was included in his *Meditationes Algebraicae* of 1770 and then 1782. Waring called these the second and third editions, despite the fact that they had been renamed. In his 1782 edition of the *Meditationes Algebraicae*, Waring presented the result only in the preface. Although Simpson's results on this question were published in 1758 and Waring's in 1762, Waring maintained that Simpson had appropriated the idea of partitioning series from him, though Waring acknowledged that de Moivre had earlier used this idea for the case of recurrent series. The English translation of the 1782 edition of his *Meditationes* expresses Waring's claim:[20]

De Moivre gives a method of generating the sum of a series of terms that are equal, or alternating, or cyclic over an interval of distance two, three, or more, $a + bx + cx^2 + \cdots$ through division of unity by a rational multinomial expression $p + qx + rx^2 + \cdots$. In 1757 I sent the first version of this work to the Royal Society of London, which Simpson read, then in 1758 he inserted in the *Philosophical Transactions* a short piece containing a rule that was in the work I submitted, *viz* let $S$ be a given function of the quantity $x$, which is expanded into a series proceeding according to the dimensions of $x$, say $a + bx + cx^2 + dx^3 + \cdots$; in $S$ now substitute for $x$ respectively $\alpha x, \beta x, \gamma x, \delta x, \ldots$ where $\alpha, \beta, \gamma, \delta, \ldots$ are roots of the equation $x^n - 1 = 0$, resulting in a total of $n$ quantities $A, B, C, D$, etc., then $\frac{A+B+C+D+\cdots}{n}$ will be the sum of the first term and of those whose position is respectively $n, 2n, 3n$, etc. beyond the first term. Nevertheless at the end of his paper he says of the series $a + bx + cx^2 + \cdots$ that he has given a solution of an example of this problem by a method which differs a little from another one where a general method is indicated. But I say that no one, before my submission to the Royal Society in 1757, had ever claimed to have devised a general method, and it was my notes that Simpson had read, in which the above method was contained.

It is difficult to judge whether Waring's accusation has merit. Simpson was a very able mathematician and he was certainly capable of executing this work on his own; it is quite possible that he worked out his results independently. Also, it is not clear that he read the communication to the Royal Society, or whether it was readily available for perusal. Indeed, a record of that communication does not now seem to exist. Note that Waring gave no proof of his proposition that given

$$f(x) = a_0 + a_1 x + a_2 x^2 + \cdots + a_n x^n + \cdots$$

---

[20] Waring (1991) pp. xlii–xliii.

and setting $\alpha_1, \alpha_2, \ldots, \alpha_n$ as the roots of the equation $z^n - 1 = 0$, then

$$\frac{f(\alpha, x) + f(\alpha_2 x) + \cdots + f(\alpha_n x)}{n} = a_0 + a_n x^n + a_{2n} x^{2n} + \cdots . \tag{6.36}$$

The question arises, whether Waring's results, as given in the *Meditationes*, allow us to reconstruct a proof of (6.36); indeed, this seems quite possible. The first chapter of his book dealt with the sums of powers of the roots of an equation. Here he made use of a recurrence formula of Newton, stating that if $\sigma_1, \sigma_2, \sigma_3, \ldots, \sigma_n$ denote the elementary symmetric functions of the roots $\beta_1, \beta_2, \ldots, \beta_n$ of any $n$th degree polynomial and if $s_k$ denotes the sum of the $k$th powers of $\beta_1, \ldots \beta_n$, then

$$s_k - \sigma_1 s_{k-1} + \sigma_2 s_{k-2} - \cdots + (-1)^{k-1}\sigma_{k-1} s_1 + (-1)^k k \sigma_k = 0, \tag{6.37}$$

for $k = 1, 2, 3, \ldots$, provided that $\sigma_j = 0$ for $j > n$. Now if we take the equation to be $x^n - 1 = 0$, then it is clear that $\sigma_1 = 0, \sigma_2 = 0, \ldots, \sigma_{n-1} = 0$ and the recurrence relation (6.37) shows us that $s_k = 0, \ k = 1, 2, \ldots, n - 1$. Therefore the sums of the $k$th powers of the $n$th roots of unity are zero when $k = 1, 2, \ldots, n - 1$. We see that this proof, similar to that given by Simpson for the case $n = 3$, could possibly have been the one conceived of by Waring. On the other hand, we observe that Waring performed several calculations with finite geometric series and with roots of unity in his *Meditationes Algebraicae*; perhaps he also used this method.

In this connection, we note that Waring was aware of Cotes's and de Moivre's result that the numbers

$$\alpha_1 = \cos\frac{2\pi}{n} + i\sin\frac{2\pi}{n}, \quad \alpha_2 = \cos\frac{4\pi}{n} + i\sin\frac{4\pi}{n}, \cdots$$

were the $n$th roots of unity; Simpson knew this too. In fact, in a footnote in his paper on the partitioning of series, Simpson wrote

If $\alpha, \beta, \gamma, \delta$, &c. be supposed to represent the cosines of the angles $\frac{360°}{n}, 2 \times \frac{360°}{n}, 3 \times \frac{360°}{n}$, &c. (the radius being unity); then the roots of the equation $z^n - 1 = 0$ (expressing the several values of $p, q, r, s$, &c.) will be truly defined by $1, \alpha + \sqrt{\alpha\alpha - 1}, \alpha - \sqrt{\alpha\alpha - 1}, \beta + \sqrt{\beta\beta - 1}, \beta - \sqrt{\beta\beta - 1}$, &c.

Thus, both Waring and Simpson understood that if

$$\alpha_1 = \cos\frac{2\pi}{n} + i\sin\frac{2\pi}{n},$$

then the $n$ $n$th roots of unity would be given by $1, \alpha_1, \alpha_1^2, \ldots, \alpha_2^{n-1}$. Moreover, if $1 \leq k < n$, then

$$1 + \alpha_1^k + \alpha_1^{2k} + \cdots + \alpha_1^{(n-1)k} = \frac{\alpha_1^{nk} - 1}{\alpha_1^k - 1} = 0 \tag{6.38}$$

and this result could then be applied to prove (6.36). Though neither Simpson nor Waring explicitly mentioned (6.38) in their work on partitioning series, we see that is likely that they had discovered it. Waring, in particular, utilized (6.38) in his proofs of some other propositions in his *Meditationes*.

Interestingly, de Moivre's result (6.25) can be applied to a problem posed by Ramanujan. Ramanujan defined the sequence of numbers $\tau(n)$ as the coefficients in the series expansion

$$x \prod_{n=1}^{\infty} (1 - x^n)^{24} = \sum_{n=1}^{\infty} \tau(n)x^n \tag{6.39}$$

and he conjectured that

$$1 + \tau(p)x + \tau(p^2)x^2 + \tau(p^3)x^3 + \cdots = \frac{1}{1 - \tau(p)x + p^{11}x^2}. \tag{6.40}$$

As one might suspect, Simpson's dissection (6.1) would play a role in the proof of (6.40). However, as Mordell realized, the theory of modular forms is a key ingredient in this proof. Note that the left-hand side of (6.39) is a cusp form of weight 12 for $\Gamma(1)$ when we set $x = e^{2\pi i \omega}$, Im $\omega > 0$. As we will discuss in our Chapter 10, Mordell utilized ideas of Kiepert, Klein, and Hurwitz on modular forms to show that the coefficients $\tau(p^n)$ in (6.40) satisfy the second-order recurrence relation

$$a_n - \tau(p)a_{n-1} + p^{11}a_{n-2} = 0. \tag{6.41}$$

Now (6.41) is equivalent to (6.24) with $\alpha_1 = -\tau(p)$ and $\alpha_2 = p^{11}$ and, since $a_0 = 1$ and $a_1 = \tau(p)$, we see that (6.40) follows from (6.25).

## 6.2  Simpson and the *Ladies Diary*

Thomas Simpson did not receive even an elementary school eduction; he is an example of the curious phenomenon in eighteenth century England of widespread interest in mathematics among the laity. One might speculate that this fascination with mathematics was awakened by Newton's monumental and celebrated work, with its applications in such influential fields as dynamics, astronomy, navigation, annuities, and insurance.

Thomas Simpson's father was a weaver who was able to teach his son to read English, but taught his son no further, because he wished Simpson to devote himself to the art of weaving. Simpson the younger wished to pursue intellectual interests; he taught himself to write, and associated with persons who could help him further his quest for knowledge. This disagreement with his father forced Simpson to leave home as a teenager and reside at the house of Mrs. Swinfield, a widow thirty years Simpson's senior who later married Simpson, bore him two children, and survived him by seventeen years. Mrs. Swinfield operated a lodging house where an itinerant peddler sometimes stayed. This peddler, impressed with Simpson's intellectual curiosity, provided him with a copy of Cocker's *Arithmetic* with an appendix on algebra. Cocker was an engraver and calligrapher, whose excellent elementary text launched Simpson on his mathematical journey, arming him with sufficient arithmetic, algebra, and geometry to comprehend some of the mathematical questions appearing in the famous *Ladies Diary*, later known as the *Ladies' Diary*, a periodical Simpson himself would later edit. Similar to other magazines of the time on intellectual topics but intended for lay people, this was journal in Latin containing articles and problems on mathematical and nonmathematical topics, with solutions provided by readers.

Simpson learned of the existence of calculus from the *Ladies Diary* but he was unable to read Latin. Fortunately, a friend then lent him a copy of the recently published *A method of fluxions, both direct and inverse* by Edmund Stone. The first part of this work was a translation into English of l'Hôpital's *Analyse des infiniment petits*, a monograph based on the calculus lectures l'Hôpital received from Johann Bernoulli in 1691–92. Note that Stone (1700–1768) himself was the son of a gardener on the estate of the second Duke of Argyll. Self-taught in mathematics, he also mastered French and Latin and performed the service of translating works from these languages into English. In 1740, he published a paper on Newton's classification of cubic curves. In 1737, Simpson published *A new treatise of fluxions*, on the calculus; like all his works, this was very clearly written and accessible to the ordinary student. Then in 1743, he was appointed instructor in mathematics at the Woolwich Military Academy, founded in 1741; this appointment spurred Simpson to pursue problems in applied mathematics and engineering.

Simpson was undoubtedly an inherently talented mathematician, but it was not unusual in England at that time for persons without the benefit of much formal education to master a good deal of mathematics. Surprisingly large numbers of persons without access to a university education pursued an active interest in mathematics. They formed clubs, such as the Philomaths of which Simpson was a member, and it was the duty of each member to help others in the club with their mathematical questions. This was the time of the penny universities, coffee shops where intellectual topics and finance were discussed by persons of largely working-class backgrounds. During this period, several popular magazines devoted to mathematical topics and problems enjoyed widespread appeal; their contributors included primarily men and women without a university education. In addition to the *Ladies Diary*, other magazines such as *Merlin's Oracle* and *Urban's Magazine* contained challenging mathematical problems. Most of these journals appeared for only a few years, but the *Ladies Diary* was founded in 1704 and was in print for more than 150 years.

The widespread interest in mathematics among a fairly large section of the general population in eighteenth century England has been termed "the democratization of mathematics." In her book, *Thomas Simpson and his times*, Frances Clarke has described this phenomenon:[21]

All this goes to show that mathematics was becoming democratic. The universities were educating a chosen few who could afford the expense; the dissenting academies often gave a more extensive course in science than the universities, but were chiefly concerned in educating dissenters; special schools, such as the military and naval academies, were primarily for sons of gentlemen; while the great public schools paid little attention to the scientific side of education. It was in the democratic movement sponsored by the self-taught and private teachers that mathematics was coming to the front.

... Simpson as an exceptional teacher as was well known in London, but Simpson as a textbook writer on such an important and advanced subject [calculus] became known throughout all England. Many wrote to him asking help on certain questions which appeared in the *Ladies' Diary* or in *Merlin's Magazine*, or on some difficulty encountered in reading a current text.

---

[21] Clarke (1928), especially p. 24.

The *Ladies Diary* had a circulation of several thousand. It contained numerous problems as a challenge to its readers, and solutions were submitted and then published. Many women proposed and solved problems. Charles Hutton, author of a history of algebra, served as editor for more than forty years and John Landen also made contributions to this magazine. Like Simpson, neither of these noted mathematicians had a university education. As examples of the kind of mathematics pursued by readers of the *Ladies Diary*, we present four problems from its issues:[22]

1.  Problem posed by Amos Fish in 1711: A ship sails north, from a certain port; after running some days in this direction she alters her course and steers west till her difference of longitude was a degree more than the difference of latitude made upon the first course. And it was also observed that a perpendicular from the point where the ship altered her course to the right line joining the ship and the port sailed from, was 150 miles. Required the distance run in all, and the direct distance from the ship to the port sailed from.

    Solution by Mrs. Barbara Sidway
    Sailed North ...............................................188.256 miles
    Sailed West ...............................................248.256 miles
    Distance Run .......................................... 436.512 miles
    From the port sailed to the ship's place .....311.555

2.  Problem posed by John Landen in 1748: Let a ball of heavy metal be laid upon one end of a horizontal plane, of an indefinite length, round which end let the plane be made to revolve downwards, with such an uniform motion, that the angle of inclination may increase at any given rate: It is required to find what length the ball will descend along the plane, before it acquires such a velocity as will cause it to fly off, and cease touching the plane?

    Landen solved the equation of motion and found an expression for the pressure exerted by the ball on the plane in terms of the angle of inclination $z$ of the plane to the horizon. The ball would leave the plane when the pressure became zero. This gave Landen the equation $2 \cos z = \cosh z$ so that he had $z = 47° \, 11' \, 54''$.

    Simpson also solved the problem; he used infinite series to solve the differential equation of motion and obtained $z = 47° \, 9'$.

3.  Problems posed by Thomas Moss in 1754 and 1755 respectively:
    (a) To determine the least equilateral triangle that can be circumscribed about a given triangle, whereof the three sides are 8, 10, and 12 inches?
    (b) In the three sides of an equi-angular field stand three trees, at the distances of 10, 12, and 16 chains from one another; to find the content of the field, it being the greatest the data will admit of?

A very elegant solution to problem (b) was offered at the time by Mr. W. Bevil, but the problem may be of interest to us in another respect. In his 1991 article, *Nonlinear Programming: a Historical Note*, Harold W. Kuhn has observed[23] that problem 3(b) is the dual to a problem posed by Fermat: For a given triangle, find the point for which the sum of its distances from the three vertices of the triangle is the least. According to Kuhn, a pair of optimization problems would be dual to each another if, based on the same data, the aim is to maximize a function $f$ in one problem and minimize a function $h$ in the other problem, such that for all feasible solutions to the two problems, $h \geq f$, with the optimal solution occurring when $h = f$. The duality of Fermat's problem and 3(b) was noted by Vecten, a professor at the Lycée de Nismes, in *Gergonne's*

---

[22] See Leybourn (1817).     [23] See Kuhn (1991) pp. 82–96, especially p. 87.

*Journal*.[24] Vecten showed that the altitude of the largest equilateral triangle circumscribing a given triangle must equal the sum of the distances of the vertices of a given triangle to a point at which this sum is least.

## 6.3   Development of Complex Variables Theory

In 1827 and 1828, Abel and Jacobi, following Euler and Laplace, used complex numbers in a formal way when they defined elliptic functions of a complex variable. This means that they applied complex substitution in an integral (initially defined over a real interval), but without an explanation of what this might mean. In a similar manner, Euler had already used a complex substitution to obtain some remarkable results on definite integrals. In a 1781 paper, he started with a gamma integral he had discovered a few years earlier

$$\Gamma(s) = \int_0^\infty e^{-t} t^{s-1} dt$$

and set $t = kx$, where $k = p + q\sqrt{-1}$ with $p > 0$. After taking real and imaginary parts, he got

$$\int_0^\infty x^{s-1} e^{-px} \cos qx \, dx = \frac{\Gamma(s)\cos s\theta}{f^n} \tag{6.42}$$

$$\int_0^\infty x^{s-1} e^{-px} \sin qx \, dx = \frac{\Gamma(s)\sin s\theta}{f^n}, \tag{6.43}$$

where $f = (p^2 + q^2)^{\frac{1}{2}}$ and $\tan\theta = \frac{q}{p}$. Euler obtained the evaluations of a number of very interesting integrals as particular and limiting cases of (6.42) and (6.43):

$$\int_0^\infty \frac{\cos x}{\sqrt{x}} dx = \sqrt{\frac{\pi}{2}}, \qquad \int_0^\infty \frac{\sin x}{\sqrt{x}} dx = \sqrt{\frac{\pi}{2}}$$

$$\int_0^\infty e^{-px} \frac{\sin qx}{x} dx = \theta, \qquad \int_0^\infty \frac{\sin x}{x} dx = \frac{\pi}{2}. \tag{6.44}$$

Euler was somewhat unsure of the validity of his methods using complex substitution. However, he had verified some of these integrals by well-established methods and he checked the second relation in (6.44) by numerical integration. A few years after Euler, Laplace rediscovered this method of complex substitution. Since this method was clearly powerful, Cauchy wrote a paper[25] in 1814 to justify it, thus initiating the theory of functions of a complex variable.

Unknown to Cauchy and to the mathematical community in general, Gauss had already taken some major steps toward such a theory. We know that Gauss, like Abel and Jacobi, had made formal use of complex variables in his initial steps toward a theory of elliptic functions. He also employed complex variables in the second part of his theory of hypergeometric functions; this work, like his theory of quartic reciprocity, went unpublished during his lifetime and it is very likely that he had wished

---

[24] Rochat et al. (1811–1812), especially pp. 91–92.   [25] Cauchy (1882–1974) vol. 1, ser. 1, pp. 329–506.

to set these on firm foundations by working out a theory of functions of a complex variable.

Gauss presented an outline of complex integration in an 1811 letter to his friend F. W. Bessel. He started by explaining the geometric representation of a complex number $a + ib$ as a point in the plane with abscissa $a$ and ordinate $b$. Gauss may have independently discovered this representation. If so, he was one of several mathematicians to do so during the late eighteenth and early nineteenth centuries. For example, in 1797, the Norwegian surveyor, Caspar Wessel, read a paper before the Danish Academy on this topic;[26] the paper was published the same year. Others who gave the geometric representation were A. Q. Buée in 1805, J. R. Argand in 1806, and J. F. Français in 1813.[27]

In his letter to Bessel,[28] after giving the geometric representation of a complex number, Gauss raised the question of the meaning of $\int \phi x \cdot dx$ (using his notation), when $x$ was complex and the integral was over any rectifiable curve in the complex plane. He wrote that this integral would be the sum of the infinitesimals $\phi x \cdot dx$, where $dx$ was an infinitesimal increment along the curve. Gauss then noted that the value of $\int \phi x \cdot dx$ would be the same along two different curves as long as $\phi x \neq \infty$ at any point in the region between the curves. As an example of a case in which $\phi x = \infty$, he defined $\log x$ as $\int \frac{1}{x} dx$ starting at 1 and ending at $x$ along a curve that did not include 0. He noted that each circuit around 0 added $\pm 2\pi i$, observing that this explained the multivaluedness of $\log x$. This letter gives clear evidence that in 1811 Gauss understood complex integration and Cauchy's theorem. He called his proof of Cauchy's theorem not difficult. It is possible that his proof employed Green's formula and the Cauchy-Riemann equations. The Cauchy-Riemann equations had appeared in a 1752 paper of d'Alembert on the resistance of fluids, and again in a 1757 paper of Euler on a plane fluid flow. In 1766, Lagrange also published these equations in a paper on hydrodynamics.[29]

Euler communicated nine papers on integrating complex functions to the Petersburg Academy. In 1777 he communicated his first paper, in which he found the Cauchy-Riemann equations;[30] he first defined the integral

$$\Delta(z) = \int Z(z)\,dz, \tag{6.45}$$

where $z = x + iy$ and $Z = M + iN$, by splitting it into its real and imaginary parts

$$\Delta(z) = \int M\,dx - N\,dy + i \int N\,dx + M\,dy \equiv P + iQ. \tag{6.46}$$

Euler assumed that the integrals in (6.46) defined functions. Now the necessary conditions under which the differentials $M\,dx - N\,dy$ and $N\,dx + M\,dy$ would be exact had

---

[26] See Smith (1959) vol. 1, pp. 55–66.    [27] See Smith (1958) vol. 2, pp. 265–266.

[28] Gauss (1863–1927) vol. 8, pp. 90–92. For a translation of the relevant portion of the letter, see Remmert (1991) pp. 167–168.

[29] See Smithies (1997) pp. 8–9.    [30] Eu.I-19, pp. 1–44.

been determined in the 1720s and 1730s by N. Bernoulli,[31] Euler,[32] and Clairaut.[33] These conditions were the Cauchy-Riemann equations:

$$\frac{\partial M}{\partial y} = -\frac{\partial N}{\partial x}, \quad \frac{\partial M}{\partial x} = \frac{\partial N}{\partial y}. \tag{6.47}$$

In a later paper in the series, Euler observed that the functions $P$ and $Q$ in (6.46) also satisfied the Cauchy-Riemann equations

$$\frac{\partial P}{\partial x} = \frac{\partial Q}{\partial y}, \quad \frac{\partial P}{\partial y} = -\frac{\partial Q}{\partial x}. \tag{6.48}$$

One might ask how Gauss may have confirmed the Cauchy-Riemann equations. Now Gauss was one of the earliest to take significant steps toward introducing stringent standards of mathematical rigor. Among other contributions to this effort, he gave an important condition for the convergence of a hypergeometric series and, in an unpublished manuscript, he defined the superior limit of a sequence. One therefore imagines that Gauss derived the Cauchy-Riemann equations in the modern way by taking the derivatives along the $x$ and $y$ axes and equating them.

In his 1814 paper, Cauchy derived two equations that can be combined to produce Cauchy's integral theorem over a class of curvilinear quadrilaterals, though Cauchy did not utilize such geometric language in this paper. He then proceeded to collapse the quadrilateral to a sector by taking two vertices to be the same. This helped him to explain how the integral $\int_0^\infty f(y)dy$ could be transformed into an integral along a ray through $p + iq$. In this way, he verified Euler's results (6.42) and (6.43).

Cauchy returned to this topic in his 1825 paper, "Mémoire sur les intégrales définies prises entre des limites imaginaires."[34] He there discussed points and curves in the complex plane, stating the theorem

$$\int_C f(z)dz = 0 \tag{6.49}$$

for $C$, a more general closed curve than in his 1814 paper. He stated that the function $f(z)$ must be "finite and continuous," apparently taking this to mean continuously differentiable. He attempted to prove (6.49) by a homotopy type of argument, but the proof was not rigorous. Perhaps he himself was unsure of his argument, for later on in the paper he tried to reformulate it in terms of the calculus of variations. Cauchy revisited this question and provided a proof much later, in his 1846 paper, "Sur les intégrales qui s'étendent à tous les points d'une courbe fermée,"[35] wherein he expressed the integral as

$$\int_C f dz = \int_C (P\,dx - Q\,dy) + i \int_C Q\,dx + P\,dy, \tag{6.50}$$

---

[31] See Engelsman (1984) Chapter 4.  [32] Eu.I-19, pp. 265–286.  [33] Clairaut (1739, 1740).
[34] Cauchy (1882–1974) ser. 2, vol. 15, pp. 41–89.  [35] Cauchy (1882–1974) ser. 1, vol. 10, pp. 70–75.

where $C$ was given as a closed curve within which there was no singularity of $f(z)$. He applied Green's theorem to each of the integrals on the right-hand side of (6.50):

$$\int_C A\,dx + B\,dy = \pm \int \int_D \left( \frac{\partial B}{\partial x} - \frac{\partial A}{\partial y} \right) dx dy, \qquad (6.51)$$

choosing the plus sign if the closed curve $C$ had a positive orientation, and with $D$ representing the region inside $C$. The Cauchy-Riemann equations (6.48) show the value of two integrals on the right-hand side of (6.50) to be zero. It seems that Cauchy was the first to publish Green's theorem; he may have been inspired by George Green's 1828 paper on the application of mathematical analysis to electrical and magnetic phenomena.[36] Though Cauchy promised a proof of Green's theorem, he did not fulfill this promise, so that Riemann's proof of Green's theorem,[37] in his 1851 dissertation on complex analysis, appears to have been the first to appear in print. We note that in 1842, Weierstrass was aware of the proof of Cauchy's theorem via Green's theorem, but such a proof required the continuity of the derivative $f'(z)$. In 1883, E. Goursat (1858–1936) discovered a new proof in which he retained this continuity; he realized soon thereafter that his proof did not require the continuity of the derivative $f'(z)$. Thus, when he republished his proof in 1900 in the *Transactions of the American Mathematical Society*, he removed the condition.

In his 1814 paper, Cauchy stated a result that can be seen as a special case of the residue theorem. Then in his 1826 paper "Sur un nouveau genre de calcul analog au calcul infinitésimal,"[38] he defined the concept of the residue and showed that residues could be used to evaluate definite integrals. In papers dating from 1826 to 1829, he gave many applications of the calculus of residues to the summation and transformation of series.

In 1831, Cauchy made further headway in the development of complex analysis by proving the famous Cauchy integral formula, deriving from it the theorem that an analytic function has a Taylor series expansion. He presented this work to the Turin Academy, since he was then in self-imposed exile from France for political reasons. In 1841, he republished significant portions of this paper with some alterations in his *Exercises d'analyse*.[39] Cauchy's statement of the integral formula took the form

$$f(x) = \frac{1}{2\pi} \int_{-\pi}^{\pi} \frac{\overline{x} f(\overline{x})}{\overline{x} - x} dp, \qquad (6.52)$$

where $\overline{x} = X e^{p\sqrt{-1}}$ and $|x| < X$; note that in this context, $\overline{x}$ did not represent the complex conjugate. He first proved a particular case of (6.52):

$$2\pi f(0) = \int_{-\pi}^{\pi} f(\overline{x})\,dp. \qquad (6.53)$$

Cauchy started with the observation that

$$D_X f(\overline{x}) = \frac{1}{X\sqrt{-1}} D_p f(\overline{x}); \qquad (6.54)$$

---

[36] Green (1970).     [37] Riemann (1990) pp. 44–46.
[38] Cauchy (1882–1974) ser. 2, vol. 6, pp. 23–87.     [39] Cauchy (1882–1974) ser. 2, vol. 12, pp. 58–112.

note that this is merely the Cauchy-Riemann equation in polar form. He then integrated (6.54) with respect to $X$ from 0 to $X$ and with respect to $p$ from $-\pi$ to $\pi$, so that the left-hand side of (6.54) was simplified to

$$\int_{-\pi}^{\pi} \left(f(\overline{x}) - f(0)\right) dp = \int_{-\pi}^{\pi} f(\overline{x})\, dp - 2\pi f(0),$$

while the right-hand side reduced to zero. This proved (6.53). Replacing $f(\overline{x})$ by $\overline{x}\frac{f(\overline{x})-f(x)}{\overline{x}-x}$, where $|x| < X$, Cauchy completed his proof of (6.52), arriving at

$$\int_{-\pi}^{\pi} \frac{\overline{x}f(\overline{x})}{\overline{x}-x}\, dp = \int_{-\pi}^{\pi} \frac{\overline{x}f(x)}{\overline{x}-x}\, dp$$

$$= f(x)\int_{-\pi}^{\pi} \left(1 + \frac{x}{\overline{x}} + \frac{x^2}{\overline{x}^2} + \cdots\right) dp = 2\pi f(x). \tag{6.55}$$

From (6.52), Cauchy derived the Maclaurin expansion of an analytic function $f(x)$: As in (6.55), he expanded $\frac{\overline{x}}{\overline{x}-x}$ as a geometric series to get

$$f(x) = \frac{1}{2\pi}\int_{-\pi}^{\pi} \frac{f(\overline{x})\,dp}{1 - \frac{x}{\overline{x}}} = \frac{1}{2\pi}\int_{-\pi}^{\pi} f(\overline{x})\left(1 + \frac{x}{\overline{x}} + \frac{x^2}{\overline{x}^2} + \cdots\right) dp$$

$$= \sum_{n=0}^{\infty} a_n x^n, \tag{6.56}$$

where

$$a_n = \frac{1}{2\pi}\int_{-\pi}^{\pi} \frac{f(\overline{x})}{\overline{x}^n}\, dp. \tag{6.57}$$

We observe that he did not justify the term-by-term integration used in obtaining (6.56). Again, note that (6.56) implies that $f(x)$ has derivatives of all orders in $|x| < X$ and that $a_n = \frac{f^{(n)}(0)}{n!}$. Cauchy, however, appears to have assumed the existence of the derivatives, and proceeded to apply repeated integration by parts to obtain

$$\frac{1}{2\pi}\int_{-\pi}^{\pi} \frac{f(\overline{x})}{\overline{x}^n}\, dp = \frac{1}{2\pi n}\int_{-\pi}^{\pi} \frac{f'(\overline{x})}{\overline{x}^{n-1}}\, dp = \cdots$$

$$= \frac{1}{2\pi n!}\int_{-\pi}^{\pi} f^{(n)}(\overline{x})\, dp$$

$$= \frac{f^{(n)}(0)}{n!}, \tag{6.58}$$

where the final step leading to (6.58) utilizes (6.53). Cauchy then observed that an analytic function $f(x)$ could be expanded as a Maclaurin series for any value of $x$ whose modulus was less than the least value for which $f(x)$ had a singularity, that is, where $f(x)$ was not finite and continuous.

We see that by 1832 Cauchy had sufficiently developed the theory of analytic functions of a complex variable to make it useful in putting the theory of elliptic functions on a more firm foundation. However, although Abel had thought highly of Cauchy's results on the foundations of analysis, Abel's premature death in 1829 prevented him from connecting his work on elliptic functions with the complex analytical ground-work provided by Cauchy. On the other hand, in the 1830s, Jacobi created a basis for the theory of elliptic functions by defining an elliptic function as the ratio of two theta functions. In 1836, he gave lectures at Königsburg on this topic; these were published by Jacobi's student Carl Borchardt in 1880 and then republished in the first volume of Jacobi's collected works.[40] However, as early as 1834, Jacobi published a paper on functions of two variables and with four periods. Note that a period $\omega$ is regarded as fundamental if there does not exist a period $\omega_1$ and an integer $n > 1$ such that $\omega = n\omega_1$. In this paper, Jacobi raised the question of whether there existed functions with two fundamental periods whose ratio was real.[41]

He argued that this was impossible: First, the ratio of the periods must be an irrational number $\lambda$; thus, the two fundamental periods could be written as $2\omega$ and $2\lambda\omega$. Secondly, since there would be rational numbers arbitrarily close to $\lambda$, one could choose integers $a$ and $b$ such that

$$2a\omega - 2\lambda b\omega \qquad (6.59)$$

could be made arbitrarily small. But (6.59) would also be a period of the given function with two fundamental periods. His argument having shown that this function would have arbitrarily small periods, Jacobi wrote at this point: "*Quod fieri non potest*," or "This cannot be." Here he did not complete his argument. But observe that if he had had the machinery to do so, he could have taken the given function (with two fundamental periods whose ratio was real) to be $f(x)$, an analytic function of a complex variable. Now such a function $f(x)$ would take on the same value at a sequence of points that has a limit point. This is manifestly impossible for a nonconstant function because it must necessarily have a power series expansion, so that points where it takes on the same value would have to be isolated. But Jacobi did not specify the class of functions he had in mind; in Jacobi's time, no one seems to have been clear on this point. Jacobi proceeded to show that if a function had three periods, then they were constructible from two of them.[42]

In the 1840s, Joseph Liouville (1809–1882) gave a different direction to the theory of elliptic functions, defining them as doubly periodic analytic functions and then deriving their properties on the basis of this definition. He appears to have become interested in developing this point of view after he found a new proof of Jacobi's theorem, using Fourier series, in 1844. Liouville recorded in his notebook[43] that the suggestion to use Fourier series in that context came from his younger friend, Charles Hermite (1823–1902). After finding his new proof of Jacobi's theorem, Liouville noticed that a similar argument would prove that a doubly periodic function without singularities must be a constant. He used this result to derive the more general theorem that a bounded entire function was a constant, now known as Liouville's theorem. In modern textbooks, this theorem is verified by the use of Cauchy's theorem. In fact, when Liouville published

---

[40] Jacobi (1969) vol. 1, pp. 499–538.  [41] Jacobi (1969) vol. 2, pp. 23–50.

[42] For a careful discussion of the points raised by Jacobi's remark, see Markushevich (1992) pp. 17–23.

[43] Lützen (1990) pp. 535–555. Lützen has described in very interesting detail the origins and development of Liouville's theory of elliptic functions.

some of his results on elliptic functions, including the particular case of Liouville's theorem, Cauchy immediately pointed out that those results could easily be obtained by the methods Cauchy had been developing for thirty years. This information did not deter Liouville from forging ahead with his own theory. In 1847, he gave lectures on his theory of doubly periodic functions, proving some important basic results:[44]

- There does not exist a nonconstant doubly periodic entire function.
- A doubly periodic function with only one simple pole in a period parallelogram (a concept due to Liouville) does not exist.
- In a period parallelogram, the sum of the orders of the zeros is equal to the sum of the orders of the poles.
- The sum of the residues of a doubly periodic function inside a period parallelogram is zero. Liouville proved this for functions with two poles.
- The sum of the zeros minus the sum of the poles in a period parallelogram is equal to a period of the function.

In proving these results, he did not employ Cauchy's theory of complex integration. Still, Liouville's original approach to elliptic functions as doubly periodic meromorphic functions opened the door to the application of Cauchy's theory of residues to this class of functions. In 1848, Hermite made use of Cauchy's theory to show that the sum of the residues of an elliptic function was zero inside a period parallelogram. Briot and Bouquet attended Liouville's 1851 lectures at the Collège de France and in 1856 they published a paper on Cauchy's theory of functions of a complex variable, applying it to derive Liouville's main results on doubly periodic functions. Perhaps independently, Riemann also derived these results by Cauchy's methods, and presented this work in a course of lectures on elliptic and Abelian functions, given at Göttingen in 1855–56 and 1861–62. Liouville was apparently not pleased with the work of Briot and Bouquet, especially when they extended it in 1859 to a full-length text on elliptic functions.[45] Borchardt, who also attended Liouville's 1847 lectures, took notes and published them in 1880 in order to preserve this approach. These methods are examples of a general principle developed by Riemann, that an analytic function is determined to a large extent by the location of its singularities: its zeros and poles. In his short note of 1844,[46] Liouville wrote that within his approach, "the integrals which have given rise to the the elliptic functions and even the moduli disappear in a way leaving only the periods and the points for which the functions become zero or infinite."[47]

In an unpublished note, Liouville offered an argument for Jacobi's theorem on the nonexistence of a nonconstant analytic function of a complex variable with two independent periods over the rational numbers. First, suppose $f$ to be an analytic function with periods $\alpha$ and $\beta$ such that $\frac{\alpha}{\beta} \notin Q$. In that case, $f$ has a Fourier series expansion

$$f(x) = \sum_{k=0}^{\infty} A_k \cos\left(\frac{2k\pi x}{\alpha} + \epsilon_k\right), \quad \text{with } \epsilon_0 = 0.$$

[44] Liouville (1880).      [45] Lützen (1990) p. 529.      [46] Liouville (1844).      [47] Lützen (1990) p. 548.

Next, since $\beta$ is a period, it is necessary that

$$A_k \cos\left(\frac{2k\pi(x+\beta)}{\alpha} + \epsilon_k\right) = A_k \cos\left(\frac{2k\pi x}{\alpha} + \epsilon_k\right), \quad k \neq 0.$$

Thus, either $A_k = 0$ or $\frac{2k\pi\beta}{\alpha} = 2m\pi$ for some integer $m$. But the latter equation implies that $\frac{\beta}{\alpha}$ is a rational number, so that $f(x) = A_0$, a constant, so that the theorem is confirmed.

In a similar fashion, Liouville proved that a doubly periodic function without a pole must be a constant. Suppose, he argued, $\psi(t)$ was a function with two periods, $\omega_1$ and $\omega_2$, whose ratio was complex. He let $\frac{z}{2\pi} = \frac{t}{\omega_1}$, with $z$ complex, and showed that $\psi(z)$ had a Fourier expansion

$$\psi(z) = \sum_{k=0}^{\infty} A_k \cos(kz + \epsilon_k).$$

Since $\cos z$ had period $2\pi$, it followed as before that $\frac{\omega_2}{\omega_1}$ was a rational real number, contradicting the original assumption.

Liouville's theorem now refers to the more general proposition that a bounded entire function is a constant, a theorem first given by Cauchy after Liouville published a statement of his own result. However, as J. Lützen argued in his book, *Joseph Liouville 1809–1882*, the general theorem turns out to be correctly named. In an unpublished note, Liouville observed that if $f$ was a bounded entire function, then $f$ composed with the Jacobi function sn $z$ would be a doubly periodic function that was bounded and therefore a constant; hence, $f$ must be a constant.

Liouville applied this theorem to prove that a doubly periodic function cannot have only one simple pole and no other singularity in a period parallelogram: Suppose $\phi(z)$ is such a function with a simple pole at $\alpha$ and with residue $G$. Liouville denoted the singular part by

$$[\phi(z)]^\alpha = \frac{G}{z-\alpha} \quad \text{or} \quad [\phi(\alpha+t)]^0 = \frac{G}{t}.$$

This implies that

$$[\phi(\alpha+t) + \phi(\alpha-t)]^0 = 0$$

and, thus, by Liouville's theorem

$$\phi(\alpha+t) + \phi(\alpha-t) = 2c,$$

where $c$ is a constant. Now let $f(t) = \phi(\alpha+t) - c$, so that $f(t) = -f(-t)$. Suppose $2\omega_1$ and $2\omega_2$ are the periods of $f$. Then

$$f(\omega_1) = -f(-\omega_1) = -f(-\omega_1 + 2\omega_1) = -f(\omega_1) = 0 \tag{6.60}$$

and, similarly,

$$f(\omega_2) = 0 \quad \text{and} \quad f(\omega_1 + \omega_2) = 0. \tag{6.61}$$

Now the function $F(t) = f(t)f(t + \omega_1)$ has no singularity because, by (6.60), the poles cancel with the zeros. Therefore, there exist constants $k_1, k_2, k_3$ such that

$$f(t)f(t + \omega_1) = k_1, \quad f(t)f(t + \omega_2) = k_2, \quad f(t)f(t + \omega_1 + \omega_2) = k_3. \quad (6.62)$$

Changing $t$ to $t + \omega_1$ in the third equation in (6.62) then gives us

$$f(t + \omega_1)f(t + \omega_2) = k_3. \tag{6.63}$$

Divide the product of the first two equations in (6.62) by (6.63) to obtain

$$\left(f(t)\right)^2 = \frac{k_1 k_2}{k_3}.$$

This means that $\phi(\alpha + t) = f(t) + c$ is a constant, proving the theorem. In this manner, Liouville proved all his basic theorems without the use of complex integration.

## 6.4  Hermite: Complex Analysis in Elliptic Functions

In 1848, Hermite published a short "Note sur la théorie des fonctions elliptiques" in the *Cambridge and Dublin Mathematical Journal*[48] in which he gave a method, avoiding inverses of integrals, for defining doubly periodic functions. Hermite gave an exposition of this method in 1862, starting with the observation that Liouville's theorem implied that a periodic entire function

$$\sum_{m=-\infty}^{\infty} A_m e^{2m \frac{i\pi x}{\omega_1}} \tag{6.64}$$

could not, without reducing to a constant, have another period $\omega_2$. This in turn implied that a nonconstant elliptic function had to be meromorphic, that is, a ratio of two entire functions. Then in his 1848 paper, he considered the situation in which the ratio of the two series

$$\Phi(x) = \sum_{m=-\infty}^{\infty} A_m e^{2m \frac{i\pi x}{\omega_1}} \quad \text{and} \quad \Pi(x) = \sum_{m=\infty}^{\infty} B_m e^{2m \frac{i\pi x}{\omega_1}}. \tag{6.65}$$

had another period, $\omega_2$. He proposed to determine the conditions on $A_m$ and $B_m$ such that

$$\sum_{m=-\infty}^{\infty} A_m e^{2m \frac{i\pi x}{\omega_1}} \sum_{m=\infty}^{\infty} B_m e^{2m \frac{i\pi x}{\omega_1}(x+\omega_2)} = \sum_{m=\infty}^{\infty} B_m e^{2m \frac{i\pi x}{\omega_1}} \sum_{m=-\infty}^{\infty} A_m e^{2m \frac{i\pi x}{\omega_1}(x+\omega_2)}, \tag{6.66}$$

reasoning that in this case, $\frac{\Phi(x)}{\Pi(x)}$ would have periods $\omega_1$ and $\omega_2$. Without loss of generality, he could assume that $\text{Im} \frac{\omega_2}{\omega_1} > 0$. He let

$$q_1 = e^{i\pi \frac{\omega_2}{\omega_1}},$$

---

[48] Hermite (1905–1917) vol. 1, pp. 71–73 or, for more details, see Hermite's 1862 exposition of this work, vol. 2, pp. 143–148. See also the treatment in Tannery and Molk (1972) vol. 2, pp. 152–158.

and equated the coefficients of $e^{2\mu \frac{i\pi x}{\omega_1}}$ of both sides in (6.66) to obtain

$$\sum_{m=-\infty}^{\infty} A_{\mu-m} B_m q_1^{2m} = \sum_{n=-\infty}^{\infty} A_{\mu-n} B_n q_1^{2(\mu-n)}. \tag{6.67}$$

Hermite next observed that equation (6.67) would be satisfied if he were to choose an integer $k$ such that

$$A_{\mu-m} B_m q_1^{2m} = A_{\mu-m-k} B_{m+k} q_1^{2(\mu-m-k)} \tag{6.68}$$

for all $m$. By setting $\mu - m - k = m'$, (6.68) was converted to

$$\frac{A_{m'+k}}{q_1^{2m'} A_{m'}} = \frac{B_{m+k}}{q_1^{2m} B_m}. \tag{6.69}$$

Since $m$ and $m'$ were independent, Hermite observed that by (6.69), $A_m$ and $B_m$ would satisfy the difference equation

$$\frac{z_{m+k}}{q_1^{2m} z_m} = \text{const.}$$

or

$$z_{m+k} - c q_1^{2m} z_m = 0, \tag{6.70}$$

whose general solution was

$$z_m = q_1^{\frac{m^2}{k} + \alpha m} u_m,$$

where $\alpha$ was dependent on the constant $c$ and $u_m$ satisfied

$$u_{m+k} = u_m.$$

Hermite noted that he could take $\alpha = 0$ without loss of generality by changing $x$ to $x + \beta$ for a suitable $\beta$. Thus, $A_m$ and $B_m$ were given by

$$A_m = a_m q_1^{\frac{m^2}{k}}, \qquad B_m = b_m q_1^{\frac{m^2}{k}}$$

where $a_m$ and $b_m$ satisfied conditions

$$a_{m+k} = a_m, \qquad b_{m+k} = b_m. \tag{6.71}$$

This implied that with $a_m$ and $b_m$ as in (6.71), he had

$$\Phi(x) = \sum_{m=-\infty}^{\infty} a_m q_1^{\frac{m^2}{k}} e^{2m \frac{i\pi x}{\omega_1}}. \tag{6.72}$$

$$\Pi(x) = \sum_{m=-\infty}^{\infty} b_m q_1^{\frac{m^2}{k}} e^{2m \frac{i\pi x}{\omega_1}}. \tag{6.73}$$

He observed that the functions $\Phi$ and $\Pi$ were periodic with respect to $\omega_1$ and quasi-periodic with respect to $\omega_2$, that is, both functions satisfied the equations

$$\psi(x + \omega_1) = \psi(x) \tag{6.74}$$

$$\psi(x + \omega_2) = \psi(x)e^{-k\frac{i\pi}{\omega_1}(2x+\omega_2)}. \tag{6.75}$$

The first equation (6.74) is an immediate consequence of (6.72) and (6.73). To prove (6.75), Hermite noted that the sum in (6.72), for example, would remain unchanged if $m$ were replaced by $m - k$ and $a_{m-k}$ by $a_m$. Thus

$$
\begin{aligned}
\Phi(x + \omega_2) &= \sum_{m=-\infty}^{\infty} a_m q_1^{\frac{m^2}{k}} e^{2m\frac{i\pi(x+\omega_2)}{\omega_1}} \\
&= \sum_{m=-\infty}^{\infty} a_m q_1^{\frac{m^2}{k}+2m} e^{2m\frac{i\pi x}{\omega_1}} \\
&= \sum_{m=-\infty}^{\infty} a_m q_1^{\frac{(m-k)^2}{k}+2(m-k)} e^{2(m-k)\frac{i\pi x}{\omega_1}} \\
&= \sum_{m=-\infty}^{\infty} a_m q_1^{\frac{m^2}{k}-k} e^{2(m-k)\frac{i\pi x}{\omega_1}} \\
&= q_1^{-k} e^{-2k\frac{i\pi x}{\omega_1}} \sum_{m=-\infty}^{\infty} a_m q^{\frac{m^2}{k}} e^{2m\frac{i\pi x}{\omega_1}} \\
&= \Phi(x)e^{\frac{-k\pi i}{\omega_1}(2x+\omega_2)}.
\end{aligned}
$$

As a particular case, Hermite took $k = 2$. He let

$$\omega_1 = 4K, \quad \omega_2 = 2iK', \quad q_1 = e^{-\frac{\pi K'}{2K}}, \quad q = e^{-\frac{\pi K'}{K}}$$

so that

$$
\begin{aligned}
\Phi(x) &= \sum a_m q_1^{\frac{m^2}{2}} e^{m\frac{i\pi x}{2K}} \\
&= a_0 \sum_{m=-\infty}^{\infty} q^{m^2} e^{m\frac{i\pi x}{K}} + a_1 \sum_{m=-\infty}^{\infty} q^{\frac{(2m+1)^2}{4}} e^{(2m+1)\frac{i\pi x}{2x}} \\
&= a_0 \left(1 + 2q \cos 2\left(\frac{\pi x}{2K}\right) + 2q^4 \cos 4\left(\frac{\pi x}{2K}\right) + 2q^9 \cos\left(\frac{\pi x}{2K}\right) + \cdots\right) \\
&\quad + a_1 \left(2q^{\frac{1}{4}} \cos\left(\frac{\pi x}{2K}\right) + 2q^{\frac{9}{4}} \cos 3\left(\frac{\pi x}{2K}\right) + 2q^{\frac{25}{4}} \cos 5\left(\frac{\pi x}{2K}\right) + \cdots\right).
\end{aligned}
$$

Hermite denoted the series of which $a_0$ was the coefficient by $\Theta_1(x)$ and the series of which $a_1$ was the coefficient by $H_1(x)$. Thus, he had

$$\Theta_1\left(\frac{2Kx}{\pi}\right) = 1 + 2q \cos 2x + 2q^4 \cos 4x + 2q^9 \cos 6x + \cdots \tag{6.76}$$

$$H_1\left(\frac{2Kx}{\pi}\right) = 2q^{\frac{1}{4}} \cos x + 2q^{\frac{9}{4}} \cos 3x + 2q^{\frac{25}{4}} \cos 5x + \cdots. \tag{6.77}$$

He noted that these series were related to Jacobi's series for the theta functions $\Theta$ and $H$, given by (3.184) and (3.185), respectively, by the equations

$$\Theta_1 \left( \frac{2Kx}{\pi} + K \right) = \Theta \left( \frac{2Kx}{\pi} \right), \quad H_1 \left( \frac{2Kx}{\pi} + K \right) = -H \left( \frac{2Kx}{\pi} \right).$$

In this manner, Hermite was able to derive from his general theory the Jacobi elliptic functions as ratios of theta functions.

In 1849, Hermite used complex integration to establish Liouville's basic theorems on doubly periodic functions. Hermite submitted a paper containing these results to the Académie des Sciences (Paris) but unfortunately it went unpublished. Only Cauchy's report on Hermite's paper was published by the academy in its *Comptes Rendus*.[49] From this report, we learn how Hermite proved that a nonconstant doubly periodic function must have more than one simple pole. He integrated the doubly periodic function $f(z)$ over the boundary of a period parallelogram $ABCD$. If we denote the point $A$ by the complex number $\zeta$, then the points $B, C, D$ will be denoted by $\zeta + \omega_1, \zeta + \omega_1 + \omega_2, \zeta + \omega_2$, respectively, where $\omega_1, \omega_2$ are the primitive (fundamental) periods of $f$. Now denoting the contour of the boundary of $ABCD$ by $L$, Hermite showed that

$$\int_L f(z) dz = 0. \tag{6.78}$$

His argument was simple; he observed that

$$\int_L f(z) dz = \int_A^B + \int_B^C + \int_C^D + \int_D^A f(z) dz. \tag{6.79}$$

From the periodicity of $f$, it followed that

$$\int_C^D f(z) dz = \int_B^A f(z + \omega_2) dz = -\int_A^B f(z) dz \tag{6.80}$$

and

$$\int_D^A f(z) dz = \int_C^B f(z + \omega_1) dx = -\int_B^C f(z) dz. \tag{6.81}$$

Substituting (6.80) and (6.81) in (6.79) gave him (6.78). Since by Cauchy's residue theorem, $\int_L f(z) dz$ was the sum of the residues of $f$ inside the period parallelogram, it followed that the nonconstant function $f$ had to have more than one simple pole.

It appears from Cauchy's report that Hermite was the first mathematician to systematically apply Cauchy's complex variables theory to the theory of elliptic functions. Hermite's approach was taken up by Briot and Bouquet in their 1859 treatise, *Théorie des fonctions elliptiques*. A number of Liouville's theorems can be verified using complex integration; they applied complex integration to prove Liouville's theorem: In its period parallelogram, a doubly periodic function has as many zeros as poles, counting multiplicity. Begin by letting $f$ be a doubly periodic function. Then $\frac{f'}{f}$ must also

---

[49] Hermite (1905–1917) vol. 1, pp. 75–83.

be doubly periodic and hence, as before, the integral of $\frac{f'}{f}$ over the boundary of the period parallelogram is zero. Moreover, if $f$ has a zero of order $n$ at some point $z_0$ in the period parallelogram, then $\frac{f'}{f}$ has residue $n$ at $z_0$. Similarly if $f$ has a pole of order $m$, then $\frac{f'}{f}$ has residue $-m$. The result now follows from Cauchy's residue theorem. Although Cauchy did not explicitly mention this theorem in his report on Hermite's work, he wrote that Hermite had derived Liouville's results by means of complex integration.

### 6.5   Riemann: Meaning of the Elliptic Integral

Riemann presented his groundbreaking forty-page work on complex analysis in this Göttingen thesis, "Grundlagen für eine allgemeine Theorie der Functionen einer Veränderlichen complexen Grösse." Riemann wrote that before composing this work, he had given careful study to the papers of Cauchy on analysis and complex variables and to Gauss's 1825 paper on conformal mappings between surfaces.[50] However, it appears that before Riemann no one had made the connection to consider complex analytic functions as conformal mappings. Recall that a conformal mapping is one that conserves angles.

Riemann established the basis for his theory in the Cauchy-Riemann equations. He was primarily interested in the geometric properties of analytic functions; he viewed a function $w = f(z) = u(x, y) + iv(x, y)$ as a map from the $z = x + iy$ plane to the $w = u + iv$ plane. In the third section of his dissertation, he showed that if $f'(z) \neq 0$, then $f$ is conformal at $z$. In the sixth section, he defined the important concept of a simply connected surface (region). In Section 21, the penultimate section, he proved his immortal mapping theorem, that two simply connected regions, each with at least two points on the boundary, were conformally equivalent. He stated his theorem without the terminology developed later:[51]

> Zwei gegebene einfach zusammenhängende ebene Flächen können stets so auf einander bezogen werden, dass jedem Punkte der einen Ein mit ihm stetig fortrükender Punkt der andern entspricht und ihre entsprechenden kleinsten Theile ähnlich sind; und zwar kann zu Einem innern Punkte und zu Einem Begrenzungspunkte der entsprechende beliebig gegeben werden; dadurch aber ist für alle Punkte die Beziehung bestimmt.
>
> Two given simply connected plane surfaces can always be related to one another in such a way that to each point on the one corresponds a point on the other, such that both points move together continuously, and their corresponding smallest parts are similar; and in fact to one interior point and to one boundary point the corresponding point can be given randomly; thereby, however, correlation is defined for all points.

Riemann's statement that "ihre entsprechenden kleinsten Theile ähnlich sind" or "their corresponding smallest parts are similar" signifies that the angles between corresponding curves are preserved, or in more modern terms, that the mapping is conformal. The theorem implies the existence of a conformal mapping from an open simply connected

---

[50] Gauss (1863–1927) vol. 4, pp. 189–216; English translation in Gauss (2005).
[51] Riemann (1990) p. 72.

region that is a proper subset of the complex plane to the interior of the unit disk. Moreover, Riemann's statement that "... zwar kann zu Einem innern Punkte und zu Einem Begrenzungspunkte der entsprechende beliebig gegeben werden; dadurch aber ist für alle Punkte die Beziehung bestimmt," or "... in fact to one interior point and to one boundary point the corresponding point can be given randomly; thereby, however, correlation is defined for all points," means that the mapping is unique when a value at an interior point and a value at a boundary point are assigned at random. The random choice of points implicitly implies that the boundary of the simply connected region contains at least two points. Riemann was well aware, for example, that the whole complex plane $\mathbb{C}$, with only one point on the boundary (at infinity), was not conformal to the unit disk, since such a function could be a constant by Liouville's theorem. In modern expositions, the reference to the boundary point is replaced by an additional condition on the interior point: For example, with $\mathbb{D}$ the unit disk, Kra and his coauthors write:[52]

> Let $D$ be a nonempty proper simply-connected open subset of $\mathbb{C}$, and let $c \in D$. Then there exists a unique conformal map $f : D \to \mathbb{D}$ with $f(c) = 0$, $f'(c) > 0$, and $f(D) = \mathbb{D}$.

Section 20 of Riemann's dissertation presents a summary of its contents, where Riemann wrote that, in contrast to previous treatments of functions of complex variables containing explicit expressions of such functions, he had defined his functions by using a small number of their general properties. Note that Cauchy had done something similar in his work on complex analysis. Riemann then took this approach a step further, with his geometric point of view and the general principle that a function was to a significant extent determined by its singularities, a principle foreshadowed by Liouville's 1844 remarks on elliptic functions, given in our Section 6.3.

The first chapter of Riemann's lectures on elliptic functions are instructive in their application of geometric ideas, such as the Riemann surface, and other general principles used in defining elliptic functions.[53] These lectures[54] were are not included in Riemann's collected mathematical works, but were published by Hermann Stahl in 1899. We first note that we shall use Riemann's notation by denoting that $f(z)$ has a pole of order $m$ at $v$ by $f(v) = \infty^m$.

Riemann began his lectures by using contour integration to establish some theorems of Liouville on doubly periodic functions, in particular, that such a function had at least two simple poles and that it had as many zeros as poles in a period parallelogram.[55] Recall that Hermite had proved Liouville's theorem in this manner; in addition, Riemann was well aware of Hermite's work.

He then considered a doubly periodic function $z = \phi(v)$ with two simple poles inside its period parallelogram and showed that such a function satisfied a differential equation of the form

$$\frac{dz}{dv} = \sqrt{(z - a_1)(z - a_2)(z - a_3)(z - a_4)}. \tag{6.82}$$

[52] Rodríguez, Kra, Gilman (2013) p. 207; also see Ahlfors (1966) p. 222.
[53] For a discussion of Riemann's view of complex analysis, see Laugwitz (1999) and Bottazzini and Gray (2013).
[54] Riemann (1899).    [55] Ibid. pp. 1–3.

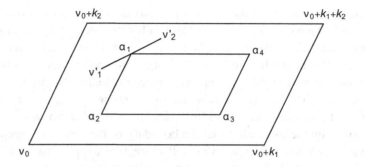

Figure 6.1. Zeros Inside Period Parallelogram.

To prove this result,[56] Riemann let $\phi(v_1') = \phi(v_2') = \infty^1$ and set the residue at $v_1'$ to be $A$; thus, since the sum of the residues must be zero, the residue at $v_2'$ would be $-A$. Then the singular part of $\phi$ at $v_1'$ is $\frac{A}{v-v_1'}$ while the singular part at $v_2'$ is $\frac{-A}{v-v_2'}$. Now the function $\psi(v) = \phi(v_1' + v_2' - v)$ again has $\psi(v_1') = \psi(v_2') = \infty^1$, and the singular part of $\psi$ at $v_2'$ is $\frac{A}{v_2'-v}$ and at $v_1'$ it is $\frac{-A}{v_1'-v}$. This means that $\phi(v) - \psi(v)$ has a vanishing singular part and is therefore a constant, that is,

$$\phi(v_1' + v_2' - v) = \phi(v) + C. \tag{6.83}$$

We may see that $C = 0$ by substituting $v = \frac{v_1' + v_2'}{2}$ in (6.82). Taking the derivative of (6.82), we obtain

$$\phi'(v_1' + v_2' - v) = -\phi'(v).$$

So $\phi'(v) = 0$ for $v_1' + v_2' - v \equiv v$ modulo the periods of $\phi$. Denote the periods by $k_1$ and $k_2$. The singular parts of $\phi'$ at $v_1'$ and $v_2'$ are $\frac{-A}{(v-v_1')^2}$ and $\frac{A}{(v-v_2')^2}$, respectively. Therefore, the sum of the poles counting multiplicity is four and $\phi'$ has four zeros in the period parallelogram. Now $\phi'(v) = 0$ at

$$v = \frac{v_1' + v_2' + mk_1 + nk_2}{2},$$

where $m$ and $n$ are integers, so that the four zeros $\alpha_1, \alpha_2, \alpha_3, \alpha_4$ can be chosen in the period parallelogram such that

$$\alpha_1 \equiv \frac{v_1' + v_2'}{2}, \quad \alpha_2 \equiv \frac{v_1' + v_2'}{2} \pm \frac{k_1}{2}, \quad \alpha_3 \equiv \frac{v_1' + v_2'}{2} \pm \frac{k_2}{2}, \quad \alpha_4 \equiv \frac{v_1' + v_2'}{2} \pm \frac{k_1}{2} \pm \frac{k_2}{2}.$$

We can choose $v_1'$ and $v_2'$ to be in the period parallelogram, so that $\alpha_1, \alpha_2, \alpha_3, \alpha_4$ will be as shown in the Figure 6.1, given here as Riemann himself presented it.

Let $\phi(\alpha_i) = a_i$, $i = 1, 2, 3, 4$. Now, with $z = \phi(v)$, the two functions $(\phi'(v))^2$ and $(z - a_1)(z - a_2)(z - a_3)(z - a_4)$ have poles of order 4 at $v_1'$ and at $v_2'$ and have zeros of order 2 at each of the four points $\alpha_1, \alpha_2, \alpha_3, \alpha_4$. This implies that each function is a

---

[56] Ibid. pp. 5–7.

constant times the other; hence

$$C\frac{dz}{dv} = \sqrt{(z - a_1)(z - a_2)(z - a_3)(z - a_4)} \tag{6.84}$$

or

$$v = \int^{z} \frac{dz}{\sqrt{(z - a_1)(z - a_2)(z - a_3)(z - a_4)}}, \tag{6.85}$$

verifying Riemann's proposition and illustrating the connection between a doubly periodic function and an elliptic integral. Riemann then remarked that a new geometric meaning of elliptic functions and integrals would be achieved by using (6.85) to map a surface onto the period parallelogram.

Let $v = \psi(z)$ be this mapping.[57] Then

$$\psi'(z) = \frac{1}{\sqrt{(z - a_1)(z - a_2)(z - a_3)(z - a_4)}}. \tag{6.86}$$

The points $a_1, a_2, a_3, a_4$ are branch points of order 2, since $\sqrt{z - a_i}$ changes sign as arg $(z - a_i)$ changes by $2\pi$. To make $\psi'(z)$ a single-valued function, Riemann considered a two-sheeted surface over the Riemann sphere, $\mathbb{C} \cup \{\infty\}$. Each sheet was cut along two lines, $a_1a_2$ and $a_3a_4$. The cut $a_1a_2$ prevented any closed path from winding around only $a_1$ or only $a_2$. Observe that a closed path winding around both $a_1$ and $a_2$ would not cause a change in $\sqrt{(z - a_1)(z - a_2)}$, since each of the factors $\sqrt{z - a_1}$ and $\sqrt{z - a_2}$ would undergo a change in sign. The two sheets were then joined together at the cuts to form a surface $T$ in such a way that when a closed path passed through either cut ($a_1a_2$ or $a_3a_4$), the path moved from one sheet to the other. Riemann called the two edges of a cut, say $a_1a_2$, the right and left of $a_1a_2$. The right edge of $a_1a_2$ (or $a_3a_4$) of the lower sheet was to be connected to the left edge of the upper sheet and the left edge of $a_1a_2$ (or $a_3a_4$) of the lower sheet would be connected to the right edge of the upper sheet. Thus was formed the surface $T$ on which $\psi'(z)$ was single-valued and analytic.

When $v = v_1'$ or when $v = v_2'$, then $z = \phi(v) = \infty$ and thus two infinitely small circles around $v_1'$ and $v_2'$ in the $v$-plane map onto infinitely large circles in each sheet of the surface $T$. Riemann next described the curves to which the border lines of the period parallelogram were mapped under $v = \psi(z)$. The line from $v_0$ to $v_0 + k_1$ corresponds to a curve $b$ on the surface $T$, a curve that starts at some point $z_0$ in the upper sheet and returns to the same point, so that it must travel around two branch points, say $a_1$ and $a_2$, because it cannot cut out any piece from $T$. The line from $v_0 + k_1$ to $v_0 + k_1 + k_2$ corresponds to a curve $a$ in $T$ that again starts at $z_0$ and returns to it without intersecting the curve $b$ at any point in between. This curve $a$ must travel around one of the branch points, either $a_1$ or $a_2$, say $a_2$, and one of the other branch points $a_3$. Clearly, $a$ must travel partly in the upper sheet and partly in the lower sheet. Note the dotted lines in the lower leaf as shown in Figure 6.2.

[57] Ibid. pp. 7–8.

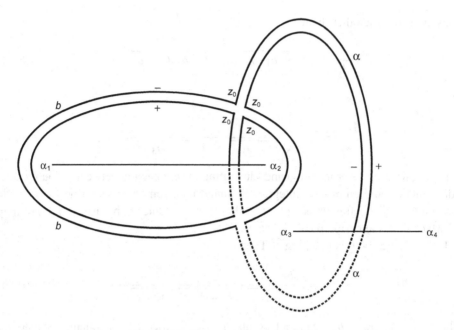

Figure 6.2. Riemann's Curves $a$ and $b$.

The line from $v_0 + k_1 + k_2$ up to $v_0 + k_2$ corresponds to the first curve $b$ on the surface, but traveling in the opposite direction. Finally, the line from $v_0 + k_2$ to $v_0$ corresponds to the second curve $a$, but moving in the opposite direction. Riemann labeled these curves with $+$ or $-$, as shown.

Since the function $\psi'(z)$ is single-valued on the surface $T$, we can integrate this function along curves on the surface. If we consider the integral (6.85) and set $v_0 = \psi(z_0)$, then the value of the integral along the closed curve $b$ changes by $k_1$ so that

$$k_1 = \int_b \psi'(z)dz.$$

Similarly,

$$k_2 = \int_a \psi'(z)dz.$$

In connection with Jacobi's elliptic integral

$$\int \frac{dz}{\sqrt{(1 - z^2)(1 - k^2 z^2)}}, \quad \text{with } 0 < k^2 < 1, \tag{6.87}$$

we remark that Riemann preferred to use the integral obtained by the substitution $x = z^2$:

$$\int \frac{dx}{2\sqrt{x(1 - x)(1 - k^2 x)}}. \tag{6.88}$$

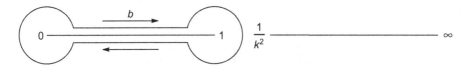

Figure 6.3. Riemann's Diagram for Circuit $b$.

He wrote the denominator of the integrand as

$$\sqrt{(x, k)} = 2\sqrt{x(1 - x)(1 - k^2 x)}. \tag{6.89}$$

The two-sheeted Riemann surface of $y = \sqrt{(x, k)}$ was obtained by making cuts on the real axis from 0 to 1 and from $k^{-2}$ to $\infty$. Riemann denoted the circuit around the cut from 0 to 1 by $b$, and the circuit going from one sheet to another by $a$, as shown in Figures 6.3 and 6.4.

As we have seen, the periods of the inverse function were

$$2K = \int_b \frac{dx}{\sqrt{(x, k)}}$$

and

$$2iK' = \int_a \frac{dx}{\sqrt{(x, k)}}.$$

Riemann next explained how the closed curve $b$ could be deformed into a circuit from 0 to 1 and then back from 1 to 0 so that

$$2K = \int_0^1 du_1 + \int_1^0 du_2.$$

Now $du_2 = -du_1$ because of the change in sign of $\sqrt{1 - x}$ as the curve moves around $x = 1$. Thus,

$$2K = 2 \int_0^1 \frac{dx}{\sqrt{(x, k)}}$$

or

$$K = \int_0^1 \frac{dx}{\sqrt{(x, k)}}.$$

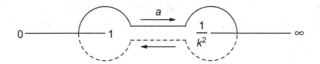

Figure 6.4. Riemann's Diagram for Circuit $a$.

Similarly,

$$iK = \int_1^{\frac{1}{k^2}} \frac{dx}{\sqrt{(x, k)}} = i \int_0^1 \frac{dx'}{\sqrt{(x', k')}},$$

where $k^2 + k'^2 = 1$ and $k^2 x^2 + k'^2 x'^2 = 1$. This also showed that when $0 < k^2 < 1$, then $K$ and $K'$ were real and positive. Perhaps Riemann preferred his own normal form of the elliptic integral (6.88) to Jacobi's form (6.87) because the periods of the inverse of his normal form were $2K$ and $2iK'$, whereas the periods of Jacobi's function, sn, were $4K$ and $2iK'$.

If we take the integral (6.85) as a starting point, then Riemann's construction of the Riemann surface corresponding to (6.85) explains the double periodicity of the inverse function. This in turn provides a complete answer to Eisenstein's question on the inverse of an integral and its periodicity. In this connection, see Chapter 4. Riemann had attended Eisenstein's 1847 lectures on elliptic functions at Berlin. According to Dedekind[58] Riemann discussed with Eisenstein the theory of elliptic functions as well as the problem of introducing complex numbers into the theory of functions; he found that they had entirely different views as to the principles that should be used as the foundations for complex analysis. Riemann felt that Eisenstein was working at the level of formal computations, whereas he himself recognized that the essential definition of a function of a complex variable lay in the differential equation. This definition was closely related to Riemann's ideas on analytic continuation and Riemann surfaces. Dedekind thought that Riemann had thoroughly worked out these ideas in the fall of 1847. In fact, in a November 1847 letter to his father, Riemann wrote that he had essentially solved a problem of Eisenstein and that he had shown this work to Jacobi, who was willing to communicate it to *Crelle's Journal*.

### 6.6   Weierstrass's Rigorization

Weierstrass first presented his theory of elliptic functions in his Berlin lectures of 1862–1863. This theory was firmly based on his theory of functions of a complex variable. He advised his students to take his function theory course before attending his lectures on elliptic and Abelian functions. In his lectures, Weierstrass gave a rigorous step-by-step development of complex variables theory, employing the idea of uniform congruence, a concept absent in Eisenstein's work. Perhaps this is why Weierstrass did not refer to Eisenstein, even though many of their results in elliptic functions were nearly identical.

Weierstrass's teacher, Christoph Gudermann (1798–1852) appears to be the first mathematician to explicitly discuss the idea of uniform convergence. In an 1838 paper on elliptic modular functions he mentioned that certain series "einen im Ganzen gleichen Grad der Convergenz haben."[59] The term "gleichmäßige Konvergenz," uniform convergence, was first employed by Weierstrass in his 1841 paper, "Zur Theorie der

---

[58]  Riemann (1990) p. 576.
[59]  Gudermann (1838) Crelle's Journal vol. 18, pp. 220–258. This quote is on p. 252.

Potenzreihen."[60] Unfortunately, this foundational paper using power series as a basis for the study of analytic functions was not published until 53 years later, when it appeared in the first volume of Weierstrass's collected papers. In fact, another 1841 paper of Weierstrass, "Darstellung einer analytischen Function einer complexen Varänderlichen, deren absoluter Betrag zwischen zwei gegebenen Grenzen liegt,"[61] suffered the same fate. In the latter paper, Weierstrass derived the Laurent series expansion in the neighborhood of a singularity two years before Pierre Laurent did so.[62] In the same paper, Weierstrass also derived the removable singularity theorem, stated in Riemann's 1851 dissertation.[63]

Weierstrass took power series as the basis for his theory of analytic functions and their analytic continuation. In developing his theory of analytic functions, he avoided the use of integration in the complex domain; in his "Zur Theorie der Potenzreihen" of 1841, he proved without integration Cauchy's result that if

$$f(z) = \sum_{n=0}^{\infty} a_n z^n \text{ in } |z| < R \text{ and if } |f(z)| \leq M \text{ on } |z| = r < R,$$

then

$$|a_n| \leq M r^n.$$

Weierstrass did not include a discussion of Laurent's theorem in his lectures, perhaps because Laurent's 1841 proof utilized integration. It appears that Weierstrass viewed the basic operations of addition and multiplication as more fundamental than integration. This perspective proved valuable in the twentieth century, when analysts considered analytic functions of $p$-adic or other types of numbers.

In his 1862–63 lectures on elliptic functions, Weierstrass introduced the absolutely convergent infinite product

$$\sigma u = u \prod_{w}' \left( 1 - \frac{u}{w} \right) e^{\frac{u}{w} + \frac{1}{2}\frac{u^2}{w^2}}, \tag{6.90}$$

where $w = m\omega_1 + n\omega_2$ and $m, n$ ran over all integers except $(m, n) = (0, 0)$. Note that Weierstrass's practice was to take the periods to be $2\omega_1, 2\omega_2$. He gave an interesting proof that

$$\sum_{w}' \frac{1}{|w|^k}$$

was convergent for all $k \geq 3$.[64] Of course, it is sufficient to prove the result for $k = 3$. Presenting the Weierstrass's proof in its essentials, let

$$\omega_1 = a + bi, \quad \omega_2 = a' + b'i, \quad \text{Im} \frac{\omega_2}{\omega_1} \neq 0, \quad w = m\omega_1 + n\omega_2.$$

---

[60] Weierstrass (1894–1927) pp. 1–49.    [61] Weierstrass (1894–1927) pp. 51–66.    [62] See Laurent (1853).
[63] Riemann (1990) p. 55.    [64] Weierstrass (1894–1927) vol. 5, pp. 116–117.

Then

$$|w|^2 = (ma + na')^2 + (mb + nb')^2,$$

and $|w|^2$ is a positive definite quadratic form in $m$ and $n$. So there exist $M$ and $N$ that are linear combinations of $m$ and $n$ and there exist positive numbers $g$ and $h$ such that

$$|w|^2 = gM^2 + hN^2.$$

subject to the requirement that $m^2 + n^2 = M^2 + N^2$. Now let $g \leq h$; in that case,

$$|w|^2 = gM^2 + hN^2 \geq g(M^2 + N^2) = g(m^2 + n^2)$$

and it follows that

$$\sum_{w}' \frac{1}{|w|^3} \leq \frac{1}{g} \sum_{m,n}' \frac{1}{(m^2 + n^2)^{\frac{3}{2}}} = \frac{4}{g} \sum_{m,n=1}^{\infty} \frac{1}{(m^2 + n^2)^{\frac{3}{2}}}.$$

Since $m^2 + n^2 > mn$, we have

$$\sum_{m,n=1}^{\infty} \frac{1}{(m^2 + n^2)^{\frac{3}{2}}} < \sum_{m,n=1}^{\infty} \frac{1}{(mn)^{\frac{3}{2}}} < \sum_{m=1}^{\infty} \frac{1}{m^{\frac{3}{2}}} \sum_{n=1}^{\infty} \frac{1}{n^{\frac{3}{2}}}.$$

In fact, if we define the exponent of convergence of the zeros of $\sigma$ as the greatest lower bound of all $\alpha$, such that $\sum_{w}' \frac{1}{|w|^\alpha}$ converges, then a modification of the argument showing convergence for $\alpha = \frac{3}{2}$ shows that $r = 2$. Thus, Weierstrass proved the exponent of convergence of the zeros of the $\sigma$ function (6.90) to be 2.

From a general result proved by Weierstrass in his lectures on the theory of functions of a complex variable, he deduced $\sigma(u)$ to be an entire function with zeros only at $m\omega_1 + n\omega_2$, where $m$ and $n$ take all possible integer values. In his lectures on elliptic and Abelian functions, Weierstrass repeatedly applied the general results from his function theory lectures. Thus, the latter course of lectures were a prerequisite for the former. It is interesting to note that in 1882, Hermite introduced Weierstrass's methods in his analysis courses in Paris.

By taking the second derivative of the logarithm of $\sigma$, Weierstrass could obtain his $\wp$ function:

$$\wp u = -\frac{d^2}{du^2} \log \sigma u.$$

## 6.7 The Phragmén-Lindelöf Theorem

This theorem is of interest in our discussion of complex analysis because Hecke applied a form of it to derive his correspondence between a type of automorphic form and a Dirichlet series satisfying a functional relation. Probably the first example of this correspondence is found in the work of Riemann who proved that the relation

$$\theta_3\left(-\frac{1}{\omega}\right) = \sqrt{\frac{\omega}{i}}\, \theta_3(\omega)$$

implies the invariance of

$$\pi^{-\frac{s}{2}} \Gamma\left(\frac{s}{2}\right) \zeta(s) \quad \text{under} \quad s \to 1 - s.$$

To prove a converse of this result, the Phragmén-Lindelöf theorem is required.

Lars Eduard Phragmén (1863–1937) was a student of Gösta Mittag-Leffler, founder of *Acta Mathematica*. Phragmén assisted his teacher by proof-reading papers for the journal and in this service, he gained fame by finding a mistake in Poincaré's paper on the three-body problem. Poincaré was led to revise his paper and thereby discover mathematical chaos.[65] Phragmén was a professor of mathematics at Stockholm until 1903, when he was appointed chief inspector of insurance, though he continued to pursue mathematics. In 1904, he published an extension of the maximum modulus principle; this result can be seen as an early version of the Phragmén-Lindelöf theorem. In 1908, he collaborated with Lindelöf to obtain the correct form of this extension, the very important and useful result now bearing their names. A student of Hjalmar Mellin, Ernst Lindelöf (1870–1946) had a deep interest in complex analysis and he founded the noted Finnish school of complex analysts which came to include Rolf and Frithiof Nevanlinna and Lars Ahlfors.[66]

The maximum modulus principle states that if a function $f$ is analytic in a region $D$, then $|f|$ cannot have a proper local maximum in $D$. In his 1850 lectures on elliptic functions, Liouville gave an early version of this principle.[67] To prove that an elliptic function without singularities was a constant, he argues that such a function would have a maximum at some point, say $z_0$. He assumed the function to be nonconstant and analytic in the disk $|z - z_0| \le r$, so that by Cauchy's integral formula,

$$A = |f(z_0)| = \frac{1}{2\pi} \left| \int_0^{2\pi} f(z_0 + re^{i\theta}) \, d\theta \right|$$

$$< \frac{1}{2\pi} \int_0^{2\pi} |f(z_0 + re^{i\theta})| \, d\theta \le A.$$

This contradiction yielded the necessary result. Liouville's formulation of the maximum principle implies that if $f(z)$ is analytic in a region $D$ and continuous on $D \cup \partial D$, where $\partial D$ is the boundary of $D$, and if $|f(z)| \le M$ for $z \in \partial D$, then $|f(z)| \le M$ for $z \in D$.

Phragmén and Lindelöf's very important extension of the maximum principle has as its basic idea that if $|f(z)| \le M$ for all $z \in \partial D$ except for one point $z_0$, then $|f(z)| \le M$ for $z \in D$, provided that $|f(z)|$ does not grow to rapidly as $z \to z_0$. Usually, the exceptional point is at infinity.

Phragmén and Lindelöf started their paper[68] with a general formulation of their idea, though their statement of (6.91) was slightly less precise. They supposed $\partial D$ to be a simple closed curve containing the region $D$ such that $f$ is analytic on $D \cup \partial D$, except

[65] See Barrow-Green (1997).

[66] For a fascinating discussion of Finnish mathematics, see Elfving (1981).

[67] Peiffer (1983), especially p. 229.     [68] Phragmén and Lindelöf (1908).

at a point $P \in \partial D$. Note, however, that they themselves allowed for more than one exceptional point.

They supposed also that there exists a function $\omega(z)$ analytic and nonzero in $D \cup \partial D$, such that $|\omega(z)| \leq 1$ in $D$. Let $\sigma > 0$, and suppose there exists a system of curves arbitrarily close to $P$ joining the two sides of $\partial D$ around $P$, on which

$$|\omega^\sigma(z) f(z)| \leq M. \qquad (6.91)$$

Then

$$|f(z)| \leq M.$$

In his interesting and informative work on the history of mathematics in Sweden, Gårding gives insight into the collaboration of Phragmén and Lindelöf: "One might ask who got the idea or had the insight to find the right formulation of Phragamén's problem [of finding the form of the extension]. Since Phragamén is not the sole author of his most important paper it is easy to guess that the decisive idea came from Lindelöf, an opinion that his compatriots could express in private."[69]

The manner in which $\omega(z)$ is chosen depends upon the region $D$. Recall that the exceptional point is usually at infinity. If $D$ were chosen to be the region $|\arg z| < \alpha < \pi$, then $\omega(z)$ could be chosen as

$$\omega(z) = \exp(-z^\delta),$$

where $0 < \delta < \frac{\pi}{2\alpha}$. In applications to Dirichlet series $\sum_{n=1}^\infty a_n n^{-s}$, the region $D \cup \partial D$ would be an infinite strip $\sigma_1 \leq \operatorname{Re} s \leq \sigma_2$. We state a form of the Phragmén-Lindelöf theorem for such a strip, useful in drawing a connection between modular forms and Dirichlet series. Let $\phi(s)$, $s = \sigma + it$, be analytic in a region containing $D_1 = D \cup \partial D$ and for $\gamma > 0$ let

$$|\phi(s)| = O(e^{|t|^\gamma}), \quad |t| \to \infty, \qquad (6.92)$$

uniformly in $D_1$. Suppose that on the vertical lines $\operatorname{Re} s = \sigma_1$ and $\operatorname{Re} s = \sigma_2$ and for a real number $\mu$

$$|\phi(s)| = O(|t|^\mu) \quad \text{as} \quad |t| \to \infty.$$

Then

$$|\phi(s)| = O(|t|^\mu) \quad \text{as} \quad |t| \to \infty,$$

uniformly in $D_1$. In this case, $\omega(s)$ can be taken as $e^{s^m}$ for a suitably chosen integer $m$.

This result is often used in conjunction with the asymptotic behavior of the $\Gamma$ function in a vertical strip: If

$$s = \sigma + it, \quad \sigma_1 \leq \sigma \leq \sigma_2, \text{ and } |t| \to \infty,$$

---

[69] Gårding (1998) p. 114.

then

$$\left|\Gamma(s)\right| = \sqrt{2\pi}\,|t|^{\sigma-\frac{1}{2}}\,e^{-\frac{\pi}{2}|t|}\left(1 + O\left(\frac{1}{|t|}\right)\right), \tag{6.93}$$

where the constant implied by $O$ depends only on $\sigma_1$ and $\sigma_2$. The formula (6.93) may be derived from Stirling's formula:[70]

$$\Gamma(s) = \sqrt{2\pi}\,s^{s-\frac{1}{2}}\,e^{-s+\mu(s)} \tag{6.94}$$

where $\mu(s) \to 0$ uniformly, in some half-plane $\sigma \geq \sigma_0 > 0$, as $|s| \to \infty$.

---

[70] A history of Stirling's formula can be found in Chapter 24 of Roy (2011).

# 7

# *Hypergeometric Functions*

## 7.1 Preliminary Remarks

The complete elliptic integral $K(k)$ may be expressed as a hypergeometric series

$$
\begin{aligned}
K(k) &= \int_0^{\frac{\pi}{2}} \frac{d\theta}{\sqrt{1 - k^2 \sin^2 \theta}} \\
&= \int_0^{\frac{\pi}{2}} \left(1 + \frac{1}{2}k^2 \sin^2 \theta + \frac{1 \cdot 3}{2 \cdot 4}k^4 \sin^4 \theta + \frac{1 \cdot 3 \cdot 5}{2 \cdot 4 \cdot 6} k^6 \sin^6 \theta + \cdots \right) d\theta \\
&= \frac{\pi}{2} \left(1 + \frac{1^2}{2^2}k^2 + \frac{1^2 \cdot 3^2}{2^2 \cdot 4^2}k^4 + \frac{1^2 \cdot 3^2 \cdot 5^2}{2^2 \cdot 4^2 \cdot 6^2} k^6 + \cdots \right),
\end{aligned}
\tag{7.1}
$$

when $|k| < 1$. Gauss found this series for $K(k)$ in the 1790s; this may have been the motivation for his general study of hypergeometric series. Examining the early development of the theory of hypergeometric series, particularly in the work of Euler, Gauss, and Kummer, may help us understand how Riemann and Schwarz utilized functions defined by these series in their study of conformal mappings.

A hypergeometric series takes the general form

$$
{}_2F_1 \left( \begin{matrix} a, & b \\ & c \end{matrix} ; x \right) = \sum_{n=0}^{\infty} \frac{(a)_n (b)_n}{n!(c)_n} x^n,
\tag{7.2}
$$

where a shifted factorial, $(e)_n$, is defined by

$$
(e)_0 = 1, \quad (e)_n = e(e+1)\cdots(e+n-1), \quad \text{for } n \geq 1.
\tag{7.3}
$$

The elementary functions and many other special functions are representable as hypergeometric series or as limiting cases of such series. In this sense, Stirling, Newton, Leibniz, and the Bernoulli brothers Jacob and Johann were familiar with several particular cases of hypergeometric series.

## 7.2 Stirling

The first mathematician to consider hypergeometric series in a somewhat general form was the Scottish mathematician James Stirling, who had a slightly unusual mathematical career. After receiving his education in Scotland, Stirling received a scholarship to Oxford in 1710. Because of his family's sympathies with the Jacobite cause, he lost his support in 1715 during the first Jacobite rebellion. He spent several years in Italy, where he was unsuccessful in securing an academic position, though during that time he published two papers in the *Philosophical Transactions* on the method of finite differences and on extending Newton's work on cubic curves. Stirling returned to Britain in 1722 and continued his mathematical research, publishing his noted *Methodus Differentialis* in 1730. His Jacobite connections continued to prevent his obtaining an academic position; in 1735, Stirling finally took the post of manager of the Leadhills Mines in Scotland, a job in which he was very successful.

In his *Methodus*[1], Stirling showed how the calculus of finite differences could be applied to the transformation and approximate summation of series. Observe that while power series work well under differentiation because $\frac{dx^n}{dx} = nx^{n-1}$, a series of shifted factorials is more effective with finite differences, because

$$\Delta(a)_n = (a+1)_n - (a)_n$$
$$= n(a+1)_{n-1}. \tag{7.4}$$

This explains why Stirling, in his *Methodus*, introduced the numbers named after him. He employed them to convert rational functions to infinite series with shifted factorial terms, and conversely. Note that, in modern terms, the definition of Stirling numbers of the first kind, $S(m, k)$, may be given as

$$x(x-1)(x-2)\cdots(x-m+1) = \sum_{k=1}^{m} S(m, k)x^k. \tag{7.5}$$

Taking the plus signs in his shifted factorials, Stirling had

$$(x)_m = x(x+1)(x+2)\cdots(x+m-1) = \sum_{k=1}^{m} |S(m, k)|\, x^k;$$

he then noted that

$$\frac{1}{x^{n+1}} = \sum_{m=n}^{\infty} \frac{|S(m, n)|}{(x)_m}. \tag{7.6}$$

He defined Stirling numbers of the second kind, now denoted by $s(m, k)$, as

$$x^m = \sum_{k=1}^{m} s(m, k)\, x(x-1)(x-2)\cdots(x-k+1). \tag{7.7}$$

---

[1] For an English translation with extensive notes by I. Tweddle, see Stirling and Tweddle (2003).

From this definition, it is not easy to determine the numbers $s(m, k)$; thus, Stirling provided a generating function:

$$\frac{1}{(x-1)(x-2)\cdots(x-m)} = \sum_{n=m}^{\infty} \frac{s(n, m)}{x^n}. \tag{7.8}$$

He stated (7.6) and (7.8) without proof; his use of these formulas suggests that he was aware of several properties of Stirling numbers not stated in his book. He applied (7.7) for the approximate summation of series, such as $\sum_{n=1}^{\infty} \frac{1}{n^2}$. This was written several years before Euler found the exact value to be $\frac{\pi^2}{6}$. Stirling also found some transformation formulas for hypergeometric series that transformed series to more rapidly convergent series. For example, in modern notation, he found the transformation formula:

$$_2F_1\left(\begin{matrix} 1, & z-n \\ & z \end{matrix}; \frac{1}{1-m}\right) = \left(1 - \frac{1}{m}\right){}_2F_1\left(\begin{matrix} 1, & n \\ & z \end{matrix}; \frac{1}{m}\right). \tag{7.9}$$

This formula contained as a particular case a formula of Newton, dating from 1684, where $n = 1$, $z = \frac{3}{2}$, and $m = \frac{t^2+1}{t^2}$:

$$\sum_{k=0}^{\infty} (-1)^k \frac{t^{2k+1}}{2k+1} = \frac{1}{1+t^2} \sum_{k=0}^{\infty} \frac{2\cdot4\cdot6\cdots(2k)}{3\cdot5\cdot7\cdots(2k+1)} \left(\frac{t^2}{1+t^2}\right)^2. \tag{7.10}$$

Because the left-hand side of (7.10) is $\arctan t$ for $|t| \leq 1$, if we take $t = 1$, the value of the left-hand side is $\frac{\pi}{4}$. Thus, (7.10) produces

$$\frac{\pi}{4} = \frac{1}{2}\left(1 + \frac{1}{3} + \frac{1\cdot2}{3\cdot5} + \frac{1\cdot2\cdot3}{3\cdot5\cdot7} + \frac{1\cdot2\cdot3\cdot4}{3\cdot5\cdot7\cdot9} + \cdots\right). \tag{7.11}$$

To find an approximate value of $\pi$ from (7.11), Stirling summed the first twelve terms of this series and then applied (7.9) to the remainder:

$$\frac{1}{2}\left(\frac{1\cdot2\cdot3\cdots12}{3\cdot5\cdot7\cdots25} + \frac{1\cdot2\cdot3\cdots13}{3\cdot5\cdot7\cdots27} + \cdots\right)$$

$$= \frac{1}{2}\cdot\frac{12!}{3\cdot5\cdots25}\,{}_2F_1\left(\begin{matrix} 1, & 13 \\ & \frac{27}{2} \end{matrix}; \frac{1}{2}\right)$$

$$= \frac{12!}{3\cdot5\cdots25}\,{}_2F_1\left(\begin{matrix} 1, & \frac{1}{2} \\ & \frac{27}{2} \end{matrix}; -1\right)$$

$$= \frac{12!}{3\cdot5\cdots25}\left(1 - \frac{1}{27} + \frac{1\cdot3}{27\cdot29} - \frac{1\cdot3\cdot5}{27\cdot29\cdot31} + \cdots\right). \tag{7.12}$$

Next, he summed the first twelve terms of the last series in (7.12) to obtain 0.7853981639 as an approximate value for $\frac{\pi}{4}$. Note that Stirling proved (7.9) by showing that both sides of the equation satisfied the same difference equation.

## 7.3  Euler and the Hypergeometric Equation

Euler introduced the general form of the hypergeometric series (7.2) in 1778[2] and it appears that he was the first to do this. Euler proved that the hypergeometric series (7.2), denoted by $s$, satisfied the differential equation

$$x(1-x)\frac{d^2y}{dx^2} + \left(c - (1+a+b)x\right)\frac{dy}{dx} - aby = 0, \qquad (7.13)$$

called the hypergeometric differential equation. Euler, Pfaff, Gauss, and Kummer used this equation to obtain transformation formulas for hypergeometric functions. Later, Riemann and Schwarz applied it to show the connection of hypergeometric functions with conformal mappings and modular functions. To prove that $s$ satisfied (7.13), Euler noted that

$$\frac{d}{dx}\left(x^c\frac{ds}{dx}\right) = abx^{c-1} + \frac{ab}{1\cdot c}(a+1)(b+1)x^c + \cdots$$

$$\frac{d}{dx}(x^a s) = ax^{a-1} + \frac{ab}{1\cdot c}(a+1)x^a + \cdots .$$

Next he had

$$\frac{d}{dx}\left(x^{b+1-a}\frac{d}{dx}(x^a s)\right) = abx^{b-1} + \frac{ab}{1\cdot c}(a+1)(b+1)x^b + \cdots$$

$$= x^{b-c}\frac{d}{dx}\left(x^c\frac{ds}{dx}\right),$$

or

$$\frac{d}{dx}\left(ax^b s + x^{b+1}\frac{ds}{dx}\right) = x^{b-c}\left(cx^{c-1}\frac{ds}{dx} + x^c\frac{d^2s}{dx^2}\right)$$

or

$$a\left(bx^{b-1}s + x^b\frac{ds}{dx}\right) + (b+1)x^b\frac{ds}{dx} + x^{b+1}\frac{d^2s}{dx^2} = cx^{b-1}\frac{ds}{dx} + x^b\frac{d^2s}{dx^2}. \qquad (7.14)$$

Euler obtained the differential equation (7.13) after dividing (7.14) by $x^{b-1}$. He applied (7.13) to find the transformation

$${}_2F_1\left(\begin{matrix}a, & b\\ & c\end{matrix}; x\right) = (1-x)^{c-a-b}\,{}_2F_1\left(\begin{matrix}c-a, & c-b\\ & c\end{matrix}; x\right). \qquad (7.15)$$

He obtained (7.15) by showing that if $s = (1-x)^n z$, then $z$ would also satisfy a differential equation of hypergeometric form when $n = c - a - b$. He began by observing that

$$\frac{1}{s}\frac{ds}{dx} = \frac{1}{z}\frac{dz}{dx} - \frac{n}{1-x}. \qquad (7.16)$$

[2] "Specimen transformationis singularis serierum" Eu.I-16-2, pp. 41–55.

Taking the derivative of (7.16) yielded

$$\frac{1}{s}\frac{d^2s}{dx^2} - \frac{1}{s^2}\left(\frac{ds}{dx}\right)^2 = \frac{1}{z}\frac{d^2z}{dx^2} - \frac{1}{z^2}\left(\frac{dz}{dx}\right)^2 - \frac{n}{(1-x)^2}. \tag{7.17}$$

Squaring (7.16), Euler arrived at

$$\frac{1}{s^2}\left(\frac{ds}{dx}\right)^2 = \frac{1}{z^2}\left(\frac{dz}{dx}\right)^2 - \frac{2n}{z(1-x)}\frac{dz}{dx} + \frac{n^2}{(1-x)^2}; \tag{7.18}$$

adding (7.18) and (7.17) produced

$$\frac{1}{s}\frac{d^2s}{dx^2} = \frac{1}{z}\frac{d^2z}{dx^2} - \frac{2n}{z(1-x)}\frac{dz}{dx} + \frac{n(n-1)}{(1-x)^2}. \tag{7.19}$$

Taking the expressions for $\frac{1}{s}\frac{ds}{dx}$ and $\frac{1}{s^2}\frac{d^2s}{dx^2}$ from (7.16) and (7.19) and substituting them into the differential equation (7.13), Euler had

$$\frac{x(1-x)}{z}\frac{d^2z}{dx^2} - \frac{2nx}{z}\frac{dz}{dx} + \frac{\left(c-(a+b+1)x\right)}{z}\frac{dz}{dx}$$
$$+ \frac{n(n-1)x}{1-x} - \frac{n\left(c-(a+b+1)x\right)}{1-x} - ab = 0. \tag{7.20}$$

On combining the two terms with the common denominator $1 - x$, he had

$$\frac{n\left((n+a+b)x - c\right)}{1-x};$$

when $n = c - a - b$, the $1 - x$ would cancel. Thus, for $n = c - a - b$, (7.20) became

$$x(1-x)\frac{d^2z}{dx^2} + [c + (a+b-2c-1)x]\frac{dz}{dx} - (c-a)(c-b)z = 0, \tag{7.21}$$

a hypergeometric equation in which $a, b, c$ were replaced by $c - a, c - b, c$ respectively. Thus, since

$$z = {}_2F_1\left(\begin{matrix} c-a, & c-b \\ & c \end{matrix}; x\right),$$

was a solution of (7.21) and the first two terms of the series on each side of (7.15) were the same, the proof of (7.15) was complete.

### 7.4  Pfaff's Transformation

In his *Disquisitiones Analyticae* of 1797, J. F. Pfaff proved a generalization of Stirling's transformation (7.9):

$${}_2F_1\left(\begin{matrix} a, & b \\ & c \end{matrix}; x\right) = (1-x)^{-a}\,{}_2F_1\left(\begin{matrix} a, & c-b \\ & c \end{matrix}; \frac{x}{x-1}\right). \tag{7.22}$$

Pfaff stated his transformation for the case $a = -n$ with $n$ a positive integer. This condition reduced the two ${}_2F_1$ functions in (7.22) to polynomials.

We note that in 1836 Kummer wrote a paper on hypergeometric functions, published in *Crelle's Journal*,[3] in which he showed Euler's transformation (7.15) to be a corollary of (7.22). An application of (7.22) to the $_2F_1$ on the right-hand side of (7.22) gives

$$_2F_1\left(\begin{matrix} c-b, & a \\ & c \end{matrix}; \frac{x}{x-1}\right) = \left(1 - \frac{x}{x-1}\right)^{-c+b} {}_2F_1\left(\begin{matrix} c-a, & c-b \\ & c \end{matrix}; x\right)$$
$$= (1-x)^{c-b} {}_2F_1\left(\begin{matrix} c-a, & c-b \\ & c \end{matrix}; x\right). \qquad (7.23)$$

Euler's transformation (7.15) follows when (7.23) is applied to the right-hand side of (7.22). We remark that formulas (7.15) and (7.22) are very important, especially for their numerous applications in analysis; we do not consider these here, since our primary focus is elliptic and modular functions.

## 7.5 Gauss and Quadratic Transformations

Gauss may have become interested in hypergeometric series through his contact with Pfaff, who was his teacher at Helmstedt and nominal dissertation advisor in the fall of 1798. In fact, Gauss resided in Pfaff's house during that period. Gauss wrote a paper on the hypergeometric function, the first part of which was published in 1812; he did not publish the second part. In this paper, Gauss defined the concept of contiguous hypergeometric functions; he desgnated two such functions contiguous if they both had the same power series variable $x$; if two of the three parameters were pairwise the same; and if the third pair differed by 1. Gauss showed that a hypergeometric function and any two functions contiguous to it are linearly related. Since there are six functions contiguous to a given $_2F_1$, we get $\binom{6}{2} = 15$ relations. Gauss derived several basic results from these fifteen contiguous relations, including the hypergeometric differential equation and an important summation formula for hypergeometric functions. The differential equation arises from the fact that

$$\frac{d}{dx} {}_2F_1\left(\begin{matrix} a, & b \\ & c \end{matrix}; x\right) = \frac{ab}{c} {}_2F_1\left(\begin{matrix} a+1, & b+1 \\ & c+1 \end{matrix}; x\right). \qquad (7.24)$$

Gauss may possibly have gleaned the idea and the importance of contiguous functions from his careful reading of Stirling's *Methodus Differentialis*. Stirling obtained his transformation formula (7.9) by substituting two contiguous functions in an appropriate difference equation.

Gauss regarded the variable $x$ and the parameters $a$, $b$, $c$ in the hypergeometric function as taking on complex values. As his 1811 letter to Bessel clearly reveals, Gauss was in the process of developing his theory of functions of complex variables; one may observe the impact of this work in the second part of his paper on hypergeometric functions. Another innovation of Gauss consisted of the quadratic transformation of hypergeometric functions. For example, Gauss found the transformation, here given in

[3] Kummer (1975) vol. 2, pp. 75–166. H. Nagaoka provided an English translation on pages 151–253 of *Memoirs on Infinite Series* published by the Tokio [sic] Mathematical and Physical Society, 1891.

modern notation,

$$
{}_2F_1 \left( \begin{matrix} a, & b \\ & a+b+\tfrac{1}{2} \end{matrix} ; \ 4x - 4x^2 \right) = {}_2F_1 \left( \begin{matrix} 2a, & 2b \\ & a+b+\tfrac{1}{2} \end{matrix} ; \ x \right), \qquad (7.25)
$$

which he proved by means of the hypergeometric differential equation. When $x$ is changed to $1 - x$, the left-hand side of (7.25) remains unchanged; this seems to imply the contradictory result that

$$
{}_2F_1 \left( \begin{matrix} 2a, & 2b \\ & a+b+\tfrac{1}{2} \end{matrix} ; \ x \right) = {}_2F_1 \left( \begin{matrix} 2a, & 2b \\ & a+b+\tfrac{1}{2} \end{matrix} ; \ 1-x \right). \qquad (7.26)
$$

Gauss called this result a paradox, and his explanation indicates that in 1812 he had conceived of analytic continuation. In the unpublished portion of his paper, he wrote:[4]

> To explain this, it ought to be remembered that proper distinction should be made between the two significations of the symbol $F$ [i.e., ${}_2F_1$], viz., whether it represents the function whose nature is expressed by the differential equation [(7.13)], or simply the sum of an infinite series. The latter is always a perfectly determinate quantity so long as the fourth element lies between $-1$ and $+1$, and care must be taken not to exceed these limits for otherwise it is entirely without any meaning. On the other hand, according to the former signification, it [$F$] represents a general function which always varies subject to the law of continuity if the fourth element vary continuously whether you attribute real values or imaginary values to it, provided you always avoid the values 0 and 1. Hence it is evident that in the latter sense, the function may for equal values of the fourth element (the passage or rather the return being made through imaginary quantities) attain unequal values of which that which the *series F* represents is only one, so that it is not at all contradictory that while some *one* value of the function $F(a, b, a + b + 1/2, 4y - 4yy)$ is equal to $F(2a, 2b, a + b + 1/2, y)$ the *other* value should be equal to $F(2a, 2b, a + b + 1/2, 1 - y)$ and it would be just as absurd to deduce thence the equalities of these values as it would be to conclude, that since Arc. sin $\tfrac{1}{2} = 30°$, Arc. sin $\tfrac{1}{2} = 150°$, $30° = 150°$. – But if we take $F$ in the less general sense, viz. simply as the sum of the series $F$, the arguments by which we have deduced (7.25), necessarily suppose $y$ to increase from the value 0 only up to the point when $x$ [$= 4y - 4yy$] becomes $= 1$, i.e., up to $y = 1/2$. At this point, indeed, the *continuity* of the series $P = F(a, b, a + b + 1/2, 4y - 4yy)$ is interrupted, for evidently $\frac{dP}{dy}$ jumps suddenly from a positive (finite) value to a negative. Thus in this sense equation (7.25) does not admit of being extended outside the limits $y = 1/2 - \sqrt{1/2}$ up to $y = 1/2$. If perferred, the same equation can also be put thus:–
>
> $$
> F\left(a, b, a + b + \frac{1}{2}, x\right) = F\left(2a, 2b, a + b + \frac{1}{2}, \frac{1 - \sqrt{1 - x}}{1}\right).
> $$

Gauss may have been motivated to study quadratic transformations because of their connection with the arithmetic-geometric mean and elliptic integrals. To see the connection with the arithmetic-geometric mean, note that Gauss proved the quadratic transformation

$$
{}_2F_1 \left( \begin{matrix} a, & b \\ & 2b \end{matrix} ; \ \frac{4x}{(1+x)^2} \right) = (1+x)^{2a} \, {}_2F_1 \left( \begin{matrix} a, & a - b + \tfrac{1}{2} \\ & b + \tfrac{1}{2} \end{matrix} ; \ x^2 \right). \qquad (7.27)
$$

---

[4] This excerpt from the translation of Gauss's Latin paper is by K. Kikuchi, and it appeared on pages 144–145 of *Memoirs on Infinite Series*, an 1891 publication of the Tokio [sic] Mathematical and Physical Society.

If we set

$$x = \frac{1 - \sqrt{1 - y^2}}{1 + \sqrt{1 - y^2}},$$

then (7.27) may be written as

$$
{}_2F_1\left(\begin{array}{cc} a, & b \\ & 2b \end{array}; y^2\right)
$$

$$
= \left(\frac{1 + \sqrt{1 - y^2}}{2}\right)^{-2a} {}_2F_1\left(\begin{array}{cc} a, & a - b + \frac{1}{2} \\ & b + \frac{1}{2} \end{array}; \left(\frac{1 - \sqrt{1 - y^2}}{1 + \sqrt{1 - y^2}}\right)^2\right). \tag{7.28}
$$

When (7.28) is then applied to the expression for the complete elliptic integral $K(k)$ in (7.1), we obtain

$$K(k) = \frac{2}{1 + k'} K\left(\frac{1 - k'}{1 + k'}\right), \tag{7.29}$$

where $k' = \sqrt{1 - k^2}$. Now equation (7.29) is equivalent to the statement that

$$I(1, k') = I\left(\frac{1 + k'}{2}, \sqrt{k'}\right), \tag{7.30}$$

where $I(a, b)$ is defined by (2.61), that is

$$I(a, b) = \frac{2}{\pi} \int_0^{\frac{\pi}{2}} \frac{d\theta}{\sqrt{a^2 \cos^2 \theta + b^2 \sin^2 \theta}}.$$

Note that (7.30) gives the connection of the elliptic integral $I(a, b)$ with the arithmetic-geometric mean of $a$ and $b$.

Although Gauss may not have arrived at the explicit concept of a regular singular point of a differential equation, he found series solutions of the hypergeometric equation in the neighborhood of its three singularities $0, 1, \infty$. He gave the two independent solutions in the neighborhood of $x = 0$ as

$$
{}_2F_1\left(\begin{array}{cc} a, & b \\ & c \end{array}; x\right) \quad \text{and} \quad x^{1-c}\,{}_2F_1\left(\begin{array}{cc} a - c + 1, & b - c + 1 \\ & 2 - c \end{array}; x\right); \tag{7.31}
$$

the solutions in the neighborhood of $x = 1$ as

$$
{}_2F_1\left(\begin{array}{cc} a, & b \\ & a + b - c + 1 \end{array}; 1 - x\right)
$$

$$
\text{and} \quad (1 - x)^{c-a-b}\,{}_2F_1\left(\begin{array}{cc} c - a, & c - b \\ & c - a - b + 1 \end{array}; 1 - x\right); \tag{7.32}
$$

and the solutions in the neighborhood of $x = \infty$ as

$$
(-x)^{-a}\,{}_2F_1\left(\begin{array}{cc} a, & a - c + 1 \\ & a - b + 1 \end{array}; \frac{1}{x}\right) \quad \text{and} \quad (-x)^{-b}\,{}_2F_1\left(\begin{array}{cc} b, & b - c + 1 \\ & b - a + 1 \end{array}; \frac{1}{x}\right). \tag{7.33}
$$

Gauss was well aware that since any three solutions of the hypergeometric equation must be linearly dependent, there exists a linear relation among these solutions; he found the relation between $_2F_1\left(\begin{smallmatrix} a, & b \\ & c \end{smallmatrix} ; x\right)$ and the two independent solutions in (7.33).

We note in particular that, by (7.1) with $x = k^2$,

$$K(k) = {}_2F_1\left(\begin{smallmatrix} \frac{1}{2}, & \frac{1}{2} \\ & 1 \end{smallmatrix} ; k^2\right),$$

(7.34)

is a solution of the hypergeometric equation

$$x(1-x)\frac{d^2y}{dx^2} + (1-2x)\frac{dy}{dx} - \frac{1}{4}y = 0.$$

(7.35)

From the second solution in (7.32), we see that

$$K'(k) = {}_2F_1\left(\begin{smallmatrix} \frac{1}{2}, & \frac{1}{2} \\ & 1 \end{smallmatrix} ; 1-k^2\right) = {}_2F_1\left(\begin{smallmatrix} \frac{1}{2}, & \frac{1}{2} \\ & 1 \end{smallmatrix} ; k'^2\right) = K(k').$$

(7.36)

Thus, the periods $2K$ and $2\sqrt{-1}K'$ of the Jacobi elliptic functions are independent solutions of the hypergeometric equation (7.35).

## 7.6  Kummer on the Hypergeometric Equation

We recall that Gauss did not publish the second part of his paper, the part treating the transformation of hypergeometric functions. This transformation theory was independently rediscovered by Ernst Kummer (1810–1893) and published in 1836. Kummer was a secondary school teacher from 1831 to 1841; after this, with the assistance of Jacobi and Dirichlet, he became a professor at Breslau. Kummer also served in the Prussian military during the year 1834. At that time, he communicated some papers in analysis to Jacobi who is reported by Emil Lampe to have commented: "There we are; now the Prussian Musketeers even enter into competition with the professors by way of mathematical works."[5] Kummer became professor at Berlin in 1855; by then, he had nearly completed his pioneering work on cyclotomic fields with applications to Fermat's last theorem.

Kummer began his study of hypergeometric functions with the consideration a general problem, given here in the hypergeometric case:

Suppose $v$ is a solution of the hypergeometric equation

$$z(1-z)\frac{d^2v}{dz^2} + \left(c' - (a'+b'+1)z\right)\frac{dv}{dz} - a'b'v = 0.$$

(7.37)

Determine $w$ and $z$ as functions of a variable $x$ such that $y = wv$ satisfies another hypergeometric equation

$$x(1-x)\frac{d^2y}{dx^2} + \left(c - (a+b+1)x\right)\frac{dy}{dx} - aby = 0.$$

(7.38)

---

[5] Lampe was a student of Weierstrass and Kummer. For this remark contained in Lampe's obituary of Kummer, see Kummer (1975) vol. 1, p. 18. Translation by Pieper: Goldstein et al. (2007) p. 214. For Kummer's paper on hypergeometric series, see vol. 2 of Kummer (1975) pp. 75–162.

In solving this problem, Kummer found that if $\dot{z}$ denoted the derivative of $z$ with respect to $x$, then $z$ must satisfy the differential equation

$$2\frac{\ddot{z}}{\dot{z}} - 3\left(\frac{\ddot{z}}{\dot{z}}\right)^2 - \frac{A'z^2 + B'z + C'}{z^2(1-z)^2} \cdot \dot{z}^2 + \frac{Ax^2 + Bx + C}{x^2(1-x)^2} = 0, \qquad (7.39)$$

where

$$A = (a-b)^2 - 1, \quad B = 4ab - 2c(a+b-1), \quad C = c(c-2) \qquad (7.40)$$

and $A'$, $B'$, $C'$ were similarly defined in terms of $a'$, $b'$, $c'$. He assumed that $a'$, $b'$, and $c'$ were linearly dependent on $a, b, c$ and $1$; he showed that in this case with $ps - qr \neq 0$

$$z = \frac{px + q}{rx + s}, \qquad (7.41)$$

a linear fractional transformation of $x$. Kummer then substituted this $z$ in (7.39), noting that

$$2\frac{\dddot{z}}{\dot{z}} - 3\left(\frac{\ddot{z}}{\dot{z}}\right)^2 = 0. \qquad (7.42)$$

The expression on the left-hand side of (7.42) is called the Schwarzian derivative of $z$. Thus, Kummer had shown that if $z$ was a fractional linear transformation of $x$, then its Schwarzian derivative would vanish. Thus, with $z$ given by (7.41), (7.39) would reduce to

$$\frac{Ax^2 + Bx + C}{x^2(1-x)^2} = \frac{A'z^2 + B'z + C'}{z^2(1-z)^2} \cdot \dot{z}^2. \qquad (7.43)$$

Kummer solved this equation, obtaining six values for $z$:

$$z = x, \quad z = 1 - x, \quad z = \frac{1}{x}, \quad z = \frac{1}{1-x}, \quad z = \frac{x}{x-1}, \quad z = \frac{x-1}{x}. \qquad (7.44)$$

By taking the first and second derivatives with respect to $x$ of the equation $y = wv$ and using (7.37) and (7.38), Kummer showed that $w$ satisfied the equation

$$2\frac{dw}{w} + \frac{\ddot{z}}{\dot{z}} + \frac{c - (a+b+1)x}{x(1-x)} dx - \frac{c' - (a'+b'+1)z}{z(1-z)} dz = 0. \qquad (7.45)$$

This differential equation was easily integrated; he found its solution to be

$$w^2 = c \cdot x^{-c}(1-x)^{c-a-b-1}z^{c'}(1-z)^{-c'+a'+b'+1}\frac{dx}{dz}. \qquad (7.46)$$

Now taking the first case in (7.44), namely $z = x$, and substituting it in (7.43) he got the relations

$$A = A', \quad B = B', \quad C = C'. \qquad (7.47)$$

Using (7.40), Kummer got

$$(a-b)^2 = (a'-b')^2, \quad 4ab - 2c(a+b-1) = 4a'b' - 2c'(a'+b'-1),$$
$$c(c-2) = c'(c'-2); \qquad (7.48)$$

he observed that the equations (7.48) could be satisfied in four different ways:

1. $a' = a$,              $b' = b$,              $c' = c$;
2. $a' = c - a$,          $b' = c - b$,          $c' = c$;
3. $a' = a - c + 1$,      $b' = b - c + 1$,      $c' = 2 - c$;
4. $a' = 1 - a$,          $b' = 1 - b$,          $c' = 2 - c$.

When these values were substituted in (7.46), Kummer perceived that $w$ could be one of four different factors:

$$w = k, \quad w = k(1-x)^{c-a-b}, \quad w = kx^{1-c}, \quad w = kx^{1-c}(1-x)^{c-a-b},$$

where $k$ represented a constant. Thus, when $z = x$, there were four solutions of the hypergeometric equation, corresponding to these four cases; these solutions were:

1. $\displaystyle {}_2F_1\left(\begin{matrix} a, & b \\ & c \end{matrix}; x\right)$,

2. $\displaystyle (1-x)^{c-a-b}\,{}_2F_1\left(\begin{matrix} c-a, & c-b \\ & c \end{matrix}; x\right)$,

3. $\displaystyle x^{1-c}\,{}_2F_1\left(\begin{matrix} a-c+1, & b-c+1 \\ & 2-c \end{matrix}; x\right)$,

4. $\displaystyle x^{1-c}(1-x)^{c-a-b}\,{}_2F_1\left(\begin{matrix} 1-a, & 1-b \\ & 2-c \end{matrix}; x\right)$.

Thus, Kummer arrived at 24 solutions, four for each of the six values of $z$ in (7.44). He then found the relations satisfied by any three of the 24 solutions.

### 7.7   Riemann and the Schwarzian Derivative

In the course of his researches, Riemann observed that many important functions of a complex variable and their properties were more or less determined by their singularities; he perceived that the types and locations of the singularities bore a close relationship with the global behavior of the function. He illustrated this in his lectures on elliptic functions as well as in his papers on Abelian functions, hypergeometric functions, and the zeta function.

We note that Kummer had derived 24 solutions of the hypergeometric differential equation and the relations among them. Riemann found a method employing analytic continuation and a notation that helped him state Kummer's results more succinctly and with a minimum of claculation. In his 1857 paper, "Beiträge zur Theorie der durch die Gauss'sche Reihe $F(\alpha, \beta, \gamma, x)$ darstellbaren Functionen," Riemann denoted the set of all solutions of the hypergeometric equation by

$$P\left\{\begin{matrix} 0 & \infty & 1 & \\ 0 & a & 0 & x \\ 1-c & b & c-a-b & \end{matrix}\right\}$$

Here

1. $0, \infty, 1$ denote the singularities of the solutions;
2. $0, 1-c$ are the exponents of $x$ at 0 (See (7.31));
3. $0, c-a-b$ are the exponents of $1-x$ at 1 (7.32);
4. $a, b$ are the exponents of $\frac{1}{x}$ at $\infty$ (7.33).

In fact, Riemann started with a set of functions with three singularities: at $\alpha, \beta, \gamma$ and with arbitrary exponents. He stated three axioms satisfied by these functions and showed that these axioms uniquely determined the differential equation whose solutions were this set of functions. He was able to obtain the relations among the solutions with minimal calculation.

In his lectures on hypergeometric functions, Riemann also discussed another topic: connecting the ratio of two independent solutions of the hypergeometric equation with the Schwarzian derivative and with modular functions.[6] More generally, he considered a second-order differential equation[7]

$$a_0(x)\frac{d^2y}{dx^2} + a_1(x)\frac{dy}{dx} + a_2(x)y = 0. \tag{7.49}$$

He denoted two independent solutions of (7.49) by $Y_1$ and $Y_2$ and let $x$ run over a closed path in the complex plane so that $a_0(x), a_1(x), a_2(x)$ would assume their original values. However, if the closed path were to contain branch points of $Y_1$ and $Y_2$, then the solutions would be given by $Y_1'$ and $Y_2'$:

$$Y_1' = \alpha Y_1 + \beta Y_2 \tag{7.50}$$
$$Y_2' = \gamma Y_1 + \delta Y_2, \tag{7.51}$$

where $\alpha, \beta, \gamma, \delta$ were constants such that $\alpha\delta - \beta\gamma \neq 0$, since $Y_1'$ and $Y_2'$ were given as independent solutions. In fact, Riemann assumed that $\alpha\delta - \beta\gamma = 1$. Now Riemann set $z = \frac{Y_1}{Y_2}$ and $z' = \frac{Y_1'}{Y_2'}$ so that he could write (7.50) and (7.51) as

$$z' = \frac{\alpha z + \beta}{\gamma z + \delta}. \tag{7.52}$$

He then considered the inverse of $z = z(x)$ and set $x = f(z)$ with the property that

$$x = f(z) = f(z') = f\left(\frac{\alpha z + \beta}{\gamma z + \delta}\right), \tag{7.53}$$

observing that the more branch points $z$ might have, the more transformations it would have of the form (7.4), with respect to which $f$ would be invariant. He also noted that $f$ would be invariant under the composition of such transformations.

Riemann next posed the problem of determining the differential equation satisfied by $x = f(z)$, a function that remained invariant under a group of transformations of the form (7.52). We note that Riemann must have been aware of the concept of a group at this time, because his friend Dedekind was then giving lectures on this topic; however, Riemann did not employ the term "group" in this context. Riemann began by taking the derivative of (7.52) to get

$$\frac{dz'}{dx} = \frac{\alpha\delta - \beta\gamma}{(\gamma z + \delta)^2} \cdot \frac{dz}{dx} = \frac{1}{(\gamma z + \delta)^2} \cdot \frac{dz}{dx}. \tag{7.54}$$

[6] Riemann (1990) pp. 667–691.    [7] Ibid. pp. 674–676.

This implied that

$$\left(\frac{dz'}{dx}\right)^{-\frac{1}{2}} = \left(\frac{dz}{dx}\right)^{-\frac{1}{2}} (\gamma z + \delta) \tag{7.55}$$

and

$$z'\left(\frac{dz'}{dx}\right)^{-\frac{1}{2}} = \left(\frac{dz}{dx}\right)^{-\frac{1}{2}} (\alpha z + \beta). \tag{7.56}$$

He noted that equations (7.55) and (7.56) meant that, under the transformation $z \to z'$, the functions $\left(\frac{dz}{dx}\right)^{-\frac{1}{2}}$ and $z\left(\frac{dz}{dx}\right)^{-\frac{1}{2}}$ would be transformed into linear expressions of these same functions. He thereupon set

$$Y_1 = \left(\frac{dz}{dx}\right)^{-\frac{1}{2}}, \quad Y_2 = z\left(\frac{dz}{dx}\right)^{-\frac{1}{2}} \tag{7.57}$$

so that $Y_1$ and $Y_2$ became two particular solutions of a second-order differential equation:

$$b_0 \frac{d^2 y}{dx^2} + b_1 \frac{dy}{dx} + b_2 y = 0. \tag{7.58}$$

To find $b_0$, $b_1$, and $b_2$, Riemann observed that in general, if $Y_1$ and $Y_2$ were solutions of (7.58), then the three expressions

$$Y_2' Y_1 - Y_1' Y_2, \quad Y_1'' Y_2 - Y_2'' Y_1, \quad Y_2'' Y_1' - Y_1'' Y_2'$$

were proportional to $b_0, b_1, b_2$, respectively; this fact follows from the equations

$$b_0 Y_1'' + b_1 Y_1' + b_2 Y_1 = 0 \tag{7.59}$$
$$b_0 Y_2'' + b_1 Y_2' + b_2 Y_2 = 0. \tag{7.60}$$

Multiply (7.59) by $Y_2$ and (7.60) by $Y_1$ and then subtract to obtain

$$\frac{b_0}{Y_2' Y_1 - Y_1' Y_2} = \frac{b_1}{Y_1'' Y_2 - Y_2'' Y_1}. \tag{7.61}$$

Similarly, by multiplying (7.59) by $Y_2'$ and (7.60) by $Y_1'$ and then subtracting, we arrive at

$$\frac{b_0}{Y_2' Y_1 - Y_1' Y_2} = \frac{b_2}{Y_2'' Y_1' - Y_1'' Y_2'}. \tag{7.62}$$

The two equations (7.61) and (7.62) imply Riemann's observation. He then noted that for $Y_1$ and $Y_2$ given by (7.57)

$$Y_2' Y_1 - Y_1' Y_2 = 1, \quad Y_1'' Y_2 - Y_2'' Y_1 = 0$$

and

$$Y_2'' Y_1' - Y_1'' Y_2' = -\left(\frac{dz}{dx}\right)^{\frac{1}{2}} \frac{d^2}{dx^2}\left(\frac{dz}{dx}\right)^{-\frac{1}{2}}. \tag{7.63}$$

Note that the quantity in (7.63) is half the Schwarzian derivative of $x = f(z)$, and in modern notation would be written as $\frac{1}{2} Sf$. Thus, $Y_1$ and $Y_2$ are solutions of the equation

$$\frac{d^2 y}{dx^2} - \frac{1}{2} Sfy = 0. \tag{7.64}$$

Note that this implies that the ratio $z = z(x)$ of any two independent solutions of (7.49) is also the ratio of two specific independent solutions of (7.64), where $Sf$ is the Schwarzian derivative of $f$, the function that is the inverse of $z = z(x)$.[8]

Riemann noted that all this work would be useful in determining the conditions under which solutions of (7.49) would be algebraic functions.

## 7.8   Riemann and the Triangle Functions

Again in his lectures, Riemann considered the case in which (7.49) was the hypergeometric equation. He let

$$P \begin{pmatrix} \alpha, & \beta, & \gamma \\ \alpha', & \beta', & \gamma' \end{pmatrix} ; x$$

denote all solutions of this equation, omitting the first row, $0, \infty, 1$. He could take $Y_1 = P^{(\alpha)}$ and $Y_2 = P^{(\alpha')}$ as two particular solutions, where $x^{-\alpha} P^{(\alpha)}$ and $x^{-\alpha'} P^{(\alpha')}$ were single-valued near $x = 0$ and nonvanishing and finite at $x = 0$. He next assumed that $\alpha, \beta, \gamma, \alpha', \beta', \gamma'$ were real, so that the coefficients of the series expansions of $P^{(\alpha)}$ and $P^{(\alpha')}$ were also real. In fact, Riemann assumed that the coefficients were all positive. Denoting the ratio of two solutions $\frac{Y_1}{Y_2} = z$, he let $\alpha > \alpha'$ so that when $x = 0$, $z = 0$ and when $x = 1$, the value of $z$, denoted by $z(1)$, would be finite. Next, he determined the behavior of $z$ when $x$ was negative. He noted that

$$z = x^{\alpha - \alpha_1} Q(x), \tag{7.65}$$

where $Q$ denoted the quotient of two power series in $x$. Since $0$ was a branch point of $z$, he had in the neighborhood of $x = 0$,

$$z = Q(re^{i\phi}) \cdot r^{\alpha - \alpha'} e^{(\alpha - \alpha')\phi i}, \tag{7.66}$$

where $-\pi < \phi \le \pi$. Observing that $x$ ran from $+r$ to $-r$ through values with positive imaginary elements, Riemann set

$$z = Q(-r) \cdot r^{\alpha - \alpha'} e^{(\alpha - \alpha')\pi i}. \tag{7.67}$$

Thus, if $\alpha - \alpha' < 1$, then for negative values of $x$, $z$ would run through values with argument $(\alpha - \alpha')\pi$. These values would lie in the $z$-plane, on a straight line making an angle $(\alpha - \alpha')\pi$ with the real axis. Riemann proceeded to investigate the behavior of $z$ as $x$ moved from $1$ to $\infty$. The solutions in the neighborhood of $x = 1$ were given by $P^{(\gamma)}$ and $P^{(\gamma')}$; hence, $P^{(\alpha)}$ and $P^{(\alpha')}$ were linear combinations of $P^{(\gamma)}$ and $P^{(\gamma')}$ and

---

[8]  For a modern treatment of this result, see Ford (1951) pp. 98–101.

the coefficients in the combinations were real. For $x > 1$,

$$\frac{P(\gamma')}{P(\gamma)} = (1-x)^{\gamma'-\gamma}(1 + A_1(1-x) + \cdots). \tag{7.68}$$

With the argument of $x$ taken as zero for $0 < x < 1$, $z$ took the form

$$z = \frac{p + p'e^{(\gamma-\gamma')\pi i}}{q + q'e^{(\gamma-\gamma')\pi i}}, \tag{7.69}$$

where $p, p', q, q'$ denoted real numbers. These values of $z$ lie on a circular arc. Thus, the real axis from $-\infty$ to $\infty$ was mapped onto a region which may be described as triangular: The line from 0 to 1 was mapped to the real axis from 0 to $z(0)$; the line from 0 to $-\infty$ was mapped to a line making an angle $(\alpha - \alpha')\pi$ with the real axis at 0; the line from 1 to $\infty$ was mapped to a circular arc making an angle $(\gamma - \gamma')\pi$ at $z(0)$ and an angle $(\beta - \beta')\pi$ at $z(\infty)$.

Riemann next asked whether the mapping $z$ was single-valued and conformal from the upper half-plane onto half of the triangular region he had constructed. For $z$ to be single-valued and conformal, $\frac{dz}{dx}$ must be nonzero and finite; calculating $\frac{dz}{dx}$, he found

$$\frac{dz}{dx} = \frac{Y_1 Y_2' - Y_2 Y_1'}{Y_2^2} = \frac{Cx^{\alpha+\alpha'-1}(1-x)^{\gamma+\gamma'-1}}{Y_2^2}. \tag{7.70}$$

Since $Y_2$ was not infinite in the upper half-plane, $\frac{dz}{dx}$ could not vanish except at the branch points $0, 1, \infty$. Riemann concluded that $x$ must be a single-valued function of $z$ inside the triangular region. The work of Schwarz showed a little more rigorously that Riemann's conclusions were correct.[9] In contrast to his findings on the hypergeometric equation, Riemann commented that he was aware that there were differential equations for which $x$ was not a single-valued function of $z$.

## 7.9  The Ratio of the Periods $\frac{K'}{K}$ as a Conformal Map

The triangular region considered by Riemann, discussed in in Section 7.8, had at least one of the angles $\alpha - \alpha' > 0$. He also showed that there was a case in which all the angles of the triangle were zero.

In the final section of his lectures,[10] Riemann considered the ratio of the periods $\frac{K'}{K}$, where

$$K(k^2) = {}_2F_1\left(\frac{1}{2}, \frac{1}{2}; k^2\right) \quad \text{and} \quad K'(k^2) = {}_2F_1\left(\frac{1}{2}, \frac{1}{2}; 1-k^2\right) \tag{7.71}$$

were independent solutions of the hypergeometric equation

$$x(1-x)\frac{d^2y}{dx^2} + (1-2x)\frac{dy}{dx} - \frac{1}{4}y = 0 \tag{7.72}$$

with $x = k^2$. Concerning the ratio $\frac{K'}{K}$, Riemann noted that it could be taken as a variable, as in the *Fundamenta Nova*, where Jacobi had expressed $k^2$ in terms of the variable

[9]  Schwarz (1972) vol. 2, pp. 221–233. Also see Riemann (1990) footnote 2, p. 692.
[10]  Riemann (1990 pp. 689–691.

$q = e^{-\frac{\pi K'}{K}}$ and also introduced the ratio $\frac{K'}{K}$ in connection with differential equations. Note the square of Jacobi's formula for $k$ (3.161):

$$k^2 = 16q \prod_{n=1}^{\infty} \left( \frac{1 + q^{2n}}{1 + q^{2n-1}} \right)^8, \quad q = e^{-\frac{\pi K'}{K}}. \tag{7.73}$$

In fact, in his paper "Suite des notices sur les fonctions elliptiques,"[11] Jacobi wrote about the relation between $k^2$ and $\frac{K'}{K}$:

Elliptic functions are essentially different from ordinary transcendental [functions]. They have a way of being virtually absolute. Their main characteristic is to embrace all that which is periodic in analysis. In fact, the trigonometric functions having a real period, the exponentials having an imaginary period, elliptic functions embrace both cases, since one has simultaneously

$$\sin \operatorname{am}(u + 4K) = \sin \operatorname{am} u$$

$$\sin \operatorname{am}(u + 2iK) = \sin \operatorname{am} u,$$

$i$ being $= \sqrt{-1}$. Moreover, one easily shows that an analytic function cannot have more than two periods, the one real and the other imaginary or both imaginary. The latter case corresponds to an imaginary modulus $k$. The quotient $\frac{K'}{K}$ of the two periods of a given function determines the modulus $k$ of elliptic functions, so that it should be expressed by means of formulas (15) and (17) [given earlier in the paper]

$$\sqrt{\frac{2K}{\pi}} = 1 + 2q + 2q^4 + 2q^9 + 2q^{16} + \cdots \tag{15}$$

$$\sqrt{\frac{2kK}{\pi}} = 2\sqrt[4]{q} + 2\sqrt[4]{q^9} + 2\sqrt[4]{q^{25}} + \cdots. \tag{17}$$

It may perhaps be appropriate to introduce into the analysis of elliptic functions the quotient $\frac{K'}{K}$ as a modulus in place of $k$. Regarding this quotient I have found:

that $k$ does not change in value if one writes in place of $\frac{K'}{K}$ the expression [with my label]

$$\frac{1}{i} \frac{bK + ib'K'}{aK + ia'K'}; \quad = \frac{KK' - i(abKK + a'b'K'K')}{aaKK + a'a'K'K'}, \tag{7.74}$$

$a, a', b, b'$ being some whole numbers, $a$ an odd number, $b$ an even number such that $ab' - a'b = 1$;

a remarkable theorem and one which should be regarded as one of the fundamental theorems in the analysis of elliptic functions.

We see from (7.74) that if we set $\tau = i\frac{K'}{K}$, then the change in $\frac{K'}{K}$ referred to by Jacobi may be written $\tau' = \frac{b + b'\tau}{a + a'\tau}$. Thus, as early as 1828, Jacobi was aware that $k^2$ was a modular function of $\frac{iK'}{K}$ with respect to the subgroup $\Gamma(2)$ of the full modular group. In his lectures on hypergeometric functions, Riemann proved Jacobi's theorem that $k^2$ was invariant with respect to $\Gamma(2)$. In the course of the proof, he found the generators of the group $\Gamma(2)$ and constructed its fundamental demain. He also considered the inverse case, $\frac{K'}{K}$ as a function of $k^2$; he demonstrated that this function was a conformal mapping from the upper half-plane to half of the fundamental domain of $\Gamma(2)$. Note

---

[11] Jacobi (1969) vol. 1, pp. 255–263, particularly pp. 262–263.

Figure 7.1. Deformed Path of Integration.

that in 1878 Émile Picard applied the function $\frac{K'}{K}$ of $k^2$ in proving his theorem that an entire function assumes every value with at most one possible exception.

Riemann began his study of the quotient $\frac{K'}{K}$ by determining how $K$ and $K'$, when regarded as functions of $k^2$, would change as $k^{-2}$ made a circuit around the branch points $0, 1, \infty$. For this purpose, he took the hypergeometric functions $K$ and $K'$ to be defined by the integrals

$$K = \int_0^1 (1 - t^2)^{-\frac{1}{2}}(1 - k^2 t^2)^{-\frac{1}{2}} \, dt \;=\; \int_0^1 \frac{1}{2} x^{-\frac{1}{2}}(1 - x)^{-\frac{1}{2}}(1 - k^2 x)^{-\frac{1}{2}} \, dx \quad (7.75)$$

and

$$K' = \int_0^1 (1 - t^2)^{-\frac{1}{2}}(1 - k'^2 t^2)^{-\frac{1}{2}} dt = -i \int_1^{k^{-2}} \frac{1}{2} x^{-\frac{1}{2}}(1 - x)^{-\frac{1}{2}}(1 - k^2 x)^{-\frac{1}{2}} dx. \tag{7.76}$$

Riemann's lectures on hypergeometric functions started with a study of the integral[12]

$$\int_0^1 s^a(1 - s)^b(1 - xs)^c \, ds \tag{7.77}$$

with branch points at $s = 0, 1, \infty, x^{-1}$. Note that, in order to avoid a discontinuous change in the value of the integral, the path of integration should not pass through these branch points. In order to determine the behavior of the integral (7.77), Riemann varied the value of $x$, producing a corresponding change in the value of the branch point $x^{-1}$. However, the path of integration from 0 to 1 must necessarily avoid the point $x^{-1}$. Felix Klein described this situation in a picturesque manner[13]:

> In order to conceive of our definite integral as a function of $x$, we must think of the path of integration as an elastic thread and the branch points as pins over which the thread cannot slide, so that if a moving branch point encounters our thread, it pushes the thread ahead of itself, without breaking it or crossing it.

Riemann next let $x^{-1}$ make a circuit around 1 and return to its original value in such a manner that the path of integration never intersected with the path of $x^{-1}$. Thus, the path of integration from 0 to 1 had to be deformed as shown in Figure 7.1, Klein's diagram of the deformed path of integration. The broken line shows the movement of $x^{-1}$ around 1 and the solid line shows the deformed path of integration from 0 to 1.

[12] Riemann (1990) pp. 667–668
[13] Klein (1933) p. 102. These lectures were given at Göttingen in 1893, but they were not published until 1933 when Otto Haupt edited them.

The altered path would proceed from 0 to 1, then from 1 to $x^{-1}$, and then go back along a circular path, around the point $x^{-1}$ in a counterclockwise direction, from $x^{-1}$ to 1. The circuit or detour around the point $x^{-1}$ caused the factor $(1 - xs)^c$ to change to $e^{2\pi ic}(1 - xs)^c$; thus, the original integral (7.77) became

$$\left( \int_0^1 + (1 - e^{2\pi ic}) \int_1^{x^{-1}} \right) \left( s^a (1 - s)^b (1 - xs)^c \, ds \right). \tag{7.78}$$

Riemann applied this technique to the integral in (7.75). He let $k^{-2}$ move in a circuit in a positive direction around 1 and then return to its original position. The path of integration in $K$ would then start at 0 and proceed positively around $k^{-2}$ and back again to 1. Riemann observed that $K$ would change to $K - 2iK'$, while $K'$ would remain unchanged. He also noted that if $k^{-2}$ made a positive circuit around 0, then $K$ would change to $3K - 2iK'$ and $iK'$ would change yo $2K - iK'$. Moreover, when $k^{-2}$ took a positive rotation around $\infty$, or a negative rotation around 0 and 1, $K$ remained unchanged and $iK'$ changed to $iK' + 2K$.

It was at this point in his lectures[14] that Riemann suggested using $\frac{K'}{K}$ as the variable; he noted that Jacobi had already done this in his *Fundamenta Nova*. Now since $k^2$ would not change as

$$K \to K - 2iK' \quad \text{and} \quad iK' \to iK', \tag{7.79}$$

it would follow that with $\tau = \frac{iK'}{K}$, the two transformations in (7.79) may be rewritten as

$$\tau \to \frac{\tau}{-2\tau + 1} \tag{7.80}$$

and hence

$$k^2 \left( \frac{\tau}{2\tau + 1} \right) = k^2(\tau). \tag{7.81}$$

We note that Riemann did not employ the symbol $\tau$, but wrote the variable as $\frac{K'}{K}$.

In a similar fashion, the rotation of $k^{-2}$ around $\infty$ showed that

$$k^2(\tau + 2) = k^2(\tau). \tag{7.82}$$

Note that (7.80) and the transformation $\tau \to \tau + 2$ produce the generators of $\Gamma(2)/\pm I$:

$$\begin{pmatrix} 1 & 0 \\ -2 & 1 \end{pmatrix} \quad \text{and} \quad \begin{pmatrix} 1 & 2 \\ 0 & 1 \end{pmatrix}. \tag{7.83}$$

After finding the generators of the group $\Gamma(2)$, Riemann raised the question: How does $\frac{K'}{K}$ behave as $k^2$ runs over all the values in the upper half-plane? He gave a very brief sketch of an answer. Seeing that the boundary of the upper half-plane was the real line, he determined the lines onto which $\frac{K'}{K}$ would map the real line. The hypergeometric representations of $K'$ and $K$ given in (7.71) showed him that as $k^2$ traveled from 0 to 1, $\frac{K'}{K}$ would take all real values from $\infty$ at $k^2 = 0$ to 0 at $k^2 = 1$.

---

[14] Ibid. p. 689.

Now observe that $K$ has branch points at 1 and $\infty$; $K'$ has branch points at 0 and $\infty$; thus, $\frac{K'}{K}$ must have branch points at 0, 1, $\infty$. Riemann remarked that $K'$ has a logarithmic singularity at 0 and has an expansion

$$K' = -\frac{1}{\pi} \log k^2 - \frac{2}{\pi} (a_0 + a_1 k^2 + \cdots). \tag{7.84}$$

He deduced from this expression that as $k^2$ went around 0 from a positive value of $k^2$ to a negative value of $k^2$, through values whose imaginary part was positive, the imaginary component of $\frac{K'}{K}$ was $-i$ for negative $k^2$. In fact, as $k^2$ moved from 0 to $-\infty$, $\frac{K'}{K}$ took values on a line parallel to the real axis going from $-i + \infty$ to $-i$.

Riemann went on to observe that if $k^2$ and $1 - k^2$ were exchanged, then $K$ and $K'$ would be exchanged. Therefore, $K$ would have a logarithmic singularity at $k^2 = 1$ and it was clear that the imaginary component of $\frac{K(k^2)}{K'(k^2)}$ would be $+i$ as $k^2$ moved from 1 to $\infty$. See exercise 6.

Riemann then noted that the values of $\frac{K'(k^2)}{K(k^2)}$, as $k^2$ moved from 1 to $\infty$, would lie in the right half-plane on the semicircle with center $-\frac{i}{2}$ and radius $\frac{1}{2}$. The truth of this observation can be verified by noting that any point $z$ on this semicircle is given by

$$z = -\frac{i}{2} + \frac{1}{2} e^{i\theta}, \quad -\frac{\pi}{2} \le \theta \le \frac{\pi}{2}$$

so that

$$\frac{1}{z} = \frac{2}{\cos\theta - i(1 - \sin\theta)} = i + \frac{\cos\theta}{1 - \sin\theta}.$$

Hence, $\frac{1}{z}$ takes all values from $i$ to $i + \infty$ as $\theta$ goes from $-\frac{\pi}{2}$ to $\frac{\pi}{2}$. These values of $\frac{1}{z}$ are identical with the values of $\frac{K'(k^2)}{K(k^2)}$ as $k^2$ moves from 1 to $\infty$. Thus, the values of $z$ on the semicircle are identical with the values of $\frac{K'(k^2)}{K(k^2)}$ as $k^2$ moves from 1 to $\infty$.

Riemann also stated that $\frac{K'(k^2)}{K(k^2)}$ took every value exactly once as $k^2$ moved around in the upper half-plane. This was a consequence of the fact that $\frac{K'}{K}$ was analytic in the upper half-plane and took each value exactly once as $k^2$ moved around the boundary (the real axis) exactly once. Riemann's statement is equivalent to the proposition that $\frac{K'}{K}$ conformally maps the upper half-plane onto the so-called triangular region enclosed by: the real axis, the line parallel to the real axis passing through $-i$, and the semicircle lying in the right half-plane with diameter extending from 0 to $-i$. Similarly, the lower half-plane is mapped conformally onto the triangular region enclosed by: the real axis, the parallel line passing through $i$, and the semicircle with 0 to $i$ as diameter. We can thus conclude that $\frac{K'}{K}$ maps the region $\mathbb{C} \setminus (-\infty, 0] \cup [1, \infty)$ conformally onto the region $R$ enclosed by: the two lines parallel to the real axis passing through $i$ and $-i$, respectively, and the two semicircles with diameters 0 to $i$ and 0 to $-i$.

Each member of the group of transformations $\Gamma(2)$, when applied to the region $R$, will yield another region bounded by four semicircles. A straight line in the right half-plane, perpendicular to the imaginary axis, can be viewed as a semicircle of infinite radius. After applying all transformations in $\Gamma(2)$, we obtain a tesselation of the right

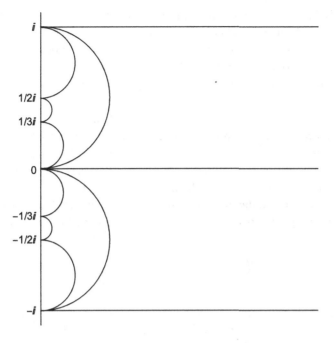

Figure 7.2. Riemann's Fundamental Domain for $\Gamma(2)$.

half-plane created by these four-sided figures. In a tesselation of the half-plane, images of $R$ under the group cover without overlapping the whole half-plane. Riemann gave a diagram[15], shown in Figure 7.2, illustrating the fundamental domain of $\Gamma(2)$ as he envisioned it.

As an example, also shown in Figure 7.2, Riemann described the region onto which the transformation

$$\frac{K'}{K} \rightarrow \frac{K'}{K - 2iK'} \tag{7.85}$$

would map $R$. Here recall that (7.85) was the transformation Riemann obtained when $k^{-2}$ made one circuit around the branch point 1. The region $R$ would be mapped onto the region bounded by four semicircles: with diameter from 0 to $i$, with diameter $i$ to $\frac{i}{2}$, with diameter from $\frac{i}{2}$ to $\frac{i}{3}$, and with diameter from $\frac{i}{3}$ to 0. The value $k^2 = 0$ corresponded to $\frac{i}{2}$ and the value $k = 1$ to $\frac{i}{3}$.

## 7.10 Schwarz: Hypergeometric Equation with Algebraic Solutions

In a justifiably famous paper, written in 1872 and published in *Crelle's Journal* in early 1873, "Ueber diejenigen Fälle, in welchen die *Gauss*iche hypergeometrische Reihen

---

[15] Riemann (1990) p. 691.

eine *algebraische* Function ihres vierten Elementes darstellt,"[16] Schwarz took up several problems that Riemann had considered in his 1858–59 lectures on the hypergeometric function. Schwarz was not directly aware of the content of Riemann's lectures. Nevertheless, it was only natural that Schwarz's deep interest in conformal mappings would cause him to consider similar problems; moreover, as a student of Weierstrass, he always sought rigorous proofs.

Hermann Schwarz (1843–1921) studied at Berlin University where the lectures of Weierstrass and Kummer drew him to mathematics. Schwarz's notes of Weierstrass's 1861 lectures are still useful today. He received his Ph.D. in 1864, on a topic in geometry, under the supervision of Kummer. Schwarz made several important contributions to the theory of conformal mappings. In an 1868 paper,[17] he showed how integrals of suitable functions mapped the upper half-plane onto simply connected regions whose boundary was a polygonal path. For example, the Legendre-Jacobi elliptic integral of the first kind

$$\omega(z) = \int_0^z \frac{dt}{\sqrt{(1 - t^2)(1 - k^2 t^2)}}, \quad 0 < k < 1,$$

maps the upper half-plane onto a rectangle in the $\omega$-plane, where the points $z = \pm 1$, $z = \pm \frac{1}{k}$, correspond to the corners of the rectangle. Thus, the corners are located at $-K, K, K + iK', -K + iK'$. This theorem on the conformal mapping of the upper half-plane onto a polygonal region was discovered at about the same time by the excellent German mathematician Elwin Bruno Christofell (1829–1900), a student of Dirichlet who made significant advances in numerical analysis, function theory, partial differential equations, and differential geometry.

Schwarz did fundamental work in minimal surfaces, conformal mappings, and potential theory. He applied potential theory to prove the Riemann mapping theorem for classes of simply connected regions bounded by analytic arcs. Note that Riemann's own proof of his theorem had gaps, as pointed out by his student Friedrich Prym, Weierstrass, and others.

In his important 1873 paper, Schwarz started off by discussing the conditions under which a hypergeometric differential equation had an algebraic solution; he then turned to the case in which all the solutions were algebraic. Note that the latter case occurs when the differential equation has two linearly independent algebraic solutions. Suppose the hypergeometric differential equation

$$x(1 - x)y'' + \left(\gamma - (\alpha + \beta + 1)x\right)y' - \alpha\beta y = 0 \qquad (7.86)$$

has an algebraic solution $y(x)$, where, as before, $x$ is a complex variable. In general, the function $y(x)$ is not single-valued, but we may assume that the logarithmic derivative $\frac{y'(x)}{y(x)}$ is single-valued and since this derivative is an algebraic function, Schwarz argued that it must, in fact, be a rational function. Next, since the expansions of $y(x)$ at 0, 1 and $\infty$ cannot have a logarithmic term, it follows that if $y(x)$ is analytic at $x = 0$ and $x = 1$, then $y(x)$ must be a polynomial. We can see that in this case, one of the parameters, $\alpha$, or

---

[16] Schwarz (1972) vol. 2, pp. 211–259.    [17] Schwarz (1972) vol. 2, pp. 84–101.

$\beta$, must be either 0 or a negative integer. Schwarz took $\alpha = -n$, $n$ being a nonnegative integer; he then wrote

$$y(x) = cF(-n, \beta, \gamma, x).$$

Schwarz next noted that in case $y(x)$ was not analytic at $x = 0$, and/or $x = 1$, there must exist rational numbers $a$ and $c$ such that

$$y(x) = x^a(1 - x)^c p(x),$$

where $p(x)$ was a polynomial. Thus,

$$y(x) = x^a(1 - x)^c \prod_{j=1}^{k} (x - a_i)^{\alpha_i}. \tag{7.87}$$

Observe that values of $a, c$ and the degree of the polynomial are determined by the indicial equations. Schwarz worked out the values of these parameters and showed that each of the cases was actually possible. He next considered the case in which all the solutions of the hypergeometric differential equation (7.86) were algebraic. Taking $y_1(x)$ and $y_2(x)$ to be independent algebraic solutions of the hypergeometric equation, he found that the ratio

$$y(x) = \frac{y_1(x)}{y_2(x)} \tag{7.88}$$

satisfied the differential equation

$$Sf(x) \equiv \frac{f'''}{f'} - \frac{3}{2}\left(\frac{f''}{f'}\right)^2 = \frac{1 - \lambda^2}{2x^2} + \frac{1 - \mu^2}{2(1 - x)^2} + \frac{1 - \lambda^2 - \mu^2 + \nu^2}{2x(1 - x)} \tag{7.89}$$

where

$$\lambda = 1 - \gamma, \quad \mu = \gamma - \alpha - \beta, \quad \nu = \alpha - \beta$$

and $Sf$ was the Schwarzian derivative of $f$; $\lambda$, $\mu$, and $\nu$ were exponent differences of two independent solutions at 0, 1, $\infty$. The solution of (7.89) then takes the form

$$\frac{ay(x) + b}{cy(x) + d}, \quad ad - bc = 1, \tag{7.90}$$

where $y(x)$ is given by (7.88). Observe that if we start at a point $x = x_0$, where $y(x)$ is analytic, and allow $y(x)$ to follow a closed path, then all the possible determinations of the solutions of (7.89) will be given by (7.90). Moreover, if all the solutions of $y(x)$ are algebraic, then the number of different expressions of the form (7.90) must be finite. This yields a finite number of 2 by 2 matrices

$$\begin{pmatrix} a & b \\ c & d \end{pmatrix}, \quad ad - bc = 1,$$

that form a group under composition. Note that Schwarz did not use the language and conceptual framework of groups. Klein was apparently the first to employ this language in this context; he showed that, except for the group consisting of only the identity element, there were, up to conjugation, five finite groups of linear transformations: cyclic, dihedral, tetrahedral, octahedral, and icosahedral.[18]

Schwarz, denoted the solution of (7.89) by $s(\lambda, \mu, \nu, x)$. He considered power series solutions in the neighborhood of the singular points at $0, 1, \infty$ to prove the theorem: When the quantities $\lambda, \mu, \nu$ are real, a particular solution $s$ of the equation (7.89) will conformally map the upper half-plane $\mathbb{H}$ onto a simply connected region whose boundary consists of three circular arcs forming a triangle. The angles at the vertices corresponding to $0, 1, \infty$ are, respectively, $\lambda\pi, \mu\pi, \nu\pi$.

Schwarz also showed that the mapping must be one-to-one and conformal under the conditions $|\lambda| \leq 2$, $|\mu| \leq 2$, and $|\nu| \leq 2$. Recall that $\lambda, \mu, \nu$ are the differences of the exponents at $0, 1, \infty$. If any one of $\lambda, \mu, \nu$ is equal to zero, then there will be a logarithmic term in the solution. This solution cannot be algebraic. Thus, for algebraic solutions $s$ of (7.89), none of $\lambda, \mu, \nu$ may be zero. Recall here that Riemann proved in his lectures that $\tau = \frac{iK'}{K}$ mapped the upper half-plane onto the interior of a circular triangle with all three angles equal to zero. The inverse of $\tau = \frac{iK'}{K}$ is $k^2(\tau)$, that maps the triangle with vertex angles $0, 0, 0$ onto the upper half-plane. Schwarz, who at the time of this work was probably not aware of Riemann's lectures since they were published much later, also mentions these facts about $\frac{K'}{K}$ and $k^2$. Also observe that the Dedekind-Klein $J$ function maps onto the upper half-plane the interior of a circular triangle with $\lambda = \frac{1}{2}, \mu = \frac{1}{3}, \nu = 0$.

Schwarz proved that the inverse of a solution $s(\lambda, \mu, \nu, x)$ was single-valued if $\lambda, \mu, \nu$ were each either zero or the reciprocal of an integer and he gave a list of values of positive $\lambda, \mu, \nu$ for which single-valued inverses existed. He considered, in addition, the cases for which $s(\lambda, \mu, \nu, x)$ is algebraic but does not have a single-valued inverse.[19]

## 7.11  Exercises

1. Prove (7.79), that is, show that as $k^{-2}$ makes one circuit around 1 in the positive direction, $K$ changes to $K - 2iK'$ and $K'$ remains unchanged.

2. Prove that as $k^{-2}$ makes a positive circuit around 0, $K$ will change to $3K - 2iK'$ and $iK'$ will change to $2K - iK'$.

3. Show that when $k^{-2}$ makes a positive circuit around $\infty$, $K$ remains unchanged and $iK$ changes to $iK' + 2K$.

4. Set

$$E = \int_0^1 \frac{1}{2} x^{-\frac{1}{2}} (1-x)^{-\frac{1}{2}} (1-k^2 x)^{\frac{1}{2}} dx$$

$$E' = \int_0^1 \frac{1}{2} x^{-\frac{1}{2}} (1-x)^{-\frac{1}{2}} (1-k'^2 x)^{\frac{1}{2}} dx.$$

---

[18]  For a proof of this result with some historical discussion, see Klein (1956) pp. 126–132, an English translation of Klein's 1884 work. For a more modern treatment, see Ford (1959) pp. 123–136. Also see McKean and Moll (1997) pp. 16–24.

[19]  Schwarz (1972) vol. 2, pp. 211–259. Gray (1986) puts Schwarz's paper into its historical context. Also see Ford (1957) pp. 284–309 and Hille (1997) pp. 374–400.

Show that

$$2\frac{dE}{dk^2} = \frac{E-K}{k^2}, \qquad\qquad 2\frac{dK}{dk^2} = \frac{E-Kk'^2}{k^2k'^2},$$

$$2\frac{dE'}{dk^2} = \frac{K'-E'}{k'^2}, \qquad\qquad 2\frac{dK'}{dk^2} = \frac{K'k^2-E'}{k^2k'^2}.$$

5. Set

$$v(\omega) = \frac{iK'(\omega)}{K(\omega)}.$$

Deduce from the results of Exercise 4 that

$$\frac{dv}{d\omega} = \frac{\pi}{4iK^2\,\omega(1-\omega)}.$$

From this differential equation, deduce that

$$v(\omega) = \frac{1}{\pi i}\,\log\omega + f(\omega), \tag{7.91}$$

where $f(\omega)$ is analytic in $|\omega| < 1$. Deduce Riemann's result (7.84) from (7.91).

6. Show that

$$\frac{d}{d\omega}\left(\frac{1}{v(\omega)}\right) = \frac{\pi}{4iK'^2\,\omega(1-\omega)}.$$

Then deduce

$$\frac{1}{v(\omega)} = \frac{1}{\pi i}\,\log(1-\omega) + g(\omega), \tag{7.92}$$

where $g(\omega)$ is analytic in $|1-\omega| < 1$.

# 8

---

## *Dedekind's Paper on Modular Functions*

### 8.1 Preliminary Remarks

In his foreword to the interesting volume published in honor of the anniversary of Richard Dedekind's one hundred fiftieth birthday,[1] Winfried Scharlau remarked that anyone attempting to definitively evaluate Dedekind's mathematical work would be struck by two overarching aspects of his thought: First, Dedekind was a pioneer in abstract mathematics, blazing a trail toward the perspective of Van der Waerden and even Bourbaki. Scharlau points out, "It is surely no coincidence that in Bourbaki's history of mathematics, no mathematician is mentioned more frequently than Dedekind." Second, Dedekind was profoundly influenced by his penetrating study of his distinguished forbears, including Gauss, Dirichlet, Kummer, Galois, and Riemann. Scharlau offers an interesting insight: "It is certainly no contradiction that one so strongly committed to the scientific tradition can at the same time pave the way to an exceptional degree for new theories and developments. [Dedekind's] place in the history of mathematics and his mathematical work must be viewed and recognized within this tension between tradition and progress."

Richard Dedekind (1831–1916) may be most famous for his work on the foundation of the theory of real numbers; almost every mathematics student has heard of or used the Dedekind cut. Dedekind developed this concept in 1858; in 1872 it was published as a short treatise, *Stetigkeit und irrationale Zahlen*. In the preface to this work, Dedekind wrote that while teaching a course in calculus at the Zürich Polytechnic School in 1858, he realized that there was no proof of the theorem that a bounded increasing sequence must converge. We note that during the period 1810–1825, Gauss, Bolzano, and Cauchy, based on an intuitive idea of real numbers, had succeeded in developing a rigorous theory of convergence of sequences and series. This project was further carried forward by Abel and Dirichlet in the 1820s and 1830s. From his close association with Dirichlet, Dedekind gained an appreciation for mathematical rigor and this led him to

---

[1] Scharlau (1981).

perceive the necessity within the foundations of analysis for a logically-satisfactory definition of an irrational number.

Dedekind employed infinite sets in his definition of an irrational number; this was of great benefit to his work on the theory of algebraic numbers, for which he defined and successfully developed the concept of ideals, and in particular prime ideals, as infinite sets. Within this theory, ideals may be seen to serve in the place of numbers; thus, Dedekind was able to establish unique factorization in terms of prime ideals. This work in algebraic numbers led to the discovery or refinement of several important concepts in the theories of groups, rings, and fields; he also conceived of the lattice structure. Supported by his association with Gauss and Riemann, Dedekind also made perhaps less famous but very significant contributions to the theory of modular functions; again, he perceived and investigated key connections between this theory and quadratic number fields within algebraic number theory.

Dedekind developed into the great mathematician we now study under the guidance and influence of three towering Göttingen mathematicians, Gauss, Dirichlet, and Riemann.[2] In a letter of 1856 to his sister Julie, Dedekind wrote prophetically concerning Riemann and Dirichlet: "The association with both of them is invaluable, and I hope will bear its fruit."[3] Dedekind attended Gauss's lectures on least squares during the winter 1850–51 and on geodesy in the following term. Gauss concluded his course on least squares with thanks to his students for being so attentive to his "rather dry" lectures. In 1901, Dedekind commented on these lectures: "A half century has now passed since then, but this allegedly dry lecture is unforgettable in memory as one of the finest I have ever heard."[4] In 1852, he wrote his Ph.D. dissertation on the Eulerian integrals, that is, the beta and gamma integrals; however, he soon realized that there were some gaps in his mathematical education. Thus, he spent two years engaged in intensive study of topics such as elliptic functions, number theory, advanced geometry, and mathematical physics. In 1854, he was appointed lecturer in mathematics at Göttingen, a few weeks after Riemann, who was five years his senior. At that time, Riemann had already done great work in laying new foundations for complex analysis, differential geometry, and trigonometric series. During the next three years, Dedekind accordingly attended Riemann's lectures on the foundations of complex analysis, on elliptic and Abelian functions, and on the hypergeometric functions. While learning from others, Dedekind's own lectures also included original material, including the idea of an abstract group in the context of Galois theory.[5] Galois himself had used only the idea of a group of permutations.

After Gauss's death in 1855, his professorship at Göttingen was awarded to Dirichlet, who had a most profound effect on Dedekind's development as a mathematician. Dedekind learned from Dirichlet through personal contact and by means of Dirichlet's lectures on number theory, partial differential equations, and definite integrals. On his early contact with Dirichlet, Dedekind wrote, "... he also recently visited me and engaged me in a mathematical discussion, which was certainly quite interesting, but for me it turned out to be somewhat humbling, although this was surely not his

---

[2] See Duke and Tschinkel (2005) pp. 31–32 and Dunnington (2004) pp. 259–261.
[3] Scharlau (1981) p. 37.    [4] Dunnington (2004) p. 261.    [5] Scharlau (1981) pp. 60–70.

intention."[6] We may contrast this with his later comments on his relationship with Dirichlet, a few months before Dirichlet's death: "I feel as indebted to him as hardly any other persons; every day, when I face my students in unshakable certainty and pre-paredness, I think to myself that I have him to thank. I have indeed greatly exerted my mind, but I nevertheless doubt whether I would have had sufficient energy for this, had this unique person not intervened in my destiny."[7]

Dedekind played an important role in editing Gauss's posthumous papers on number theory. He compiled and edited Dirichlet's *Vorlesungen über Zahlentheorie* and added several supplements of his own, including one on ideal theory. Dedekind also partici-pated in the editing of Riemann's mathematical papers. This work led him to write his two papers on modular forms, and led to the concept of Dedekind sums. It is interesting to see Dedekind's motivation in his work on modular forms as an attempt to free it from the theory of elliptic functions.

In his paper "Schreiben an Herr Borchardt über die Theorie der elliptischen Mod-ulfunktionen," Dedekind remarked that he applied the theory of elliptic functions to derive the theorem that for integers $\alpha, \beta, \gamma, \delta$ with $\beta$ and $\gamma$ even and $\alpha\delta - \beta\gamma = 1$,

$$k^2 \left( \frac{\gamma + \delta\omega}{\alpha + \beta\omega} \right) = k^2(\omega). \tag{8.1}$$

Of course, this means that $k^2$ is invariant under the action of $\Gamma(2)$. Recall that Jacobi knew this result in 1828 and that Riemann had derived this theorem in his 1858–59 lectures on hypergeometric functions. At the time of those lectures, Dedekind had left Göttingen for Zürich; in any case, Dedekind's proof is different from Riemann's.

Dedekind wrote in his paper to Borchardt that this theorem pushed him to explore the connection among the quantities $k^2$, $K$, and the ratio of the periods, separately from the theory of elliptic functions. He further stated in this connection that Hermite made a striking remark in his overview of elliptic functions, "Note sur la theorie des fonctions elliptiques" published as an addition to Lacroix's 1862 sixth edition of *Traité élémentaire de calcul différential et de calcul intégral*[8]:

> It is mainly in algebra, in the theory of equations, and in arithmetic, in the theory of binary quadratic forms, that the consideration of these functions [i.e., the functions $k^2(\omega)$, $k'^2(\omega)$] has suggested itself of completely new points of view and opened fertile pathways, where one has obtained the most interesting results. The limits of this note do not permit us to enter into this already so extensive field of nice researches.... it does not appear to be possible, within the actual state of our knowledge of analysis, to arrive at all their properties solely on the basis of their definition [given, e.g., in (8.8)] as the quotient of the series given earlier [(8.8)].

[In a footnote to this, Hermite wrote:]

Poisson and Cauchy arrived, by two different methods, at this identity:

$$\sqrt{-i\omega}\left(1 + 2e^{i\pi\omega} + 2e^{4i\pi\omega} + 2e^{9i\pi\omega} + \cdots\right) = \left(1 + 2e^{-\frac{i\pi}{\omega}} + 2e^{-\frac{4i\pi}{\omega}} + 2e^{-\frac{9i\pi}{\omega}} + \cdots\right), \tag{8.2}$$

---

[6] In a letter to his sister Julie, Scharlau (1981) p. 34.
[7] in a letter to his sister Mathilde, Scharlau (1981) p. 47.
[8] This note is given in Hermite's *Oeuvres* (1905–1917) vol. 2, pp. 125–238.

which leads to the fundamental properties of the moduli considered as functions of $\omega$. But it is possible to extract from this only the transformation of the first order, and no way of arriving at *modular equations* has yet been offered other than that which was given by the founders of the theory of elliptic functions.[9]

Observe that the identity mentioned by Hermite can be written in the form

$$\sqrt{\frac{2K(\omega)}{\pi}} = \sqrt{\frac{i}{\omega} \cdot \frac{2K\left(-\frac{1}{\omega}\right)}{\pi}}. \tag{8.3}$$

Also recall than an $n$th order transformation is given by

$$\omega' = \frac{a\omega + b}{c\omega + d}, \quad \text{with } a, b, c, d \text{ integers such that } ad - bc = n,$$

and a modular equation of order $n$ is the algebraic equation satisfied by, for example, $k^2(\omega)$ and $k^2(n\omega)$. Hermite had used the modular equation of order 5, found by Jacobi, to solve an irreducible fifth-degree equation. Now Jacobi had obtained this modular equation with the use of elliptic integrals (or elliptic functions) and Hermite felt that the use of elliptic functions was unavoidable there. Dedekind perceived this as a gap in the theory of modular functions and he set out to fill it.

Dedekind started this effort in his 1877 paper by defining the modular group $\Gamma = \Gamma(1)$ and obtaining its fundamental domain. He then defined the valency function $v$; this conformally mapped the interior of this fundamental domain onto $\mathbb{C} \setminus (-\infty, 0) \cup (1, \infty)$. The inverse of the function $v$ had branch points at $0, 1, \infty$. A year later, Klein denoted the function $v$ by $J$ and that is the symbol we still employ. In fact, we now define the basic function

$$j(\omega) = 1728J(\omega).$$

The factor 1728 serves to provide that $j(\omega)$ has integer coefficients when expanded in powers of $q = e^{\pi\omega i}$. Since $v(\omega)$ and $k^2(\omega)$ are both invariant under action of the group $\Gamma(2)$ and behave suitably at the cusps $0, 1, \infty$, there must be an algebraic relation between these two functions. Dedekind found this relation:

$$v \equiv J = \frac{4}{27} \frac{(1 - \lambda + \lambda^2)^2}{\lambda^2(1 - \lambda)^2}, \quad \lambda = k^2. \tag{8.4}$$

Note here that Dedekind himself wrote $k$ for $k^2$, but we follow the earlier notation of Jacobi. Recall that in 1858 Hermite arrived at the invariant function $j(\omega)$ and the relation (8.4) in the course of his work on binary quadratic forms.

Dedekind wished to create a theory of modular forms independent of elliptic and theta functions; thus, he defined $k(\omega)$ in terms of $v$. He set

$$\eta(\omega) = \text{const. } v^{-\frac{1}{6}}(1 - v)^{-\frac{1}{8}}\left(\frac{dv}{d\omega}\right)^{\frac{1}{4}} \tag{8.5}$$

[9] Hermite (1905–1917) vol. 2, p. 163.

and considered the second-order transformations of this function:

$$\eta_1(\omega) = \eta(2\omega), \quad \eta_2(\omega) = \eta\left(\frac{\omega}{2}\right), \quad \eta_3(\omega) = \eta\left(\frac{1+\omega}{2}\right). \tag{8.6}$$

He then defined the functions

$$\sqrt[8]{\lambda(\omega)} = \sqrt[4]{k(\omega)} = \phi(\omega) = e^{\frac{2\pi i}{48}}\sqrt{2}\,\frac{\eta_1(\omega)}{\eta_3(\omega)}$$

$$\sqrt[8]{1-\lambda} = \sqrt[4]{k'} = \psi(\omega) = e^{\frac{2\pi i}{48}}\frac{\eta_2(\omega)}{\eta_3(\omega)},$$

and showed that these functions could be written as ratios of theta functions. He next employed a known result on theta functions to show that

$$\eta(\omega) = q^{\frac{1}{12}}\prod_{n=1}^{\infty}(1-q^{2n}), \quad q = e^{\pi\omega i}, \tag{8.7}$$

where the constant in (8.5) was chosen appropriately. Dedekind wrote that he had not yet succeeded in deriving (8.7) without the use of $\theta$-functions; Hurwitz achieved this goal by employing Eisenstein series.

With the use of his valency function $v$, in the final section of his paper, Dedekind obtained a theory of modular equations, thus successfully producing a theory of modular equations independent of the theory of elliptic functions. He wrote that he wrote his paper with applications to the theory of "singular moduli" in his mind, but that he would present those results at a later time; unfortunately, he did not return to this topic in writing.

## 8.2 Dedekind's Approach

Dedekind started with the well-known series formulas for the four theta functions. Note that we must take $x = \pi z$ in (3.186) through (3.189) to get these series.

$$\theta_0(z, \omega) = \sum_{s=-\infty}^{\infty} e^{2\pi i\left(s^2\frac{\omega}{2}+s(z-\frac{1}{2})\right)}$$

$$\theta_1(z, \omega) = \sum_{s=-\infty}^{\infty} e^{2\pi i\left((s+\frac{1}{2})^2\frac{\omega}{2}+(s+\frac{1}{2})(z-\frac{1}{2})\right)}$$

$$\theta_2(z, \omega) = \sum_{s=-\infty}^{\infty} e^{2\pi i\left((s+\frac{1}{2})^2\frac{\omega}{2}+(s+\frac{1}{2})z\right)}$$

$$\theta_3(z, \omega) = \sum_{s=-\infty}^{\infty} e^{2\pi i\left(s^2\frac{\omega}{2}+sz\right)}.$$

For the sake of brevity, in his paper Dedekind actually wrote $e^{2\pi iz}$ as $1^z$. Following Hermite, he set

$$\sqrt{k} = \frac{\theta_2(0, \omega)}{\theta_3(0, \omega)} = \phi^2(\omega), \quad \sqrt{k'} = \frac{\theta_0(0, \omega)}{\theta_3(0, \omega)} = \psi^2(\omega) \tag{8.8}$$

so that $k^2 + k'^2 = 1$ by (3.197). He also set

$$\sqrt{x} = \frac{1}{\sqrt{k}} \frac{\theta_1(z, \omega)}{\theta_0(z, \omega)} = \text{sn}(2Kz, k), \tag{8.9}$$

allowing him to write

$$\sqrt{1 - x} = \sqrt{\frac{k'}{k}} \frac{\theta_2(z, \omega)}{\theta_0(z, \omega)} = \text{cn}(2Kz, k), \tag{8.10}$$

$$\sqrt{1 - k^2 x} = \sqrt{k'} \frac{\theta_3(z, \omega)}{\theta_0(z, \omega)} = \text{dn}(2Kz, k), \tag{8.11}$$

$$\frac{d\sqrt{x}}{dz} = 2K\sqrt{1 - x}\sqrt{1 - k^2 x}, \tag{8.12}$$

where

$$2K = \frac{\theta_3(0, \omega)\,\theta_1'(0, \omega)}{\theta_0(0, \omega)\,\theta_2(0, \omega)} = \pi\theta_3^2(0, \omega).$$

Dedekind observed that it was known that the zeros of $\theta_0$, $\theta_1$, $\theta_2$, and $\theta_3$ occurred respectively at the values $z = m + (n + \frac{1}{2})\omega$, $z = m + n\omega$, $z = m + \frac{1}{2} + n\omega$, and $z = m + \frac{1}{2} + (n + \frac{1}{2})\omega$, where $m$ and $n$ were integers. This is easy to check using the infinite product expansions (3.214) through (3.217) for these $\theta$-functions. He then used formulas (8.9) through (8.12) to prove that for those $\omega$ and $\omega_1$ with positive imaginary parts

$$\omega_1 = \frac{a\omega + b}{c\omega + d},$$

where $a$, $b$, $c$, $d$ were integers with $b$ and $c$ even and $ad - bc = 1$, if and only if $k^2(\omega) = k^2(\omega_1)$.[10] Following Dedekind, let $k^2(\omega) = k^2(\omega_1)$ and set

$$2K_1 = \pi\,\theta_3^2(0, \omega_1)$$

and define a new variable $z_1$ by $z_1 = \frac{K}{K_1}z$. Then with

$$\sqrt{x_1} = \frac{1}{\sqrt{k}} \frac{\theta_1(z_1, \omega_1)}{\theta_0(z_1, \omega_1)}$$

and similar relations for $\sqrt{1 - x_1}$ and $\sqrt{1 - k^2 x_1}$, it follows from (8.12) that

$$\left(\frac{d\sqrt{x}}{dz}\right)^2 = 4K^2(1 - x)(1 - k^2 x) \tag{8.13}$$

and

$$\left(\frac{d\sqrt{x_1}}{dz}\right)^2 = 4K^2(1 - x_1)(1 - k^2 x_1). \tag{8.14}$$

---

[10] Dedekind (1930–) vol. 1, pp. 175–177.

Taking the derivatives of (8.13) and (8.14), obtain

$$\frac{1}{4K^2} \frac{d^2 \sqrt{x}}{dz^2} = \sqrt{x} \left( -(1 + k^2) + 2k^2 x \right)$$

$$\frac{1}{4K^2} \frac{d^2 \sqrt{x_1}}{dz^2} = \sqrt{x_1} \left( -(1 + k^2) + 2k^2 x_1 \right).$$

Therefore

$$\frac{1}{4K^2} \left( \sqrt{x_1} \frac{d^2 \sqrt{x}}{dz^2} - \sqrt{x} \frac{d^2 \sqrt{x_1}}{dz^2} \right) = 2k^2 \sqrt{xx_1}(x - x_1)$$

or

$$\frac{1}{4K^2} \frac{d}{dz} \left( \sqrt{x_1} \frac{d \sqrt{x}}{dz} - \sqrt{x} \frac{d \sqrt{x_1}}{dz} \right) = 2k^2 \sqrt{xx_1} \, (x - x_1). \tag{8.15}$$

Note also that it follows directly from (8.13) and (8.14) that

$$\frac{1}{4K^2} \left( x_1 \left( \frac{d \sqrt{x}}{dz} \right)^2 - x \left( \frac{d \sqrt{x_1}}{dz} \right)^2 \right) = (x_1 - x)(1 - k^2 xx_1). \tag{8.16}$$

By dividing (8.15) by (8.16) arrive at

$$\frac{\frac{d}{dz} \left( \sqrt{x_1} \frac{d \sqrt{x}}{dz} - \sqrt{x} \frac{d \sqrt{x_1}}{dz} \right)}{\sqrt{x_1} \frac{d \sqrt{x}}{dz} - \sqrt{x} \frac{d \sqrt{x_1}}{dz}} = \frac{2k^2 \sqrt{xx_1} \left( \sqrt{x_1} \frac{d \sqrt{x}}{dz} + \sqrt{x} \frac{d \sqrt{x_1}}{dz} \right)}{k^2 xx_1 - 1}. \tag{8.17}$$

Integrating (8.17) and exponentiating the result then yields

$$\sqrt{x_1} \frac{d \sqrt{x}}{dz} - \sqrt{x} \frac{d \sqrt{x_1}}{dz} = C(1 - k^2 xx_1), \tag{8.18}$$

where $C$ is a constant. Since $z_1 = 0$ when $z = 0$, we see that both $\sqrt{x_1}$ and $\sqrt{x}$ vanish when $z = 0$ and thus $C = 0$. Substituting (8.18) in (8.15) then shows that $x$ is identical with $x_1$. Similarly, $\sqrt{1 - x} \equiv \sqrt{1 - x_1}$ and $\sqrt{1 - k^2 x} \equiv \sqrt{1 - k^2 x_1}$. Equations (8.9), (8.10), and (8.11) then imply that each of the four theta functions

$$\theta_0(z, \omega), \ \theta_1(z, \omega), \ \theta_2(z, \omega), \ \theta_3(z, \omega)$$

vanishes when, of the theta functions

$$\theta_0(z_1, \omega_1), \ \theta_1(z_1, \omega_1), \ \theta_2(z_1, \omega_1), \ \theta_3(z_1, \omega_1),$$

the corresponding function vanishes, and conversely. As we mentioned earlier, $\theta_1(z, \omega)$ is zero if and only if $z = m + n\omega$, where $m$ and $n$ are integers. So if we take the zeros $z_1 = 1$ and $z_1 = \omega_1$ of $\theta_1(z_1, \omega_1)$, the corresponding zeros of $\theta_1(z, \omega)$ must be

$$z = \frac{K_1}{K} = c\omega + d \quad \text{and} \quad z = \frac{K_1}{K}\omega_1 = a\omega + b, \tag{8.19}$$

where $a, b, c, d$ are integers. Thus

$$\omega_1 = \frac{a\omega + b}{c\omega + d}, \quad (c\omega + d)z_1 = z. \tag{8.20}$$

Conversely, taking

$$z = 1 \quad \text{gives} \quad z_1 = \frac{K}{K_1} = c_1\omega_1 + d_1$$

$$z = \omega \quad \text{gives} \quad z_1 = \frac{K}{K_1}\omega = a_1\omega_1 + b_1,$$

where $a_1, b_1, c_1, d_1$ are integers and

$$\omega = \frac{a_1\omega_1 + b_1}{c_1\omega_1 + d_1}. \tag{8.21}$$

From (8.20) and (8.21) we see that $(ad - bc)(a_1d_1 - b_1c_1) = 1$ and hence $ad - bc = \pm 1$. Since the imaginary parts of $\omega$ and $\omega_1$ are both positive, it follows that $ad - bc = 1$. Moreover, since $\theta_0(z_1, \omega_1)$ vanishes at precisely $z_1 = r + (s + \frac{1}{2})\omega_1$, it follows that

$$\text{when} \quad z = \frac{\omega}{2}, \quad \text{we have} \quad z_1 = r + \left(s + \frac{1}{2}\right)\omega_1$$

for some integers $r$ and $s$. Equation (8.20) then implies

$$(c\omega + d)r + (a\omega + b)\left(s + \frac{1}{2}\right) = \frac{\omega}{2},$$

so that $dr + b(s + \frac{1}{2}) = 0$ and $b \equiv 0 \pmod 2$. In a similar manner, using the zeros of $\theta_2(z, \omega)$ and $\theta_2(z_1, \omega_1)$, one may show that

$$c \equiv 0 \pmod 2.$$

This verifies the theorem, since the converse was proved by Jacobi and Hermite, as shown by equation (5.29). After presenting the proof in his paper, Dedekind noted that the proof was clearly based on the fact that the elliptic functions $\text{sn}(u, k^2)$, $\text{cn}(u, k^2)$, and $\text{dn}(u, k^2)$ were single-valued functions of the modular function $k^2$. The proof of this theorem motivated Dedekind in his effort to develop a theory of modular functions independent of elliptic functions. He noted that he was intrigued that Charles Hermite remarked, "... there is no way leading to these modular functions, other than that which the founders of the theory of elliptic functions have pursued."[11]

## 8.3 The Fundamental Domain for SL$_2$ ($\mathbb{Z}$)

The first step in Dedekind's construction of a theory of modular functions was setting up the fundamental domain for the action of $SL_2(\mathbb{Z})$, also denoted by $\Gamma(1)$, on $S = \mathbb{H} \cup \{\infty\} \cup \mathbb{Q}$, where $\mathbb{H}$ denoted the upper half-plane and $\mathbb{Q}$ the set of rational numbers.

---

[11] Dedekind (1930–) vol. 1, p. 178.

Two points $\omega$ and $\omega_1$ in $S$ would be equivalent if

$$\omega_1 = \frac{a\omega + b}{c\omega + d}, \quad \text{with } a, b, c, d \in \mathbb{Z} \text{ and } ad - bc = 1.$$

Note that the point at infinity is equivalent to every rational number $\frac{a}{c}$. The fundamental domain for $\Gamma(1)$ consists of a connected region with exactly one point in each equivalence class; if the full boundary of the region is included in the fundamental domain, then there are pairs of equivalent points on the boundary. Dedekind showed that points $\omega_0$ in the fundamental domain (including boundary) satisfied the three conditions[12]

$$|\omega_0 - 1| \ge |\omega_0|, \quad |\omega_0 + 1| \ge |\omega_0|, \quad \text{and} \quad |\omega_0| \ge 1. \tag{8.22}$$

He noted that if $\omega_0 = x_0 + i y_0$, then the first condition would be reduced to

$$-2x_0 + 1 \ge 0 \quad \text{or} \quad x_0 \le \frac{1}{2}$$

and the second condition to

$$2x_0 + 1 \ge 0 \quad \text{or} \quad x_0 \ge -\frac{1}{2}.$$

Thus, the region called the fundamental domain would encompass points

$$-\frac{1}{2} \le \operatorname{Re} \omega_0 \le \frac{1}{2} \quad \text{and} \quad |\omega_0| \ge 1. \tag{8.23}$$

However, the points on the boundary $|\omega_0| = 1$, $0 \le \operatorname{Re} \omega \le \frac{1}{2}$ are equivalent to the points on $|\omega_0| = 1$, $-\frac{1}{2} \le \operatorname{Re} \omega_0 \le 0$ under $\omega_0 \to -\frac{1}{\omega_0}$; points on $\operatorname{Re} \omega_0 = \frac{1}{2}$ are equivalent to the points on $\operatorname{Re} \omega_0 = -\frac{1}{2}$ under $\omega_0 \to \omega_0 + 1$. Consider the set $D$ that consists of the points

$$-\frac{1}{2} \le \operatorname{Re} \omega_0 < \frac{1}{2}, \quad |\omega_0| \ge 1, \quad |\omega_0| = 1 \text{ with } -\frac{1}{2} \le \operatorname{Re} \omega_0 \le 0. \tag{8.24}$$

For convenience, we will continue to call $D$ a fundamental domain. While no two points in $D$ are equivalent, any point in the upper half-plane $\mathbb{H}$ is equivalent to exactly one point in $D$. In fact, Dedekind proved only that every point in the upper half-plane was equivalent to a point in the region $D$ as defined by (8.24). He noted that the proof applied the same method used to prove that a binary quadratic form of negative determinant was equivalent to one and only one reduced form. In his proof, he let $\omega \in \mathbb{H}$ and let $|c\omega + d|$ denote the least value among all pairs of relatively prime numbers $c$ and $d$. Then there exist integers $a$ and $b$ such that $ad - bc = 1$. Also note that $a' = a + mc$ and $b' = b + md$ satisfy $a'd - b'c = 1$, where $m$ is an integer. Choose $m$ such that

$$\left| \frac{a\omega + b}{c\omega + d} \right| = \left| \frac{a'\omega + b'}{c\omega + d} - m \right|$$

is the smallest possible value, so that the number equivalent to $\omega$, given by

$$\omega_0 = \frac{a\omega + b}{c\omega + d}$$

[12] Ibid. pp. 179–180.

has the properties delineated in (8.22). The third condition in (8.22) follows from the definition of the integers $c, d$; the first and second conditions follow from the definition of $m$.

Dedekind observed that the proof showing that reduced binary quadratic forms of a given negative determinant were inequivalent could be applied, except in exceptional cases, to demonstrate that two points of $D$ were inequivalent. In this connection, he referred to Section 65 of Dirichlet's lectures on number theory. The proof along these lines is fairly easy. Another simple proof of this fact is done by contradiction. Suppose $z, z' \in D$ are such that

$$z' = \frac{az + b}{cz + d}, \quad ad - bc = 1, \quad z \neq z'.$$

The construction of $D$ shows that the transformations $z' = -\frac{1}{z}$ and $z' = z + m$ are not possible. So $c \neq 0$ and

$$z' - \frac{a}{c} = \frac{1}{c(cz + d)}$$

or

$$\left| z' - \frac{a}{c} \right| \left| z + \frac{d}{c} \right| = \frac{1}{c^2}. \tag{8.25}$$

Since the distance between any point in $D$ and the real axis is at least $\frac{\sqrt{3}}{2}$, it follows that $c = 1$ and (8.25) then becomes

$$|z' - a||z + d| = 1. \tag{8.26}$$

Moreover, the distance between any point in $D$ and any integer is at least 1, so that

$$|z' - a| = 1 \quad \text{and} \quad |z + d| = 1. \tag{8.27}$$

Now the integers lying at a distance 1 from points in $D$ are limited to $-1$, 0, and 1. Thus, $a = 0, 1$ and $d = 0, -1$. One need check only that these values of $a$ and $d$ do not lead to equivalent points $z$ and $z'$. For example, $a = 0$ and $d = 0$ are excluded because that results in $z' = -\frac{1}{z}$. Then again, $a = 1, d = -1$ gives

$$z' = -\frac{1}{2} + \frac{\sqrt{3}}{2}i \quad \text{and} \quad z = -\frac{1}{2} + \frac{\sqrt{3}}{2}i, \quad \text{that is, } z = z'.$$

The other cases can similarly be excluded so that we have shown that $z$ and $z'$ cannot be equivalent unless they are identical.

Dedekind was familiar with the fundamental domain for $\Gamma(2)$; the third volume of Gauss's *Werke*, published in 1866, contains a figure of such a domain on pages 477–478. Then again, Riemann dealt with this fundamental domain in his 1858–59 lectures on hypergeometric functions. Since he had already departed for Zürich, Dedekind did not attend these lectures; nevertheless, it is very possible that Riemann discussed these ideas with his close friend Dedekind.

### 8.4  Tesselation of the Upper Half-plane

Set $V \in \Gamma = SL_2(\mathbb{Z})$, and let $V(D)$ be the set of all points $Vz$, $z \in D$, where $D$ is the domain described by (8.24). Then $V(D)$ can also be a fundamental domain of $\Gamma$, since every point in $\mathbb{H}$ is equivalent to a point in $V(D)$ and no two distinct points in $V(D)$ are equivalent. Now the boundary of $D$ consist of three lines: a vertical line starting at $z = -\frac{1}{2} + \frac{\sqrt{3}}{2}i$ and another starting at $z = \frac{1}{2} + \frac{\sqrt{3}}{2}i$, meeting each other at $i\infty$; and an arc of the circle $|z| = 1$ joining these two $z$-values. The region $D$ may be viewed as a triangular region with the point at infinity as one of the vertices. If $V(\omega) = \omega + n$, with $n$ an integer, the transformed region $V(D)$ will again be a similar triangular region with infinity as a vertex; on the other hand, if

$$V(\omega) = \frac{a\omega + b}{c\omega + d}, \quad \text{with} \quad ad - bc = 1 \quad \text{and} \quad c \neq 0,$$

then $V(D)$ will be a triangular region with a cusp at the rational point $\frac{a}{c}$. In this way, if $V_1, V_2, V_3, \ldots$ is a list of all members of $\Gamma$, then the upper half-plane may be understood as being divided into nonoverlapping triangular regions $V_i(D)$, $i = 1, 2, 3, \ldots$. This is a tesselation of the upper half-plane.[13] In speaking of the fundamental domain or region of $\Gamma = SL_2(\mathbb{Z})$, we mean the region $D$ described by (8.24).

### 8.5  Dedekind's Valency Function

Let $D$ be the region defined by (8.24). Then the closure of $D$, denoted by $\overline{D}$, is symmetric about the imaginary axis. Call $D_1$ the interior of the left-hand region, that is

$$D_1 : -\frac{1}{2} < \operatorname{Re}\omega < 0, \quad |\omega| > 1.$$

Since this is a simply connected region not equal to the whole plane, Dedekind applied the Riemann mapping theorem to map it conformally onto the upper half-plane such that the boundary of $D_1$ was mapped to the real axis. The interior of the right half of the region

$$D_2 : 0 < \operatorname{Re}\omega < \frac{1}{2}, \quad |\omega| > 1$$

could then be mapped to the lower half-plane by conjugation. He denoted by $v$ (for valency) the function that mapped $D$ onto the whole complex plane in a one-to-one manner, such that

$$v(x + iy) = \overline{v(-x + iy)}, \quad \text{when } x + iy \in D_2.$$

Let $v_1$ be another conformal mapping of $D_1$ onto the upper half-plane, so that $v_1 \circ v^{-1}$ is a conformal mapping of the upper half-plane onto itself, and is therefore a linear fractional transformation. Thus

$$v_1 = \frac{Av + B}{Cv + D} \quad \text{with } AD - BC \neq 0 \text{ and } A, B, C, D \text{ real.} \tag{8.28}$$

---

[13] Dedekind (1930–) vol. 1, pp. 181–183.

Now (8.28) is completely determined by its values at three points. Dedekind chose these values as

$$v(\infty) = \infty, \quad v\left(-\frac{1}{2} + \frac{\sqrt{3}}{2}i\right) = 0, \quad v(i) = 1.$$

In this situation, the function $v$ can be extended uniquely to all $z \in \mathbb{H} \cup \{\infty\} \cup \mathbb{Q}$, where $\mathbb{Q}$ is the set of rational numbers, by setting

$$v\left(\frac{a\omega + b}{c\omega + d}\right) = v(\omega), \tag{8.29}$$

where $\left(\begin{smallmatrix} a & b \\ c & d \end{smallmatrix}\right) \in \Gamma$. Thus, $v(\omega) = v(\omega_1)$ if and only if

$$\omega_1 = \frac{a\omega + b}{c\omega + d}, \tag{8.30}$$

where $a, b, c, d$ are integers such that $ad - bc = 1$. Note that $v(\omega)$ is analytic for $\omega \in D$, because $v$ is a conformal mapping on $D$. Moreover, $v(\omega)$ remains analytic on the boundary points of $D$ given by $\omega = -\frac{1}{2} + iy$ for $-\frac{\sqrt{3}}{2} < y < \infty$. This is because the right-hand side of a small neighborhood of a point on the boundary will be mapped by $v$ into points in the upper half-plane and the left-hand side will be mapped into conjugate poiints in the lower half-plane. A similar argument can be made about points on the boundary of $D$ given by $|\omega| = 1$ from $\rho = -\frac{1}{2} + \frac{\sqrt{3}}{2}i$ to $i$, but not including the end points. Dedekind showed that $i$ and $\rho$ were fixed points of the transformations $S\omega = -\frac{1}{\omega}$ and $ST\omega = -\frac{1}{\omega+1}$, respectively, where $T\omega = \omega + 1$. Observe that $S^2 = I$ and $(ST)^3 = I$, so that $\omega = i$ is a fixed point of order 2 and $\omega = \rho$ is a fixed point of order 3.

## 8.6 Branch Points

Dedekind proved that $\omega = i$, $\omega = \rho$, $\omega = \infty$ and all points equivalent to these were the only fixed points of the transformations in $\Gamma$.[14] Let $\tau$ be a fixed point of the linear fractional transformation (8.30). Then

$$a\tau + b = \tau(c\tau + d)$$

or

$$(2c\tau + d - a)^2 = (a + d)^2 - 4. \tag{8.31}$$

Observe that $\tau$ must either be rational or have a positive ordinate, so that only three cases are possible: (1) $a + d = 0$, (2) $(a + d)^2 = 1$, and (3) $(a + d)^2 = 4$. The first possibility implies

$$d^2 + 1 = (d + i)(d - i) = -bc$$

---

[14] Ibid. pp. 184–187.

and we may assume $c$ to be positive. From the factorization theory of Gaussian integers, it follows that

$$-d + i = (d_1 - c_1 i)(b_1 + a_1 i) = b_1 d_1 + a_1 c_1 + (a_1 d_1 - b_1 c_1)i \qquad (8.32)$$

$$b = -(b_1 + a_1 i)(b_1 - a_1 i) = -(b_1^2 + a_1^2) \qquad (8.33)$$

$$c = (d_1 + c_1 i)(d_1 - c_1 i) = d_1^2 + c_1^2 \qquad (8.34)$$

$$d + i = -(d_1 + c_1 i)(b_1 - a_1 i) = -(b_1 d_1 + a_1 c_1) + (a_1 d_1 - b_1 c_1)i. \qquad (8.35)$$

By (8.32) we see that $a_1 d_1 - b_1 c_1 = 1$; hence, the Gaussian integers $b_1 + a_1 i$ and $d_1 + c_1 i$ are relatively prime. By (8.31) through (8.35) we may conclude that

$$\tau = \frac{-d + i}{c} = \frac{b}{d + i} = \frac{a_1 i + b_1}{c_1 i + d_1}.$$

Thus, $\tau$ is equivalent to $i$ and $v(\tau) = v(i) = 1$. Taking $\tau = i$, we find $\tau$ to be a fixed point of order 2 of the transform $S\omega = -\frac{1}{\omega}$. Therefore, as $\omega$ moves around $i$ once, $v(\omega)$ moves around $v(i) = 1$ twice so that we obtain

$$1 - v(\omega) = (\omega - i)^2 g(\omega), \quad \text{where} \quad g(i) \neq 0.$$

It follows that, in a neighborhood of $\omega = i$,

$$\frac{d}{d\omega} \sqrt{1 - v} = \frac{-1}{2\sqrt{1 - v}} \frac{dv}{d\omega} \neq 0, \qquad (8.36)$$

and the same holds at any point equivalent to $i$.

In a similar way, Dedekind showed that the second case implies that $\tau$ is equivalent to $\rho = -\frac{1}{2} + \frac{\sqrt{3}}{2}i$. Since $\rho$ is a fixed point of order 3, it follows that in a neighborhood of $\rho$

$$v(\omega) = (\omega - \rho)^3 h(\omega),$$

where $h(\rho) \neq 0$. Thus, in a neighborhood of $\omega = \rho$

$$\frac{d}{d\omega} v^{\frac{1}{3}} = \frac{1}{3} v^{-\frac{2}{3}} \frac{dv}{d\omega} \neq 0. \qquad (8.37)$$

In the third case, the fixed point is given by

$$2c\tau + d - a = 0,$$

meaning that $\tau$ is a rational number when $c \neq 0$ and $\tau = \infty$ when $c = 0$. Observe that all these rational points are equivalent to $\infty$.

Because $v(\omega + 1) = v(\omega)$, $v(\omega)$ has a Fourier series expansion. More explicitly, observe that $q^2 = e^{2\pi i \omega}$ maps the region $-\frac{1}{2} \leq \text{Re}\,\omega < \frac{1}{2}$ in the upper half-plane conformally onto the unit disc with the point at $\infty$ going to 0. Since $v$ is one-to-one in $-\frac{1}{2} \leq \text{Re}\,\omega < \frac{1}{2}$ and $v(\infty) = \infty$, we may assert that

$$v(\omega) = \frac{a_{-1}}{q^2} + \sum_{n=0}^{\infty} a_n q^{2n}. \qquad (8.38)$$

Therefore, in the neighborhood of infinity, $q^2 v$ is finite and nonzero, implying that at infinity

$$v^{-1} \frac{dv}{d\omega} = \frac{d \log v}{d\omega} \approx -2\pi i. \tag{8.39}$$

Dedekind was well aware of the Fourier expansion for $v(\omega)$ in (8.38) from Hermite's 1859 paper, "Sur la théorie des équations modulaires"[15]:

$$1728 v(\omega) = \frac{1}{q^2} - 744 + 196884 q^2 + \cdots . \tag{8.40}$$

We note parenthetically that all the coefficients in the series (8.40) are integers, and these integers bear a direct relationship to the integers that arise in the character table for the finite group known as the Monster.[16]

Now suppose that $f$ is an analytic function defined on the upper half-plane and is invariant under the action of the modular group $\Gamma = SL_2(\mathbb{Z})$ so that, in particular,

$$f(\omega + 1) = f(\omega), \quad f\left(-\frac{1}{\omega}\right) = f(\omega). \tag{8.41}$$

Note that the first equation in (8.41) shows that $f$ has a Fourier expansion; this is given by

$$f(\omega) = \sum_{n=-\infty}^{\infty} a_n q^{2n}, \quad q = e^{\pi \omega i}.$$

Suppose that $f(\omega) \to \infty$ as $y \to \infty$ in $\omega = x + iy$; then $f(q^2)$ has a pole of order, say, $m$ at $q = 0$ and

$$f(\omega) = \sum_{n=-m}^{\infty} a_n q^{2n}, \quad q = e^{\pi \omega i}. \tag{8.42}$$

It is said that $f$ is meromorphic at infinity if (8.42) holds. Thus, $v$ has a pole of order 1 at infinity.

## 8.7 Differential Equations

To find the differential equation involving $v$, Dedekind started with the Schwarzian derivative of $v$.[17] Recall from our discussion of hypergeometric functions that Riemann and Schwarz had taken a similar approach in their study of triangle functions; note that $v$ is a triangle function. Dedekind had gleaned this from Schwarz's 1872 paper on the hypergeometric function. Also recall that Dedekind may well have discussed these

---

[15] Hermite (1905–1917) vol. 2, p. 60. This paper contains a typographical error: The coefficient of $q^2$ is given as 196880 rather than 196884.

[16] See Gannon (2006) or, for a popular account, Ronan (2006).     [17] Dedekind (1930–) vol. 1, pp. 187–190.

ideas with Riemann. Dedekind began by setting

$$[v, \omega] = -\frac{4}{\sqrt{\frac{dv}{d\omega}}} \frac{d^2}{dv^2} \sqrt{\frac{dv}{d\omega}}. \tag{8.43}$$

Observe that (8.43) is twice the Schwarzian derivative of $\omega$ with respect to $v$, denoted by $\{\omega, v\}$:

$$[v, \omega] = 2\frac{\omega'''}{\omega'} - 3\left(\frac{\omega''}{\omega'}\right)^2 \equiv 2\{\omega, v\}, \quad \omega' = \frac{d\omega}{dv}.$$

Recall that the Schwarzian derivative $\{f, z\}$ has the following property:

$$\{f \circ \zeta, z\} = \{f, \zeta\}(\zeta'(z))^2 + \{\zeta, z\}. \tag{8.44}$$

Now if $f$ is a fractional linear transformation

$$f(\zeta) = \frac{A\zeta + B}{C\zeta + D},$$

then $\{f, \zeta\} = 0$ and by (8.44)

$$\{f \circ \zeta, z\} = \{\zeta, z\}. \tag{8.45}$$

On the other hand, if $\zeta$ is a fractional linear transformation of $z$, then

$$\{f \circ \zeta, z\} = \{f, \zeta\}(\zeta'(z))^2. \tag{8.46}$$

It is clear that $[v, \omega]$ is a single-valued function of $\omega$. Dedekind employed (8.45) to show that $[v, \omega]$ was a single-valued function of $v$, that is, that for a given value of $v$, there would be only one value of $[v, \omega]$. Observe that $v$ has the same value at $\omega$ and $\omega_1$ if and only if

$$\omega_1 = \frac{a\omega + b}{c\omega + d},$$

$a, b, c, d$ integers and $ad - bc = 1$. Thus, $[v, \omega]$ is a single-valued function of $v$ if

$$[v, \omega] = [v, \omega_1]; \tag{8.47}$$

and (8.47) is true by (8.45). Dedekind could then set

$$[v, \omega] = F(v), \tag{8.48}$$

where $F(v)$ denoted a single-valued function of $v$. To determine the function $F(v)$, he had to study its behavior at the points $i, \rho, \infty$. He first noted that (8.36), (8.37) and (8.39), implied that the functions

$$(1 - v)^{-\frac{1}{2}} \frac{dv}{d\omega}, \quad v^{-\frac{2}{3}} \frac{dv}{d\omega}, \quad \text{and} \quad v^{-1} \frac{dv}{d\omega}$$

were finite and nonzero at $i$, $\rho$, and $\infty$, respectively. He then observed that

$$(1-v)^2 F(v) = \frac{3}{4}, \quad v^2 F(v) = \frac{8}{9}, \quad \text{and} \quad v^2 F(v) = 1 \qquad (8.49)$$

at $i$, $\rho$, and $\infty$, respectively. Note that the values in (8.49) are limits of the left-hand sides as $\omega$ tends to $i$, $\rho$, and $\infty$, respectively, or as $v$ tends to 1, 0, and $\infty$, respectively. We may verify (8.49) by a rough calculation: From (8.36), we can conclude that in a sufficiently small neighborhood of $v = 1$, and for some constant $C$,

$$\sqrt{\frac{dv}{d\omega}} = C(1-v)^{\frac{1}{4}} + \text{ higher powers of } 1-v. \qquad (8.50)$$

Hence

$$\frac{d^2}{dv^2}\left(\sqrt{\frac{dv}{d\omega}}\right) = -\frac{3C}{16}(1-v)^{-\frac{7}{4}} + \cdots . \qquad (8.51)$$

Dividing (8.51) by (8.50) and then multiplying the resulting equation by $-4$ gives us

$$F(v) = \frac{3}{4}(1-v)^{-2} + \cdots .$$

The first result in (8.49) then follows; the other two may be derived similarly.

Thus, from (8.49) Dedekind saw that $v^2(1-v)^2 F(v)$ was an analytic and single-valued function of $v$ that behaved like $v^2$ at $v = \infty$. This implied that

$$v^2(1-v)^2 F(v) = Av^2 + Bv + C$$

was a polynomial of degree 2. The values of $A, B$ and $C$ could be determined from (8.49), allowing Dedekind to write

$$F(v) = \frac{36v^2 - 41v + 32}{36v^2(1-v)^2}. \qquad (8.52)$$

Dedekind therefore concluded that $v = v(\omega)$ was a solution of the third order equation

$$[u, \omega] = F(u). \qquad (8.53)$$

He also proved that the general solution of (8.53) was given by

$$u = v\left(\frac{C + D\omega}{A + B\omega}\right),$$

where $A, B, C, D$ were arbitrary constants. Here Dedekind referred to Sections 32 and 33 of Jacobi's *Fundamenta Nova*, where Jacobi implicitly considered the Schwarzian derivative and obtained analogous results for the modular function $k^2$.

It is clear from (8.48) and (8.43) that $u = \sqrt{\frac{dv}{d\omega}}$ is a solution of the second order linear differential equation

$$4\frac{d^2u}{dv^2} + F(v)u = 0. \qquad (8.54)$$

After observing this fact, Dedekind remarked that it would be more appropriate to consider the function

$$w = \text{const.} \, v^{-\frac{1}{3}} (1 - v)^{-\frac{1}{4}} \left( \frac{dv}{d\omega} \right)^{\frac{1}{2}}. \tag{8.55}$$

After applying (8.52) and (8.54), we see with Dedekind that $w$ satisfies the hypergeometric equation

$$v(1 - v) \frac{d^2 w}{dv^2} + \left( \frac{2}{3} - \frac{7}{6} v \right) \frac{dw}{dv} - \frac{1}{144} w = 0. \tag{8.56}$$

He noted that the general solution of the equation would be

$$w = C_1 F \left( \frac{1}{12}, \frac{1}{12}, \frac{2}{3}, v \right) + C_2 F \left( \frac{1}{12}, \frac{1}{12}, \frac{1}{2}, 1 - v \right), \tag{8.57}$$

where $F$ denoted Gauss's hypergeometric function. Note the modern notation for the first function on the right-hand side of (8.57):

$$_2F_1 \left( \begin{matrix} \frac{1}{12}, \frac{1}{12} \\ \frac{2}{3} \end{matrix}; v \right).$$

Later in his paper, Dedekind showed that

$$w = \eta^2(\omega) = q^{\frac{1}{6}} \prod_{n=1}^{\infty} (1 - q^{2n})^2 = (2\pi)^{-1} \sqrt[12]{\Delta(\omega)}, \quad q = e^{\pi \omega i}, \tag{8.58}$$

where $(2\pi)^{-12} \Delta(\omega)$ is the discriminant function:

$$(2\pi)^{-12} \Delta(\omega) = q^2 \prod (1 - q^{2n})^{24}, \quad q = e^{\pi \omega i}. \tag{8.59}$$

Thus, the twelfth root of the discriminant function can be expressed as a sum of a hypergeometric function of $v = J$ and a hypergeometric function of $1 - J$.

## 8.8   Dedekind's $\eta$ Function

In his theory of elliptic modular functions, Dedekind repeatedly utilized the result: If $f$ is an analytic (holormophic) function on the upper half-plane such that

$$f(\omega + 1) = f(\omega) = f \left( -\frac{1}{\omega} \right)$$

and $f$ tends to a finite value as $q^2 \to 0$ (or as $v \to \infty$), then $f$ is a constant. Dedekind did not indicate a proof of this result, taking it to be generally known. There are several possible arguments for this result. A process outlined by Klein[18], in a paper published a year after Dedekind's paper on modular functions, requires the identification of the

---

[18]  Klein (1921–1923) vol. 3, pp. 35–37.

boundary of the fundamental domain of $\Gamma = SL_2(\mathbb{Z})$ by means of the transformations $\omega \to \omega + 1$ and $\omega \to -\frac{1}{\omega}$. If the point at infinity is included in the boundary, this results in a surface of genus 0. A function defined on such a surface, topologically equivalent to the Riemann sphere, will be a constant if it has it has no poles. These ideas can be traced back to Riemann so that Dedekind was familiar with them. Dedekind could possibly have provided an argument by integrating the function $\frac{f'(z)}{f(z)}$ around the boundary of the fundamental domain. This argument was given in this context by Hurwitz.[19] In any case, the result was well known.[20]

Dedekind also often made use of two more general facts, obtainable from the particular case: (a) that an analytic function invariant under $\Gamma$ and satisfying (8.42) is a polynomial of degree $m$ in $v$, the Dedekind valency function defined by (8.40); (b) that if $f$ is meromorphic in the upper half-plane, invariant under $\Gamma$, and satisfies (8.42), then $f$ must be a rational function in the valency function $v$.

Dedekind introduced the function now known as Dedekind's $\eta$ function, as the square root of $w$ in (8.58):[21]

$$\eta(\omega) = \text{const. } v^{-\frac{1}{6}} (1 - v)^{-\frac{1}{8}} \left(\frac{dv}{d\omega}\right)^{\frac{1}{4}}. \tag{8.60}$$

From the behavior of $v$ at infinity, as given in (8.38) and (8.39), Dedekind had

$$e^{-\frac{2\pi i\omega}{24}} \eta(\omega) = \text{const. [taken to be 1]} \quad \text{when} \quad \omega = i\infty. \tag{8.61}$$

Dedekind took the constant in (8.61) to be 1 so as to fully define $\eta(\omega)$. Note that (8.56) implies that $\eta^2(\omega)$ is a single-valued function of $v(\omega)$ for $\omega$ in the upper half-plane. Taking the derivative of (8.29) yields

$$(c\omega + d)^{-2} v' \left(\frac{a\omega + b}{c\omega + d}\right) = v'(\omega), \quad \text{when} \quad ad - bc = 1. \tag{8.62}$$

From (8.29) and (8.60) one obtains

$$\eta\left(\frac{a\omega + b}{c\omega + d}\right) = u(c\omega + d)^{\frac{1}{2}} \eta(\omega), \tag{8.63}$$

where $u^{24} = 1$. In particular, (8.63) and (8.61) imply that

$$\eta(1 + \omega) = e^{\frac{\pi i}{12}} \eta(\omega) \tag{8.64}$$

and

$$\eta\left(-\frac{1}{\omega}\right) = u_1 \omega^{\frac{1}{2}} \eta(\omega). \tag{8.65}$$

---

[19] Hurwitz (1962) vol. 1, pp. 589–590.

[20] Rademacher (1974) vol. 2, pp. 358–360. Rademacher gave a simple proof of this result by using Jensen's inequality in complex analysis. Jensen proved his theorem on meromorphic functions in 1899, but Jacobi had found a form of this theorem in 1827. See Roy (2011) pp. 892–894.

[21] Dedekind (1930–) vol. 1, pp. 190–194.

By setting $\omega = i$ in (8.65), we see that $u_1 = i^{-\frac{1}{2}} = e^{-\frac{\pi i}{4}}$, because $\eta(i)$ cancels. Thus,

$$\eta\left(-\frac{1}{\omega}\right) = \sqrt{\frac{\omega}{i}}\,\eta(\omega). \tag{8.66}$$

The $\eta$ function is uniquely defined by relations (8.61), (8.64), and (8.66). To see how Dedekind verified this, first assume that $f$ is a function different from $\eta$ satisfying these relations. Then $\frac{f(\omega)}{\eta(\omega)} = g(\omega)$ satisfies

$$g(\omega + 1) = g(\omega) \quad \text{and} \quad g\left(-\frac{1}{\omega}\right) = g(\omega) \tag{8.67}$$

and $g$ is analytic in a neighborhood of infinity with $g(i\infty) = 1$. Now since $S\omega = \omega + 1$ and $T\omega = -\frac{1}{\omega}$ generate the modular group, $g(\omega)$ is invariant under the action of the modular group and is bounded at $i\infty$. Thus, Dedekind concluded that $g$ was a constant equal to 1; this implied that $f(\omega) = \eta(\omega)$ so that the result was proved.

In order to establish the connection of the $\eta$ function with the elliptic modular function $k^2$, Dedekind considered the three second-order transformations of the $\eta$ function:

$$\eta_1(\omega) = \eta(2\omega); \quad \eta_2(\omega) = \eta\left(\frac{\omega}{2}\right); \quad \eta_3(\omega) = \eta\left(\frac{1+\omega}{2}\right). \tag{8.68}$$

These functions possess properties as consequences of (8.64) and (8.66):

$$\eta_1(1+\omega) = e^{\frac{\pi i}{6}}\,\eta_1(\omega), \quad \eta_1\left(-\frac{1}{\omega}\right) = e^{-\frac{\pi i}{4}}\left(\frac{\omega}{2}\right)^{\frac{1}{2}}\eta_2(\omega), \tag{8.69}$$

$$\eta_2(1+\omega) = \eta_3(\omega), \quad \eta_2\left(-\frac{1}{\omega}\right) = e^{-\frac{\pi i}{4}}(2\omega)^{\frac{1}{2}}\eta_1(\omega), \tag{8.70}$$

$$\eta_3(1+\omega) = e^{\frac{\pi i}{12}}\,\eta_2(\omega), \quad \eta_3\left(-\frac{1}{\omega}\right) = e^{-\frac{\pi i}{4}}\omega^{\frac{1}{2}}\eta_3(\omega). \tag{8.71}$$

We present proofs of (8.69) and (8.71); (8.70) is proved in a similar manner. First, by (8.64), we have

$$\eta_1(1+\omega) = \eta(2+2\omega) = e^{\frac{2\pi i}{12}}\eta(2\omega) = e^{\frac{\pi i}{6}}\eta_1(\omega).$$

Next, by (8.66) we may write

$$\eta_1\left(-\frac{1}{\omega}\right) = \eta\left(-\frac{2}{\omega}\right) = \eta\left(-\frac{1}{\frac{\omega}{2}}\right) = e^{-\frac{\pi i}{4}}\left(\frac{\omega}{2}\right)^{\frac{1}{2}}\eta\left(\frac{\omega}{2}\right) = e^{-\frac{\pi i}{4}}\left(\frac{\omega}{2}\right)^{\frac{1}{2}}\eta_2(\omega).$$

Again using (8.64),

$$\eta_3(1+\omega) = \eta\left(\frac{2+\omega}{2}\right) = \eta\left(1+\frac{\omega}{2}\right) = e^{\frac{\pi i}{12}}\eta\left(\frac{\omega}{2}\right) = e^{\frac{\pi i}{12}}\eta_2(\omega).$$

We are now in a position to prove the second equation given in (8.71). Note that by (8.64) and (8.66)

$$
\eta_3\left(-\frac{1}{\omega}\right) = \eta\left(\frac{\omega-1}{2\omega}\right) = e^{-\frac{\pi i}{4}}\left(\frac{2\omega}{1-\omega}\right)^{\frac{1}{2}}\eta\left(\frac{2\omega}{1-\omega}\right)
$$

$$
= e^{-\frac{\pi i}{4}}\left(\frac{2\omega}{1-\omega}\right)^{\frac{1}{2}}\eta\left(-2+\frac{2}{1-\omega}\right)
$$

$$
= e^{-\frac{\pi i}{4}}e^{-\frac{\pi i}{6}}\left(\frac{2\omega}{1-\omega}\right)^{\frac{1}{2}}\eta\left(-\frac{2}{\omega-1}\right)
$$

$$
= e^{-\frac{\pi i}{4}}e^{-\frac{\pi i}{6}}\left(\frac{2\omega}{1-\omega}\right)^{\frac{1}{2}}e^{-\frac{\pi i}{4}}\left(\frac{\omega-1}{2}\right)^{\frac{1}{2}}\eta\left(\frac{\omega-1}{2}\right)
$$

$$
= e^{-\frac{\pi i}{4}}e^{-\frac{\pi i}{6}}\omega^{\frac{1}{2}}e^{-\frac{\pi i}{4}}e^{\frac{\pi i}{2}}e^{-\frac{\pi i}{12}}\eta\left(\frac{\omega+1}{2}\right)
$$

$$
= e^{-\frac{\pi i}{4}}\omega^{\frac{1}{2}}\eta_3(\omega).
$$

From (8.61) we can deduce that as $\omega \to i\infty$

$$
e^{-\frac{2\pi i\omega}{12}}\eta_1(\omega) \to 1\,; \quad e^{-\frac{2\pi i\omega}{48}}\eta_2(\omega) \to 1\,; \quad e^{-\frac{2\pi i\omega}{48}}\eta_3(\omega) \to e^{\frac{2\pi i}{48}}. \tag{8.72}
$$

From (8.69) through (8.71) it follows that

$$
f(\omega) = \frac{\eta_1(\omega)\,\eta_2(\omega)\,\eta_3(\omega)}{\eta^3(\omega)} \tag{8.73}
$$

has the properties

$$
f(1+\omega) = f(\omega) \quad \text{and} \quad f\left(-\frac{1}{\omega}\right) = f(\omega). \tag{8.74}
$$

As we saw before, this means that $f$ is invariant under $\Gamma = SL_2(\mathbb{Z})$ and can be seen to be a single-valued, analytic, bounded function, and hence a constant. In fact, when $\omega = i\infty$, we may conclude from (8.61) and (8.72) that

$$
f(\omega) = e^{\frac{2\pi i}{48}} \cdot e^{\frac{2\pi i\omega}{12} + \frac{2\pi i\omega}{48} + \frac{2\pi i\omega}{48} - \frac{6\pi i\omega}{24}}
$$

$$
= e^{\frac{2\pi i}{48}}.
$$

Thus,

$$
\eta_1(\omega)\,\eta_2(\omega)\,\eta_3(\omega) = e^{\frac{2\pi i}{48}}\eta^3(\omega). \tag{8.75}
$$

Similarly, it can be shown that

$$
f_1(\omega) = \frac{2^4\,\eta_1^8(\omega) + \eta_2^8(\omega) + e^{\frac{2\pi i}{3}}\eta_3^8(\omega)}{\eta^8(\omega)}
$$

satisfies the properties

$$
f_1(1+\omega) = e^{\frac{2\pi i}{3}}f_1(\omega) \quad \text{and} \quad f_1\left(-\frac{1}{\omega}\right) = f_1(\omega). \tag{8.76}
$$

Relations (8.76) imply that $f_1^3(\omega)$ is invariant with respect to the modular group; its behavior at infinity shows that it is a polynomial in the valency function $v$. In fact, we can go further and say that $f_1(\omega)e^{\frac{2\pi i\omega}{6}} \to 0$ as $\omega \to i\infty$, or $f_1^3(\omega)v^{-\frac{1}{2}} \to 0$ as $v \to \infty$. Recall that $v \to \infty$ as $\omega \to i\infty$. Thus, $f_1^3(\omega)$ cannot become infinitely large, because it cannot be of order $v^s$, where $s$ is a fraction $< 1$. This in turn implies that $f_1^3$ is a constant, and hence so is $f_1$. The first relation in (8.76) shows that this constant must be 0. We may therefore conclude that

$$2^4\, \eta_1^8(\omega) + \eta_2^8(\omega) + e^{\frac{2\pi i}{3}}\, \eta_3^8(\omega) = 0. \qquad (8.77)$$

This relation led Dedekind to introduce functions corresponding to those given by Hermite:[22]

$$\phi(\omega) = e^{\frac{2\pi i}{48}}\sqrt{2}\,\frac{\eta_1(\omega)}{\eta_3(\omega)} = \sqrt[8]{k^2} \qquad (8.78)$$

$$\psi(\omega) = e^{\frac{2\pi i}{48}}\,\frac{\eta_2(\omega)}{\eta_3(\omega)} \quad = \sqrt[8]{k'^2} = \phi\left(-\frac{1}{\omega}\right) \qquad (8.79)$$

$$\chi(\omega) = e^{\frac{2\pi i}{48}}\sqrt[6]{2}\,\frac{\eta(\omega)}{\eta_3(\omega)}. \qquad (8.80)$$

Note that (8.78) and (8.79) give us the definitions of $k^2$ and $k'^2$. Now it follows from (8.77) that

$$\phi^8(\omega) + \psi^8(\omega) = 1, \text{ that is, } k^2 + k'^2 = 1, \qquad (8.81)$$

and from (8.75) that

$$\phi(\omega)\,\psi(\omega) = \chi^3(\omega). \qquad (8.82)$$

Dedekind defined the quantities $K$ and $K'$

$$\sqrt{\frac{2K}{\pi}} = e^{-\frac{2\pi i}{24}}\,\frac{\eta_3^2(\omega)}{\eta(\omega)}, \quad K'i = K\omega. \qquad (8.83)$$

An application of formulas (8.69) and (8.71) shows that for $\lambda = \phi^8$

$$\phi^8(1+\omega) = -2^4\,\frac{\eta_1^8(\omega)}{\eta_3^8(\omega)} = \frac{\lambda}{\lambda-1}, \qquad (8.84)$$

$$\phi^8\left(-\frac{1}{\omega}\right) = e^{\frac{2\pi i}{6}}\,\frac{\eta_2^8(\omega)}{\eta_3^8(\omega)} = 1 - \lambda, \qquad (8.85)$$

and by (8.72)

$$\lambda e^{-\pi i\omega} \to 2^4 \text{ as } \omega \to i\infty. \qquad (8.86)$$

At this juncture, taking $\rho = e^{\frac{2\pi i}{3}}$, Dedekind defined the function

$$f_2(\omega) = \frac{(\lambda+\rho)^3(\lambda+\rho^2)^3}{\lambda^2(1-\lambda)^2} = \frac{(\lambda^2-\lambda+1)^3}{\lambda^2(1-\lambda)^2} \qquad (8.87)$$

[22] Hermite (1905–1917) vol. 2, pp. 7–8.

Note that it is easy to check that

$$f_2(1 + \omega) = f_2(\omega), \quad f_2\left(-\frac{1}{\omega}\right) = f_2(\omega);$$

hence, $f_2$ is a single-valued function of $v = \text{val}(\omega)$. It follows from (8.87) that $f_2$ can be infinite only when $\lambda = 0, 1,$ or $\infty$. Since $\lambda$ and $1 - \lambda$ are quotients of the $\eta$-functions, $\lambda$ can take the value $0, 1,$ or $\infty$ only when $v = \infty$, that is, for example, when $\omega = i\infty$. Relation (8.86) implies that $\lambda$ behaves like $e^{\pi i \omega}$ as $\omega \to i\infty$; from (8.87) it then follows that $f_2(\omega)$ becomes infinitely large and is of the order of $v$ as $\omega \to i\infty$. Thus, $f_2(\omega)$ is a polynomial of degree 1 in $v$ and

$$f_2(\omega) = \frac{(\lambda^2 - \lambda + 1)^3}{\lambda^2(1 - \lambda)^2} = av(\omega) + b. \tag{8.88}$$

To find $b$, we set $\omega = \rho$; observing that $v(\rho) = 0$, we have $b = f_2(\rho)$. Since $\rho^2 + \rho + 1 = 0$, we may write $\frac{1+\rho}{2} = -\frac{1}{2\rho}$. Therefore

$$\eta_3(\rho) = \eta\left(\frac{1+\rho}{2}\right) = \eta\left(-\frac{1}{2\rho}\right)$$

and by (8.66)

$$\eta_3^8(\rho) = \eta^8\left(-\frac{1}{2\rho}\right) = (2\rho)^4 \eta^8(2\rho) = (2\rho)^4 \eta_1^8(\rho).$$

So by the definition of $\lambda = k^2$ in (8.78), we have

$$\lambda(\rho) = \phi^8(\rho) = e^{\frac{2\pi i}{6}} 2^4 \frac{\eta_1^8(\rho)}{\eta_3^8(\rho)} = -\rho.$$

Thus, by (8.87), $f_2(\rho) = 0$, and we get $b = 0$. Next, to determine the value of $a$ in (8.88), we set $\omega = i$ and since $v(i) = 1$, we have $a = f_2(i)$. Now by (8.79) and (8.81)

$$\phi^8(i) = \phi^8\left(-\frac{1}{i}\right) = 1 - \phi^8(i);$$

therefore $\lambda(i) = \phi^8(i) = \frac{1}{2}$. Substituting this value in (8.87), we have

$$f_2(i) = \frac{\left(\frac{1}{4} - \frac{1}{2} + 1\right)^3}{\frac{1}{4} \cdot \frac{1}{4}} = \frac{27}{4} = a.$$

We finally obtain Dedekind's formula[23]

$$v = \text{val}(\omega) = \frac{4}{27} \frac{(\lambda + \rho)^3 (\lambda + \rho^2)^3}{\lambda^2 (1 - \lambda)^2} = \frac{4}{27} \frac{(\lambda^2 - \lambda + 1)^3}{\lambda^2 (1 - \lambda)^2}. \tag{8.89}$$

Note that (8.86) and (8.89) imply that with $q = e^{\pi i \omega}$

$$q^2 v \to \frac{1}{1728} \quad \text{as} \quad \omega \to i\infty. \tag{8.90}$$

[23] Ibid. p. 193.

## 8.9 The Uniqueness of $k^2$

Recall that Dedekind proved that the function $\eta(\omega)$ was completely determined by (8.61), (8.64), and (8.66). He went on to say that it could be similarly proved that $\phi^8(\omega) = k^2(\omega)$ was uniquely determined by (8.83), (8.84), and (8.85).

Note that (8.84) and (8.85) together imply that $k^2(\omega)$ is invariant with respect to the subgroup $\Gamma(2)$ of the modular group. As Dedekind observed, $\Gamma(2)$ was of index 6 in $\Gamma(1)$. He remarked that, just as the Schwarzian derivative of $\omega$ with respect to the invariant of the modular group, $v$, could be expressed as a rational function of $v$ as in (8.52), the Schwarzian derivative of $\omega$ with respect to the invariant of $\Gamma(2)$, $\lambda = k^2$, could be expressed as a rational function of $\lambda$:

$$[\lambda, \omega] = \frac{(\lambda + \rho)(\lambda + \rho^2)}{\lambda^2(1 - \lambda)^2} = \frac{\lambda^2 - \lambda + 1}{\lambda^2(1 - \lambda)^2}, \quad \rho = e^{\frac{2\pi i}{3}}. \tag{8.91}$$

He noted that one could show that

$$\frac{d\lambda}{d\omega} = -\frac{4}{\pi i} \lambda(1 - \lambda)K^2. \tag{8.92}$$

Since this implied that

$$K = \text{const.} \, \lambda^{-\frac{1}{2}}(1 - \lambda)^{-\frac{1}{2}} \sqrt{\frac{d\lambda}{d\omega}},$$

by (8.91), $K$ must satisfy the hypergeometric equation

$$\lambda(1 - \lambda)\frac{d^2K}{d\lambda^2} + (1 - 2\lambda)\frac{dK}{d\lambda} - \frac{1}{4}K = 0. \tag{8.93}$$

Therefore, $K$ was seen to be uniquely defined as a single-valued function of $\lambda$.

## 8.10 The Connection of $\eta$ with Theta Functions

We have seen that Dedekind obtained the elliptic modular functions $k, K, K'$ without using elliptic function theory. He regretted the fact that in his derivation of the infinite product expression for $\eta(\omega)$, he required some properties of the $\theta$ function. It can be shown, using Hermite's transformation (5.7) through (5.9), and equations (8.84) and (8.85), that the function $k^2 = \phi^8(\omega)$, on one hand, and the ratio of the theta functions

$$\frac{\theta_2^4(0, \omega)}{\theta_3^4(0, \omega)},$$

on the other hand, are transformed in the same way under the action of the modular group and that their behavior at $\infty$ is the same and that they must be identical. Dedekind pointed this out, without referring to Hermite. Dedekind next observed that it then followed that

$$\theta_0(0, \omega) = \frac{\eta_2^2(\omega)}{\eta(\omega)}, \quad \theta_1'(0, \omega) = 2\pi \, \eta^3(\omega), \tag{8.94}$$

$$\theta_2(0, \omega) = \frac{2\eta_1^2(\omega)}{\eta(\omega)}, \quad \theta_3(0, \omega) = e^{-\frac{\pi i}{12}} \frac{\eta_3^2(\omega)}{\eta(\omega)}; \tag{8.95}$$

for proofs one may use (8.75), (8.78), (8.79), and (8.83).

Next Dedekind employed Jacobi's formula (3.201)

$$\theta_1'(0, \omega) = 2\pi\, e^{\frac{\pi i \omega}{4}} \prod_{n=1}^{\infty} (1 - e^{2\pi i n \omega})^3$$

to arrive at

$$\eta(\omega) = q^{\frac{1}{12}} \prod_{n=1}^{\infty} (1 - q^{2n}), \quad q = e^{\pi i \omega}. \tag{8.96}$$

Dedekind wrote that he had not succeeded in deriving this representation of $\eta(\omega)$ without the help of the theory of theta functions.

## 8.11  Hurwitz's Infinite Product for $\eta(\omega)$

In his 1881 doctoral dissertation, Hurwitz gave a derivation of (8.96) without utilizing the properties of theta functions,[24] thereby completing the work begun by Dedekind to construct a theory of modular functions independent of elliptic and theta functions. Hurwitz therefore titled the first part of his dissertation "Foundation of an independent theory of elliptic modular functions." Following Klein, Hurwitz denoted Dedekind's valency function $v$ by the symbol $J$. He proved that for some constant $C$

$$q^{\frac{1}{3}} \prod_{n=1}^{\infty} (1 - q^{2n})^4 = C J^{-\frac{2}{3}} (1 - J)^{-\frac{1}{2}} \frac{dJ}{d\omega}. \tag{8.97}$$

Comparing (8.97) with (8.60), we see that (8.96) follows. In this section and the next, we sketch Hurwitz's derivation of (8.97). Recall that Hurwitz proved that

$$\Delta(\omega) \equiv (2\pi)^{12} q^2 \prod_{n=1}^{\infty} (1 - q^{2n})^{24}, \quad q = e^{i\pi \omega} \tag{8.98}$$

was a cusp form of weight 12. See Section 4.11 in this connection. He then had to prove that

$$\Delta(\omega) = g_2^3(\omega) - 27 g_3^2(\omega). \tag{8.99}$$

Recall that Hurwitz could not draw upon the modern argument showing that the dimension of the space of cusp forms of weight 12 must be 1. Note that in this section, following Hurwitz's reasoning, $\Delta(\omega)$ is defined by (8.98) rather than (8.99). Recall that

$$g_2(\omega) = 60 \sideset{}{'}\sum_{m,n} \frac{1}{(m\omega + n)^4}, \quad g_3(\omega) = 140 \sideset{}{'}\sum_{m,n} \frac{1}{(m\omega + n)^6},$$

where $m$ and $n$ run over all integers except $(m, n) = (0, 0)$. Hurwitz observed that $g_2^3(\omega)$ was also a modular form of weight 12, and hence $f_1 \equiv \frac{g_2^3}{\Delta}$ was a modular function of weight 0 in $\omega$ and was thus a single-valued function of $J(\omega)$. Since $\Delta$ did not vanish

---

[24]  Hurwitz (1932) pp. 23–30.

at any point in $\mathbb{H}$, $f_1$ was holomorphic on $\mathbb{H}$. Moreover, $f_1$ had a pole of order 1 at $i\infty$ because

$$q^2 \frac{g_2^3}{\Delta} \to \text{ a nonzero finite number} \quad \text{as } \omega \to i\infty.$$

Therefore, $f_1$ was a linear function of $J$. Similarly, by the same reasoning, the modular function of weight 0, $f_2 = \frac{g_3^2}{\Delta}$, must be a linear function of $J$. In addition,

$$\frac{g_2}{\sqrt[3]{\Delta}} \quad \text{and} \quad \frac{g_3}{\sqrt{\Delta}}$$

were single-valued analytic functions of $\omega$ and were also functions of $J$. Hurwitz argued[25] that branch points of functions of $J(\omega)$ that were single-valued functions of $\omega$ would be of order 2 at $J = 1$ and of order 3 at $J = 0$. It followed that

$$\frac{g_2^3}{\Delta} = aJ, \quad \frac{g_3^2}{\Delta} = a_1(J - 1). \tag{8.100}$$

From the $q$-series expansions of $g_2$ and $g_3$ in (4.173) and using (8.98), Hurwitz determined that $a = 1$ and $a_1 = \frac{1}{27}$, that is

$$\frac{g_2^3}{\Delta} = J, \quad \frac{27g_3^2}{\Delta} = J - 1, \quad \text{and hence } \Delta = g_2^3 - 27g_3^2. \tag{8.101}$$

Hurwitz noted that though it could be directly shown that $g_2(\rho) = 0$ and $g_3(i) = 0$, these facts were consequences of (8.100).

## 8.12   Algebraic Relations among Modular Forms

Hurwitz pointed out[26] that the method used to obtain the algebraic relationship

$$\Delta(\omega) = g_2^3(\omega) - 27g_3^2(\omega)$$

could be adapted to discover all the relationships of this kind. More generally, he considered forms with a multiplier system $V(\gamma)$, $\gamma \in \Gamma$, where $V(\gamma)$ was a root of unity. Suppose $V(\gamma)$ is an $n$th root of unity, that $f$ is a meromorphic modular form of weight $k$ with multiplier system $V(\gamma)$, and that $f$ is also meromorphic at infinity. Under those conditions, the function

$$g(\omega) = \left( \frac{f(\omega)}{\left( \sqrt[12]{\Delta} \right)^k} \right)^{12n}$$

is a modular function of weight 0 with a finite number of poles in the fundamental region, including the point $i\infty$; thus, $g$ is a rational function of $J$. Furthermore, since $g^{\frac{1}{12n}}(\omega)$ is a single -valued function of $\omega$ as well as a function of $J$, it can have a branch point of order 2 at $J(\omega) = 1$, or it can have a branch point of order 3 at $J(\omega) = 0$. It

---

[25] Ibid. p. 28.      [26] Ibid. p. 29–30.

follows that

$$\frac{f(\omega)}{\left(\sqrt[12]{\Delta}\right)^k} = \sqrt[3]{J^a}\,\sqrt{(J-1)^b}\,R(J), \tag{8.102}$$

where $a$ and $b$ are integers and $R$ is a rational function. By (8.101), we may write (8.102) as

$$f(\omega) = \left(\sqrt[12]{\Delta}\right)^r \frac{h_n(g_2, g_3)}{h_m(g_2, g_3)}, \tag{8.103}$$

where $h_n$ and $h_m$ are polynomials in $g_2$, $g_3$ and hence are modular forms of weights, say, $n$ and $m$. This means that $k = r + n - m$. Hurwitz noted the particular case of (8.103) where $f$ had no poles in the fundamental domain of $\Gamma(1)$, except at $i\infty$. In this case,

$$f(\omega) = \left(\sqrt[12]{\Delta}\right)^r h_n(g_2, g_3). \tag{8.104}$$

In particular, (8.104) implies that each Eisenstein series $G_k(\omega)$, $k \geq 4$, is a polynomial in $g_2$ and $g_3$. For example, Hurwitz noted that there existed a constant $c$ such that $G_8 = cG_4^2$; or, with an application of (4.173), such that

$$1 + 480 \sum_{n=1}^{\infty} \sigma_7(n) q^{2n} = \left(1 + 240 \sum_{n=1}^{\infty} \sigma_3(n) q^{2n}\right)^2. \tag{8.105}$$

Equating the coefficients of $q^{2p}$ in (8.105), with $p$ an odd prime, Hurwitz obtained the number theoretic result:

$$p^7 - p^3 = 240 \sum_{a=1}^{\frac{p-1}{2}} \sigma_3\left(\frac{p^2 - (2a-1)^2}{4}\right).$$

Observe that since a bounded modular form of weight zero is a constant, Hurwitz's (8.104) implies that the number of linearly independent modular forms of weight $k$ is $1 + \lfloor \frac{k}{12} \rfloor$. This result was noted explicitly by Hecke in a paper of 1935.[27]

Hurwitz noted another consequence of (8.103): If $V(\gamma)$ is an $n$th root of unity, then $n$ is a divisor of 12. To derive (8.97), he observed that the derivative of a modular function was a meromorphic modular form of weight 2, since

$$\frac{dJ(\gamma\,\omega)}{d\omega}\gamma'(\omega) = \frac{dJ(\omega)}{d\omega}, \quad \gamma\,\omega = \frac{a\omega+b}{c\omega+d}, \quad ad - bc = 1,$$

and since $q^2 \frac{dJ(\omega)}{d\omega}$ was nonzero and finite as $\omega \to i\infty$. Hence, by Hurwitz's (8.104) and (8.100), there is a modular form of weight 14 such that

$$\frac{dJ(\omega)}{d\omega} = \frac{h_{14}(g_2, g_3)}{\Delta} = C\frac{g_2^2 g_3}{\Delta} \tag{8.106}$$

$$= CJ^{\frac{2}{3}}(1-J)^{\frac{1}{2}}\sqrt[6]{\Delta}. \tag{8.107}$$

Note that we require $\Delta$ in the denominator of (8.106) because $\frac{dJ}{d\omega}$ has a pole of order 1 at $i\infty$. This completes the proof of (8.97). Observe that Hurwitz could have worked

[27] Hecke (1959) pp. 577–590, especially p. 582.

with $\left(\sqrt[24]{\Delta}\right)^k$ in (8.102), but in the paper under discussion, he did not consider modular forms of half-integral weight.

## 8.13   The Modular Equation

Recall that Hermite was of the view that the theory of modular equations could not be developed separately from the theory of elliptic functions. Indeed, Dedekind wrote his paper on modular functions to show that, contrary to Hermite's opinion, this could be achieved. Of course, Dedekind took his basic modular function of weight 0 to be $v \equiv J$, instead of $k^2$, since $v$ was invariant with respect to the full modular group $\Gamma = \Gamma(1)$, whereas $k^2$ was invariant only under the action of the subgroup $\Gamma(2)$. The modular equation is defined here as the polynomial equation satisfied by the distinct functions

$$v_n = v\left(\frac{A\omega + B}{C\omega + D}\right) \tag{8.108}$$

where $n$ is a positive integer and $A, B, C, D$ are integers with no common factor such that

$$AD - BC = n. \tag{8.109}$$

The coefficients of the polynomial satisfied by $v_n$ are polynomials in $v(\omega)$. We first verify that there is a finite number of functions $v_n$. Begin by setting $M_n$ as the set of all $2 \times 2$ matrices with integer entries $A, B, C, D$ with no common factor and with determinant $n$; this means that (8.109) is satisfied. Then, as Dedekind proved,[28]

$$M_n = \bigcup_\lambda \Gamma\lambda \tag{8.110}$$

where $\lambda$ runs over all integer matrices $\left(\begin{smallmatrix} a & b \\ 0 & d \end{smallmatrix}\right)$ with $ad = n$, $d > 0$, $0 \le b < d$, and $a, b, d$ relatively prime. Exercise 7 at the end of this chapter gives a guide to recreating Dedekind's argument. Moreover, $\Gamma\lambda_1 \ne \Gamma\lambda_2$ when $\lambda_1 \ne \lambda_2$, and the number of distinct $\lambda$, where $p$ is prime, is expressed by

$$\psi(n) = n \prod_{p|n} \left(1 + \frac{1}{p}\right). \tag{8.111}$$

In order to deduce (8.111) from (8.110), first consider the matrix

$$\lambda = \begin{pmatrix} a & b \\ 0 & d \end{pmatrix} \tag{8.112}$$

with $d > 0$, $0 \le b < d$, and $a, b, d$ relatively prime. Suppose $e = \gcd(a, d)$. Taking $\phi$ to be the Euler totient function, the number of possible distinct values $b$ can take in (8.112) is given by $\frac{d}{e}\phi(e)$, where $\phi(e)$ represents the number of integers $b$ relatively

---

[28] Dedekind (1930–) vol. 1, p. 196.

prime to $e$ and satisfying $0 \leq b < d$. Summing over all $d$ then yields

$$\psi(n) = \sum_{d|n} \frac{d}{e} \phi(e) \tag{8.113}$$

as the number of distinct $\lambda$ in (8.110). From (8.113) we may then infer that if $n_1$ and $n_2$ are relatively prime, then

$$\psi(n_1 n_2) = \psi(n_1)\psi(n_2). \tag{8.114}$$

Observe that (8.111) follows from (8.114) if we can demonstrate that

$$\sum_{d|n} \frac{d}{e} \phi(e) = p^m + p^{m-1} = p^m \left(1 + \frac{1}{p}\right) \tag{8.115}$$

when $n = p^m$. To check this, consider the case $m = 2l$ and note that the different cases in the sum (8.113) arise from factorizations

$$1 \cdot p^{2l}, \; p \cdot p^{2l-1}, \; p^2 \cdot p^{2l-2}, \ldots, \; p^l \cdot p^l, \; p^{l+1} \cdot p^{l-1}, \ldots, \; p^{2l-1} \cdot p, \; p^{2l} \cdot 1.$$

The values of the pair $(d, e)$ in these cases is in turn given by

$$(1, 1), \; (p, p), \; (p^2, p^2), \ldots, \; (p^l, p^l), \; (p^{l+1}, p^{l-1}), \ldots, \; (p^{2l}, 1);$$

the sum in (8.113) can thus be written as

$$\phi(1) + \phi(p) + \phi(p^2) + \cdots + \phi(p^l) + p^2\phi(p^{l-1}) + p^4\phi(p^{l-2}) + \cdots + p^{2l}\phi(1)$$

$$= 1 + p - 1 + p^2 - p + \cdots + p^l - p^{l-1} + p^2(p^{l-1} - p^{l-2}) + \cdots + p^{2l}$$

$$= p^{2l-1} + p^{2l} = p^{2l} \left(1 + \frac{1}{p}\right).$$

By taking $m = 2l + 1$, a similar result is obtained. The proof of (8.111) is thus complete.

Note that for given $A, B, C, D$ as in (8.108), there exist relatively prime integers $a, b, d$ with $ad = n$, $0 \leq b < d$ and

$$v_n = v\left(\frac{A\omega + B}{C\omega + D}\right) = v\left(\frac{a\omega + b}{d}\right). \tag{8.116}$$

We may thus infer that there are $v = \psi(n)$ distinct functions $v_n$, one for each matrix (8.112). Denote these $v$ distinct functions by

$$f_1(\omega), f_2(\omega), \ldots, f_v(\omega),$$

where we can take $f_1(\omega) = v(n\omega)$. Letting $\sigma$ be a variable independent of $\omega$, consider the polynomial

$$F_n(\sigma, v) = \prod_{k=1}^{v} (\sigma - f_k(\omega)). \tag{8.117}$$

Dedekind defined the modular equation by $F_n(\sigma, v) = 0$ and demonstrated some essential properties of $F_n(\sigma, v)$:

- $F_n(\sigma, v)$ is also a polynomial in $v$ of degree $v = \psi(n)$. See (8.111).
- $F_n(\sigma, v)$ is symmetric in $\sigma$ and $v$, meaning that $F_n(\sigma, v) = F_n(v, \sigma)$.
- As a polynomial in $\sigma$, $F_n(\sigma, v)$ is irreducible.
- The coefficients of $F_n(\sigma, v)$ are integers.

Note that Jacobi's modular equations, such as (3.16) and (3.17), have these properties; as we shall see, this was proved by Jacobi and Sohnke for Jacobi modular equations.

The coefficients of the polynomial (8.117) with degree $v$ are the elementary symmetric functions of $f_k(\omega)$, with $k = 1, 2, \ldots, v$. Dedekind observed that these symmetric functions would remain unchanged under the action of the full modular group, since the $f_k(\omega)$ were permuted under this action. Hence, the coefficients of the various powers of $\sigma$ in $F_n$ were weakly modular of weight zero. He next noted that (8.40) implied that

$$1728v \left( \frac{a\omega + b}{d} \right) = e^{-\frac{2\pi i(a\omega+b)}{d}} + 744 + \cdots ;$$

thus, $F_n$ would have a pole at $i\infty$ of order

$$\sum \frac{a}{d} \cdot \frac{d}{e} \phi(e) = \sum \frac{a}{e} \phi(e) = \sum \frac{d}{e} \phi(e) = \psi(n) = v.$$

From this he had that

$$F_n(\sigma, v) = \sigma^v + V_1 \sigma^{v-1} + V_2 \sigma^{v-2} + \cdots + V_v$$

was also a polynomial in $v$ of degree $v$. He pointed out that

$$-V_1 = \sum v_n = 1728^{n-1} v^n + \cdots$$

was of a polynomial of degree $n$ in $v$.

We now present Dedekind's proof[29] that $F_n(\sigma, v)$ is irreducible as a polynomial in $\sigma$. For this purpose, suppose that $G(\sigma, v)$ is any polynomial in $\sigma$ whose coefficients are polynomials in $v$ and that $G(v_n, v) \equiv 0$ when $v_n = v(n\omega)$. The action of $\Gamma(1)$ on $G(v_n(\omega), v(\omega))$ leaves it unchanged. It follows that for any four integers $\alpha, \beta, \gamma, \delta$ such that $\alpha\delta - \beta\gamma = 1$ we have

$$G\left( v \left( n \frac{\alpha\omega + \beta}{\gamma\omega + \delta} \right), v(\omega) \right) \equiv 0. \tag{8.118}$$

Now given a matrix $\lambda$, as in (8.112), Dedekind showed[30] that there exist eight integers $\alpha, \beta, \gamma, \delta, \alpha', \beta', \gamma', \delta'$ such that $\alpha\delta - \beta\gamma = 1$, $\alpha'\delta' - \beta'\gamma' = 1$, and

$$\begin{pmatrix} n & 0 \\ 0 & 1 \end{pmatrix} \begin{pmatrix} \alpha & \beta \\ \gamma & \delta \end{pmatrix} = \begin{pmatrix} \alpha' & \beta' \\ \gamma' & \delta' \end{pmatrix} \begin{pmatrix} a & b \\ 0 & d \end{pmatrix} ; \tag{8.119}$$

---

[29] Ibid. p. 200.     [30] Ibid. p. 199.

in this connection, see Exercise 8 at the end of this chapter. When (8.119) is used in (8.118), the result is

$$G\left(v\left(\frac{a\omega+b}{d}\right), v(\omega)\right) \equiv 0.$$

We have seen that if any one of the functions $f_k(\omega)$, $k = 1, \ldots, \nu$, is a solution of the equation $G(\sigma, v(\omega)) = 0$, then all of these functions are solutions of that equation.

This means that $F_n(\sigma, v)$ divides $G(\sigma, v)$ and that $F_n(\sigma, v)$ is irreducible.

We finally prove that $F_n(\sigma, v)$ is symmetric in $\sigma$ and $v$ for $n > 1$. Observe that, since we may replace $\omega$ by $\frac{\omega}{n}$, the identity

$$F_n(v(n\omega), v(\omega)) \equiv 0$$

implies

$$F_n\left(v(\omega), v\left(\frac{\omega}{n}\right)\right) \equiv 0.$$

Moreover, because $v(\frac{\omega}{n})$ is one of the $\nu$ functions $f_k(\omega)$, the irreducibility of $F_n(\sigma, v)$ implies that $F_n(v, \sigma)$ must be divisible by $F_n(\sigma, v)$ and therefore

$$F_n(v, \sigma) = \pm F_n(\sigma, v). \tag{8.120}$$

In addition, since using the negative sign would imply that $\sigma - v$ is a factor of the irreducible polynomial $F_n(\sigma, v)$, the positive sign must be chosen.

Dedekind[31] employed differential equations to deduce that the coefficients $a_{rs}$ in $F_n(\sigma, v) = \sum a_{rs}\sigma^r v^s$ were integers. We present a proof that is useful in computing the modular equation; this proof was given by Dedekind's friend Heinrich Weber in his algebra book of 1908.[32] We will prove this proposition for the case in which $n = p$ is prime and leave the general case as Exercise 9 at the end of this chapter.

Weber used the function $j(\omega) = 1728v(\omega)$ and its expansion as a series in $q^2 = e^{2\pi \omega i}$, given by (5.35) in which $\omega = \frac{1+i\sqrt{\Delta}}{2}$. Recall that this series was first published by Hermite in 1859. The coefficients of the powers of $q$ are all integers. Weber showed that

$$F_p(\sigma, j) = \sum b_{rs}\sigma^r j^s$$

where $b_{rs}$ were integers; this implied the result for $F_p(\sigma, v)$. Weber started by observing that

$$j(\omega) = q^{-2} + \sum_{h=0}^{\infty} a_h q^{2h}, \tag{8.121}$$

where $a_h$ were integers. Then by Fermat's theorem that $a^p \equiv a \pmod{p}$ for any integer $a$, it followed that

$$j^p(\omega) = q^{-2p} + \sum_{h=0}^{\infty} a_h q^{2ph} + pq^{-2(p-1)} \sum_{h=0}^{\infty} b_h q^{2h}, \tag{8.122}$$

---

[31] Dedekind (1930–) vol. 1, p. 201.    [32] Weber (1894–1908) vol. 3, pp. 243–245.

where $b_n$ were also integers. He replaced $\omega$ by $p\omega$ in (8.121), set

$$j(\omega) = u, \quad j(p\omega) = v,$$

to obtain

$$v = j(p\omega) = q^{-2p} + \sum_{h=0}^{\infty} a_h q^{2ph}, \tag{8.123}$$

so that (8.122) implied

$$u^p - v = pq^{-2(p-1)} \sum_{h=0}^{\infty} b_h q^{2h}. \tag{8.124}$$

Now the lowest power of $q$ in $v^p$ was $q^{-2p^2}$ and (8.124) implied

$$(u^p - v)(u - v^p) = pq^{-2(p^2+p-1)} \sum_{h=0}^{\infty} c_h q^{2h}, \tag{8.125}$$

where $c_h$ were integers. Weber next observed that since $F_p(x, y)$ had the terms $x^{p+1} + y^{p+1}$, he could set

$$F_p(x, y) = (x^p - y)(x - y^p) - \sum_{h,k=0}^{p} c_{h,k} x^h y^k; \tag{8.126}$$

the symmetry in $x$ and $y$ implied that

$$c_{h,k} = c_{k,h}. \tag{8.127}$$

Because $F_p(u, v) = 0$, Weber concluded that

$$(u^p - v)(u - v^p) = \sum_{h,k}^{p} c_{h,k} u^h v^k. \tag{8.128}$$

Thus, $c_{p,p} = 0$. To see this, note that if $c_{p,p} \neq 0$, then there would be a term involving $q^{-2(p^2+p)}$, whereas (8.125) showed that the lowest power of $q$ had to be $q^{-2(p^2+p-1)}$. Therefore, applying (8.127), he could write (8.128) as

$$(u^p - v)(u - v^p) = \sum_{h=0}^{p} \sum_{k=0}^{h-1} c_{h,k}(u^h v^k + u^k v^h) + \sum_{h=0}^{p-1} c_{h,h} u^h v^h. \tag{8.129}$$

He then inserted into (8.129) the expansions of $u$ and $v$ from (8.121) and (8.123); he observed that the lowest power of $q$ in $u^h v^k + u^k v^h$ was $q^{-2(hp+k)}$ with coefficient one and the lowest power of $q$ in $u^h v^h$ was $q^{-2h(p+1)}$ with coefficient one. Thus, since

$$hp + k = h'p + k'$$

implied that $k = k'$ and $h = h'$, the right-hand side would contain not more than one term whose expansion began with a given power of $q$.

Equating powers of $q$ in (8.129) with the powers of q in (8.125) then yielded the result that $c_{h,k}$ were integers divisible by $p$. Weber thus arrived at the result:

$$F_p(x, y) = (x^p - y)(x - y^p) - p \sum_{h,k=0}^{p} a_{h,k} x^h y^k, \tag{8.130}$$

where $a_{h,k}$ were integers satisfying $a_{h,k} = a_{k,h}$, $a_{p,p} = 0$. The relation (8.130) is called Kronecker's congruence when written in the form

$$F_p(x, y) \equiv (x^p - y)(x - y^p) \mod p\mathbb{Z}[x, y], \tag{8.131}$$

where $\mathbb{Z}[x, y]$ denotes the ring of polynomials in $x$ and $y$ with integer coefficients. Kronecker proved[33] this congruence in 1886 in connection with his work on complex multiplication and singular moduli. He used Jacobi's modular function $k^2$ rather than the Dedekind-Klein $J$ function.

## 8.14 Singular Moduli and Quadratic Forms

Dedekind referred to quadratic forms and to singular moduli of elliptic functions for which complex multiplication holds at the beginning as well as the end of his paper on modular forms. Indeed, one of his motivations for developing the theory of modular equations was to understand singular moduli. He concluded his paper with a promise to address this area at a later time. Unfortunately, he apparently did not do this. However, we present some results on the topic in order to illustrate close connections with Dedekind's paper.

Recall that an elliptic function with periods $\omega_1$ and $\omega_2$ is said to admit of complex multiplication if there exist a complex number $\alpha$ and integers $a, b, c, d$ such that

$$\alpha \omega_1 = a \omega_1 + b \omega_2$$
$$\alpha \omega_2 = c \omega_1 + d \omega_2.$$

It follows that $\omega = \frac{\omega_1}{\omega_2}$ satisfies the quadratic equation

$$\omega = \frac{a\omega + b}{c\omega + d} \tag{8.132}$$

and the equation

$$\alpha = c\omega + d. \tag{8.133}$$

Note that (8.132) is equivalent to

$$c\omega^2 + (d - a)\omega - b = 0. \tag{8.134}$$

Next, let $\rho = \gcd(c, d - a, b)$ and set $c = \rho\gamma$, $d - a = \rho\beta$, $-b = \rho\alpha$. Then

$$\alpha + \beta \omega + \gamma \omega^2 = 0,$$

---

[33] Kronecker (1968) vol. 4, pp. 389–471, especially equation (66) on p. 453.

where

$$\alpha > 0, \quad \beta^2 - 2\alpha\gamma = -n < 0,$$

and

$$\omega = \frac{-\beta + \sqrt{-n}}{2\gamma}. \tag{8.135}$$

Formula (8.135) for $\omega$ shows that $\omega \in \mathbb{H}$, $\mathbb{H}$ being the upper half-plane. The numbers in $\mathbb{H}$ satisfying a quadratic equation with coefficients in $\mathbb{Z}$, as in (8.134), are designated *CM* points and are related to positive definite binary quadratic forms in an interesting way.

A binary quadratic form is an expression in two variables $x$, $y$ of the form

$$f(x, y) = ax^2 + bxy + cy^2. \tag{8.136}$$

Such forms were initially studied by Euler and Lagrange, who took $a$, $b$, $c$ to be integers, as we shall see. They were interested in determining the integers representable by a given form. Now an integer $n$ can represented by the form (8.136) if there exist integers $x_0$ and $y_0$ such that

$$ax_0^2 + bx_0y_0 + cy_0^2 = n.$$

We consider only primitive forms, that is, forms satisfying (8.136) for which gcd $(a, b, c) = 1$. A form must fall in one of two groups: positive definite or indefinite. A form is designated "positive definite" if for every pair of integers, $(x, y) \neq (0, 0))$, the value of the expression (8.136) is a positive integer. The reader may be aware that Fermat determined the integers representable by the forms $x^2 + y^2, x^2 + 2y^2, x^2 + 3y^2$. For example, he discovered that primes of the form $4m + 1$ could be expressed in the form $x^2 + y^2$, whereas primes of the form $4m + 3$ could not be so expressed. The three forms considered by Fermat were obviously positive definite. It is not difficult to show that $ax^2 + bxy + cy^2$ is positive definite if and only if $a > 0$ and $4ac - b^2 > 0$. Each positive definite form can be identified with a complex number in the upper half-plane $\mathbb{H}$ as the solution of the equation

$$a\omega^2 + b\omega + c = 0, \tag{8.137}$$

so that

$$\omega = \frac{-b + \sqrt{b^2 - 4ac}}{2a} \in \mathbb{H}. \tag{8.138}$$

In matrix notation

$$f(x, y) = (x \quad y) \begin{pmatrix} a & \frac{b}{2} \\ \frac{b}{2} & c \end{pmatrix} \begin{pmatrix} x \\ y \end{pmatrix}.$$

Next consider the effect of the change in variables

$$\begin{pmatrix} x \\ y \end{pmatrix} = \begin{pmatrix} \alpha & \beta \\ \gamma & \delta \end{pmatrix} \begin{pmatrix} x' \\ y' \end{pmatrix}, \tag{8.139}$$

where $\alpha, \beta, \gamma, \delta$ are integers such that $\alpha\delta - \beta\gamma = 1$. Then we have

$$f(x, y) = (x' \quad y') \begin{pmatrix} \alpha & \gamma \\ \beta & \delta \end{pmatrix} \begin{pmatrix} a & \frac{b}{2} \\ \frac{b}{2} & c \end{pmatrix} \begin{pmatrix} \alpha & \beta \\ \gamma & \delta \end{pmatrix} \begin{pmatrix} x' \\ y' \end{pmatrix}$$

$$= (x' \quad y') \begin{pmatrix} a' & \frac{b'}{2} \\ \frac{b'}{2} & c' \end{pmatrix} \begin{pmatrix} x' \\ y' \end{pmatrix},$$

where

$$a = a'\alpha^2 + b'\alpha\gamma + c'\gamma^2, \quad b = 2a'\alpha\beta + b'(\beta\gamma + \alpha\delta) + 2c'\gamma\delta, \quad c$$
$$= a'\beta^2 + b'\beta\delta + c'\delta^2. \tag{8.140}$$

It is not difficult to check that

$$(a', b', c') = 1 \quad \text{and} \quad b^2 - 4ac = b'^2 - 4a'c' = D. \tag{8.141}$$

The quantity $D = b^2 - 4ac$ is called the discriminant of the quadratic form given by (8.136); it is a negative integer when the form is positive definite. A form $ax^2 + bxy + cy^2$ is said to be equivalent to the form $a'x^2 + b'xy + c'y^2$ if $a, b, c$ are related to $a', b', c'$ by means of (8.140), where $\alpha, \beta, \gamma, \delta$ are defined by (8.139). It is clear that two equivalent forms represent the same set of integers. Let $\omega' \in \mathbb{H}$ be a solution of

$$a'\omega'^2 + b'\omega' + c' = 0; \tag{8.142}$$

then (8.139) implies that

$$\omega' = \frac{\alpha\omega + \beta}{\gamma\omega + \delta}, \quad \alpha\delta - \beta\gamma = 1. \tag{8.143}$$

Thus, we see that if $\omega$ and $\omega' \in \mathbb{H}$ correspond to two equivalent forms, then there exists $g \in \Gamma$, the full modular group, such that $\omega' = g\omega$. This in turn implies that

$$J(\omega') = J(\omega). \tag{8.144}$$

Conversely, if (8.144) holds and if $\omega$ and $\omega'$ are solutions of (8.137) and (8.142) respectively, then the forms

$$ax^2 + bxy + cy^2 \quad \text{and} \quad a'x'^2 + b'x'y' + c'y'^2$$

are equivalent. Checking this is straightforward, since (8.144) implies (8.143). Thus, it follows that the number of inequivalent forms of a given discriminant $b^2 - 4ac = D = -n, n > 0$, is equal to the number of *CM* points $\omega$ of the form (8.138) that lie in the fundamental domain of the modular group $\Gamma = \Gamma(1)$. Recall that a complex number $\omega$ lies in this domain if and only if

$$-\frac{1}{2} \leq \text{Re}\,\omega \leq \frac{1}{2} \quad \text{and} \quad |\omega| \geq 1. \tag{8.145}$$

By (8.138), the first condition in (8.145) translates to the condition that $|b| \leq a$, and the second condition translates to $a < c$. It is possible to see that there exist only a finite number of integers $a, b, c$ with $|b| \leq a \leq c$ and $4ac - b^2 = n$, with $n$ a given positive integer. A form $ax^2 + bxy + cy^2$ that satisfies the condition $|b| \leq a < c$ or $0 \leq b \leq a = c$ is called a reduced form. We have shown that the number of reduced forms of a given discriminant is finite. This number is termed the class number of the discriminant $D = -n$, and is often denoted by $h(D) \equiv h(-n)$.

Let the *CM* points corresponding to the reduced forms of a discriminant $-n$ be denoted by $\omega_k$, with $k = 1, 2, \ldots, h(-n)$. Then it can be demonstrated that the class polynomial

$$H_D(x) = \prod_{k=1}^{h(D)} \left( x - j(\omega_k) \right), \quad \text{where} \quad j(\omega) = 1728J(\omega),$$

is irreducible with coefficients in $\mathbb{Z}$. The numbers $j(\omega_k)$ are therefore algebraic integers, once called singular invariants, though now they are called singular moduli along with $k^2(\omega_k)$. Note also the interesting result that $H_D(x)$ is a factor of $F_n(x, x)$, where $F_n(x, j) = 0$ is Dedekind's modular equation.

We note in passing that Dedekind referred[34] to a page in Gauss's collected papers,[35] and wrote that Gauss too had the intention of introducing a function like $J(\omega)$, in connection with binary quadratic forms. On the page referred to by Dedekind, Gauss wrote:

Let $\alpha, \beta, \gamma, \delta$ be whole real numbers, $\alpha\delta - \beta\gamma = 1$, and

$$\frac{\alpha t - \beta i}{\delta + \gamma t i} = t'.$$

Here $t = \dfrac{\sqrt{d} + bi}{a}, \quad t' = \dfrac{\sqrt{d} + b'i}{a'}, \quad -d = bb - ac = b'b' - a'c',$

so that the form $(a, b, c)$ carries over into $(a', b', c')$ by the transformation

$$\begin{pmatrix} \delta & -\beta \\ -\gamma & \alpha \end{pmatrix}.$$

The relationship between the forms of negative determinant $-p$ and the summing function.

In particular, if the forms $(a, b, c)$ and $(A, B, C)$ are equivalent, so then set $f(t) = f(u)$ if $\frac{t-u}{i}$ is a whole number or if $t = \frac{1}{u}$. Every class must then correspond to a definite value of $f\left(\frac{\sqrt{p}+b}{a}\right)$.

Note that for Gauss, $(a, b, c)$ denoted the form $ax^2 + 2bxy + cy^2$. Also, he took the right half-plane instead of the upper half-plane as the region in which his function $f$ was defined. So he took the point corresponding to the form $(a, b, c)$ to be $t = \frac{\sqrt{d}+bi}{a}$. Note here also that the modular group is generated by the transformations $t = u - i$ and $t = \frac{1}{u}$.

Kronecker published the results given in this section without proof in his famous 1857 paper, "Über die elliptische Funktionen für welche Komplexe Multiplikation stattfindet"[36] [On elliptic functions for which complex mulitiplication holds]. In this and later papers, Kronecker founded the theory of complex multiplication, showing the deep interconnections of elliptic functions with algebra and with number theory. As Kronecker noted in his 1857 paper, Abel had glimpsed some of these connections, especially between elliptic functions and the theory of equations. In this connection, we

---

[34] Dedekind (1930–) vol. 1, p. 193.    [35] Gauss (1863–1927) vol. 3, p. 386.
[36] Kronecker (1968) vol. 4, pp. 177–183.

here present an English translation of the initial portion of Kronecker's 1857 paper:[37]

> In a paper by *Abel* (volume I, p. 272 of the collected works)[38] we find the remark that the moduli of those elliptic functions, for which complex multiplication holds, are all representable by radicals. Lacking, however, is any hint on the manner by which *Abel* discovered this remarkable property of that special class of elliptic functions. That this appeared *after* the composition of the paper "Recherches sur les fonctions elliptiques" [Researches on elliptic functions] is apparent from a passage found in the same place (volume I, p. 248 of the collected works or volume III, p. 183 of *Crelle*'s Journal)[39], which still contains doubt about the solvability of the equations through which the above-mentioned moduli are determined. – Inspired by the first remark of *Abel* cited above, and with the intention of seeking a proof of it, this past winter I occupied myself with the investigation of those elliptic functions for which complex multiplication holds, and in this process, in addition to the proof I sought, I have found many other interesting results, a few of which I wish to here briefly communicate.
>
> If $n$ denotes a positive odd number which is larger than 3, if one further denotes by $\chi$ the module of the elliptic function, and by $k$ its square, then the number of different values of $k$, for which multiplication of the elliptic function by $\sqrt{-n}$ is possible, that is for which $\sin^2 \operatorname{am}(\sqrt{-n} \cdot u, \chi)$ [$\operatorname{sn}^2(\sqrt{-n} \cdot u, \chi)$] can be expressed as a rational function of $\sin^2 \operatorname{am}(u, \chi)$ [$\operatorname{sn}^2(u, \chi)$] and $\chi$, equals six times the number of different classes of quadratic forms of determinant $-n$. All these values of $k$ are explicit algebraic functions of any one of them(selves); they are, moreover, roots of an equation with whole-number [coefficients], whose degree equals the number of those values [of $k$] and which can be factored into as many whole-number factors as the number of different orders of determinant $-n$. To each of these orders corresponds a definite factor of that equation whose degree equals exactly six times the number of classes in the corresponding order. The factor belonging to the proper primitive order is finally factorable into six factors of the same degree whose coefficients contain only whole numbers and $\sqrt{-n'}$.[40] The degree of each one of these six factor-equations is then equal to the number of proper primitive classes of determinant $-n$, and it is one of these factor-equations which most clearly reveals the property of solvability. The roots themselves have in particular the property that each can be represented as a rational function (containing only integer coefficients) of one of the roots and that for every two such functions $\phi(k)$ and $\psi(k)$, the equation $\phi\big(\psi(k)\big) = \psi\big(\phi(k)\big)$ holds true.[41]

Observe that Kronecker denoted by $k$ the quantity denoted by Jacobi as $k^2$. The reason Kronecker obtained six times the number of different classes of quadratic forms was that he worked with the modular function $k$ rather than $j$. Recall that $j$ is an invariant of the full modular group, whereas $k$ is invariant with respect to $\Gamma(2)$, a subgroup of index 6. It is also interesting to note that Kronecker saw fit to publish such important results without proofs; it may be that the circumstances of his life made this necessary.

Leopold Kronecker (1823–1891) was a student during the 1830s at Leignitz Gymnasium where he studied under Ernst Kummer, who became a lifelong friend. Kronecker matriculated from the University of Berlin in 1841 where he attended lectures by Lejeune Dirichlet and Jacob Steiner. Four years later, he received his Ph.D. under Dirichlet's supervision, though he also spent two semesters in 1844 with Kummer, then professor at the University of Breslau. Kronecker worked during the next ten years running the family business in Leignitz, so that when he finally settled in Berlin in 1855, he had

---

[37] Kronecker (1968) vol. 4, pp. 179–180.
[38] [editor's footnote:] *Abel*, Oeuvres (new edition), volume I, p. 426. H
[39] [editor's footnote:] *Abel*, Oeuvres, volume I, p. 383. H
[40] [editor's footnote:] See Supplement 39 at the end of this volume. H
[41] [editor's footnote:] See Supplement 40 at the end of this volume. H

become independently wealthy without an urgent need of employment. However, he had continued to do mathematics even while engaged in his business. It was perhaps during this period that, due to a shortage of time, he developed the habit writing up his results without full proofs. Kronecker's correspondence with Kummer and Dirichlet seems to confirm this. He was finally appointed professor of mathematics at Berlin after Kummer's retirement in 1883; this appears to be the only academic position he ever held. But he continued the practice of publishing results with proofs in greatly abbreviated form, even after he had more leisure. Thus, he often gave the particulars of his proofs many years after announcing the results. For example, Kummer announced in a paper of 1859[42] that Kronecker had generalized Kummer's theory of ideal numbers for cyclotomic fields to algebraic number fields. Kronecker published this generalization only twenty-three years later.[43] In the meanwhile, in 1871, Dedekind had published his very different version of this generalization, called ideal theory. This was a source of some tension between Kronecker and Dedekind; Kronecker and his friends in Berlin felt that Dedekind should have not have published his theory before Kronecker.

As we have noted, Emmy Noether greatly admired Dedekind's work and she was a very influential mathematician. Thus, Dedekind's ideas and approach have become much better known than Kronecker's. However, some mathematicians, including Hermann Weyl and Martin Eichler, have preferred or appreciated Kronecker's point of view. In his 1950 lecture at the International Mathematics Congress,[44] André Weil gave some insightful remarks on this matter:

> There appears to have been a certain feeling of rivalry, both scientific and personal, between Dedekind and Kronecker during their life-time; this developed into a feud between their followers, which was carried on until the partisans of Dedekind, fighting under the banner of "purity of algebra," seemed to have won the field, and to have exterminated or converted their foes. Thus many of Kronecker's far-reaching ideas and fruitful results now lie buried in the impressive but seldom opened volumes of his Complete Works. While each line of Dedekind's XI th Supplement, in its three successive and increasingly "pure" versions, has been scanned and analyzed, axiomatized and generalized, Kronecker's once famous *Grundzüge* are either forgotten, or are thought of merely as presenting an inferior (and less pure) method for achieving part of the same results, viz., the foundation of ideal-theory and of the theory of algebraic number-fields. In more recent years, it is true, the fashion has veered to a more multiplicative and less additive approach than Dedekind's, to an emphasis on valuations rather than ideals; but, while this trend has taken us back to Kronecker's most faithful disciple, Hensel, it has stopped short of the master himself.
>
> Now it is time for us to realize that, in his *Grundzüge*, Kronecker did not merely intend to give his own treatment of the basic problems of ideal-theory which form the main subject of Dedekind's life-work. His aim was a higher one. He was, in fact, attempting to describe and to initiate a new branch of mathematics, which would contain both number-theory and algebraic geometry as special cases. This grandiose conception has been allowed to fade out of our sight, partly because of the intrinsic difficulties of carrying it out, partly owing to historical accidents and to the temporary successes of the partisans of purity and of Dedekind. It will be the main purpose of this lecture to try to rescue it from oblivion, to revive it, and to describe the few modern results which may be considered as belonging to the Kroneckerian program.

---

[42] Kummer (1975) vol. 1, pp. 699–839.     [43] Kronecker (1969) vol. 2, pp. 239–387.
[44] Weil (1980) vol. 1, p. 442.

Kronecker's *Grundzüge* is difficult to read. One might wish to start with H. M. Edwards's *Divisor Theory* before tackling Kronecker. One may also read Edwards's *Essays in Constructive Mathematics* that takes the view that in this age of the computer, we might prefer the algorithmic and constructive approach of Kronecker to the nonconstructive and infinite set-theoretic approach of Dedekind.

## 8.15 Exercises

1.  Derive the hypergeometric differential equation (8.56). This differential equation was published in 1875 by Heinrich Bruns in the *Dorpat Festschrift*, in honor of Viktor Buniakovski. The paper was republished in 1886.[45]
2.  Verify equations (8.91) and (8.92).
3.  Show that the value of $C$ in (8.107) is $\frac{\sqrt{3}}{\pi i}$. See Hurwitz (1962) vol. 1, p. 34.
4.  Prove (8.91), (8.92), and (8.93).
5.  Show that $w$ in (8.55) is a solution of (8.56) and that the general solution of (8.56) is given by (8.57).
6.  Prove (8.94) and (8.95).
7.  Prove (8.110) as follows: Let $A, B, C, D$ be integers as in (8.109) and let $a = \gcd(A, C)$ so that $A = a\alpha$, $C = a\gamma$ for some integers $\alpha$ and $\gamma$. Note that if $d = \alpha D - \gamma B$, then $ad = n$. Prove that there exist integers $\beta$ and $\delta$ so that $\alpha\gamma - \beta\delta = 1$ and at the same time the number $b = B\delta - \beta D$ satisfies $0 \le b < d$. Thus

    $$\begin{pmatrix} A & B \\ C & D \end{pmatrix} = \begin{pmatrix} \alpha & \beta \\ \gamma & \delta \end{pmatrix} \begin{pmatrix} a & b \\ 0 & d \end{pmatrix},$$

    completing the proof of (8.110). See Dedekind (1931) p. 196.
8.  Prove equation (8.119). See Dedekind (1931) vol. 1, pp. 199–200.
9.  (a) Take $p$ to be a prime and $m$ a positive integer. Let $n = p^m$ so that the degree of $F_n(x, j(\omega))$ is $p^{m-1}(p+1)$ and the degree of $F_{p^{m-1}}(x, j(\omega))$ is $v' = p^{m-2}(p+1)$. Denote the roots of $F_{p^{m-1}}(x, j(\omega)) = 0$ by $x_1, x_2, \ldots, x_{v'}$.
    *   Show that the coefficients of the polynomial

        $$P(x) = F_p(x, x_1) F_p(x, x_2) \cdots F_p(x, x_{v'})$$

        are rational functions of $j(\omega)$ and that $P(x) = 0$ for $x = j\left(\frac{\omega}{p^m}\right)$.
    *   Deduce that $F_n(x, j(\omega))$ divides $P$ and that

        $$F_n\big(x, j(\omega)\big) = \frac{P}{\big(F_{p^{m-2}}(x, j(\omega))\big)^p} \quad \text{when } m > 2.$$

    *   Show also that for $m = 2$

        $$F_{p^2}(x, j(\omega)) = \frac{F_p(x, x_1)F_p(x, x_2) \cdots F_p(x, x_{p+1})}{(x - j(\omega))^{p+1}}.$$

    (b) Let $n$ be a positive integer and let $n = n_1 n_2$ where $n_1$ and $n_2$ are relatively prime and greater than one. Let $x_1, x_2, \ldots, x_{v_2}$, where $v_2 = \psi(n_2)$

---

[45] Bruns (1886). See especially p. 238.

(as defined by (8.111)) denote the zeros of the equation $F_{n_2}(x, j(\omega)) = 0$. Prove that

$$F_n(x, j(\omega)) = F_{n_1}(x, x_1) F_{n_1}(x, x_2) \cdots F_{n_1}(x, x_{\nu_2}).$$

(c) Using the result proved in the text that $F_p(x, j(\omega))$ has integer coefficients, deduce from (a) and (b) that $F_n(x, j(\omega))$ has the same property. See Weber (1894–1908) vol. 3, pp. 241–242.

# 9

---

# *The η Function and Dedekind Sums*

## 9.1 Preliminary Remarks

Dedekind came upon the sums now named after him in the course of his analysis of the function:

$$\eta(\omega) = q^{\frac{1}{12}} \prod_{n=1}^{\infty} (1 - q^{2n}), \tag{9.1}$$

as $\omega$ tends to a rational number $\frac{m}{n}$, with $q = e^{\pi i \omega}$, and $\omega \in \mathbb{H}$, the upper half-plane. Now known as the Dedekind $\eta$ function, (9.1) was in fact implicitly contained in the work of Jacobi. He presented products for the modular functions $k^2, k'^2$ and for the complete elliptic integal $K$; Dedekind observed that these functions were expressible in terms of the more basic function $\eta(\omega)$. For example, as may be verified from (3.161):

$$\sqrt[4]{k} = e^{\frac{2\pi i}{48}} \sqrt{2} \frac{\eta(2\omega)}{\eta\left(\frac{1+\omega}{2}\right)}. \tag{9.2}$$

In this context, Dedekind wished to determine the behavior of $\eta(\tau)$ under the action of full the modular group $\Gamma(1)$. He applied a known formula for a theta function to $\eta(\tau)$ in order to show that if $a, b, c, d$ were integers with $ad - bc = 1$, then there existed an integer-valued function of $d$ and $c$, denoted by Dedekind as $(d, c)$, such that for $c \neq 0$

$$\log \eta\left(\frac{a\omega + b}{c\omega + d}\right) = \log \eta(\omega) + \frac{1}{4} \log\left(-(c\omega + d)^2\right) + \frac{a+d}{12c}\pi i - \frac{(d, c)}{6c}\pi i. \tag{9.3}$$

It can be shown that when $c > 0$, (9.3) gives rise to an alternate form of a transformation formula for the $\eta$ function:

$$\eta\left(\frac{a\omega + b}{c\omega + d}\right) = e^{\pi i \left(\frac{a+d-2(d,c)}{12c} - \frac{1}{4}\right)} \sqrt{c\omega + d}\, \eta(\omega). \tag{9.4}$$

One will have a similar formula for the case $c < 0$. Dedekind expressed $\frac{(d,c)}{6c}$, when $c > 0$, as a finite sum of rational numbers, now called a Dedekind sum. Note that when $c = 0$, Dedekind defined $(\pm 1, 0) = \pm 1$, so that the numerators of the coefficients of

251

$\pi i$ would cancel in (9.3). He went on to show, after several pages of work, that

$$\frac{(d,c)}{6c} = \sum_{\sigma=1}^{c-1} \frac{\sigma}{c} \left( \left( \frac{\sigma d}{c} \right) \right), \tag{9.5}$$

where

$$((x)) = \begin{cases} x - [x] - \frac{1}{2} & \text{if } x \text{ is not an integer,} \\ 0 & \text{if } x \text{ is an integer.} \end{cases} \tag{9.6}$$

Observe that $B_1(x) = x - [x] - \frac{1}{2}$ is the first Bernoulli polynomial, restricted to the interval $(0, 1)$. The formula (9.5) is now taken as the definition of a Dedekind sum, a purely arithmetical definition often presented in the form

$$\frac{(d,c)}{6c} \equiv s(d,c) = \sum_{\sigma=1}^{c-1} \left( \left( \frac{\sigma}{c} \right) \right) \left( \left( \frac{\sigma d}{c} \right) \right). \tag{9.7}$$

Now (9.7) is equivalent to (9.5), since it can be shown that

$$\sum_{\sigma=1}^{c-1} \left( \left( \frac{\sigma d}{c} \right) \right) = 0. \tag{9.8}$$

For clarity, Rademacher employed the notation $s(c, d)$ for the Dedekind sum instead of using $(c, d)$; we shall use Rademacher's notation.

Dedekind applied the transformation $\omega \to -\frac{1}{\omega}$ in (9.3) to obtain a relation between $s(c, d)$ and $s(d, c)$, known as the reciprocity formula for Dedekind sums: For any two relatively prime integers $c$ and $d$,

$$s(c,d) + s(d,c) = -\frac{1}{4} + \frac{1}{12} \left( \frac{c}{d} + \frac{1}{cd} + \frac{d}{c} \right). \tag{9.9}$$

Of course, in addition to the analytic proofs of Dedekind and Rademacher and others, there exist completely arithmetical proofs of this result.[1]

Dedekind sums occur in several contexts other than the study of the $\eta$ function. For example, consider the permutation

$$c, 2c, 3c, \ldots, (d-1)c \pmod{d} \tag{9.10}$$

of the integers $1, 2, 3, \ldots, d - 1$. Let $I(c, d)$ denote the number of inversions in the permutaion (9.10), that is, the number of times a larger entry precedes a smaller one. In 1957, Curt Meyer expressed $I(c, d)$ in terms of Dedekind sums:[2]

$$I(c,d) = -3d\, s(c,d) + \frac{1}{4}(d-1)(d-2). \tag{9.11}$$

Even earlier, in 1872, the noted Russian number theorist Zolotarev (also Zolotareff) related the number of inversions $I(c, d)$ to the Jacobi symbol $(\frac{c}{d})$, showing[3] that for

---

[1]  For these proofs and references, see Rademacher and Grosswald (1972).
[2]  Rademacher and Grosswald (1972) pp. 35–37.     [3]  Ibid. p. 38.

odd $d$ and gcd $(c, d) = 1$

$$(-1)^{l(c,d)} = \left(\frac{c}{d}\right). \tag{9.12}$$

Zolotarev's result, combined with (9.11) and (9.9), may be used to prove the law of quadratic reciprocity.[4] Dedekind sums also occur in counting the lattice points inside certain triangles, parallelepipeds, pyramids, and tetrahedrons. Another interesting situation in which generalized Dedekind sums have proved useful is in testing series correlation in a sequence of pseudorandom numbers; in this connection, Donald Knuth defined a generalized Dedekind sum[5] as

$$\sigma(d, c, f) = 12 \sum_{j=0}^{c-1} \left(\left(\frac{j}{c}\right)\right)\left(\left(\frac{dj+f}{c}\right)\right), \tag{9.13}$$

where $0 \le f < c$. He then proved a reciprocity formula for integers $d, c, f$, with $0 \le f < c$ and $0 < d < c$, and gcd $(c, d) = 1$:

$$\sigma(d, c, f) + \sigma(c, d, f) = \frac{d}{c} + \frac{c}{d} + \frac{1}{cd} + \frac{6f^2}{cd} - 6\left\lfloor\frac{c}{d}\right\rfloor - 3e(d, f), \tag{9.14}$$

where $e(d, f) = [f = 0] + [f \bmod d \ne 0]$. Here $\lfloor \ \rfloor$ denotes the greatest integer function; $[ \ ]$ denotes the character function of the condition within the square brackets.

Instead of Dedekind sums, most contemporary mathematicians employ the Jacobi symbol to express transformation formulas. Of course, there is a connection between Dedekind sums and Jacobi symbols; Dedekind used this connection to show that quadratic reciprocity could be obtained from the reciprocity formula for Dedekind sums. Dedekind sums are of intrinsic importance, and it is worthwhile to understand how Dedekind derived his transformation formula for the $\eta$ function in terms of his sums, because they arose so naturally in this context.

Dedekind's research on Dedekind sums arose within his explanation of certain formulas of Riemann, stated in some fragmentary notes on modular functions. Upon his death, Riemann's scientific papers were left in the care of his devoted friend Dedekind; some of the papers were complete or nearly so, but there were also several pages of formulas without explanation. Dedekind, five years younger than Riemann, became his close friend when they were both in Göttingen together. Along with Gauss and Dirichlet, Riemann was greatly admired by Dedekind. Dedekind thought as an algebraist and taught courses on algebraic topics, in contrast to Riemann's analytic approach.

The great Emmy Noether viewed Dedekind as a brilliant algebraist, instructing her students to read each of the four editions of Dedekind's work on algebraic number theory, published as a long supplement in *Vorlesungen über Zahlentheorie*.[6] Indeed, Noether perceived the origins of her own innovative ideas to lie in Dedekind. This

---

[4] Ibid. p. 39   [5] Knuth (1998) pp. 83–84.

[6] This supplement, numbered as 10 and then 11, was Dedekind's addition to his presentation of Dirichlet's lectures on number theory.

led her to regularly remark, "It is all already in Dedekind."[7] Despite their contrasting outlooks, Dedekind valued Riemann's work very highly, attending his lectures on the foundations of complex analysis, on elliptic functions, and on hypergeometric functions.

Perhaps because of his algebraic point of view, or perhaps for some other reason, Dedekind did not feel fully competent to edit Riemann's papers single-handedly. He first sought the help of Alfred Clebsch (1833–1872), a teacher of Felix Klein and Max Noether, but Clebsch tragically died soon thereafter. Dedekind then turned to his friend Heinrich Weber (1842–1913); together, they edited Riemann's papers which appeared in 1876. Among these papers were two notes, one containing 68 formulas written in Latin and the other containing eight formulas on a single sheet of paper. Dedekind edited these himself and wrote extensive notes on how the eight formulas could be derived.[8]

Riemann's cryptic eight formulas determined the behavior of the elliptic modular functions $\log k(\omega)$, $\log k'(\omega)$, and $\log \frac{2K}{\pi}$ as $\omega \to \frac{h}{k}$, with $h$ and $k$ relatively prime. He started out with Jacobi's formulas, such as

$$\log k = \log 4\sqrt{q} + \sum_{l=1}^{\infty} (-1)^l \frac{4}{l} \frac{q^l}{1+q^l}, \quad q = e^{i\pi\omega}, \quad \text{Im } \omega > 0. \tag{9.15}$$

Note that (9.15) can be obtained from (3.161) by taking the logarithm of both sides of the equation, expanding the logarithms as series, and then summing after changing the order of summation. This change in order of summation may be justified by the absolute convergence of the series. Riemann set $\omega = \frac{h}{k} + iy$, $y > 0$, and let $y \to 0$. Observe that the real part of $\log k \to \infty$ as $y \to 0$; however, he was interested in determining the exact value of the imaginary part. After finding these values, he noted the results separately for the cases in which: (i) $h$ was even and $k$ was odd, (ii) $h$ and $k$ were both even or both odd, and (iii) $h$ was odd and $k$ even. For example, when $h = 2m$ and $k = n$ (i.e., odd), Riemann wrote the formula for the imaginary part of $\log k(\omega)$ as $\omega \to \frac{2m}{n}$:

$$\pi \left( \frac{2m}{2n} + 2\sum_{\sigma=1}^{n-1} (-1)^\sigma \frac{2m\sigma}{2n} - 2\sum_{\sigma=1}^{n-1} (-1)^\sigma \left[ \frac{2m\sigma}{2n} + \frac{1}{2} \right] \right). \tag{9.16}$$

We remark that Riemann himself used the symbol $E(x)$, as opposed to $[x]$, to denote the greatest integer in $x$.

Dedekind derived (9.16) by finding the behavior of $\log \eta(\omega)$ as $\omega \to \frac{2m}{n}$ and then using (9.2) to determine the behavior of $\log k(\omega)$. He began by observing that when $\omega = x + iy$ and $y \to \infty$,

$$q = e^{i\pi\omega} \to 0,$$

[7] "Es steht alles schon bei Dedekind – Emmy Noether" is given as the motto at the beginning the 1975 article by Noether's student B. L. Van der Waerden, "On the Sources of My Book *Modern Algebra*."

[8] Riemann (1990) pp. 498–510.

so that from (9.1), it followed that

$$\log \eta(\omega) - \frac{\omega \pi i}{12} \approx 0 \quad \text{for large } y. \tag{9.17}$$

Dedekind then determined the behavior of $\log \eta(\omega)$ as $\omega \to \frac{m}{n}$, gcd $(m, n) = 1$, from the behavior of this function as $\omega \to i\infty$ by means of a modular transformation. Note with $m$ and $n$ given as relatively prime, there exist integers $m'$ and $n'$ such that $m'm - n'n = 1$. Thus, the transformation

$$\omega_1 = \frac{m'\omega - n'}{n\omega - m}$$

must be a modular transformation such that $\omega_1 \to i\infty$ as $\omega \to \frac{m}{n}$. Dedekind could conclude from (9.17) that

$$\log \eta(\omega_1) - \frac{\omega_1 \pi i}{12} \approx 0$$

with $\omega$ close to $\frac{m}{n}$. An application of (9.3) then gave him

$$\log \eta(\omega) + \frac{1}{4} \log \left( - (n\omega - m)^2 \right) + \frac{(m' - m)\pi i}{12n} - s(-m, n)\pi i$$

$$= \log \eta(\omega_1)$$

$$\approx \frac{\pi i}{12} \omega_1$$

$$= \frac{\pi i}{12} \left( \frac{m'}{n} - \frac{1}{n(n\omega - m)} \right). \tag{9.18}$$

Observing that $s(-m, n) = -s(m, n)$, Dedekind could rewrite (9.18) as

$$\log \eta(\omega) + \frac{\pi i}{12n(n\omega - m)} + \frac{1}{4} \log \left( - (n\omega - m)^2 \right) \approx \pi i \left( \frac{m - 12n\, s(m, n)}{12n} \right). \tag{9.19}$$

Formula (9.19) formed the basis for Dedekind's derivation of Riemann's eight results.[9] However, although he had in his possession a complete derivation of Riemann's formulas, Dedekind did not present it in his paper included in Riemann's collected works. The complete derivation, using Dedekind's ideas, was given by Rademacher and Whiteman in their 1940 paper, "Theorems on Dedekind Sums."[10]

We shall present Dedekind's derivation of (9.19) and of the reciprocity formula for Dedekind sums, (9.9). We will also discuss Rademacher's representation of the Dedekind sums in terms of trigonometric functions, and his derivation of the reciprocity formula from this representation. Rademacher was perhaps the first mathematician to recognize the importance of these sums; he wrote many papers and a short book on this topic.

In 1917, Hans Rademacher (1892–1969) wrote his Ph.D. dissertation, on a topic in real analysis, under Carathéodory at the University of Berlin. From 1923 onwards, however, he worked mainly in complex analysis and number theory. For example,

[9] Riemann (1990) pp. 493–510.    [10] See Rademacher (1974) vol. 2, pp. 220–250.

in 1928 he presented the paper, "Zur Theorie der Modulfunktionen" to the International Mathematical Congress at Bologna. Three years later, he published details of the proofs.[11] Note that Dedekind had given the transformation formula for the $\eta$ function, from which follows:

Set:

$$\text{sign } c = \begin{cases} \frac{c}{|c|}, & c \neq 0; \\ 0, & c = 0. \end{cases}$$

Then

$$\eta\left(\frac{a\omega+b}{c\omega+d}\right) = \begin{cases} e^{\pi i\left(\frac{a+d}{12c}-(\text{sign } c)\left(s(d,|c|)-\frac{1}{4}\right)\right)}\sqrt{c\omega+d}\,\eta(\omega), & c \neq 0; \\ e^{\frac{\pi ib}{12d}}\eta(\omega), & c = 0, \end{cases} \tag{9.20}$$

where $s(d, |c|)$ denotes the Dedekind sum. Note that when $c = 0$, $d = \pm 1$, since $ad - bc = 1$.

Rademacher defined

$$\Phi\begin{pmatrix} a & b \\ c & d \end{pmatrix} = \begin{cases} \frac{b}{d} & \text{for } c = 0; \\ \frac{a+d}{c} - 12(\text{sign } c)\,s(d,|c|) & \text{for } c \neq 0. \end{cases} \tag{9.21}$$

He proved that $\eta(\omega)$ was a modular form of weight $\frac{1}{2}$ with the multiplier system $v_\eta$ given by

$$v_\eta\begin{pmatrix} a & b \\ c & d \end{pmatrix} = \frac{\pi i}{12}\,\Phi\begin{pmatrix} a & b \\ c & d \end{pmatrix} - \frac{\pi i}{4}(\text{sign } c); \tag{9.22}$$

note that the definition of the multiplier system is given by (1.50) and (1.53).

Rademacher achieved this very important result by showing that if $U_1, U_2 \in \Gamma(1)$, $U_3 = U_1 U_2$, and $c_1, c_2, c_3$ were the corresponding entries in the second row and first column of $U_1, U_2, U_3$, then

$$\Phi(U_3) = \Phi(U_1) + \Phi(U_2) - 3\big(\text{sign }(c_1 c_2 c_3)\big). \tag{9.23}$$

Some of Rademacher's most important work was in the area of modular functions – in particular, on the Fourier coefficients of modular forms. Perhaps his best-known result in this field is a formula related to Hardy and Ramanujan's asymptotic formula for the number of partitions $p(n)$ of a positive integer $n$. Recall that $p(n)$ represents the number of ways $n$ can be written as an ordered sum of positive integers, that is

$$n = n_1 + n_2 + \cdots + n_k, \quad 1 \le n_1 \le n_2 \le \cdots \le n_k \le n.$$

Now in 1741, Euler gave the generating function for $p(n)$:[12]

$$\frac{1}{(1-x)(1-x^2)(1-x^3)\cdots} = 1 + \sum_{n=1}^{\infty} p(n)x^n. \tag{9.24}$$

[11] Rademacher (1974) vol. 1, pp. 652–676.

[12] See Eu.I-2, pp. 163–193. Also see Chapter 16 of Euler (1988), an English translation of Euler's 1747 *Introductio in analysin infinitorum*.

Note that except for the factor $x^{\frac{1}{24}}$, the denominator of the left-hand side of (9.24) is Dedekind's $\eta$ function. Hardy and Ramanujan used this fact and their so-called circle method to find the asymptotic formula:

$$p(n) = \frac{1}{2\pi\sqrt{2}} \sum_{k \leq \alpha\sqrt{n}} A_k(n) k^{\frac{1}{2}} \frac{d}{dn}\left(\frac{e^{\frac{C\lambda_n}{k}}}{\lambda_n}\right) + O\left(n^{-\frac{1}{4}}\right), \qquad (9.25)$$

with $\alpha$ an arbitrary positive constant, $C = \pi\sqrt{\frac{2}{3}}$, $\lambda_n = \sqrt{(n - \frac{1}{24})}$, and

$$A_k(n) = \sum_{\substack{h \bmod k \\ (h, k) = 1}} \exp\left(\pi i \left(-\frac{2hn}{k} + s(h, k)\right)\right). \qquad (9.26)$$

Formula (9.25) for $p(n)$ gives its exact value, for sufficiently large $n$, as the integer nearest the sum. In fact, the first six terms give the correct value of $p(100)$ and eight terms yield the correct value of $p(200)$. Hardy and Ramanujan were unable to determine, however, whether the series

$$\frac{1}{2\pi\sqrt{2}} \sum_{k=1}^{\infty} A_k(n) k^{\frac{1}{2}} \frac{d}{dn}\left(\frac{e^{\frac{C\lambda_n}{k}}}{\lambda_n}\right)$$

converged or diverged. In 1937, D. H. Lehmer showed that the series diverged[13]. A year earlier, Rademacher had found a convergent series for $p(n)$[14]:

$$p(n) = \frac{1}{\pi\sqrt{2}} \sum_{k=1}^{\infty} A_k(n) k^{\frac{1}{2}} \frac{d}{dn}\left(\frac{\sinh \frac{C\lambda_n}{k}}{\lambda_n}\right). \qquad (9.27)$$

We note that Rademacher also gave the expression (9.26) for $A_k(n)$ in terms of Dedekind sums, in contrast to Hardy and Ramanujan's equivalent but different expression.[15] See the exercises at the end of this chapter.

In 1937, Atle Selberg (1917–2007) independently obtained the convergent series (9.27) for $p(n)$. He also found an expression for $A_k(n)$ that was simpler than the one given in (9.26), as treated in the exercises at the end of this chapter. Selberg's very interesting remarks on how he made these discoveries are contained in his entertaining article, "Reflections around the Ramanujan centenary."[16] The reader might also wish to study Jan Hendrik Bruinier and Ken Ono's derivation of a formula for $p(n)$ as a finite sum of algebraic numbers.[17] Of additional interest, Michael Dewar and M. Ram Murty showed how to recover the Hardy-Ramanujan asymptotic formula from the formula of Bruinier and Ono.[18]

[13] Lehmer (1937) pp. 171–176.   [14] Rademacher (1974) vol. 2, pp. 100–106.
[15] For a good treatment of the mathematical work of Rademacher, see Berndt (1992).
[16] Berndt and Rankin (2001) pp. 203–213.   [17] Brunier and Ono (2013).   [18] Dewar and Murty (2013).

## 9.2 Riemann's Notes

As we have seen, among the treasures left behind by Riemann were two very brief notes on modular functions, based on the formulas in Jacobi's *Fundamenta Nova*.[19] The first note, probably dating from 1852, contained 68 formulas. The second note, undated but perhaps from about the same time, consisted of one barely legible page of eight formulas with a very few words in German. In this note, Riemann considered these formulas of Jacobi:[20]

$$\log k = \log 4\sqrt{q} - \frac{4q}{1+q} + \frac{4q^2}{2(1+q^2)} - \frac{4q^3}{3(1+q^3)} + \frac{4q^4}{4(1+q^4)} - \cdots \quad (9.28)$$

$$-\log k' = \frac{8q}{1-q^2} + \frac{8q^3}{3(1-q^6)} + \frac{8q^5}{5(1-q^{10})} + \frac{8q^7}{7(1-q^{14})} + \cdots \quad (9.29)$$

$$\log \frac{2K}{\pi} = \frac{4q}{1+q} + \frac{4q^3}{3(1+q^3)} + \frac{4q^5}{5(1+q^5)} + \frac{4q^7}{7(1+q^7)} + \cdots . \quad (9.30)$$

Observe that these series are the result of taking the logarithms of the infinite products for $k$, $k'$, and $\frac{2K}{\pi}$. The logarithms are converted into infinite series by using the power series expansion

$$\log(1+x) = x - \frac{x^2}{2} + \frac{x^3}{3} - \frac{x^4}{4} + \cdots, \quad \text{when } |x| < 1;$$

changing the order of summation produces the necessary results. Note that Jacobi took $q = e^{i\pi\omega}$, where $\omega = x + iy$ with $y > 0$, so that $|q| < 1$. Riemann inquired into what might happen when $\omega = \frac{m}{n} + iy \to \frac{m}{n}$, with $m$ and $n$ integers, that is, when $q^{2n} = 1$. Under these conditions, as $y \to 0$, the real parts of $\log k$, $-\log k'$, and $\log \frac{2K}{\pi}$ depend on $y$ and so tend to infinity; Riemann ignored these real parts. Yet as $y \to 0$, the imaginary parts of these series tend toward finite limits depending on $\frac{m}{n}$. Riemann computed these limits.

Judging from the date on the first note, September 1852, Dedekind observed that it was very likely that in both notes, Riemann had been working out examples of functions having an infinite number of discontinuities within a finite interval. In his famous 1854 paper on trigonometric series, Riemann gave examples of such functions that were integrable.

Dedekind's commentary on these fragmentary notes of Riemann, published in the collected papers of Riemann,[21] began by setting

$$\eta(\omega) = q^{\frac{1}{12}} \prod_{n=1}^{\infty} (1 - q^{2n}), \qquad q = e^{i\pi\omega}, \quad (9.31)$$

so that

$$\eta(2\omega)\,\eta\left(\frac{\omega}{2}\right)\eta\left(\frac{1+\omega}{2}\right) = e^{\frac{2\pi i}{48}}\eta^3(\omega) \quad (9.32)$$

[19] Riemann (1990) pp. 487–497.   [20] Jacobi (1969) vol. 1, p. 159.   [21] Riemann (1990) pp. 487–497.

$$\sqrt[4]{k} = e^{\frac{2\pi i}{48}} \sqrt{2} \, \frac{\eta(2\omega)}{\eta(\frac{1+\omega}{2})} \tag{9.33}$$

$$\sqrt[4]{k'} = e^{\frac{2\pi i}{48}} \, \frac{\eta(\frac{\omega}{2})}{\eta(\frac{1+\omega}{2})} \tag{9.34}$$

$$\sqrt{\frac{2K}{\pi}} = e^{-\frac{2\pi i}{24}} \, \frac{\eta^2(\frac{1+\omega}{2})}{\eta(\omega)}. \tag{9.35}$$

Taking $\omega = x + iy$, he took that branch of $\log \eta(\omega)$ for which

$$\lim_{y \to \infty} \frac{\log \eta(\omega)}{\omega} = \frac{\pi i}{12},$$

or, as he wrote it,

$$\log \eta(\omega) - \frac{\omega \pi i}{12} = 0 \tag{9.36}$$

where $\omega$ was infinitely large. Thus, for example, by (9.33)

$$\log k = \log 4 + 4 \log \eta(2\omega) - 4 \log \eta \left( \frac{1+\omega}{2} \right) + \frac{\pi i}{6} \tag{9.37}$$

so that, using (9.36), Dedekind had

$$\log k - \log 4 - \frac{\omega \pi i}{2} \to 0 \quad \text{as } y \to \infty.$$

To be able to derive Riemann's formulas, Dedekind required that when $ad - bc = 1$, $\eta(\frac{a\omega+b}{c\omega+d})$ be expressible in terms of $\eta(\omega)$ and a factor that depended upon $\frac{a\omega+b}{c\omega+d}$. To this end, he began with a formula that, as he remarked, was generally known:

$$\theta_1 \left( z, \frac{a\omega+b}{c\omega+d} \right) = \alpha \sqrt{c\omega+d} \, e^{i\pi c(c\omega+d)z^2} \, \theta_1((c\omega+d)z, \omega), \tag{9.38}$$

where $a, b, c, d$ were integers with $ad - bc = 1$ and $\alpha$, an eighth root of unity, was a function of $a, b, c, d$. In fact, Hermite had explicitly computed the constant $\alpha$ in a paper of 1858.[22] See formula (5.11). Dedekind took the derivative of equation (9.38) with respect to $z$ and then set $z = 0$ to obtain

$$\theta_1' \left( 0, \frac{a\omega+b}{c\omega+d} \right) = \alpha(c\omega+d)^{\frac{3}{2}} \theta_1'(0, \omega). \tag{9.39}$$

Then, by Jacobi's formulas (3.201) and (3.215) through (3.217),

$$\theta_1'(0, \omega) = 2\pi \, \eta^3(\omega), \tag{9.40}$$

so that

$$\eta \left( \frac{a\omega+b}{c\omega+d} \right) = \alpha^{\frac{1}{3}}(c\omega+d)^{\frac{1}{2}} \eta(\omega), \tag{9.41}$$

[22] Hermite (1905) vol. 1, pp. 482–486.

where $\alpha^{\frac{1}{3}}$ was a 24th root of unity depending on $a, b, c, d$. Dedekind's paper is devoted to the determination of $\alpha^{\frac{1}{3}}$ in a computable form. The case $c = 0$ was an immediate consequence of the definition (9.31) of $\eta(\omega)$:

$$\log \eta(\omega + 1) = \log \eta(\omega) + \frac{\pi i}{12}, \tag{9.42}$$

and, more generally, for any integer $n$,

$$\log \eta(\omega + n) = \log \eta(\omega) + \frac{n\pi i}{12}. \tag{9.43}$$

Note that if $c \neq 0$, the quantity $-(c\omega + d)^2$ could not be negative, so that Dedekind could define its logarithm uniquely by requiring that its imaginary part lie between $\pm\pi$, and then rewrite (9.41) in the form

$$\log \eta \left(\frac{a\omega + b}{c\omega + d}\right) = \log \eta(\omega) + \frac{1}{4} \log(-(c\omega + d)^2) + (a, b, c, d)\frac{\pi i}{12}, \tag{9.44}$$

where $(a, b, c, d)$ denoted an integer depending on the integers $a, b, c, d$. Clearly,

$$(-a, -b, -c, -d) = (a, b, c, d). \tag{9.45}$$

In his commentary on Riemann's fragments, Dedekind's primary aim was to completely determine the integer $(a, b, c, d)$. He first proved that there existed an integer-valued function of $c$ and $d$, denoted by Dedekind as $(d, c)$, such that

$$c(a, b, c, d) = a + d - 2(d, c). \tag{9.46}$$

Note that when $c > 0$, we now call the function $\frac{(d,c)}{6c}$ a Dedekind sum. In order to prove (9.46) when $c \neq 0$, Dedekind noted that, under the given condition that $ad - bc = 1$, and if there were integers $a', b'$ such that $a'd - b'c = 1$, then $a' = a + nc$ and $b' = b + nd$ for some integer $n$. Hence by (9.43),

$$\log \eta \left(\frac{a'\omega + b'}{c\omega + d}\right) = \log \left(\frac{a\omega + b}{c\omega + d} + n\right) = \log \left(\frac{a\omega + b}{c\omega + d}\right) + \frac{n\pi i}{12};$$

it followed from (9.44) and the relation $a' - a = nc$ that

$$(a', b', c, d) - \frac{a'}{c} = (a, b, c, d) - \frac{a}{c}$$

depended only on the two relatively prime integers $c, d$. This implied the existence of a function $(d, c)$ such that (9.46) would hold true. Dedekind could therefore rewrite (9.44) as

$$\log \eta \left(\frac{a\omega + b}{c\omega + d}\right) = \log \eta(\omega) + \frac{1}{4} \log \left(-(c\omega + d)^2\right) + \frac{a + d - 2(d, c)}{12c} \pi i; \tag{9.47}$$

his problem was thus reduced to deriving properties of $(c, d)$ when $c$ and $d$ were relatively prime. For example, by changing $a, b, c, d$ to $-a, -b, -c, -d$, he had

$$(-d, -c) = -(d, c). \tag{9.48}$$

Next, by replacing all the terms in (9.47) with the corresponding conjugate values, he obtained

$$\log \eta \left( -\frac{a\overline{\omega}+b}{c\overline{\omega}+d} \right) = \log \eta \, (-\overline{\omega}) + \frac{1}{4} \log \left( -(c\overline{\omega}+d)^2 \right) - \frac{a+d-2(d,c)}{12c} \pi i.$$

$$(9.49)$$

He applied (9.47) after rewriting the left-hand side of (9.49) as

$$\log \eta \left( \frac{a(-\overline{\omega})-b}{-c(-\overline{\omega})+d} \right) = \log \eta \, (-\overline{\omega}) + \frac{1}{4} \log \left( -(c\overline{\omega}+d)^2 \right) + \frac{a+d-2(d,-c)}{12(-c)} \pi i.$$

$$(9.50)$$

Comparison of (9.49) and (9.50) yielded

$$(d, -c) = (d, c) \qquad (9.51)$$

and an application of (9.48) gave

$$(-d, c) = -(d, c). \qquad (9.52)$$

Dedekind extended the definition of $(d, c)$ to the case $c = 0$ by remarking that in this case $a = d = \pm 1$ and thus (9.47) suggested that $(\pm 1, 0) = \pm 1$. On the other hand, (9.52) gave him the value $(0, \pm 1) = 0$; so that by taking $a = 0, b = -1, c = 1, d = 0$ in (9.47), he arrived at the important transformation formula

$$\log \eta \left( -\frac{1}{\omega} \right) = \log \eta(\omega) + \frac{1}{4} \log(-\omega^2). \qquad (9.53)$$

Recall that $\log(-\omega^2)$ is defined by its principal value, and since $0 < \arg \omega < \pi$, it follows that $\log(-\omega^2) = 2 \log \omega - \pi i$. Thus, though Dedekind did not write it explicitly, clearly

$$\eta \left( -\frac{1}{\omega} \right) = \sqrt{\frac{\omega}{i}} \, \eta(\omega). \qquad (9.54)$$

Dedekind next verified two basic properties of $(d, c)$:

$$(d + c, c) = (d, c) \qquad (9.55)$$

$$2d(d, c) + 2c(c, d) = 1 + d^2 + c^2 - 3|dc|. \qquad (9.56)$$

These formulas are clearly true when $c = 0$ or $d = 0$. So to prove (9.55), Dedekind took $c \neq 0$, replaced $\omega$ by $\omega + 1$ in (9.47) and then applied (9.42) to get

$$\begin{aligned}
\log \eta & \left( \frac{a\omega + a + b}{c\omega + c + d} \right) \\
&= \log \eta(\omega + 1) + \frac{1}{4} \log(-(c\omega + c + d)^2) + \frac{a+d-2(d,c)}{12c} \pi i \\
&= \log \eta(\omega) + \frac{1}{4} \log(-(c\omega + c + d)^2) + \frac{a+d+c-2(d,c))}{12c} \pi i. \qquad (9.57)
\end{aligned}$$

On the other hand, by (9.47),

$$\log \eta \left( \frac{a\omega + a + b}{c\omega + c + d} \right)$$
$$= \log \eta(\omega) + \frac{1}{4} \log(-(c\omega + c + d)^2) + \frac{a + d + c - 2(d + c, c)}{12c} \pi i. \quad (9.58)$$

Equation (9.55) resulted upon equating (9.57) and (9.58). In a similar way, to prove (9.56), he replaced $\omega$ by $-\frac{1}{\omega}$ in (9.47) and applied (9.53) to obtain

$$\log \eta \left( \frac{a\left(-\frac{1}{\omega}\right) + b}{c\left(-\frac{1}{\omega}\right) + d} \right)$$
$$= \log \eta(\omega) + \frac{1}{4} \left( \log(-\omega^2) + \log\left( -\left( c\left(-\frac{1}{\omega}\right) + d \right)^2 \right) \right) + \frac{a + d - 2(d, c)}{12c} \pi i. \quad (9.59)$$

The left-hand side of (9.59) could also be written as

$$\log \eta \left( \frac{b\omega - a}{d\omega - c} \right) = \log \eta(\omega) + \frac{1}{4} \log(-(d\omega - c)^2) + \frac{b - c - 2(-c, d)}{12d} \pi i. \quad (9.60)$$

Taking $c$ and $d$ positive, equating (9.59) and (9.60), and using (9.52) then yielded

$$-\frac{i\pi}{4} cd + \frac{d\left(a + d - 2(d, c)\right) \pi i}{12} = \frac{c\left((b - c) + (c, d)\right)}{12} \pi i,$$

or, since $ad - bc = 1$,

$$1 + d^2 + c^2 - 3cd = 2d(d, c) + 2c(c, d), \quad \text{when } c > 0, \ d > 0. \quad (9.61)$$

Note that when $c < 0$, it is clear that the $c$ on the left-hand side of (9.61) must be replaced by $|c|$ and a similar applies for the case $d < 0$. Thus, (9.56) is verified. We remark that (9.61) is a form of the reciprocity formula for Dedekind sums.

Dedekind found another expression for $\frac{(d,c)}{c}$ by studying the behavior of $\log \eta(\omega)$ as $\omega = x + iy \to \frac{d}{c}$. Recall that Riemann applied this method to determine the value of the imaginary part of $\log k$ in (9.28) when $\omega \to \frac{d}{c}$. So Dedekind proceeded by supposing that $\omega = x + iy$ approached $-\frac{d}{c}$ in such a manner that $d + cx$ was an infinitesimal of order higher than $\sqrt{y}$. It is interesting to note here that, although he knew very well how to express this idea in terms of limits, Dedekind found it convenient to use the language of the infinitesimal. We shall follow him in this respect. Thus, the ordinate of

$$\omega_1 = \frac{a\omega + b}{c\omega + d} = \frac{a}{c} - \frac{1}{c(c\omega + d)} \quad (9.62)$$

was positive and approached infinity as $\omega$ approached $-\frac{d}{c}$, so that by (9.36),

$$\log \eta(\omega_1) - \frac{\omega_1 \pi i}{12} = 0. \quad (9.63)$$

Dedekind remarked that with $\omega_1$ infinitely large, (9.63) would hold; in contemporary notation, we would write

$$\lim_{\omega_1 \to \infty} \frac{\log \eta(\omega_1)}{\omega_1} = \frac{\pi i}{12}.$$

Using (9.47), (9.62), and (9.63), Dedekind concluded that

$$\lim_{\omega \to -\frac{d}{c}} \left( \log \eta(\omega) + \frac{\pi i}{12c(c\omega + d)} + \frac{1}{4} \log \left( -(c\omega + d)^2 \right) \right) = \frac{2(d, c) - d}{12c} \pi i.$$

$$(9.64)$$

In order to present his result in the notation used by Riemann in his notes, Dedekind replaced $d$, $c$ by $-m$, $n$, respectively, in (9.64). Applying (9.52), he could rewrite (9.64) as

$$\lim_{\omega \to \frac{m}{n}} \left( \log \eta(\omega) + \frac{\pi i}{12n(n\omega - m)} + \frac{1}{4} \log \left( -(n\omega - m)^2 \right) \right) = \frac{m - 2(m, n)}{12n} \pi i,$$

$$(9.65)$$

with gcd $(m, n) = 1$. Note that since $\omega = x + iy$ approached $\frac{m}{n}$, both $y$ and $mx - n$ would tend to 0. Dedekind then observed that if one applied the more stringent condition, that $mx - n$ be an infinitesimal of order higher than $y^2$, then the imaginary parts of the second and third terms on the left-hand side of (9.65) would vanish; indeed, this can be easily checked. Thus, upon subtracting the formula obtained by replacing $\omega$ by $-\overline{\omega}$ in (9.65), he arrived at the limit formula

$$\log \eta(\omega) - \log \eta(-\overline{\omega}) = \frac{m - 2(m, n)}{6n} \pi i, \qquad (9.66)$$

as $\omega = x + iy \to \frac{m}{n}$, as described. He noted that the conditions for convergence would be satisfied if $\omega = \frac{m}{n} + iy$ and $y \to 0$, since in this case $mx - n$ was in fact 0. Hence, when gcd $(m, n) = 1$ for $\omega = \frac{m}{n} + iy$,

$$\lim_{\omega \to \frac{m}{n}} (\log \eta(\omega) - \log \eta(-\overline{\omega})) = \frac{m - 2(m, n)}{6n} \pi i. \qquad (9.67)$$

To evaluate the left-hand side of (9.67) in another way, Dedekind applied the definition (9.31) of $\eta(\omega)$ to obtain

$$\log \eta(\omega) = \frac{\omega \pi i}{12} + \sum_{\nu=1}^{\infty} \log \left( 1 - e^{2\pi i \omega \nu} \right). \qquad (9.68)$$

Dedekind noted that the logarithms on the right-hand side of (9.68) were taken to vanish as $e^{2\pi i \omega \nu} \to 0$. Consequently,

$$\log(1 - e^{2\pi i \omega \nu}) = -\sum_{\mu=1}^{\infty} \frac{e^{2\pi i \omega \nu \mu}}{\mu}.$$

Since $\omega = \frac{m}{n} + iy$, $y > 0$, this series converges absolutely and

$$\log \eta(\omega) = \frac{\omega \pi i}{12} - \sum_{v=1}^{\infty} \sum_{\mu=1}^{\infty} \frac{e^{2\pi i \omega v \mu}}{\mu}$$

$$= \frac{\omega \pi i}{12} - \sum_{\mu=1}^{\infty} \frac{1}{\mu} \sum_{v=1}^{\infty} e^{2\pi i \omega v \mu}$$

$$= \frac{\omega \pi i}{12} - \sum_{\mu=1}^{\infty} \frac{1}{\mu} \cdot \frac{e^{2\pi i \omega \mu}}{1 - e^{2\pi i \omega \mu}}.$$

Dedekind could therefore write

$$\log \eta(\omega) - \log \eta(-\overline{\omega}) = \frac{(\omega + \overline{\omega})\pi i}{12} - \sum_{\mu=1}^{\infty} \frac{1}{\mu} \left( \frac{e^{2\pi i \omega \mu}}{1 - e^{2\pi i \omega \mu}} - \frac{e^{-2\pi i \overline{\omega} \mu}}{1 - e^{-2\pi i \overline{\omega} \mu}} \right)$$

$$= \frac{(\omega + \overline{\omega})\pi i}{12} - \sum_{\mu=1}^{\infty} \frac{a_\mu}{\mu}, \tag{9.69}$$

where

$$a_\mu = \frac{1}{1 - e^{2\pi i \omega \mu}} - \frac{1}{1 - e^{-2\pi i \overline{\omega} \mu}}. \tag{9.70}$$

When $\omega = \frac{m}{n} + iy$,

$$\frac{(\omega + \overline{\omega})\pi i}{12} = \frac{m \pi i}{6n} \tag{9.71}$$

and $a_\mu$ could be written as

$$a_\mu = \frac{1}{1 - \theta^\mu r^\mu} - \frac{1}{1 - \theta^{-\mu} r^\mu}, \tag{9.72}$$

where

$$\theta = e^{\frac{2m\pi i}{n}} \quad \text{and} \quad r = e^{-2\pi y}.$$

Thus, Dedekind had

$$\log \eta(\omega) - \log \eta(-\overline{\omega}) = \frac{m \pi i}{6n} - \sum_{\mu=1}^{\infty} \frac{a_\mu}{\mu}, \tag{9.73}$$

where $a_\mu$ was given by (9.72). He employed (9.73) to obtain his famous result (9.5).

### 9.3   Dedekind Sums in Terms of a Periodic Function

Recall that Dedekind sought to compute the imaginary part of the limit of $\log k(\omega)$ as $\omega = \frac{m}{n} + iy$ tends to $\frac{m}{n}$, where $n$ denotes a positive integer. To apply (9.33), it was necessary to find the corresponding limit of $\log \eta(\omega)$. Since $\log \eta(\omega) - \log \eta(-\overline{\omega})$ was purely imaginary, Dedekind needed the limit of (9.73) as $y \to 0$ or, equivalently, as $r \to 1$. Using a theorem he had published in the ninth supplement to Dirichlet's *Lectures on*

*Number Theory*, he was able to conclude that the sum on the right-hand side of (9.73) was continuous in $r \in (0, 1]$. Thus, taking $r = 1$ in the series in (9.73) gave him the limit. The theorem from the ninth supplement stated:[23]

If the absolute value of the sum

$$A_n = a_1 + a_2 + \cdots + a_n$$

remains bounded as $n$ tends to infinity, and if the sum $\beta$ of absolute values of the differences

$$b_1 - b_2, b_2 - b_3, b_3 - b_4, \ldots$$

is bounded, and moreover $b_n$ tends to zero as $n$ tends to infinity, then the series

$$\gamma = a_1 b_1 + a_2 b_2 + a_3 b_3 + \cdots$$

converges, and its sum varies continuously with the quantities in the sequence $(b)$ $[(b_n)]$, and hence with the sum $\beta$.

Dedekind wrote that the proof of this theorem employed summation by parts in the manner used by Abel to prove his continuity theorem on power series:

$$\text{If } \sum_{n=0}^{\infty} a_n \text{ converges, then } \sum_{n=0}^{\infty} a_n x^n \text{ converges for } |x| < 1$$

$$\text{and } \lim_{x \to 1^-} \sum_{n=0}^{\infty} a_n x^n = \sum_{n=0}^{\infty} a_n.$$

The method of summation by parts is often attributed to Abel, but in fact it had been known for more than a hundred years before Abel. In a paper of 1717, Brook Taylor, of Taylor series fame, described the method of summation by parts as an analog of integration by parts. Abel's contribution lay in the insightful application of this method to the study of convergence of series.[24] We also mention that the theorem presented by Dedekind in Dirichlet's lectures is now known as Dirichlet's theorem, though only the particular case for $b_n = \frac{1}{n^s}$ appears to have been explicitly stated by Dirichlet. The general theorem may well be due to Dedekind, who applied it to the series $\sum \frac{a_\mu}{\mu}$, given in (9.73). He noted that the absolute value of the sum

$$A_\mu = a_1 + a_2 + \cdots + a_\mu, \tag{9.74}$$

where $a_\mu(r)$ was defined by (9.72), was bounded for $0 < r = e^{-y} \le 1$. Observe that when $r = 1$, if we set $c_\mu = a_\mu(1)$, we have

$$c_\mu = \lim_{r \to 1^-} a_\mu(r) = \begin{cases} 0, & n \mid \mu \\ \frac{1}{1-\theta^\mu} - \frac{1}{1-\theta^{-\mu}}, & n \nmid \mu \end{cases} \tag{9.75}$$

with $\theta = e^{\frac{2\pi i m}{n}}$. To show that

$$c_1 + c_2, + \cdots + c_\mu \tag{9.76}$$

---

[23] English translation by John Stillwell. Dirichlet and Dedekind (1999) p. 261.

[24] Abel (1965) vol. 1, pp. 219–250.

is bounded, note that when $n \nmid \mu$

$$c_\mu = \frac{\theta^\mu - \theta^{-\mu}}{2 - (\theta^\mu + \theta^{-\mu})} = 2i \frac{\sin \frac{2m\mu\pi}{n}}{2 - 2\cos \frac{2m\mu\pi}{n}} = i \cot \frac{m\mu\pi}{n}. \qquad (9.77)$$

Since gcd $(m, n) = 1$, we then have

$$\sum_{\mu=1}^{n-1} c_\mu = i \sum_{\mu=1}^{n-1} \cot \frac{\mu m\pi}{n} = 0, \qquad (9.78)$$

showing the boundedness of the sum in (9.76). Dedekind had observed that the boundedness of $A_\mu$ in (9.74) would be evident by adding the two terms $a_{ns+v}$ and $a_{n(s+1)-v}$, where $0 < v < \frac{n}{2}$. In fact,

$$a_{ns+v} + a_{n(s+1)-v} = \frac{r^{ns}(\theta^v r^v - \theta^{-v} r^{n-v})}{1 - r^{ns}(\theta^v r^v - \theta^{-v} r^{n-v}) + r^{n(2s+1)}}. \qquad (9.79)$$

The reader may check that equation (9.79) yields the boundedness of $A_\mu$ for $r < 1$; also, (9.78) may be verified by adding the terms $c_v$ and $c_{n-v}$.

It is interesting to note that Dedekind's theorem from the ninth supplement does not actually apply directly here, since $a_\mu$ is a function of $r$ and $b_\mu = \frac{1}{\mu}$. However, the method used to prove the continuity of the earlier sum is also applicable to the sum $\sum \frac{a_\mu}{\mu}$. Applying summation by parts and taking $A_0 = 0$, we have

$$\sum_{\mu=1}^{\infty} \frac{a_\mu}{\mu} = \lim_{n \to \infty} \sum_{\mu=1}^{n} \frac{A_\mu - A_{\mu-1}}{\mu} = \lim_{n \to \infty} \left( \sum_{\mu=1}^{n-1} A_\mu \left( \frac{1}{\mu} - \frac{1}{\mu+1} \right) + \frac{A_n}{n} \right).$$

Since $|A_\mu| \le K$ with $K$ a constant, for $0 \le r \le 1$ for all $\mu \ge 0$, we see with Dedekind that

$$\sum_{n=1}^{\infty} \frac{a_\mu}{\mu} = \sum_{\mu=1}^{\infty} A_\mu \left( \frac{1}{\mu} - \frac{1}{\mu+1} \right). \qquad (9.80)$$

Dedekind concluded that the series on the right-hand side of (9.80) converged uniformly on $[0, 1]$ and hence represented a continuous function of $r = e^{-y}$. It followed from (9.66), (9.69), (9.71), and (9.75) that

$$\frac{(m, n)\pi i}{3n} = \sum_{\mu=1}^{\infty} \frac{c_\mu}{\mu}. \qquad (9.81)$$

To find another expression for the left-hand side of (9.81), Dedekind expressed $c_\mu$ as

$$c_\mu = \frac{1}{n} \sum_{\sigma=1}^{n-1} \sigma (\theta^{-\mu\sigma} - \theta^{\mu\sigma}) \qquad (9.82)$$

and noted that (9.82) in fact followed from the identity

$$\frac{1}{1 - \theta^\mu} = -\frac{1}{n} \sum_{\sigma=1}^{n-1} \sigma \theta^{-\mu\sigma}, \quad n \nmid \mu. \qquad (9.83)$$

He did not offer any hint on a proof of (9.83), since it is very easy to prove; start with the identity

$$(1 - x) \sum_{\sigma=0}^{n-1} x^\sigma = 1 - x^n.$$

Take the derivative with respect to $x$ to obtain

$$-\sum_{\sigma=0}^{n-1} x^\sigma + (1 - x) \sum_{\sigma=1}^{n-1} \sigma x^{\sigma-1} = -nx^{n-1}. \tag{9.84}$$

Multiply (9.84) by $x$ and set $x = \theta^\mu = e^{\frac{2\pi i m \mu}{n}}$, where $n \nmid \mu$, to conclude the proof:

$$(1 - \theta^\mu) \sum_{\sigma=1}^{n-1} \sigma \theta^{\mu\sigma} = -n.$$

Dedekind could therefore write (9.81) as

$$\frac{(m, n)\pi i}{6n} = \frac{1}{n} \sum_{\mu=1}^{\infty} \sum_{\sigma=1}^{n-1} \frac{\sigma}{\mu} (\theta^{-\mu\sigma} - \theta^{\mu\sigma})$$

$$= \sum_{\sigma=1}^{n-1} \frac{\sigma}{n} \sum_{\mu=1}^{\infty} \frac{\theta^{-\mu\sigma} - \theta^{\mu\sigma}}{\mu}. \tag{9.85}$$

To sum the series (9.85), he employed the Fourier series expansion of the periodic function $((x))$, defined by:

$$((x)) = \begin{cases} x - [x] - \frac{1}{2} & \text{if } x \text{ is not an integer,} \\ 0 & \text{if } x \text{ is an integer,} \end{cases} \tag{9.86}$$

with $[x]$ the greatest integer not exceeding $x$. Note that here we have taken Rademacher's definition of $((x))$; Dedekind's definition would correspond to $((x + \frac{1}{2}))$.

The Fourier series of (9.86) was originally given by Euler. It is often said that Euler believed that a so-called arbitrary function could not be expanded as a Fourier series. However, as early as 1752, Euler expanded the first two Bernoulli polynomials as Fourier series. In the 1770s, he published a more complete paper[25] in which he gave a method for obtaining the Fourier series for the successive Bernoulli polynomials, based upon the work of Daniel Bernoulli[26]. In his 1752 work,[27] Euler published the formula

$$\frac{x}{2} = \sin x - \frac{\sin 2x}{2} + \frac{\sin 3x}{3} - \cdots, \tag{9.87}$$

obtained by integrating the divergent series

$$\cos x - \cos 2x + \cos 3x - \cdots = \frac{1}{2}. \tag{9.88}$$

---

[25] Eu.I-15, pp. 446–450.    [26] Bernoulli (1982–1996) vol. 2, pp. 119–134, especially pp. 120–121.
[27] Eu.I-14, p. 584.

Note that the right-hand side of (9.88) is the Abel or Cesàro sum of the series on the left-hand side. Dedekind noted that (9.87) could be derived from the two formulas

$$\log(1 + e^{i\theta}) = \sum_{\mu=1}^{\infty} (-1)^{\mu-1} \frac{e^{i\mu\theta}}{\mu}$$

$$\log(1 + e^{i\theta}) = \log\left(2\left(1 + \cos\theta\right)\right) + \frac{i\theta}{2},$$

where $\theta \neq \pi$ and where Dedekind chose the principal value of the logarithm. Euler gave no range of validity for (9.87), but two decades later Daniel Bernoulli specified the range of validity as $-\pi < x < \pi$. Now if we replace $x$ by $2\pi(x - \frac{1}{2})$, then we may rewrite (9.87) as

$$\pi\left(x - \frac{1}{2}\right) = -\left(\sin 2\pi x + \frac{1}{2}\sin 4\pi x + \frac{1}{3}\sin 6\pi x + \cdots\right) \quad \text{for } 0 < x < 1. \quad (9.89)$$

By periodicity, the function $((x))$ extends the function $f(x) = x - \frac{1}{2}, 0 < x < 1$, to all values of $x$ except the integers. Thus, from (9.89) we have

$$2\pi i((x)) = \sum_{\mu=1}^{\infty} \frac{e^{-2\pi i\mu x} - e^{2\pi i\mu x}}{\mu}$$

and therefore

$$2\pi i\left(\left(\frac{\sigma m}{n}\right)\right) = \sum_{\mu=1}^{\infty} \frac{\theta^{-\mu\sigma} - \theta^{\mu\sigma}}{\mu}. \quad (9.90)$$

So Dedekind could conclude from (9.90) and (9.85) that

$$\frac{(m, n)}{6n} = \sum_{\sigma=1}^{n-1} \frac{\sigma}{n}\left(\left(\frac{\sigma m}{n}\right)\right). \quad (9.91)$$

Moreover, by changing $\sigma$ to $n - \sigma$ and applying the periodicity of $((x))$, he observed that

$$\frac{1}{2}\sum_{\sigma=1}^{n-1}\left(\left(\frac{\sigma m}{n}\right)\right) = 0. \quad (9.92)$$

Subtracting (9.92) from (9.91) gave the final expression for the Dedekind sum:

$$\frac{(m, n)}{6n} = \sum_{\sigma=1}^{n-1}\left(\left(\frac{\sigma}{n}\right)\right)\left(\left(\frac{\sigma m}{n}\right)\right) = \sum_{\sigma=1}^{n}\left(\left(\frac{\sigma}{n}\right)\right)\left(\left(\frac{\sigma m}{n}\right)\right). \quad (9.93)$$

Denoting the right-hand side of (9.93) as $s(m, n)$ and by combining (9.93) with (9.61), taking $m$ and $n$ to be positive integers, we arrive at the famous reciprocity formula for Dedekind sums:

$$s(m, n) + s(n, m) = -\frac{1}{4} + \frac{1}{12}\left(\frac{m}{n} + \frac{n}{m} + \frac{1}{mn}\right). \quad (9.94)$$

Dedekind noted that (9.94) implied the quadratic reciprocity law. He provided a few more details on this topic in his paper on modular functions, written as a letter to the long-time editor of *Crelle's Journal*, Carl Borchardt. In that paper, he observed a connection between $(m, n)$ and the Legendre-Jacobi symbol $\left(\frac{m}{n}\right)$:

$$\left(\frac{m}{n}\right) \equiv \frac{n+1}{2} - (m, n) \pmod 4. \tag{9.95}$$

Combining (9.94) and (9.95) then produced

$$\left(\frac{m}{n}\right) + \left(\frac{n}{m}\right) \equiv 2 \left(1 + \frac{m-1}{2} \cdot \frac{n-1}{2}\right) \pmod 4. \tag{9.96}$$

Equation (9.96) is an alternate formulation of the law of quadratic reciprocity, usually written in the form

$$\left(\frac{m}{n}\right) \left(\frac{n}{m}\right) = (-1)^{\frac{m-1}{2} \cdot \frac{n-1}{2}}. \tag{9.97}$$

## 9.4 Rademacher

Rademacher defined the Dedekind sum by means of the expression on the right-hand side of (9.93) and denoted it by $s(n, m)$. In this way, the reciprocity law (9.96) for Dedekind sums could be stated as a completely arithmetical formula. He then freed the Dedekind sum from the modular transformation of the $\eta$ function by constructing proofs of the reciprocity formula without the use of transformation theory. In this connection, Rademacher obtained various expressions for the Dedekind sum; of particular interest was the expression

$$s(m, n) = \frac{1}{4n} \sum_{k=1}^{n-1} \cot \frac{\pi k}{n} \cot \frac{\pi mk}{n}. \tag{9.98}$$

Rademacher first presented this result in a 1933 paper, written in Hungarian with a three-page German summary, published in the Hungarian journal *Matematikai és Fizikai Lapok*.[28] In fact, this paper was an excerpt from a lecture given by Rademacher in October 1929 at the Pázmány Péter Catholic University in Budapest. In this paper, he proved (9.98) by using definite integrals, with an approach similar to that used by Dirichlet in calculating special values of some $L$-functions.[29] It is interesting to note that Dedekind remarked, before presenting his proof of (9.93), that he had accomplished it without the use of definite integrals. We note that Rademacher later gave an alternate proof of (9.98) that avoided the use of definite integrals; it is presented in his book with his student Grosswald.

Rademacher proved (9.94) by applying Cauchy's residue theorem to get

$$\frac{1}{4n} \sum_{k=1}^{n-1} \cot \frac{\pi k}{n} \cot \frac{\pi mk}{n} + \frac{1}{4m} \sum_{j=1}^{m-1} \cot \frac{\pi j}{m} \cot \frac{\pi nj}{m} = -\frac{1}{4} + \frac{1}{12} \left(\frac{m}{n} + \frac{n}{m} + \frac{1}{mn}\right). \tag{9.99}$$

---

[28] Rademacher (1974) vol. 2, pp. 26–36.
[29] See Dirichlet (1969) vol. 1, p. 421 and p. 480; compare the integrals on these pages with (9.104) and (9.105).

Rademacher started his proof of (9.98) by taking the right-hand side of (9.91) as the definition of $s(m, n)$:

$$s(m, n) = \sum_{\sigma=1}^{n} \frac{\sigma}{n} \left( \left( \frac{\sigma m}{n} \right) \right);$$  (9.100)

from (9.89) he had

$$((x)) = -\frac{1}{\pi} \sum_{k=1}^{\infty} \frac{\sin 2\pi k x}{k}$$

and therefore

$$s(m, n) = -\frac{1}{n\pi} \sum_{k=1}^{\infty} \frac{1}{k} \sum_{\sigma=1}^{n} \sigma \sin \left( \frac{2\pi m k \sigma}{n} \right).$$  (9.101)

For the inner sum, he noted that

$$S_j = \sum_{\sigma=1}^{n} \sigma \sin \left( \frac{2\pi j \sigma}{n} \right) = \begin{cases} 0, & j \equiv 0 \pmod{n} \\ -\frac{n}{2} \cot \frac{\pi j}{n}, & j \not\equiv 0 \pmod{n} \end{cases}.$$  (9.102)

Although this is not difficult to verify, Rademacher provided a proof, letting

$$U = \sum_{\sigma=1}^{n} \sigma e^{\frac{2\pi i \sigma j}{n}}$$

so that $S_j = \text{Im}(U)$. Then he had

$$U = \sum_{\sigma=1}^{n} (\sigma - 1) e^{\frac{2\pi i \sigma j}{n}} = e^{\frac{2\pi i j}{n}} \sum_{\nu=0}^{n-1} \nu e^{\frac{2\pi i \nu j}{n}}$$

$$= e^{\frac{2\pi i j}{n}} (U - n e^{2\pi i j}) = e^{\frac{2\pi i j}{n}} (U - n);$$

$$U = \frac{n}{1 - e^{\frac{-2\pi i j}{n}}} = \frac{n(1 - e^{\frac{2\pi i j}{n}})}{|1 - e^{\frac{2\pi i j}{n}}|^2}.$$  (9.103)

Taking the imaginary part of (9.103) yielded (9.102) and Rademacher could thus rewrite (9.101) as

$$s(m, n) = \frac{1}{2\pi} \sum_{t=0}^{\infty} \sum_{l=1}^{n-1} \frac{1}{tn + l} \cot \frac{\pi l m}{n}$$

$$= \frac{1}{2\pi} \sum_{t=0}^{\infty} \sum_{l=1}^{n-1} \int_{0}^{1} x^{tn+l-1} \cot \frac{\pi l m}{n} dx$$

$$= \frac{1}{2\pi} \int_{0}^{1} \frac{1}{1 - x^n} \sum_{l=1}^{n-1} x^{l-1} \cot \frac{\pi l m}{n} dx.$$  (9.104)

Applying partial fractions and the identity $\sum_{l=1}^{n-1} \cot \frac{\pi l m}{n} = 0$, he then got

$$s(m, n) = -\frac{1}{2n\pi} \int_{0}^{1} \sum_{\lambda=1}^{n-1} \sum_{l=1}^{n-1} \frac{\rho^{\lambda l} \cot \frac{\pi l m}{n}}{x - \rho^{\lambda}} dx,$$  (9.105)

with $\rho = e^{\frac{2\pi i}{n}}$. He observed that

$$\sigma(\lambda) \equiv \sum_{l=1}^{n-1} \rho^{\lambda l} \cot \frac{\pi l m}{n}$$

$$= \sum_{l=1}^{n-1} \rho^{\lambda(n-l)} \cot \frac{\pi(n-l)m}{n}$$

$$= -\sum_{l=1}^{n-1} \rho^{-\lambda l} \cot \frac{\pi l m}{n} = -\overline{\sigma(\lambda)}; \qquad (9.106)$$

hence, $\sigma(\lambda)$ was purely imaginary and could be written as

$$\sigma(\lambda) = i \sum_{l=1}^{n-1} \sin \frac{2\pi \lambda l}{n} \cot \frac{\pi l m}{n}. \qquad (9.107)$$

By evaluating the integral in (9.105), he obtained

$$s(m, n) = -\frac{1}{2n\pi} \sum_{\lambda=1}^{n-1} \sigma(\lambda) \log(1 - \rho^{-\lambda}) \qquad (9.108)$$

with

$$\left| \mathrm{Im}(\log(1 - \rho^{-\lambda})) \right| < \frac{\pi}{2}.$$

Taking the complex conjugate of (9.108) and using (9.106) gave Rademacher

$$s(m, n) = \frac{1}{2n\pi} \sum_{\lambda=1}^{n-1} \sigma(\lambda) \log(1 - \rho^{\lambda}); \qquad (9.109)$$

adding (9.108) and (9.109) then yielded

$$s(m, n) = \frac{1}{4n\pi} \sum_{\lambda=1}^{n-1} \sigma(\lambda) \log \frac{1 - \rho^{\lambda}}{1 - \rho^{-\lambda}}. \qquad (9.110)$$

Next, note that since

$$-\pi < \mathrm{Im}\left( \log\left( \frac{1 - \rho^{\lambda}}{1 - \rho^{-\lambda}} \right) \right) < \pi$$

and

$$\frac{1 - \rho^{\lambda}}{1 - \rho^{-\lambda}} = -\rho^{\lambda} = -e^{\frac{2\pi i \lambda}{n}},$$

he arrived at

$$\log \frac{1 - \rho^{\lambda}}{1 - \rho^{-\lambda}} = \frac{2\pi i \lambda}{n} - \pi i$$

and

$$s(m, n) = \frac{i}{4n} \sum_{\lambda=1}^{n-1} \sigma(\lambda) \left( \frac{2\lambda}{n} - 1 \right). \tag{9.111}$$

Moreover, since $\sigma(0) = 0$,

$$\sum_{\lambda=1}^{n-1} \sigma(\lambda) = \sum_{\lambda=0}^{n-1} \sigma(\lambda)$$

$$= \sum_{l=1}^{n-1} \cot \frac{\pi l m}{n} \sum_{\lambda=0}^{n-1} \rho^{\lambda l} = 0.$$

Thus, by (9.107) and (9.111)

$$s(m, n) = \frac{i}{2n^2} \sum_{\lambda=1}^{n-1} \lambda \sigma(\lambda)$$

$$= -\frac{1}{2n^2} \sum_{l=1}^{n-1} \cot \frac{\pi l m}{n} \sum_{\lambda=1}^{n-1} \lambda \sin \frac{2\pi \lambda l}{n} \, ;$$

finally, by (9.102) he completed his proof of (9.98):

$$s(m, n) = \frac{1}{4n} \sum_{l=1}^{n-1} \cot \frac{n l m}{n} \cot \frac{\pi l}{n}. \tag{9.112}$$

To verify the reciprocity formula (9.94), Rademacher computed the integral

$$J = \frac{1}{2\pi i} \int_R \pi \cot(\pi z) \cot\left( \frac{\pi z}{n} \right) \cot\left( \frac{\pi m z}{n} \right) dz$$

in two different ways, over the rectangular contour $R$ joining the points $n - \epsilon + Ni$, $-\epsilon + Ni$, $-\epsilon - Ni$, $n - \epsilon - Ni$, where $0 < \epsilon < \frac{1}{m}$. To follow his calculations, first observe that the integrand has, inside $R$, a triple pole at $z = 0$ and has simple poles at $z = 1, 2, \ldots, n-1$ and $z = \frac{n}{m}, \frac{2n}{m}, \cdots, \frac{(m-1)n}{m}$. The residue at the pole $z = 0$ can be found by taking the Laurent expansion $\cot w = \frac{1}{w} - \frac{w}{3} \cdots$. Then

$$\pi \cot(\pi z) \cot\left( \frac{\pi z}{n} \right) \cot\left( \frac{\pi m z}{n} \right)$$

$$= \pi \left( \frac{1}{\pi z} - \frac{\pi z}{3} - \cdots \right) \left( \frac{n}{\pi z} - \frac{\pi z}{3n} - \cdots \right) \left( \frac{n}{\pi m z} - \frac{\pi m z}{3n} - \cdots \right).$$

The coefficient of $\frac{1}{z}$ in this expansion can then be seen to be

$$R_0 = -\frac{n^2}{3m} - \frac{1}{3m} - \frac{m}{3}. \tag{9.113}$$

The residue at a simple pole $z = l$ is $\cot \frac{\pi l}{n} \cot \frac{\pi l m}{n}$ and the sum of the residues at $z = 1, 2, \ldots, n - 1$ is therefore

$$\sum_{l=1}^{n-1} \cot \frac{\pi l}{n} \cot \frac{\pi l m}{n}.$$

Similarly, the residue at $z = \frac{jn}{m}$ is $\frac{n}{m} \cot \frac{\pi j}{m} \cot \frac{\pi jn}{m}$ and the sum of the residues at $z = \frac{n}{m}, \frac{2n}{m}, \ldots, \frac{(m-1)n}{m}$ is

$$\frac{n}{m} \sum_{j=1}^{m-1} \cot \frac{\pi j}{m} \cot \frac{\pi jn}{m}.$$

By Cauchy's residue theorem

$$J = \sum_{l=1}^{n-1} \cot \frac{\pi l}{n} \cot \frac{\pi l m}{n} + \frac{n}{m} \sum_{j=1}^{m-1} \cot \frac{\pi j}{m} \cot \frac{\pi jn}{m} + R_0, \tag{9.114}$$

where $R_0$ is given by (9.113).

Rademacher computed the value of the integral $J$ in another way. Begin by noting that the integral along the two sides or lines of $R$ parallel to the imaginary axis cancel one another, inasmuch as the value of the integrand is the same along each of those lines, but the lines run in opposite directions. Moreover, the value of $J$ remains unchanged if $N$ is increased and, indeed if we let $N \to \infty$. Thus

$$\frac{1}{2\pi i} \int_R = \frac{1}{2\pi i} \lim_{N \to \infty} \left( \int_{n-\epsilon+Ni}^{-\epsilon+Ni} + \int_{-\epsilon-Ni}^{n-\epsilon-Ni} \right). \tag{9.115}$$

Since $z = x + Ni$, with $x$ lying between $n - \epsilon$ and $-\epsilon$, in the first integral on the right-hand side of (9.115), and since

$$\cot z = i \frac{e^{iz} + e^{-iz}}{e^{iz} - e^{-iz}} \to -i \quad \text{as} \quad N \to \infty,$$

it follows that the contribution of the first integral on the right-hand side of (9.115) is

$$-\frac{n}{2i}(-i)^3 = -\frac{n}{2}. \tag{9.116}$$

Similarly, the contribution of the second integral on the right-hand side of (9.115) is

$$\frac{n}{2i}(i)^3 = -\frac{n}{2}; \tag{9.117}$$

hence, by (9.115), (9.116), and (9.117)

$$J = -n. \tag{9.118}$$

Taking (9.113), (9.114), and (9.118) together, we obtain

$$-n = \sum_{l=1}^{n-1} \cot \frac{\pi l}{n} \cot \frac{\pi l m}{n} + \frac{n}{m} \sum_{j=1}^{m-1} \cot \frac{\pi j}{m} \cot \frac{\pi jn}{m} - \frac{n^2}{3m} - \frac{1}{3m} - \frac{m}{3}. \tag{9.119}$$

Finally, dividing (9.119) by $4n$ and using (9.112) we arrive at (9.94), the reciprocity of Dedekind sums.

## 9.5   Exercises

1.  Show that when $c$ and $d$ are relatively prime integers,

$$\sum_{\sigma=1}^{c-1}\left(\left(\frac{\sigma d}{c}\right)\right)=\sum_{\sigma=1}^{c-1}\left(\left(\frac{\sigma}{c}\right)\right)=0,$$

where $((x))$ is defined by (9.6).

2.  Prove Knuth's formula (9.14) by reducing it to

$$\sigma(d,c,f)+\sigma(c,d,f)$$
$$=\sigma(d,c,0)+\sigma(c,d,0)+\frac{6f^2}{cd}-6\left[\frac{f}{d}\right]-3e(d,f)+3.$$

Note that $\sigma(d,c,0)=s(d,c)$. See Knuth (1998) p. 84.

3.  Use (3.160) through (3.162) to demonstrate that

$$\log k=\log 4\sqrt{q}+\sum_{n=1}^{\infty}(-1)^n\frac{4q^n}{n(1+q^n)} \tag{9.120}$$

$$-\log k'=8\sum_{n=1}^{\infty}\frac{q^{2n-1}}{(2n-1)(1-q^{4n-2})} \tag{9.121}$$

$$\log\frac{2K}{\pi}=4\sum_{n=1}^{\infty}\frac{q^{2n-1}}{(2n-1)(1+q^{2n-1})}. \tag{9.122}$$

Riemann studied the behavior of functions (9.120) through (9.122) as $\omega\to\frac{m}{n}$, where gcd $(m,n)=1$, and $q=e^{i\pi\omega}$. These three formulas were derived by Jacobi in his *Fundamenta Nova*; see Jacobi (1969) vol. 1, p. 159.

4.  Let $\omega_{h,k}$ be a $24k$th root of unity, with $h$ and $k$ relatively prime integers, given for odd $h$ by

$$\omega_{h,k}=\left(-\frac{k}{h}\right)\exp\left(-\pi i\left(\frac{1}{4}(2-hk-h)+\frac{1}{12}\left(k-\frac{1}{k}\right)(2h-h'+h^2h')\right)\right)$$

and for odd $k$ by

$$\omega_{h,k}=\left(-\frac{k}{h}\right)\exp\left(-\pi i\left(\frac{1}{4}(k-1)+\frac{1}{12}\left(k-\frac{1}{k}\right)(2h-h'+h^2h')\right)\right),$$

where $\left(\frac{a}{b}\right)$ denotes the Jacobi symbol and $h'$ is any solution of

$$hh'\equiv-1\pmod{k}.$$

Hardy and Ramanujan gave the expression[30]

$$A_k(n) = \sum_{\substack{h \bmod k \\ (h,k)=1}} \omega_{h,k} e^{-\frac{2\pi i h n}{k}}.$$

Show that this expression is equivalent to Rademacher's expression in (9.26). See Rademacher (1974) vol. 2, pp. 3–8.

5. Prove Selberg's formula

$$A_k(n) = \sqrt{\frac{k}{3}} \sum_{\substack{l(\bmod 2k) \\ \frac{3l^2-l}{2} \equiv -n(\bmod k)}} (-1)^l \cos \frac{6l-1}{6k} \pi.$$

See Berndt and Rankin (2001) p. 213. For a proof, see Rademacher (1974) vol. 2, pp. 460–474.

6. Check that $\sum_{l=1}^{n-1} \cot \frac{\pi l m}{n} = 0$ when $m$ and $n$ are relatively prime integers.

7. Use (9.55) and (9.56) to prove Dedekind's result that when $cd \geq 0$,

$$(d, c+d) = (d, c) - (c, d) + d - c.$$

8. Show that the reciprocity formula for Dedekind sums defined by (9.98) can be proved by integrating the function

$$F(z) = \pi \cot \pi z \cot \pi m z \cot \pi n z$$

around a contour obtained from a rectangle with vertices $1 + iY, iY, -iY, 1 - iY$, $Y > 0$, by making two small, semicircular, parallel indentations so that $z = 0$ is kept inside and $z = 1$ is kept outside the contour. Let $Y \to \infty$ to get the result. See Rademacher and Grosswald (1972) pp. 21–22.

9. Read and verify Rademacher's proof of (9.23) in J. Reine Angew. Math <u>167</u> (1931), pp. 312–336 or in Rademacher (1974) vol. 1, pp. 652–676.

---

[30] Ramanujan (2000) p. 284.

# 10

---

# *Modular Forms and Invariant Theory*

## 10.1  Preliminary Remarks

Invariant theory played an important role in the development of modular functions; tracing the beginnings of invariant theory from 1840 to 1855 shows how George Boole defined an invariant and gives insight into the methods used by Boole, Arthur Cayley, and J. J. Sylvester to compute basic invariants, including those used in modular function theory. In the early 1840s, Boole and Cayley determined that the modular forms $g_2$ and $g_3$ were algebraic invariants of the quartic

$$P(x) = a_0x^4 + 4a_1x^3 + 6a_2x^2 + 4a_3x + a_4.$$

Boole was the son of a cobbler and did not receive a college education, but, like his father, he was greatly interested in science and mathematics. After opening his own school in 1834, he independently studied the works of Laplace and Lagrange. His study of Lagrange led him to the discovery of algebraic invariance. He found that if a linear fractional transformation $y = \frac{ax+b}{cx+d}$, with $ad - bc \neq 0$, was applied to

$$f(x) = a_0x^n + \binom{n}{1}a_1x^{n-1} + \binom{n}{2}a_2x^{n-2} + \cdots + \binom{n}{n}a_n, \tag{10.1}$$

and the result was

$$g(x) = b_0x^n + \binom{n}{1}b_1x^{n-1} + \cdots + \binom{n}{n}b_n = (cx + d)^n f(y), \tag{10.2}$$

then there must exist homogeneous polynomials in the coefficients $G(a_0, a_1, \ldots, a_n)$ such that for some integer $\delta$

$$G(b_0, b_1, \ldots, b_n) = (ad - bc)^\delta G(a_0, a_1, \ldots, a_n). \tag{10.3}$$

In particular, Boole's method produced an invariant that he denoted by $\theta(f)$, related to the discriminant of the polynomial $f$. The polynomials $G(a_0, a_1, \ldots, a_n)$ are called algebraic invariants. In the paper[1] containing the statement of his discovery of these

---

[1] Boole (1841).

276

invariants, Boole accurately predicted, "To those who may be disposed to engage in the investigation [of questions related to invariants], it will, I believe, present an ample field of research and discovery."

Cayley very soon took up this challenge and found an invariant for the quartic polynomial

$$g_2 = a_0 a_4 - 4a_1 a_3 + 3a_2^2. \tag{10.4}$$

Boole then communicated to Cayley another invariant for the quartic:

$$g_3 = a_0 a_2 a_4 - a_1^2 a_4 - a_0 a_3^2 - a_2^3 + 2a_1 a_2 a_3. \tag{10.5}$$

He also wrote to Cayley that for a quartic $P$, he had found by trial and error that

$$\theta(P) = g_2^3 - 27 g_3^2. \tag{10.6}$$

It can be shown that the value of $\delta$ in (10.3) for $g_2$, $g_3$, and $\theta(f)$, respectively, is 4, 6, and 12. It follows that $\delta = 0$ for the rational function

$$J = \frac{g_2^3}{\Delta} = \frac{g_2^3}{g_2^3 - 27 g_3^2}, \tag{10.7}$$

where $\Delta$ denotes Boole's $\theta(P)$. Note that $J$ is called an absolute invariant because $\delta = 0$ and because the value of $J$ remains the same when the quartic $P$ is multiplied by a constant. To see a connection of invariant theory and elliptic functions, consider the elliptic differential $\frac{dx}{\sqrt{P(x)}}$. When a fractional linear transformation $y = \frac{ax+b}{cx+d}$, with $m = ad - bc \neq 0$, is applied to this differential, we obtain

$$\frac{dx}{\sqrt{P(x)}} = m^{-1} \frac{dy}{\sqrt{Q(y)}}, \tag{10.8}$$

where

$$(cx + d)^4 Q\left(\frac{ax+b}{cx+d}\right) = P(x).$$

In an 1828 paper,[2] Abel presented the different values of $k^2$ that resulted when a linear fractional transformation $y = \frac{ax+b}{cx+d}$, with $m = ad - bc \neq 0$, produced the Legendre form of an elliptic integral

$$\frac{dx}{\sqrt{P(x)}} = m^{-1} \frac{dy}{\sqrt{(1 - y^2)(1 - k^2 y)}}. \tag{10.9}$$

Abel found that the six different values of $k^2$ could be represented in terms of any one of them:

$$k^2, \quad \frac{1}{k^2}, \quad \left(\frac{1+\sqrt{k}}{1-\sqrt{k}}\right)^4, \quad \left(\frac{1-\sqrt{k}}{1+\sqrt{k}}\right)^4, \quad \left(\frac{i+\sqrt{k}}{i-\sqrt{k}}\right)^4, \quad \left(\frac{i-\sqrt{k}}{i+\sqrt{k}}\right)^4. \tag{10.10}$$

A detailed proof of this theorem was presented in 1846 by Cayley. Arthur Cayley (1821–1895) studied at Cambridge University and was a prolific mathematician who

[2] Abel (1965) vol. 1, pp. 457–465, especially pp. 458–459.

contributed to many areas of mathematics, including theories of invariants, partitions, elimination, non-Euclidean geometry, and groups. Some of his contributions to elimination theory and on hyperdeterminants have gained prominence in the work of I. M. Gelfand and his collaborators. Concerning Cayley's 1848 paper "On the Theory of Elimination," they have written,[3]

> In his breathtaking 1848 note ... he outlined a general method of writing down the resultant of several polynomials in several variables. We were very surprised to find that Cayley introduced in this note several fundamental concepts of homological algebra: complexes, exactness, Koszul complexes, and even the invariant now sometimes called the Whitehead torsion or Reidemeister–Franz torsion of an exact complex. The latter invariant is a natural generalization of the determinant of a square matrix (which itself was a rather recent discovery back in 1848!), so we prefer to call it the determinant of a complex. Using this terminology, Cayley's main result is that the resultant is the determinant of the Koszul complex.

Cayley's work on invariants was closely intertwined with that of his older colleqgue James Joseph Sylvester (1814–1897). Both Cayley and Sylvester practiced law in London for many years before taking academic positions. Sylvester is particularly noted for his elaboration of the graphical method in the theory of partitions and for creating much of the older terminology in the theory of invariants.

To prove Abel's result (10.10), Cayley showed that

$$J = \frac{(k^4 + 14k^2 + 1)^3}{108k^2(1 - k^2)^4}. \tag{10.11}$$

Thus, for a given value of $J$, $k^2$ would satisfy an equation of degree 6; Cayley proved that the six roots were given by (10.10).

In an 1877 letter[4] to F. Brioschi, Felix Klein pointed out the role of $J$, given in (10.11), in the theory of the quintic equation. In his paper on the transformation of elliptic functions[5] he gave more details and showed that when an elliptic integral $\int \frac{dx}{\sqrt{P(x)}}$ is transformed into the Riemann normal form $\int \frac{dx}{\sqrt{x(1-x)(1-k^2x)}}$ by means of a fractional linear transformation, then

$$g_2 = \frac{k^4 - k^2 + 1}{12},$$

$$\Delta = \frac{k^4(1 - k^2)^2}{256},$$

$$J = \frac{4(k^4 - k^2 + 1)^3}{27k^4(1 - k^2)^2}.$$

Klein observed that $\delta = 1$ for $\sqrt[4]{g_2}$ and $\sqrt[12]{\Delta}$. Moreover, he noted that since the periods $\omega_1$, $\omega_2$ of an elliptic function could be expressed in terms of the elliptic integral, it followed from (10.9) that $\omega_1$, $\omega_2$ could be viewed as transcendental invariants associated

---

[3] Gelfand et al. (1994) p. 4.     [4] Klein (1921–1923) vol. 3, pp. 10–12.
[5] Ibid. pp. 13–75, especially pp. 18–20.

with $P(x)$ with $\delta = -1$. Their transcendental nature was a consequence of the many-valuedness of $\omega_1$, $\omega_2$ because $\omega_1'$, $\omega_2'$ could be taken to be the fundamental periods of

$$\omega_1' = m\,\omega_1 + n\,\omega_2,$$
$$\omega_2' = p\,\omega_1 + q\,\omega_2,$$

with $mq - np = 1$. The periods are chosen in the order $\omega_1$, $\omega_2$ so that $\text{Im }\frac{\omega_1}{\omega_2} > 0$. This order is preserved in the transformed periods $\omega_1'$, $\omega_2'$. Now $\omega_i\sqrt[4]{g_2}$ and $\omega_i\sqrt[12]{\Delta}$ may be taken to be absolute invariants because $\delta = 0$ when these quantities are linearly transformed. Klein used this fact and combined it with formulas in Jacobi's *Fundamenta Nova* to obtain the key formulas

$$\Delta = \left(\frac{2\pi}{\omega_2}\right)^{12} q^2 \prod_{n=1}^{\infty}(1 - q^{2n})^{24}, \quad q = e^{\pi \omega i}, \quad \omega = \frac{\omega_1}{\omega_2}$$

and

$$g_2 = \left(\frac{2\pi}{\omega_2}\right)^4 \frac{1}{12}\left(1 + 240\sum_{n-1}^{\infty}\frac{n^3 q^{2n}}{1 - q^{2n}}\right).$$

Felix Klein had an extremely broad knowledge of pure and applied mathematics, reflecting and contributing to his efforts to search for themes that would unify disparate disciplines within mathematics. His work on invariant theory and modular functions reflects his ability to perceive and pursue such themes. His 1884 work, *Vorlesungen über das Ikosaeder und die Auflösung der Gleichung fünften Grades*, developed connections among group theory, invariants, modular forms, and the theory of equations. His famous correspondence with Henri Poincaré during 1881–1882 led to the founding of the theory of automorphic functions. His breadth of knowledge and his talent as a teacher contributed greatly to all levels of mathematics instruction in Germany.

## 10.2 The Early Theory of Invariants

In his 1841 paper, "Exposition of a General Theory of Linear Transformation,"[6] Boole sketched out a definition of invariance through examples. His method of producing invariants of algebraic forms in many variables may perhaps be best understood by means of his simplest examples. Boole let

$$Q = ax_1^2 + 2bx_1x_2 + cx_2^2 \tag{10.12}$$

be a binary quadratic form and then set the partial derivatives of $Q$ with respect to $x_1$ and $x_2$ equal to zero, thereby eliminating $x_1$ and $x_2$:

$$2ax_1 + 2bx_2 = 0 \quad \text{and} \quad 2bx_1 + 2cx_2 = 0$$

so that

$$\theta(Q) = b^2 - ac = 0.$$

[6] Boole (1841).

He then applied a linear transformation

$$x_1 = py_1 + qy_2, \quad x_2 = ry_1 + sy_2 \tag{10.13}$$

where $p, q, r, s$ were constants, with the determinant $ps - qr \neq 0$, to obtain a new quadratic form

$$R = Ay_1^2 + 2By_1y_2 + Cy_2^2. \tag{10.14}$$

Boole showed that

$$\theta(R) = B^2 - AC = (ps - qr)^2(b^2 - ac) = \delta^2\theta(Q), \tag{10.15}$$

where $\delta$ denoted the determinant of the linear transformation (10.13). Now (10.15) is the result of an easy calculation, made even simpler if we employ the properties of matrices. The matrix notation was not available to Boole; Cayley invented it in 1858. But in terms of matrices, observe that

$$Q = (x_1 x_2) \begin{pmatrix} a & b \\ b & c \end{pmatrix} \begin{pmatrix} x_1 \\ x_2 \end{pmatrix}$$

$$\begin{pmatrix} x_1 \\ x_2 \end{pmatrix} = \begin{pmatrix} p & q \\ r & s \end{pmatrix} \begin{pmatrix} y_1 \\ y_2 \end{pmatrix},$$

so that

$$Q = (y_1 \ y_2) \begin{pmatrix} p & r \\ q & s \end{pmatrix} \begin{pmatrix} a & b \\ b & c \end{pmatrix} \begin{pmatrix} p & q \\ r & s \end{pmatrix} \begin{pmatrix} y_1 \\ y_2 \end{pmatrix}$$

$$= (y_1 \ y_2) \begin{pmatrix} A & B \\ C & D \end{pmatrix} \begin{pmatrix} y_1 \\ y_2 \end{pmatrix}.$$

It follows that

$$\begin{pmatrix} A & B \\ C & D \end{pmatrix} = \begin{pmatrix} p & r \\ q & s \end{pmatrix} \begin{pmatrix} a & b \\ b & c \end{pmatrix} \begin{pmatrix} p & q \\ r & s \end{pmatrix};$$

taking the determinant of each side, we get (10.15). More generally, Boole considered a homogeneous polynomial $Q$ of degree $n$ in $m$ variables. By $\theta(Q)$ he denoted the expression obtained by eliminating the variables $x_1, x_2, \ldots, x_m$ from the polynomials

$$\frac{\partial Q}{\partial x_1}, \frac{\partial Q}{\partial x_2}, \ldots, \frac{\partial Q}{\partial x_m}. \tag{10.16}$$

To eliminate a variable $x$ from two polynomials of degree $n$, Boole employed the Euclidean algorithm. Observe that in (10.16) he initially had $m$ polynomials in $m$ variables. He could eliminate a variable, say $x_1$, from the first two polynomials, then eliminate the same variable from the second and third, and so on until he had $m - 1$ polynomials in $m - 1$ variables. Using this procedure, all the variables could eventually be eliminated and he was left with an expression only the coefficients in $Q$: $\theta(Q)$. Boole applied a linear transformation (here given in the later matrix notation)

$$\begin{pmatrix} x_1 \\ x_2 \\ \vdots \\ x_m \end{pmatrix} = \begin{pmatrix} a_{11} & a_{12} & \cdots & a_{1m} \\ a_{21} & a_{22} & \cdots & a_{2m} \\ \vdots & \vdots & \vdots & \vdots \\ a_{m1} & a_{m2} & \cdots & a_{mm} \end{pmatrix} \begin{pmatrix} y_1 \\ y_2 \\ \vdots \\ y_m \end{pmatrix} \tag{10.17}$$

to convert $Q$ into a homogeneous polynomial $R$ of degree $n$ in the $m$ variables $y_1, y_2,$ $\ldots, y_m$. He then proved that if $\theta(R)$ denoted the result of eliminating $y_1, y_2, \ldots, y_m$ from

$$\frac{\partial R}{\partial y_1}, \frac{\partial R}{\partial y_2}, \ldots, \frac{\partial R}{\partial y_m}$$

and $\delta$ denoted the determinant of the matrix in (10.17), then

$$\theta(R) = \delta^{\frac{\gamma n}{m}} \theta(Q) \tag{10.18}$$

with $\gamma$ representing the degree of $\theta(R)$ and $\theta(Q)$.

In his 1841 paper, Boole did not prove this result, but verified it for a few special cases. A few years later, he published an indication of a proof[7] in which he wrote that

$$\gamma = m(n-1)^{m-1}. \tag{10.19}$$

Boole's result (10.18) certainly suggests the definition of an invariant, for which we here provide the definition only for the case of polynomials of two variables. Cayley and Sylvester considered mainly this case and this is the case needed for the theory of elliptic functions where the polynomial is of degree 4. Thus, let

$$f(x_1, x_2) = a_0 x_1^n + \binom{n}{1} a_1 x_1^{n-1} x_2 + \binom{n}{2} a_2 x_1^{n-2} x_2^2 + \cdots + a_n x_2^n. \tag{10.20}$$

Suppose that the linear transformation

$$\begin{pmatrix} x_1 \\ x_2 \end{pmatrix} = \begin{pmatrix} s_{11} & s_{12} \\ s_{21} & s_{22} \end{pmatrix} \begin{pmatrix} y_1 \\ y_2 \end{pmatrix} \tag{10.21}$$

with determinant $\delta = s_{11}s_{22} - s_{12}s_{21} \neq 0$ converts $f(x_1, x_2)$ into

$$A_0 y_1^n + \binom{n}{1} A_1 y_1^{n-1} y_2 + \binom{n}{2} A_2 y_1^{n-2} y_2^2 + \cdots + A_n y_2^n. \tag{10.22}$$

An invariant $I$ of $f(x_1, x_2)$ is then a polynomial in the coefficients $a_0, a_1, \ldots, a_n$; this invariant is denoted by $I(a_0, a_1, \ldots, a_n)$ such that for some integer $p$

$$I(A_0, A_1, \ldots, A_n) = \delta^p I(a_0, a_1, \ldots, a_n), \tag{10.23}$$

where $A_0, A_1, \ldots, A_n$ are given by (10.22). It can be shown, by taking both $s_{12}$ and $s_{21}$ to be zero, that $I(a_0, a_1, \ldots, a_n)$ is a homogeneous polynomial in $a_0, a_1, \ldots, a_n$. Moreover, if we assign the weight $k$ to the coefficient $a_k$, then each term of $I$ in $a_0, a_1, \ldots, a_n$ has weight $p$ and the invariant $I$ is said to have weight $p$; note that $p$ is the power of

----

[7] Boole (1844).

the discriminant $\delta$ in (10.23). It follows from (10.15) that $b^2 - ac$ is an invariant of $ax_1^2 + 2bx_1x_2 + cx_2^2$ and that it has degree 2 and weight 2. Similarly, the weight of $\theta(Q)$ in (10.18) is $n(n-1)^{m-1}$, while its degree is $m(n-1)^{m-1}$.

At the end of the second part of his 1841 paper, Boole wrote that invariant theory should prove to be a fruitful area of research. Cayley read this paper in 1844 and wholeheartedly agreed. In 1843, Cayley had published a paper[8] in which he introduced the idea of a multidimensional determinant or hyperdeterminant. After reading Boole's paper, Cayley used hyperdeterminants to show that[9]

$$g_2 = a_0a_4 - 4a_1a_3 + 3a_2^2 \tag{10.24}$$

was an invariant of the binary quartic

$$u \equiv a_0x_1^4 + 4a_1x_1^3x_2 + 6a_2x_1^2x_2^2 + 4a_3x_1x_2^3 + a_4x_2^4. \tag{10.25}$$

Note that (10.24) is an invariant of degree 2 and weight 4. Cayley communicated this result to Boole who pointed out that the quartic also had an invariant of degree 3 (and, we add, weight 6):

$$g_3 = a_0a_2a_4 - a_1^2a_4 - a_0a_3^2 - a_2^3 + 2a_1a_2a_3. \tag{10.26}$$

Cayley then wrote a paper deriving these two invariants by a different method. He also mentioned that Boole had communicated to him a result arrived at by trial and error:

$$\theta(u) = g_2^3 - 27g_3^2, \tag{10.27}$$

where $\theta(u)$ represented Boole's invariant in (10.18) when $Q$ is replaced by $u$ in (10.25). We note here that we have used Weierstrass's notation, $g_2$ and $g_3$, to denote the invariants in (10.24) and (10.26); see equation (4.100) in Chapter 4. By doing this, we can make it clear from the outset the significance of Klein's 1878 remark that the modular forms $g_2$ and $g_3$ were invariants of the quartic defining an elliptic integral. Also note that $g_2^3 - 27g_3^2$ is the discriminant of the quartic.

It is not difficult to check that (10.24), (10.26), and (10.27) are invariants of the quartic. It is also not difficult to show that invariants are homogeneous polynomials in which each term is of the same weight. Finally, to check whether a given expression is an invariant, first observe that any linear transformation is composed of the three simpler transformations

$$\begin{pmatrix} x_1 \\ x_2 \end{pmatrix} = \begin{pmatrix} c & 0 \\ 0 & d \end{pmatrix} \begin{pmatrix} y_1 \\ y_2 \end{pmatrix} \tag{10.28}$$

$$\begin{pmatrix} x_1 \\ x_2 \end{pmatrix} = \begin{pmatrix} 1 & e \\ 0 & 1 \end{pmatrix} \begin{pmatrix} y_1 \\ y_2 \end{pmatrix} \tag{10.29}$$

$$\begin{pmatrix} x_1 \\ x_2 \end{pmatrix} = \begin{pmatrix} 1 & 0 \\ f & 1 \end{pmatrix} \begin{pmatrix} y_1 \\ y_2 \end{pmatrix}. \tag{10.30}$$

[8] Cayley (1889–1898) vol. 1, pp. 63–79.    [9] Cayley (1889–1898) vol. 1, pp. 80–94.

Any expression invariant with respect to each of these simpler transformations must then be invariant with respect to the general linear transformation. These three transformations can also be used to prove general properties of invariants. For example, the special transformation (10.28) implies that invariants must be homogeneous and isobaric (each term having the same weight). Therefore, invariants must have this property. Applying (10.28) to (10.20), we obtain

$$a_0 c^n y_1^n + \binom{n}{1} a_1 c^{n-1} d y_1^{n-1} y_2 + \cdots + \binom{n}{k} a_k c^{n-k} d^k y_1^{n-k} y_2^k + \cdots + a_n d^n y_2^n$$

$$= A_0 y_1^n + \binom{n}{1} A_1 y_1^{n-1} y_2 + \cdots + \binom{n}{k} A_k y_1^{n-k} y_2^k + \cdots + A_n y_2^n.$$

Thus, $A_0 = a_0 c^n$, $A_1 = a_1 c^{n-1} d$, ... and $A_k = a_k c^{n-k} d^k$, ..., $A_n = a_n d^n$.

Now let

$$I(a_0, a_1, \ldots, a_n) = \sum \alpha_{l_0, l_1, \ldots, l_n} a_0^{l_0} a_1^{l_1} \cdots a_n^{l_n}$$

be an invariant of (10.20). Then $\delta = cd$ is the determinant of (10.28) and by (10.23)

$I(A_0, A_1, \ldots, A_n)$

$\quad = I(a_0 c^n, a_1 c^{n-1} d, \ldots, a_n d^n)$

$\quad = \sum \alpha_{l_0 l_1 \cdots, l_n} a_0^{l_0} c^{n l_0} a_1^{l_1} c^{(n-1) l_1} d^{l_1} \cdots a_k^{l_k} c^{(n-k) l_k} d^{k l_k} \cdots a_n^{l_n} d^{n l_n}$

$\quad = \sum \alpha_{l_0 l_1 \cdots, l_n} a_0^{l_0} a_1^{l_1} \cdots a_n^{l_n} c^{n l_0 + (n-1) l_1 + \cdots + (n-k) l_k + \cdots + l_{n-1}} d^{l_1 + \cdots + k l_k + \cdots + n l_n} \qquad (10.31)$

$\quad = \delta^p I(a_0, a_1, \ldots, a_n) = c^p d^p \sum \alpha_{l_0 l_1 \cdots, l_n} a_0^{l_0} a_1^{l_1} \cdots a_n^{l_n}. \qquad (10.32)$

Equating the coefficients of $a_0^{l_0} a_1^{l_1} \cdots a_n^{l_n}$ in (10.31) and (10.32) yields

$$n l_0 + (n-1) l_1 + \cdots + (n-k) l_k + \cdots + l_{n-1} = p \qquad (10.33)$$

$$l_1 + 2 l_2 + \cdots + k l_k + \cdots + n l_n = p. \qquad (10.34)$$

Adding (10.33) and (10.34), we arrive at

$$n(l_0 + l_1 + \cdots + l_n) = 2p. \qquad (10.35)$$

Observe that (10.35) implies that an invariant $I$ is homogeneous of degree

$$\theta = l_0 + l_1 + \cdots + l_n = \frac{2p}{n}$$

and equation (10.34) reveals that it is isobaric of weight

$$0 \cdot l_0 + 1 \cdot l_1 + 2 \cdot l_2 + \cdots + n \cdot l_n = p.$$

More difficult than determining the elementary properties of invariants is developing a systematic procedure to produce invariants. In December 1851, Cayley wrote Sylvester that, in general, invariants satisfied two partial differential equations, without indicating a derivation of the equations. Sylvester gave a derivation for the differential equations and these can be used to show that, for example, (10.24) and (10.26) are invariants of the binary quartic (10.25). In his 1852 paper, "On the Principles of the Calculus of

Forms,"[10] Sylvester first noted that when transformation (10.29) was applied to (10.20), then

$$a_0(y_1 + ey_2)^n + \binom{n}{1} a_1(y_1 + ey_2)^{n-1} y_2 + \cdots$$

$$+ \binom{n}{k} a_k(y_1 + ey_2)^{n-k} y_2^k + \cdots + a_n y_2^n$$

$$= A_0 y_1^n + \binom{n}{1} A_2 y_1^{n-1} y_2 + \cdots + \binom{n}{k} A_k y_1^{n-k} y_2^k + \cdots + A_n y_2^n,$$

where

$$A_k = a_k + \binom{k}{1} e a_{k-1} + \binom{k}{2} e^2 a_{k-2} + \cdots + e^k a_0, \quad k = 0, 1, 2, \ldots, n. \quad (10.36)$$

Since the determinant of transformation (10.29) was 1, Sylvester had

$$I(A_0, A_1, \ldots, A_n) = I(a_0, a_1, \ldots, a_n). \quad (10.37)$$

Next, note that (10.36) implies that

$$\frac{\partial A_k}{\partial e} = k \left( a_{k-1} + \binom{k-1}{1} a_{k-2} + \cdots + e^{k-1} a_0 \right)$$

$$= k A_{k-1}, \quad \text{with} \quad k = 1, 2, \ldots, n.$$

Thus, when Sylvester took the derivative of (10.37) with respect to $e$, he had

$$0 = \frac{d}{de} I(A_0, A_1, \ldots, A_n)$$

$$= \frac{\partial I}{\partial A_0} \frac{\partial A_0}{\partial e} + \frac{\partial I}{\partial A_1} \frac{\partial A_1}{\partial e} + \cdots + \frac{\partial I}{\partial A_n} \frac{\partial A_n}{\partial e}$$

$$= A_0 \frac{\partial I}{\partial A_1} + 2A_1 \frac{\partial I}{\partial A_2} + \cdots + nA_{n-1} \frac{\partial I}{\partial A_n}.$$

$I$ was therefore a solution of the differential equation

$$\Omega I = \left( A_0 \frac{\partial}{\partial A_1} + 2A_1 \frac{\partial}{\partial A_2} + \cdots + nA_{n-1} \frac{\partial}{\partial A_n} \right) I = 0.$$

Note that instead of (10.20), we could have started with

$$A_0 y_1^n + \binom{n}{1} A_1 y_1^{n-1} y_2 + \cdots + A_n y_2^n$$

and moved to the form (10.20) by taking the inverse of (10.29). So we can see that $\Omega$ can be taken to be

$$\Omega = a_0 \frac{\partial}{\partial a_1} + 2a_1 \frac{\partial}{\partial a_2} + \cdots + na_{n-1} \frac{\partial}{\partial a_n}. \quad (10.38)$$

[10] Sylvester (1973) vol. 1, pp. 328–363, especially pp. 352–353.

In a similar way, by applying (10.30), Sylvester had the operator

$$O = na_1 \frac{\partial}{\partial a_0} + (n-1)a_2 \frac{\partial}{\partial a_1} + \cdots + a_n \frac{\partial}{\partial a_{n-1}} \qquad (10.39)$$

such that, if $I$ was invariant, then

$$OI = 0.$$

We now show how the invariants $g_2$ and $g_3$ can be obtained using the differential operators. For example, since $g_2$ is of degree 2, by (10.35), it is of weight 4. There are three monomials of degree 2 and weight 4: $a_0a_4$, $a_1a_3$, $a_2^2$. Now let $I$ be an invariant defined by

$$I = Aa_0a_4 + Ba_1a_3 + Ca_2^2. \qquad (10.40)$$

We may find $A$, $B$, $C$ by solving the equation $\Omega I = 0$. Observe that

$$\Omega a_0 a_4 = 4a_0 a_3, \quad \Omega a_1 a_3 = a_0 a_3 + 3a_1 a_2, \quad \Omega a_2^2 = 4a_1 a_2.$$

Thus, for $I$ in (10.40), $\Omega I = (4A + B)a_0a_3 + (3B + 4C)a_1a_2 = 0$. This implies that $B = -4A$ and $C = 3A$ and that

$$I = A(a_0a_4 - 4a_1a_3 + 3a_2^2) = Ag_2,$$

so that the invariant $g_2$ has been produced. Observe that $g_3$ may be obtained in a similar manner. Note that $g_2$ and $g_3$ are of weights 4 and 6, respectively, so that

$$g_2(A_0, A_1, A_2, A_3, A_4) = \delta^4 g_2(a_0, a_1, a_2, a_3, a_4) \qquad (10.41)$$

$$g_3(A_0, A_1, A_2, A_3, A_4) = \delta^6 g_3(a_0, a_1, a_2, a_3, a_4) \qquad (10.42)$$

and

$$\frac{g_2^3}{g_2^3 - 27g_3^2}(A_0, A_1, A_2, A_3, A_4) = \frac{g_2^3}{g_2^3 - 27g_3^2}(a_0, a_1, a_2, a_3, a_4).$$

Cayley called $\frac{g_2^3}{g_2^3 - 27g_3^2}$ an absolute invariant, because the power of $\delta$ was zero. Klein denoted this absolute invariant by $J$.

## 10.3 Cayley's Proof of a Result of Abel

In an 1828 paper, "Sur les nombre des transformations différentes...,"[11] Abel stated a result on the number of different values taken by $k^2$ when a given elliptic integral $\int \frac{dy}{\sqrt{P(y)}}$, with $P(y)$ a polynomial of degree 4, was transformed into the Legendre normal form

$$\int \frac{dx}{\sqrt{(1-x^2)(1-qx^2)}} \qquad (10.43)$$

[11] Abel (1965) vol. 1, pp. 457–465, especially p. 459.

by means of the fractional linear transformation

$$y = \frac{ex + f}{gx + h}, \quad \text{where} \quad \delta = eh - fg \neq 0.$$

In a paper of 1846,[12] Cayley applied invariant theory to establish Abel's result. Cayley proved that, given a value of

$$J = \frac{g_2^3}{g_2^3 - 27g_3^2},$$

there were six different values of $\beta^4 = q$:

$$\beta^4, \quad \frac{1}{\beta^4}, \quad \left(\frac{1-\beta}{1+\beta}\right)^4, \quad \left(\frac{1+\beta}{1-\beta}\right)^4, \quad \left(\frac{1-\beta i}{1+\beta i}\right)^4, \quad \left(\frac{1+\beta i}{1-\beta i}\right)^4.$$

Cayley started his proof by noting that if $P(x_1, x_2)$ denoted the homogeneous quartic (10.25), and the linear transformation (10.21) was applied to this quartic, obtaining $P'(y_1, y_2)$, then

$$\frac{x_1 dx_2 - x_2 dx_1}{\sqrt{P(x_1, x_2)}} = \frac{\delta(y_1 dy_2 - y_2 dy_1)}{\sqrt{P'(y_1, y_2)}}, \tag{10.44}$$

where $\delta$ represented the determinant of the linear transformation. He then set

$$x = \frac{x_2}{x_1}, \quad y = \frac{y_2}{y_1}$$

so that he could rewrite (10.44) as

$$\frac{dx}{\sqrt{P(1, x)}} = \delta \frac{dy}{\sqrt{P'(1, y)}}. \tag{10.45}$$

He then supposed that

$$
\begin{aligned}
P'(1, y) &= A_0(1 - y^2)(1 - qy^2) \\
&= A_0(1 - (1+q)y^2 + qy^4).
\end{aligned}
$$

Now in the notation of the previous section

$$A_0 = A_0, \quad A_1 = 0, \quad 6A_2 = -A_0(1+q), \quad A_3 = 0, \quad A_4 = A_0 q.$$

Thus, from (10.24) and (10.26), Cayley had

$$g_2(A_0, A_1, A_2, A_3, A_4) = \frac{1}{12} A_0^2 (1 + 14q + q^2)$$

$$g_3(A_0, A_1, A_2, A_3, A_4) = \frac{1}{216} A_0^3 (1 + q)(1 - 34q + q^2)$$

and

$$J = \frac{(q^2 + 14q + 1)^3}{108q(1 - q)^4}. \tag{10.46}$$

---

[12] Cayley (1889–1898) vol. 1, pp. 224–227.

This demonstrated that, given a value of $J$, the values of $q$ could be obtained from a sixth degree equation. If $q = \beta^4$ were a value of $q$ so that

$$\frac{(q^2 + 14q + 1)^3}{q(1 - q)^4} = \frac{(\beta^8 + 14\beta^4 + 1)^3}{\beta^4(1 - \beta^4)^2}, \tag{10.47}$$

then

$$q = \left(\frac{1 - \beta}{1 + \beta}\right)^4 \quad \text{and} \quad q = \left(\frac{1 - \beta i}{1 + \beta i}\right)^4$$

would also satisfy (10.47). Observe that (10.46) remains unchanged when $q$ is replaced by $\frac{1}{q}$; thus for a given $J$, the six different values of $q$ would be

$$\beta^4, \quad \frac{1}{\beta^4}, \quad \left(\frac{1 - \beta}{1 + \beta}\right)^4, \quad \left(\frac{1 + \beta}{1 - \beta}\right)^4, \quad \left(\frac{1 - \beta i}{1 + \beta i}\right)^4, \quad \left(\frac{1 + \beta i}{1 - \beta i}\right)^4,$$

proving Abel's result.

## 10.4   Reduction of an Elliptic Integral to Riemann's Normal Form

In his 1855–1856 lectures on elliptic and Abelian functions, Riemann introduced the normal form for an elliptic integral:

$$\int \frac{dz}{\sqrt{z(1 - z)(1 - k^2 z)}}. \tag{10.48}$$

Riemann's lectures were published in 1899 by Hermann Stahl and contain his method for reducing a general elliptic integral $\int \frac{dx}{\sqrt{P(x)}}$, where $P(x)$ is a polynomial of degree four with distinct roots, to the form given in (10.48). A similar treatment[13] was presented by Weierstrass in his Berlin lectures, given in approximately 1862. For Riemann's method,[14] let

$$P(x) = a_0(x - c_1)(x - c_2)(x - c_3)(x - c_4) \tag{10.49}$$

and set

$$z = \frac{x - c_1}{x - c_4} \frac{c_2 - c_4}{c_2 - c_1} \tag{10.50}$$

so that the fractional linear transformation (10.50) will map $x = c_1, c_2, c_3, c_4$ to $z = 0, 1, \frac{1}{k^2}, \infty$, where

$$\frac{1}{k^2} = \frac{c_3 - c_1}{c_3 - c_4} \cdot \frac{c_2 - c_4}{c_2 - c_1}. \tag{10.51}$$

[13] See Weierstrass (1894–1927) vol. 6, pp. 136–137.    [14] See Riemann (1899) pp. 14–16.

A simple calculation reveals that

$$1 - z = \frac{c_1 - c_4}{c_1 - c_2} \cdot \frac{x - c_2}{x - c_4}$$

$$1 - k^2 z = \frac{c_1 - c_4}{c_1 - c_3} \cdot \frac{x - c_3}{x - c_4}$$

and hence

$$P(x) = a_0 \cdot \frac{c_3 - c_1}{c_2 - c_4} \cdot \frac{(c_2 - c_1)^2}{(c_1 - c_4)^2} (x - c_4)^4 z(1 - z)(1 - k^2 z). \tag{10.52}$$

By taking the derivative of (10.50), we may write

$$\frac{dx}{dz} = \frac{c_2 - c_1}{(c_1 - c_4)(c_2 - c_4)} (x - c_4)^2. \tag{10.53}$$

Next, combining (10.53) with (10.52) yields

$$P(x) = a_0(c_2 - c_4)(c_3 - c_1) \cdot z(1 - z)(1 - k^2 z) \left(\frac{dx}{dz}\right)^2$$

or

$$\frac{dx}{\sqrt{P(x)}} = \frac{1}{\sqrt{a_0(c_2 - c_4)(c_3 - c_1)}} \frac{dz}{\sqrt{z(1 - z)(1 - k^2 z)}}. \tag{10.54}$$

Thus, the inverse of the linear transformation (10.50) reduces the general elliptic integral to Riemann's normal form.

In homogeneous form, the polynomial $P(x)$ can be written as

$$P(x_1, x_2) = a_0(x_2 - c_1 x_1)(x_2 - c_2 x_1)(x_2 - c_3 x_1)(x_2 - c_4 x_1), \tag{10.55}$$

and the linear transformation (10.50) as

$$x_1 = \frac{z_1}{(c_1 - c_4)(c_2 - c_1)} + \frac{z_2}{(c_1 - c_4)(c_4 - c_2)}$$

$$x_2 = \frac{c_1 z_1}{(c_1 - c_4)(c_2 - c_1)} + \frac{c_4 z_2}{(c_1 - c_4)(c_4 - c_2)}.$$

$P(x_1, x_2)$ is thereby transformed into

$$P'(z_1, z_2) = \delta^2 a_0 (c_3 - c_1)(c_2 - c_4) z_1 z_2 (z_1 - z_2)(z_1 - k^2 z_2), \tag{10.56}$$

where $\delta$ is the determinant of the linear transformation and is given by

$$\delta^{-1} = (c_1 - c_4)(c_2 - c_1)(c_2 - c_4).$$

Now observe that if we write (10.55) and (10.56) in the form

$$P(x_1, x_2) = a_0 x_2^4 + 4a_1 x_2^3 x_1 + 6a_2 x_2^2 x_1^2 + 4a_3 x_2 x_1^3 + a_4 x_1^4 \tag{10.57}$$

$$P'(z_1, z_2) = A_0 z_1^4 + 4A_1 z_1^3 z_2 + 6A_2 z_1^2 z_2^2 + 4A_3 z_1 z_2^3 + A_4 z_2^4, \tag{10.58}$$

then

$$A_0 = 0, \quad 4A_1 = \delta^2 a_0 (c_3 - c_1)(c_2 - c_4),$$

$$6A_2 = -4A_1(1 + k^2), \quad 4A_3 = 4A_1 k^2, \quad A_4 = 0.$$

The invariant $g_2$ in (10.24) satisfies (10.41) and, therefore, with $a_i$ and $A_i$ defined by (10.57) and (10.58), we have

$$g_2(A_0, \ldots, A_4) = -4A_1A_3 + 3A_2^2$$

$$= -4A_1^2k^2 + \frac{4}{3}A_1^2(1 + k^2)^2$$

$$= \frac{(4A_1)^2}{12}(1 - k^2 + k^4)$$

$$= \delta^4 g_2(a_0, \ldots, a_4).$$

Hence

$$g_2(a_0, \ldots, a_4) = \frac{a_0^2(c_3 - c_1)^2(c_2 - c_4)^2}{12}(1 - k^2 + k^4). \qquad (10.59)$$

Using similar reasoning, we may deduce that

$$g_3(a_0, \ldots, a_4) = \frac{a_0^3(c_3 - c_1)^3(c_2 - c_4)^3}{432}(k^2 + 1)(2k^2 - 1)(k^2 - 2) \qquad (10.60)$$

and with the use of (10.59) and (10.60) we may show that

$$\Delta = g_2^3 - 27g_3^2 = \frac{a_0^6(c_3 - c_1)^6(c_2 - c_4)^6}{256}k^4(1 - k^2)^2 \qquad (10.61)$$

$$J = \frac{g_2^3}{\Delta} = \frac{4(k^4 - k^2 + 1)^3}{27k^4(1 - k^2)^2}. \qquad (10.62)$$

Observe that (10.62) is exactly the expression given by Dedekind for his valency function $v(\omega)$. Note also that, unlike Weierstrass, Riemann did not consider the invariants $g_2$, $g_3$, and $\Delta$. The values (10.59) through (10.62) of these invariants were calculated by Klein.[15]

## 10.5 The Weierstrass Normal Form

Additional insight may be gained by studying the Weierstrass normal form, also useful in areas of number theory. The Weierstrass normal form was first introduced by Eisenstein in 1847, and later discussed by Weierstrass in his lectures on elliptic functions dating from 1862 onwards. In Weierstrass's notation, the normal form of an elliptic integral took the form

$$\int \frac{ds}{\sqrt{4s^3 - g_2s - g_3}} = \int \frac{ds}{\sqrt{4(s - e_1)(s - e_2)(s - e_3)}}, \qquad (10.63)$$

where $e_1, e_2, e_3$ are distinct. The transformation

$$z = \frac{e_3 - e_2}{e_1 - e_2} \frac{s - e_1}{s - e_3} \qquad (10.64)$$

[15] Klein (1921–1923) vol. 3, p. 18.

takes the Riemann normal form (as a differential)

$$\frac{dz}{2\sqrt{e_3 - e_2}\sqrt{z(1 - z)(1 - k^2 z)}}, \quad \text{with} \quad k^2 = \frac{e_1 - e_2}{e_3 - e_2} \tag{10.65}$$

to the Weierstrass normal form (as a differential)

$$\frac{ds}{\sqrt{4(s - e_1)(s - e_2)(s - e_3)}}. \tag{10.66}$$

Now if (10.64) is written as the linear transformation

$$z_2 = (e_3 - e_2)(s_2 - e_1 s_1), \quad z_1 = (e_1 - e_2)(s_2 - e_3 s_1), \tag{10.67}$$

then the polynomial $P(z_1, z_2) = 4(e_3 - e_2) z_1 z_2 (z_1 - z_2)(z_1 - k^2 z_2)$ is transformed to

$$P'(s_1, s_2) = 4\delta^2(s_2 - e_1 s_1)(s_2 - e_2 s_1)(s_2 - e_3 s_1)s_1 = \delta^2 s_1^4 (4s^3 - g_2 s - g_3)$$

where $\delta$ is the determinant of the transformation (10.67): $\delta = (e_1 - e_2)(e_1 - e_3)(e_3 - e_2)$. Using the notation of the previous section, we may verify that

$$g_2(A_0, A_1, \dots, A_4) = \delta^4 g_2$$

$$g_3(A_0, A_1, \dots, A_4) = \delta^6 g_3$$

$$g_2(a_0, a_1, \dots, a_4) = \frac{4}{3}(e_3 - e_2)^2(1 - k^2 + k^4)$$

$$g_3(a_0, a_1, \dots, a_4) = -\frac{4}{27}(e_3 - e_2)^3(1 + k^2)(2 - k^2)(2k^2 - 1).$$

Therefore, the invariance relations (10.41) and (10.42) produce

$$g_2 = \frac{4}{3}(e_3 - e_2)^2(1 - k^2 + k^4) \tag{10.68}$$

$$g_3 = -\frac{4}{27}(e_3 - e_2)^3(1 + k^2)(2 - k^2)(2k^2 - 1). \tag{10.69}$$

As Weierstrass showed in one of his lectures,[16] relations (10.68) and (10.69) could also be obtained directly without the use of invariant theory: From the relation

$$4s^3 - g_2 s - g_3 = 4(s - e_1)(s - e_2)(s - e_3)$$

it follows that

$$e_1 + e_2 + e_3 = 0 \tag{10.70}$$

$$e_1 e_2 + e_2 e_3 + e_3 e_1 = -\frac{1}{4}g_2 \tag{10.71}$$

$$e_1 e_2 e_3 = \frac{1}{4}g_3 \tag{10.72}$$

and from (10.70) and (10.71) we have

$$e_1^2 + e_2^2 + e_3^2 = \frac{1}{2}g_2. \tag{10.73}$$

---

[16] Weierstrass (1894–1927) vol. 5, pp. 46–47.

Next, taking $e_1 - e_2 = \alpha$ and $e_3 - e_2 = \beta$, we know from (10.70) that

$$e_1 = \frac{2\alpha - \beta}{3}, \quad e_2 = -\frac{\alpha + \beta}{3}, \quad e_3 = \frac{2\beta - \alpha}{3}. \tag{10.74}$$

Combining (10.73) and (10.74), it follows that

$$g_2 = \frac{4}{3}(\alpha^2 + \beta^2 - \alpha\beta); \quad g_3 = -\frac{4}{27}(\alpha + \beta)(2\alpha - \beta)(2\beta - \alpha).$$

Using these values of $g_2$ and $g_3$ and (10.65), we obtain (10.68) and (10.69).

## 10.6  Proof of the Infinite Product for Δ

In his 1878–79 paper,[17] Felix Klein published a number of insightful comments on the rational invariant $J$ and the periods of elliptic functions. In this paper, entitled "Über die Transformation der elliptischen Funktionen und die Auflösungen der Gleichungen fünften Grades," he observed that (10.23) implied that an elliptic integral could be viewed as an invariant, because by (10.44) and (10.45)

$$\int \frac{dy}{\sqrt{P'(1, y)}} = \delta^{-1} \int \frac{dx}{\sqrt{P(1, x)}}. \tag{10.75}$$

Moreover, since the periods of elliptic functions could be expressed as integrals, Klein pointed out that they too were invariants. For example, the periods of the Jacobi functions sn, cn, and so on would be the sum of even-integer multiples of both

$$K = \int_0^1 \frac{dx}{\sqrt{(1 - x^2)(1 - k^2x^2)}}, \quad iK' = \int_0^1 \frac{dx}{\sqrt{(1 - x^2)(1 - k'^2x^2)}}. \tag{10.76}$$

In general, if $\omega_1$ and $\omega_2$ were the periods, then, considering the absolute invariants $\omega_i \sqrt[4]{g_2}$, $\omega_i \sqrt[6]{g_3}$, and $\omega_i \sqrt[12]{\Delta}$, one could regard the normal form of an elliptic integral to be

$$\int \frac{\sqrt[12]{\Delta}dx}{\sqrt{P(x)}}. \tag{10.77}$$

Note that the normal form is determined up to a twelfth root of unity. Klein denoted $\omega_1$ and $\omega_2$ the transcendental invariants, because they could take on infinitely many values. Observe that if $\omega_1$ and $\omega_2$ generated the period lattice $m\omega_1 + n\omega_2$, with $m$ and $n$ arbitrary integers, then the pair $\omega_1'$ and $\omega_2'$ would generate the same lattice when

$$\omega_1' = a\omega_1 + b\omega_2; \quad \omega_2' = c\omega_1 + d\omega_2$$

with $ad - bc = 1$. Klein also took $\omega = \frac{\omega_1}{\omega_2}$ as an absolute transcendental invariant, such that

$$\omega' = \frac{\omega_1'}{\omega_2'} = \frac{a\omega + b}{c\omega + d}.$$

He also observed that, while the normal form of the elliptic integral was taken to be Legendre's form (10.43) with $q = k^2$, the developments of the theory of elliptic functions

---

[17] See Klein (1923) vol. 3, pp. 13–89.

following Jacobi's approach were all related to the integral (10.48). Now the periods of the function arising from (10.48) would be $4K$ and $4iK'$, where

$$2K = \int_0^1 \frac{dx}{\sqrt{x(1-x)(1-k^2x)}},$$

$$2iK' = \int_1^{\frac{1}{k^2}} \frac{dx}{\sqrt{x(1-x)(1-k^2x)}}) = i \int_0^1 \frac{dx'}{\sqrt{x'(1-x')(1-k'^2x')}}$$

with $k'^2 = 1 - k^2$, while the function sn arising from Legendre's form has periods $4K$ and $2iK'$ as defined by (10.76). Klein pointed out that Jacobi's work made it clear that the absolute transcendental invariant $\frac{4iK'}{4K}$, rather than $\frac{iK'}{2K}$, was the correct invariant to consider.

Klein identified Jacobi's $\frac{iK'}{k}$ with his own $\omega = \frac{\omega_1}{\omega_2}$ and utilized some formulas drawn from Jacobi's *Fundamenta Nova* to represent the rational invariants $g_2$, $\Delta$, and $J$ by means of the periods. Note that by (10.54), (10.59), and (10.61), the absolute invariants $\omega_i \sqrt[12]{\Delta}$ and $\omega_i \sqrt[4]{g_2}$ do not contain the factor $\sqrt{a_0(c_3-c_1)(c_2-c_4)}$ because of cancellation. Thus, Klein had[18]

$$\sqrt[12]{\Delta}\,\omega_1 = 4iK' \sqrt[3]{\frac{kk'}{4}},$$

$$\sqrt[12]{\Delta}\,\omega_2 = 4K \sqrt[3]{\frac{kk'}{4}},$$

$$\sqrt[4]{g_2}\,\omega_1 = 4iK' \sqrt[4]{\frac{k^4-k^2+1}{12}},$$

$$\sqrt[4]{g_2}\,\omega_2 = 4K \sqrt[4]{\frac{k^4-k^2+1}{12}}.$$

Recall that Jacobi had expressed $k$, $k'$ and $K$, as infinite products in $q = e^{-\pi \frac{K'}{K}}$; see (3.160), (3.161), and (3.162). These imply that

$$\{(1-q^2)(1-q^4)(1-q^6)\cdots\}^6 = \frac{2kk'K^3}{\pi^3\sqrt{q}}, \qquad (10.78)$$

and hence Klein could conclude that

$$\omega_2 \sqrt[12]{\Delta} = 2\pi \cdot q^{\frac{1}{6}} \prod_{n=1}^{\infty} (1-q^{2n})^2, \qquad q = e^{\pi \omega i}, \qquad \omega = \frac{iK'}{K}. \qquad (10.79)$$

In addition, from Jacobi's formula[19]

$$\frac{1-k^2+k^4}{15}\left(\frac{2K}{\pi}\right)^4 = \frac{1}{15} + 16\left(\frac{q^2}{1-q^2} + \frac{2^3q^4}{1-q^4} + \frac{3^3q^6}{1-q^6} + \frac{4^3q^8}{1-q^8} + \cdots\right),$$

Klein obtained

$$g_2 \cdot \left(\frac{\omega_2}{2\pi}\right)^4 = \frac{1}{12} + 20\left(\frac{q^2}{1-q^2} + \frac{2^3q^4}{1-q^4} + \frac{3^3q^6}{1-q^6} + \frac{4^3q^8}{1-q^8} + \cdots\right). \qquad (10.80)$$

---

[18] Klein (1921–1923) vol. 3, p. 21.     [19] Jacobi (1969) vol. 1, p. 169.

Thus

$$\Delta = \left(\frac{2\pi}{\omega_2}\right)^{12} \cdot q^2 \prod_{n=1}^{\infty} (1 - q^{2n})^{24}, \tag{10.81}$$

$$g_2 = \left(\frac{2\pi}{\omega_2}\right)^4 \left(\frac{1}{12} + 20\left(\frac{q^2}{1-q^2} + \frac{2^3 q^4}{1-q^4} + \frac{3^3 q^6}{1-q^6} + \cdots\right)\right), \tag{10.82}$$

where $q = e^{\pi i \omega}$ and $\Delta, g_2$ could be calculated from the periods $\omega_1$ and $\omega_2$. Since (10.81) follows immediately from (10.78), a formula given by Jacobi,[20] the result (10.81) is sometimes attributed to Jacobi. However, Jacobi did not connect (10.78) with the discriminant function; rather this derivation of (10.81) is due to Klein. Recall that an earlier derivation was given by Weierstrass by means of the sigma function. The absolute invariant $J$ could then be found from the formula

$$J = \frac{g_2^3}{\Delta} = \frac{\left(\frac{1}{12} + 20\left(\frac{q^2}{1-q^2} + \frac{2^3 q^4}{1-q^4} + \frac{3^3 q^6}{1-q^6} + \cdots\right)\right)^3}{q^2 \prod_{n=1}^{\infty} (1 - q^{2n})^{24}}.$$

Note here that as $q \to 0$, one obtains with Klein $J \sim \frac{1}{1728 q^2}$ as a first approximation. Note also that by using the geometric series expansion and changing the order of summation, we can rewrite (10.80) as

$$\left(\frac{\omega_2}{2\pi}\right)^4 g_2 = \frac{1}{12}\left(1 + 240 \sum_{n=1}^{\infty} \sigma_3(n) q^{2n}\right), \quad \sigma_3(n) = \sum_{d|n} d^3. \tag{10.83}$$

Observe that (10.83) is a particular case of (4.173), a formula derived by Hurwitz in his 1881 thesis.

## 10.7 The Multiplier in Terms of $\sqrt[12]{\Delta}$

In a section of his 1878–79 paper, Klein discussed the multiplier for an elliptic integral normalized by means of $\sqrt[12]{\Delta}$.[21] Recall that this integral is given by (10.77) where

$$\Delta = g_2^3 - 27 g_3^2 \tag{10.84}$$

and $g_2$ and $g_3$ are defined in terms of the coefficients of the quartic $P(x)$ as in (10.24) and (10.26). Now suppose a $p$th-order transformation ($p$ prime) is applied to the elliptic integral (10.77), thereby producing the elliptic integral

$$\int \frac{\sqrt[12]{\Delta_1}\, dx_1}{\sqrt{P_1(x_1)}}. \tag{10.85}$$

Here $x_1$ is a rational function of $x$ producing the $p$th-order transformation; $P_1(x_1)$ is the resulting quartic; and $\Delta_1$ is obtained from (10.84) by replacing the coefficients of $P(x)$ in $g_2$ and $g_3$ by the corresponding coefficients in $P_1(x_1)$. We thus obtain the multiplier

[20] Ibid. p. 146.     [21] Klein (1921–1923) vol. 3, pp. 47–48.

$M$ defined by

$$M \int \frac{\sqrt[12]{\Delta}\, dx}{\sqrt{P(x)}} = \int \frac{\sqrt[12]{\Delta_1}\, dx_1}{\sqrt{P_1(x_1)}}. \tag{10.86}$$

Now the $p$-th order transformation can be given in terms of the periods of the elliptic function, defined by the elliptic integral (10.77). The periods $\omega_1$ and $\omega_2$ can be so chosen that the $p+1$ inequivalent transformations of the $p$th order can be written in the form

$$\omega_1' = \omega_1 + d\,\omega_2, \quad \omega_2' = p\,\omega_2,, \tag{10.87}$$

with $0 \le d < p$, and

$$\omega_1' = p\,\omega_1, \qquad \omega_2' = \omega_2. \tag{10.88}$$

We choose the path of integration in the left-hand side of the integral in (10.86) such that the integral represents a period, say, $\omega_2$. Then, depending on the type of transformation with respect to the given period, we obtain

$$M = \frac{\sqrt[12]{\Delta_1} \cdot \omega_2'}{\sqrt[12]{\Delta} \cdot \omega_2} \quad \text{or} \quad = \frac{1}{p} \frac{\sqrt[12]{\Delta_1} \cdot \omega_2'}{\sqrt[12]{\Delta} \cdot \omega_2}.$$

It follows from (10.79) that

$$\omega_2 \sqrt[12]{\Delta} = 2\pi q^{\frac{1}{6}} \prod_{n=1}^{\infty} (1 - q^{2n})^2, \quad q = e^{\pi \omega i}, \quad \omega = \frac{\omega_2}{\omega_1}$$

and

$$\omega_2' \sqrt[12]{\Delta_1} = 2\pi q_1^{\frac{1}{6}} \prod_{n-1}^{\infty} (1 - q_1^{2n})^2, \quad q_1 = e^{\pi \omega' i}, \quad \omega' = \frac{\omega_2'}{\omega_1'},$$

where $q_1$ takes any one of the values

$$q^{\frac{1}{n}}, \ \alpha q^{\frac{1}{n}}, \ \alpha^2 q^{\frac{1}{n}}, \dots, \ \alpha^{n-1} q^{\frac{1}{n}}, \ q^n,$$

with $\alpha$ a primitive $n$th root of unity. Here see (3.228). The $p+1$ values of $M$ are then given by[22]

$$M_k = \frac{1}{p} \frac{\alpha^{\frac{k}{6}} q^{\frac{1}{6n}} \prod_{m=1}^{\infty} \left(1 - \alpha^{2km} q^{\frac{2m}{n}}\right)^2}{q^{\frac{1}{6}} \prod_{m=1}^{\infty} (1 - q^{2m})^2}, \quad k = 0, 1, \dots, p-1$$

$$M_\infty = \frac{q^{\frac{n}{6}} \prod_{m=1}^{\infty} (1 - q^{2mn})^2}{q^{\frac{1}{6}} \prod_{m=1}^{\infty} (1 - q^{2m})^2}.$$

The values of $M_k$, $k = 0, 1, \dots, p-1$ are defined up to a sixth root of unity. These $p+1$ values of $M$ are the roots of the multiplier equation of order $p$.

---

[22] Klein (1921–1923) vol. 3, p. 48.

# 11

---

# *The Modular and Multiplier Equations*

## 11.1 Preliminary Remarks

We recall that, as presented by Jacobi, a transformation of prime order $p$ was given by $y = \frac{U(x)}{V(x)}$, $U$ and $V$ being polynomials of degree $p$ and $p - 1$, respectively, when $y$ satisfied the differential equation

$$\frac{dy}{\sqrt{(1 - y^2)(1 - \lambda^2 y^2)}} = \frac{dx}{M\sqrt{(1 - x^2)(1 - k^2 x^2)}}, \tag{11.1}$$

where $M$ depended only on $k$ and $\lambda$ and where $k^2 \neq 0$, 1. We remark that Legendre and the earlier work of Jacobi took $0 < k^2 < 1$. Corresponding to a value of $k$ were $p + 1$ distinct values of $\lambda$, given by Jacobi in terms of the values of the elliptic function sn. In this connection, please see our Section 3.4. Let $\omega$ denote any of the $p + 1$ values

$$\frac{K}{p}, \frac{iK'}{p}, \frac{K + iK'}{p}, \frac{K + 2iK'}{p}, \ldots, \frac{K + (p-1)iK'}{p}, \tag{11.2}$$

where $K$ and $K'$ are given by

$$K = \int_0^1 \frac{dx}{\sqrt{(1 - x^2)(1 - k^2 x^2)}}, \quad K' = \int_0^1 \frac{dx}{\sqrt{(1 - x^2)(1 - k'^2 x^2)}}, \quad k^2 + k'^2 = 1; \tag{11.3}$$

then the $p + 1$ values of $\lambda$ would be

$$\lambda = k^p \prod_{s=0}^{\frac{p-1}{2}} \mathrm{sn}^4(K - 4\omega). \tag{11.4}$$

Jacobi gave an alternative and more striking form of these $p + 1$ values of $\lambda$. As we have seen, Jacobi proved that

$$\sqrt{k} = \frac{2q^{\frac{1}{4}} + 2q^{\frac{9}{4}} + 2q^{\frac{25}{4}} + \cdots}{1 + 2q + 2q^4 + 2q^9 + \cdots}, \quad q = e^{-\frac{\pi K'}{K}}. \tag{11.5}$$

295

Now in a note of 1828,[1] Jacobi wrote, without giving a proof, that the $p + 1$ values of $\sqrt{\lambda}$ could be obtained from (11.5) if one replaced $q$ by each of the $p + 1$ values

$$q^p, q^{\frac{1}{p}}, \alpha q^{\frac{1}{p}}, \ldots, \alpha^{p-1} q^{\frac{1}{p}}, \quad \text{where } \alpha^p = 1, \ \alpha \neq 1; \tag{11.6}$$

obviously, Jacobi would have been capable of proving this result. In any case, in 1837, L. A. Sohnke, Professor of mathematics at the University of Halle, provided a proof in a paper on modular equations.[2] The $p + 1$ values of $\lambda$ corresponding to the value $k(q)$, as given by (11.5), would then be

$$k(q^p), k(q^{\frac{1}{p}}), k(\alpha^j q^{\frac{1}{p}}), \quad j = 1, 2, \ldots, p - 1. \tag{11.7}$$

Since $q = e^{i\pi \omega}$, Im $\omega > 0$, we will sometimes write $k(q)$ as $k(\omega)$. Then the values in (11.7) would be given as

$$k(p\omega), \quad k\left(\frac{\omega}{p}\right), \quad k\left(\frac{\omega + 2j}{p}\right), \quad j = 1, 2, \ldots, p - 1. \tag{11.8}$$

Since $k(\omega) = \frac{\theta_2^2(\omega)}{\theta_3^2(\omega)}$, the functions in (11.8) may be written in terms of theta functions. Recall that Jacobi observed[3] that $k^2(\omega)$ was invariant under the transformation

$$\omega \to \frac{a\omega + b}{c\omega + d}, \tag{11.9}$$

where $b$ and $c$ were even and $ad - bc = 1$. Note that the transformations in (11.9) constitute the group $\Gamma(2)$ and that the $p + 1$ functions

$$k^2(p\omega), k^2\left(\frac{\omega}{p}\right), k^2\left(\frac{\omega + 2j}{p}\right), \quad j = 2, 3, \ldots, p - 1 \tag{11.10}$$

are permuted by the action of any element of $\Gamma(2)$. Thus, the elementary symmetric functions of the $p + 1$ functions in (11.10) are invariant under $\Gamma(2)$, and it can be shown that these symmetric functions are polynomials in $k^2(\omega)$. Therefore

$$F(\lambda^2, k^2) = \left(\lambda^2 - k^2(p\omega)\right) \prod_{j=0}^{p-1} \left(\lambda^2 - k^2\left(\frac{\omega + 2j}{p}\right)\right)$$

$$= \lambda^{2(p+1)} + C_1 \lambda^{2p} + C_2 \lambda^{2(p-1)} + \cdots + C_{p+1}, \tag{11.11}$$

where $C_1, C_2, \ldots, C_{p+1}$ are polynomials in $k^2(\omega)$. The equation

$$F(\lambda^2, k^2) = 0 \tag{11.12}$$

is called a modular equation of order $p$. Jacobi and Sohnke did not employ this type of argument to show the existence of the modular equation (11.12), but established it by the use of the series and product expansions of $k^2(\omega)$ and $\lambda^2(\omega)$. As we have seen, such an argument in the context of the full modular group was used in the work of Dedekind and Klein, facilitated by the papers of Hermite. Instead of $k^2$ and $\lambda^2$, Jacobi and Sohnke constructed the modular equation in terms of $u = \sqrt[4]{k}$ and $v = \sqrt[4]{\lambda}$.

---

[1] Jacobi (1969) p. 252.  [2] Sohnke (1837).  [3] Jacobi (1969) p. 263.

Denoting the $p + 1$ values of $v = \sqrt[4]{\lambda}$, corresponding to a given value of $u = \sqrt[4]{k}$, by $v_1, v_2, \ldots, v_{p+1}$, Sohnke showed that the polynomial

$$F(u, v) = (v - v_1)(v - v_2) \cdots (v - v_{p+1})$$

would have the form

$$v^{p+1} + v^p u^s (\alpha + \beta u^8 + \gamma u^{16} + \cdots) + v^{p-1} u^{s'} (\alpha' + \beta' u^8 + \cdots)$$

$$+ \cdots + (-1)^{\frac{p^2-1}{8}} u^{p+1} = 0, \tag{11.13}$$

where $s$ and $s'$ were positive integers, $\alpha, \alpha', \beta, \beta', \ldots$ were constants, and the powers of $u$ would not exceed $p + 1$. Then, to determine the values of $\alpha, \alpha', \beta, \beta', \ldots$ for small values of $p$, he showed that

$$F(1, v) = (v - 1)^{p+1} \qquad \text{when } p = 8\mu \pm 1;$$
$$F(1, v) = (v - 1)(v + 1)^p \quad \text{when } p = 8\mu \pm 3.$$

In a note written on July 21, 1828, Jacobi noted that in the transformation of elliptic functions, the multiplier $M$ in (11.1) satisfied an algebraic equation.[4] In his *Fundamenta Nova*, he showed[5] that $M$ was a rational function of $k$ and $\lambda$:

$$M^2 = \frac{1}{p} \frac{\lambda(1 - \lambda^2)}{k(1 - k^2)} \frac{dk}{k\lambda} = \frac{1}{p} \frac{u(1 - v^8)}{v(1 - u^8)} \frac{du}{dv}. \tag{11.14}$$

In his 1828 note, he observed that by eliminating $\lambda$ (or $v$) from (11.14) and the modular equation in terms of $k$ and $\lambda$ (or $u$ and $v$), an algebraic equation involving $M$ and $k^2$ (or $M$ and $u$) would be obtained. He called this *Die Multiplicatorgleichung*, the multiplier equation. For example, for a transformation of order 3, the modular equation is

$$v^4 - u^4 - 2uv(1 - u^2 v^2) = 0; \tag{11.15}$$

using (3.35) and (3.41) we can verify that the multiplier is

$$\frac{1}{M} = \frac{v + 2u^3}{v} \quad \text{or} \quad \frac{1}{M} - 1 = \frac{2u^3}{v}. \tag{11.16}$$

Next, rewrite (11.15) as

$$\frac{1}{u^8} \frac{u^{12}}{v^4} - 1 + \frac{2u}{v^3} - \frac{2u^3}{v} = 0$$

or

$$\frac{1}{16u^8} \left(\frac{1}{M} - 1\right)^4 + \frac{2u}{v^3} - \left(\frac{1}{M} - 1\right) - 1 = 0$$

or

$$\left(\frac{1}{M} - 1\right)^4 + 32\frac{u^9}{v^3} - 16u^8 \frac{1}{M} = 0. \tag{11.17}$$

---

[4] Jacobi (1969) p. 361.　　[5] Ibid. p. 130.

Set $x = \frac{1}{M}$ and simplify to write (11.17) as

$$x^4 - 6x^2 + 8(1 - 2k^2)x - 3 = 0. \tag{11.18}$$

Equation (11.18) is denoted the multiplier equation of order 3. For $p = 5$, Jacobi gave an equation for $x = \frac{\lambda}{M}$:

$$x^6 - 10kx^5 + 35k^2x^4 - 60k^3x^3 + 55k^4x^2 - \left(26k^5 + 256(k - k^3)\right)x + 5k^6 = 0, \tag{11.19}$$

from which the multiplier equation of the fifth order can be easily obtained. To obtain the general multiplier equation, analogous to (11.11) or (11.13), recall that

$$\frac{1}{M} = (-1)^{\frac{p-1}{2}} \prod_{j=1}^{\frac{p-1}{2}} \frac{\text{sn}^2 4\,\omega}{\text{sn}^2(K - 4\omega)}.$$

Denote the values in the list (11.7) or (11.8) by $\lambda_1, \lambda_2, \ldots, \lambda_{p+1}$. A calculation similar to that for $\lambda$ shows that the $p + 1$ values of $\frac{1}{M}$ are given by

$$(-1)^{\frac{p-1}{2}} p\frac{\Lambda_1}{K}, \frac{\Lambda_2}{K}, \ldots, \frac{\Lambda_{p+1}}{K}, \tag{11.20}$$

where

$$\Lambda_i = \int_0^1 \frac{dx}{\sqrt{(1 - x^2)(1 - \lambda_i^2 x^2)}}. \tag{11.21}$$

The values of $\frac{1}{M}$, just as those of $k^2$, can be given in terms of theta functions. Since $K = \theta_3^2(\omega)$, we perceive that the values of $\Lambda_1, \Lambda_2, \ldots, \Lambda_{p+1}$ are, respectively,

$$\theta_3^2(p\omega), \theta_3^2\left(\frac{\omega}{p}\right), \theta_3^2\left(\frac{\omega + 2}{p}\right), \ldots, \theta_3^2\left(\frac{\omega + 2(p - 1)}{p}\right). \tag{11.22}$$

Though evidently known to Jacobi, the list in (11.20) and (11.21) was explicitly given by Francesco Brioschi (1824–1897), who was a student of the applied mathematician Antonio Bordoni (1788–1860) at the University of Pavia. Brioschi taught at Pavia, where his students included Casorati, Cremona, and Beltrami. After a visit to Göttingen where he met with Riemann and saw the significance of his work, Brioschi initiated translations of Riemann's works into Italian and lectured on them. Brioschi was an algebraist who contributed to the theory of algebraic equations, determinants, invariant theory, and elliptic and Abelian functions; he was also an applied mathematician, and made key contributions to hydraulic projects in Italy.

In 1858, Brioschi wrote the paper[6] containing (11.20) and (11.21) in order to give a proof of Jacobi's improved theorem that there existed $\frac{p+1}{2}$ quantities $A, A_1, \ldots, A_{\frac{p-1}{2}}$

---

[6] Brioschi (1901) vol. 1, pp. 321–324.

such that

$$\frac{1}{\sqrt{M_1}} = \sqrt{(-1)^{\frac{p-1}{2}} p A},$$

$$\frac{1}{\sqrt{M_2}} = A + \sum_{j=1}^{\frac{p-1}{2}} A_j,$$

$$\frac{1}{\sqrt{M_3}} = A + \sum_{j=1}^{\frac{p-1}{2}} \alpha^{j^2} A_j,$$

$$\frac{1}{\sqrt{M_4}} = A + \sum_{j=1}^{\frac{p-1}{2}} \beta^{j^2} A_j, \text{ etc.,} \tag{11.23}$$

where $\alpha, \beta, \cdots$ were the complex roots of $\alpha^p = 1$. Here $M_1, M_2, \cdots, M_{p+1}$ denoted the $p + 1$ roots of the multiplier equation given in the list (11.20).

In another paper of 1858[7], Brioschi applied the multiplier equation to solve, in terms of modular functions, an equation of the fifth degree that had been reduced to the form

$$\theta^5 - \frac{5}{2}\theta^4 - a = 0.$$

Also in 1858, the Jesuit mathematician Charles Joubert proved, for multiplier equations, some analogs of Sohnke's results on modular equations; details of his proofs appeared in his short book of 1875. He first showed that setting $k = 0$ in the multiplier equation reduces it to either

$$f(x) = (x - 1)^p(x - p) \text{ or } f(x) = (x + 1)^p(x - p),$$

according as $p = 4m + 1$ or $p = 4m + 3$. He then proved that the multiplier equation of order $p = 4m + 1$ had the form

$$f(x) + k^2 k'^2 x F(k^2 k'^2, x) = 0$$

where

$$F(k^2 k'^2, x) = Ax^{p-5} + Bx^{p-6} + Cx^{p-7} + Dx^{p-8} + (A' + A_1' k^2 k'^2)x^{p-9}$$
$$+ (B' + B_1' k^2 k'^2)x^{p-10} + \cdots, \tag{11.24}$$

where $A, A', B, B'$, etc., were constants, and the degree of $k^2 k'^2$ in the coefficients would increase by one degree after every four terms. Joubert had a similar result for $p = 4m + 3$. In a paper of 1874, "A Memoir on the Transformation of Elliptic Functions,"[8] Cayley succinctly described Joubert's results on the multiplier equation:

> The last term of the equation is constant, and the other coefficients are rational and integral functions of $u^8$, of a degree not exceeding $\frac{1}{2}(p - 1)$; and not only so but they are, $p \equiv 1$ (mod 4), rational and integral functions of $u^8(1 - u^8)$, and $p \equiv 3$ (mod 4), alternately of this form, and of the same form multiplied by the factor $(1 - 2u^8)$.

---

[7] Ibid, pp. 335–341.    [8] Cayley (1874), especially p. 420.

Note that we present Cayley's remarks in terms of $p$ instead of his $n$, in conformity with our notation.

Recall that Weierstrass began lecturing on the topic of elliptic functions in 1862. When his students then did work on this topic, they naturally used the Weierstrassian functions and notation, within which the periods of an elliptic function were denoted by $2\omega_1, 2\omega_2$. However, as in our discussion of Eisenstein, we here denote the periods by $\omega_1, \omega_2$.

In 1878, Ludwig Kiepert (1846–1934) recast Jacobi's work in terms of Weierstrass's elliptic functions. Kiepert was Weierstrass's student at Berlin and attended Weierstrass's lectures from 1865 through 1871, obtaining his doctoral degree in 1970. He later became very interested in actuarial mathematics. Kiepert described the roots of the multiplier equation in terms of the invariant $\Delta(\omega_1, \omega_2)$ that played an important role in Weierstrass's theory. In fact, he showed by means of a fairly long computation that the multiplier could be expressed as

$$M = \sqrt[12]{\frac{\Delta\left(\omega_1, \frac{\omega_2}{p}\right)}{\Delta(\omega_1, \omega_2)}}. \tag{11.25}$$

In a paper of 1879,[9] Kiepert showed that the function

$$g^2(\omega_1) = \frac{1}{\left(\wp\left(\frac{\omega_1}{5}\right) - \wp\left(\frac{2\omega_1}{5}\right)\right)^2} = e^{-\frac{2\eta\omega_1}{5}}\sigma^2\left(\frac{\omega_1}{5}\right)\sigma^2\left(\frac{2\omega_1}{5}\right)$$

was a root of the sixth degree equation

$$x^6 + \frac{10}{\Delta}x^3 - \frac{12g_2}{\Delta^2}x + \frac{5}{\Delta^2} = 0 \tag{11.26}$$

and he then used equation (11.26), whose form suggested a connection with the multiplier equation of order 5, to solve an equation of the fifth degree. More generally, Kiepert proved[10] that for $p = 6m \pm 1$, the function

$$g(\omega_1) = e^{-\frac{\eta\omega_1(p^2-1)}{24p}}\sigma\left(\frac{\omega_1}{p}\right)\sigma\left(\frac{2\omega_1}{p}\right)\cdots\sigma\left(\frac{(p-1)\omega_1}{2p}\right) \tag{11.27}$$

was the root of an equation of degree $p + 1$ with coefficients that were polynomials in $g_2$ and $g_3$ divided by $\Delta$ raised to an integer power. Recall that $g_2$ and $g_3$ are defined in (4.100) and $\Delta$ in (4.155). He proved that the $p + 1$ roots satisfied $\frac{p+1}{2}$ linear relations of the type that result from Jacobi's (11.23). Kiepert wrote the $p + 1$ roots in an alternative form:

$$\left(\frac{\Delta(p\omega_1, \omega_2)}{\Delta^p(\omega_1, \omega_2)}\right)^{\frac{1}{12}}, (-1)^{\frac{p-1}{2}}\left(\frac{\Delta(\omega_1 + 12j\omega_2, p\omega_2)}{\Delta^p(\omega_1, \omega_2)}\right)^{\frac{1}{12}}, \quad j = 0, 1, \ldots, p - 1. \tag{11.28}$$

Here we note that for $p = 6m \pm 1$, any matrix with integer entries

$$\begin{pmatrix} a & b \\ c & d \end{pmatrix}, \quad ad - bc = p \tag{11.29}$$

---

[9] Kiepert (1879a).   [10] Kiepert (1879b).

is equivalent modulo the full modular group $\Gamma(1)$ to one of the matrices

$$\begin{pmatrix} p & 0 \\ 0 & 1 \end{pmatrix}, \quad \begin{pmatrix} 1 & 12j \\ 0 & p \end{pmatrix}, \quad j = 0, 1, \dots, p - 1. \tag{11.30}$$

If we call the matrix in (11.29) a transformation of order $p$, then the numerators in (11.28), that is,

$$\Delta^{\frac{1}{12}}(p\,\omega_1, \omega_2), (-1)^{\frac{p-1}{2}} \Delta^{\frac{1}{12}}(\omega_1 + 12j\,\omega_2, p\,\omega_2), \quad j = 0, 1, \dots, p - 1, \tag{11.31}$$

give all the transformed values of $\Delta^{\frac{1}{12}}(\omega_1, \omega_2)$ under a transformation of order $p = 6m \pm 1$. The theory of transformations of elliptic integrals can thus be treated more generally by the theory of transformations of theta functions and modular forms. We have seen that Jacobi himself played an important role in the transition to this more general framework. Since Eisenstein defined elliptic functions in terms of their periods, rather than as inverses of elliptic integrals, it was natural for him to discuss transformations in terms of (11.29). Later mathematicians also found it convenient to continue this approach.

In the late 1870s, Felix Klein brought the idea of the group to the forefront; he synthesized the ideas of the previous half-century in terms of the group $\Gamma(1)$ and its subgroups, in particular in terms of the concept of a subgroup of $\Gamma(1)$ of level $N$. Recall also that in Chapter 10 we saw how Klein was able to show[11] that the function $\Delta^{\frac{1}{12}}$, used by Kiepert, could itself be regarded as a multiplier of an elliptic integral. Thus, it satisfied a multiplier equation. He also produced the multiplier equations of orders 5 and 7 for $\Delta^{\frac{1}{12}}(\omega_1, \omega_2)$. In his paper, "Über Multiplikatorgleichungen"[12], Klein presented the form of the general $p$th-order multiplier equation for $\Delta^{\frac{1}{12}}$. Since $\Delta^{\frac{1}{12}}$ is a modular form of weight one with a multiplier system, the values $\Delta^{\frac{1}{12}}(a\,\omega_1, +b\,\omega_2, c\,\omega_1 + d\,\omega_2)$, $ad - bc = p$, must be multiplied by appropriate twelfth roots of unity, so that they will be roots of the multiplier equation. Klein pointed this out and also worked out the multiplier equation of order 11. In a later paper, "Zur [Systematik der] Theorie der elliptischen Modulfunktionen,"[13] he defined a multiplier equation more generally, as the polynomial relation between a modular form $F(\omega_1, \omega_2)$ of weight $k$ and the form $F(p\,\omega_1, \omega_2)$. When $k = 0$, the result is a modular function, and thus the multiplier equation is seen to contain the modular equation as a particular case. In this paper, Klein also defined the associated idea of modular correspondence.[14]

In his Ph.D. dissertation of 1881, Klein's student Adolf Hurwitz (1859–1919) worked out the details of the multiplier equation arising from $\sqrt[12]{\Delta(\omega_1, \omega_2)}$. In his history of nineteenth-century mathematics,[15] Klein wrote about his distinguished pupil: "Hurwitz has been called an aphorist. In total mastery of the subjects he was concerned with, he sought out here and there an important problem, always advancing it by a significant step. Each of his works is complete in itself." Examples of such accomplishments are his thesis; his 1881 paper in which he gave a method for deriving the functional equation for the general Dirichlet $L$-function by means of his Hurwitz zeta-function;[16] his 1896 number theory paper on the application of quaternions to the

---

[11] Klein (1921–1923) vol. 3, pp. 46–51.    [12] Klein (1921–1923) vol. 3, pp. 137–139.
[13] Ibid. pp. 169–178.    [14] See also Klein and Fricke (1890–1892) vol. 2, pp. 596–634.
[15] Klein (1979) p. 309.    [16] Hurwitz (1962) vol. 1, pp. 72–88.

Fermat-Lagrange theorem on sums of four squares;[17] his 1897 paper which gave the first definition of an invariant measure on a group;[18] and his definitive 1904 paper on elliptic modular functions, the translation of which is included in this volume.

Hurwitz determined that the $p + 1$ roots of the multiplier equation for $\Delta^{\frac{1}{12}}$, when $p \nmid 6$, were given by

$$(-1)^{\frac{p-1}{2}} \sqrt[12]{\Delta(p\,\omega_1, \omega_2)}, \quad e^{-\frac{kp\pi i}{6}} \sqrt[12]{\Delta(\omega_1 + k\,\omega_2, p\,\omega_2)}, \quad k = 0, 1, \ldots, p - 1. \tag{11.32}$$

This result meant, in particular, that the symmetric functions of these roots change by the same factor of the twelfth root of unity as $\Delta^{\frac{1}{12}}$ changes under the action of the full modular group. This fact in turn implies that if the functions in (11.32) are each raised to a power $a$ that is a divisor of 12, then the symmetric polynomials of these functions also change by the same roots of unity as $\Delta^{\frac{a}{12}}$ under the action of the full modular group. If the inequivalent representatives of transformations of order $p$ are taken to be given by (11.30), then the functions (11.32) can be replaced by those in (11.31). Since $\Delta^{\frac{1}{12}}$ can be written in terms of Dedekind's $\eta$ function,

$$\Delta^{\frac{1}{12}}(\omega_1, \omega_2) = \frac{2\pi}{\omega_2}\, \eta^2(\omega), \quad \text{Im}\,\omega = \text{Im}\,\frac{\omega_1}{\omega_2} > 0, \tag{11.33}$$

the functions in (11.31) may be written as[19]

$$\frac{2\pi}{\omega_2}\, \eta^2(p\,\omega), \quad (-1)^{\frac{p-1}{2}} \frac{2\pi}{p\,\omega_2}\, \eta^2\left(\frac{\omega + 12j}{p}\right), \quad j = 0, 1, \ldots, p - 1. \tag{11.34}$$

These results of Kiepert, Klein, and Hurwitz[20] were taken in a new direction by Louis Mordell. Mordell was inspired by the 1916 conjectures of Ramanujan concerning the coefficients of the series expansion of $\eta^{2a}\left(\frac{12\omega}{a}\right)$ where $a$ is a divisor of 12 in powers of $q^2 = e^{2\pi i \omega}$. Now Hurwitz had proved that, when $p \nmid 6$,

$$\left((-1)^{\frac{(p-1)a}{2}} \Delta^{\frac{a}{12}}(p\,\omega_1, \omega_2) + \sum_{j=0}^{p-1} \Delta^{\frac{a}{12}}(\omega_1 + 12j\,\omega_2, p\,\omega_2)\right) \div \Delta^{\frac{a}{12}}(\omega_1, \omega_2) \tag{11.35}$$

was invariant under the action of the full modular group. Mordell's important contribution was to observe that the quantity in (11.35) was bounded in the fundamental domain and hence was a constant, and that this implied Ramanujan's conjectures. However, Mordell was not inclined to generalize Ramanujan's work to other modular forms. It was Hecke who took this step in 1935 when he developed the concept of what are now known as Hecke operators.

It is noteworthy that Ramanujan derived more modular equations than any previous mathematician. He published a few of these without proofs and recorded the rest, again without proofs, in his notebooks. Bruce Berndt and his collaborators[21] have given

[17] Ibid. vol. 2, pp. 303–330.    [18] Ibid. vol. 2, pp. 546–564.    [19] Weber (1908) vol. 3, p. 252.

[20] For a good discussion of the work of these mathematicians in this area, see Klein and Fricke (1890–1892) vol. 2, pp. 62–82.

[21] Berndt (1985–1998) vol. 3, pp. 220–493; vol. 4, pp. 138–244; vol. 5, pp. 353–408.

proofs; in his paper, "Ramanujan's theory of theta-functions,"[22] Berndt notes, "Unfortunately, many of Ramanujan's modular equations have resisted all of our efforts at proving them by using the classical analysis of theta-functions." He then offers two examples of these resistant modular equations found in Ramanujan: With $\alpha = k^2$, $\beta = \lambda^2$, and $M$ the multiplier, a modular equation of order 11:

$$\left(\frac{(1-\beta)^3}{1-\alpha}\right)^{\frac{1}{8}} - \left(\frac{\beta^3}{\alpha}\right)^{\frac{1}{8}} - \left(\frac{\beta^3(1-\beta)^3}{\alpha(1-\alpha)}\right)^{\frac{1}{8}}$$
$$= \frac{M}{\sqrt{2}}\left(1 + \sqrt{\alpha\beta} + \sqrt{(1-\alpha)(1-\beta)}\right)^{\frac{1}{2}}; \tag{11.36}$$

and a modular equation of order 17:

$$M = \left(\frac{\beta}{\alpha}\right)^{\frac{1}{4}} + \left(\frac{1-\beta}{1-\alpha}\right)^{\frac{1}{4}} + \left(\frac{\beta(1-\alpha)}{\alpha(1-\alpha)}\right)^{\frac{1}{4}}$$
$$- 2\left(\frac{\beta(1-\beta)}{\alpha(1-\alpha)}\right)^{\frac{1}{8}}\left(1 + \left(\frac{\beta}{\alpha}\right)^{\frac{1}{8}} + \left(\frac{1-\beta}{1-\alpha}\right)^{\frac{1}{8}}\right). \tag{11.37}$$

## 11.2 Jacobi's Multiplier Equation

Continuing our discussion of the multiplier equation, suppose we have a $p$th-order transformation; that is, let

$$\frac{dy}{\sqrt{(1-y^2)(1-\lambda^2 y^2)}} = \frac{dx}{M\sqrt{(1-x^2)(1-k^2 x^2)}} \tag{11.38}$$

be the differential equation satisfied by $y$, a rational function of $x$ in which the powers of $x$ take values up through $p$. For simplicity, take $p$ to be prime. Recall that the moduli $k$ and $\lambda$ satisfy a polynomial equation of degree $p + 1$ in $k$ as well as in $\lambda$; this polynomial equation is the modular equation. We can determine all the values $\sqrt{\lambda}$ by substituting into the equation

$$\sqrt{k} = \frac{2q^{\frac{1}{4}} + 2q^{\frac{9}{4}} + 2q^{\frac{25}{4}} + \cdots}{1 + 2q + 2q^4 + 2q^9 + \cdots}, \quad q = e^{-\frac{\pi K'}{K}} \tag{11.39}$$

the $p + 1$ values $q^p, q^{\frac{1}{p}}, \epsilon q^{\frac{1}{p}}, \epsilon^2 q^{\frac{1}{p}}, \ldots, \epsilon^{p-1} q^{\frac{1}{p}}$, where $\epsilon^p = 1$. Note that for $0 < k^2 < 1$, $K$ and $K'$ represent the complete integrals

$$K = \int_0^1 \frac{dx}{\sqrt{(1-x^2)(1-k^2 x^2)}}, \quad K' = \int_0^1 \frac{dx}{\sqrt{(1-x^2)(1-k'^2 x^2)}}, \quad k^2 + k'^2 = 1. \tag{11.40}$$

The multiplier $M$ is a function of $\lambda$ and $k$. In a paper of 1828,[23] Jacobi demonstrated that $M$ satisfied an algebraic equation of degree $p + 1$, whose coefficients were rational

---

[22] Berndt (1993).    [23] Jacobi (1969) vol. 1, pp. 255–263, especially pp. 260–261.

functions of $k$. In 1878, Klein called this equation a multiplier equation (*Mulitplikatorgleichung*) of level 1.[24] In his 1828 paper, Jacobi in fact took $\mathfrak{M} = \frac{1}{M}$ and wrote that $\mathfrak{M}$ satisfied an equation of degree $p + 1$; to derive this equation, he first observed that a rational expression for $\mathfrak{M}$ in terms of $k$ and $\lambda$ could be determined from the equation

$$\mathfrak{M}^2 = \frac{p(k - k^3)}{(\lambda - \lambda^3)} \frac{d\lambda}{dk}. \tag{11.41}$$

He did not derive this relation in his paper, but in the *Fundamenta Nova* he gave a derivation using a differential equation, due to Legendre, between $K$ and $k$.[25] Jacobi next noted that eliminating $\lambda$ from the modular equation and from (11.41) would give him an equation satisfied by $\mathfrak{M}$ of the same degree as the modular equation. He then gave without proof a "remarkable property" of the solutions of this equation:[26] There exist $\frac{n+1}{2}$ quantities $A_i$ such that the square roots of the solutions can be expressed linearly in terms of the $A_i$. Thus, if $\mathfrak{M}, \mathfrak{M}_1, \mathfrak{M}_2, \ldots \mathfrak{M}_p$ denoted the roots of the equation, then Jacobi could write

$$\sqrt{\mathfrak{M}} = \sqrt{(-1)\frac{p-1}{2} p A_0},$$

$$\sqrt{\mathfrak{M}_1} = A_0 + A_1 + A_2 + \cdots + A_{\frac{p-1}{2}},$$

$$\sqrt{\mathfrak{M}_2} = A_0 + \alpha A_1 + \alpha^4 A_2 + \cdots + \alpha^{(\frac{p-1}{2})^2} A_{\frac{p-1}{2}},$$

$$\sqrt{\mathfrak{M}_3} = A_0 + \beta A_1 + \beta^4 A_2 + \cdots + \beta^{(\frac{p-1}{2})^2} A_{\frac{p-1}{2}},$$

$$\text{etc.,} \tag{11.42}$$

where $\alpha$, $\beta$, etc. were roots of the equation $x^p - 1 = 0$. Although he did not provide a proof of this result in this or his subsequent papers, Jacobi noted that this theorem was of great importance in the algebraic theory of the transformation and division of elliptic functions.

## 11.3  Sohnke's Paper on Modular Equations

In an 1837 paper published in *Crelle's Journal*,[27] L. A. Sohnke, Professor at Halle, gave the details of the proof of Jacobi's assertion that the $p + 1$ values of $\sqrt{\lambda}$ that could be obtained in (11.38) could be derived from (11.39) by replacing $q$ successively by $q^p$, $q^{\frac{1}{p}}, \alpha q^{\frac{1}{p}}, \ldots, \alpha^{p-1} q^{\frac{1}{p}}$, where $\alpha^p = 1$. Recall that if

$$\omega = \frac{mK + im'K'}{p}, \tag{11.43}$$

---

[24]  Klein (1921–1923) vol. 3, pp. 50–51 and pp. 137–139.
[25]  See Jacobi (1969) vol. 1, pp. 130–131. For a more direct proof, see Borwein and Borwein (1987) p. 137.
[26]  Ibid. p. 261.    [27]  Sohnke (1837).

and $p$ does not divide both $m$ and $m'$, then the modulus $\lambda$ and the multiplier $M$ in (11.1) are given by (3.103) and (3.104):

$$\lambda = k^p \prod_{s=1}^{\frac{p-1}{2}} \text{sn}^4(K - 4s\,\omega), \quad M = (-1)^{\frac{p-1}{2}} \prod_{s=1}^{\frac{p-1}{2}} \frac{\text{sn}^2(K - 4s\,\omega)}{\text{sn}^2 4s\,\omega}. \tag{11.44}$$

Sohnke showed that, in fact, all the $p + 1$ values of $\lambda$ could be obtained by taking $\omega$ to be successively

$$\frac{K}{p}, \frac{iK'}{p}, \frac{K + ilK'}{p}, \quad l = 1, 2, \ldots, p-1; \tag{11.45}$$

and that when $\omega$ was successively assigned the values listed in (11.45), then the corresponding values of $\lambda$ would be

$$k(q^p), \ k\big(q^{\frac{1}{p}}\big), \ k\big(\alpha^l q^{\frac{1}{p}}\big), \quad l = 1, 2, \ldots, p-1 \ \text{ and } \ \alpha^p = 1. \tag{11.46}$$

With $q = e^{\frac{-\pi K'}{K}}$, we present the details of the first two cases given in the list (11.46) and leave the remaining cases as an exercise. Sohnke started by noting that

$$\text{sn}\left(K - \frac{2Kx}{\pi}\right) = \frac{\text{cn}\left(\frac{2Kx}{\pi}\right)}{\text{dn}\left(\frac{2Kx}{\pi}\right)} = \frac{2q^{\frac{1}{4}}}{\sqrt{k}} \cos x \prod_{n=1}^{\infty} \frac{1 + 2q^{2n} \cos 2x + q^{4n}}{1 + 2q^{2n-1} \cos 2x + q^{4n-2}}. \tag{11.47}$$

Note that formulas (11.47) follow from (3.62), (3.158), and (3.159). Thus, he could write $\lambda$ in (11.44) as

$$\lambda = 2^{2(p-1)} k\, q^{\frac{p-1}{2}} \prod_{s=1}^{\frac{p-1}{2}} \left\{ \cos^4 \frac{2s\pi}{p} \prod_{n=1}^{\infty} \left( \frac{1 + 2q^{2n} \cos \frac{4s\pi}{p} + q^{4n}}{1 + 2q^{2n-1} \cos \frac{4s\pi}{p} + q^{4n-2}} \right)^4 \right\}. \tag{11.48}$$

Sohnke then observed that a formula of Cotes, (6.14), implied that

$$\prod_{t=1}^{\frac{p-1}{2}} \left( 1 + 2y \cos \frac{4t\pi}{p} + y^2 \right) = \frac{1 + y^p}{1 + y}; \tag{11.49}$$

this in turn implied

$$\prod_{t=1}^{\frac{p-1}{2}} \cos^4 \frac{2t\pi}{p} = \frac{1}{2^{2(p-1)}}. \tag{11.50}$$

Thus, (11.48) simplified to

$$\lambda = k\, q^{\frac{p-1}{2}} \prod_{n=1}^{\infty} \left( \frac{(1 + q^{2pn})(1 + q^{2n-1})}{(1 + q^{p(2n-1)})(1 + q^{2n})} \right)^4. \tag{11.51}$$

then utilized the formula for $k$ from (3.161) to see that

$$\lambda = 4q^{\frac{p}{2}} \prod_{n=1}^{\infty} \left( \frac{1 + q^{2pn}}{1 + q^{p(2n-1)}} \right)^4 = k(q^p).$$

Since Jacobi and Sohnke had found the modular equation in terms of $u = \sqrt[4]{k}$ and $v = \sqrt[4]{\lambda}$, Sohnke had to calculate the fourth root of the product in (11.50). He noted that

$$\prod_{t=1}^{\frac{p-1}{2}} \cos \frac{4t\pi}{p} = \prod_{t=1}^{\frac{p-1}{2}} \cos \frac{2t\pi}{p} = \pm \frac{1}{2^{\frac{p-1}{2}}}, \tag{11.52}$$

where the $+$ sign would be chosen when $p = 8m \pm 1$ and the $-$ sign when $p = 8m \pm 3$. We can verify (11.52) by taking $y = i$ in (11.49) to obtain

$$\prod_{t=1}^{\frac{p-1}{2}} \cos \frac{2t\pi}{p} = \frac{i^{-\frac{p-1}{2}}}{2^{\frac{p-1}{2}}} \frac{1 + i^p}{1 + i} = (-1)^{\frac{p^2-1}{8}}. \tag{11.53}$$

It may be checked that (11.53) is equivalent to (11.52). We also note that

$$\left(\frac{2}{p}\right) = (-1)^{\frac{p^2-1}{8}}, \tag{11.54}$$

where the Legendre symbol $\left(\frac{a}{p}\right)$ is $\pm 1$, depending on whether or not $a$ is congruent to a square modulo $p$.

Sohnke proved (11.46) for the general case, but to simplify the calculations, we will briefly indicate how, in particular, $\omega = \frac{iK'}{p}$ leads to the value $\lambda = k(q^{\frac{1}{p}})$. Here we have to determine the value of

$$\lambda = k^p \prod_{s=1}^{\frac{p-1}{2}} \mathrm{sn}^4 \left( K - \frac{4siK'}{p} \right) \tag{11.55}$$

so we have to set $x = \frac{2si\pi K'}{pK}$ in (11.47). Then

$$\frac{1 + 2q^{2n} \cos 2x + q^{4n}}{1 + 2q^{2n-1} \cos 2x + q^{4n-2}} = \frac{\left(1 + e^{-\frac{2n\pi K'}{K} - \frac{4s\pi K'}{pK}}\right)\left(1 + e^{-\frac{2n\pi K'}{K} + \frac{4s\pi K'}{pK}}\right)}{\left(1 + e^{-\frac{(2n-1)\pi K'}{K} - \frac{4s\pi K'}{pK}}\right)\left(1 + e^{-\frac{(2n-1)\pi K'}{K} + \frac{4s\pi K'}{pK}}\right)}$$

$$= \frac{\left(1 + \left(q^{\frac{1}{p}}\right)^{2pn-4s}\right)\left(1 + \left(q^{\frac{1}{p}}\right)^{2pn+4s}\right)}{\left(1 + \left(q^{\frac{1}{p}}\right)^{p(2n-1)-4s}\right)\left(1 + \left(q^{\frac{1}{p}}\right)^{p(2n-1)+4s}\right)}.$$

When these values are substituted in (11.46) and (11.48), and observing that

$$k(q) = k\left(\left(q^{\frac{1}{p}}\right)^p\right),$$

then the expression for $\lambda$ simplifies to

$$\lambda = 4q^{\frac{1}{2p}} \prod_{n=1}^{\infty} \left( \frac{1 + \left(q^{\frac{1}{p}}\right)^{2n}}{1 + \left(q^{\frac{1}{p}}\right)^{2n-1}} \right)^4 = k(q^{\frac{1}{p}}). \tag{11.56}$$

We now turn to Sohnke's two methods[28] for finding the modular equations connecting $u = \sqrt[4]{k}$ and $v = \sqrt[4]{\lambda}$. Recall that Jacobi stated his modular equations of orders 3 and 5 in such a form. By (11.51), (11.53), and (11.56), it follows that for a given $u$ there are $p + 1$ values of $v$:

$$u = \sqrt{2}\, q^{\frac{1}{8}} \prod_{n=1}^{\infty} \frac{1+q^{2n}}{1+q^{2n-1}}, \quad v_1 = (-1)^{\frac{p^2-1}{8}} \sqrt{2}\, q^{\frac{p}{8}} \prod_{n=1}^{\infty} \frac{1+q^{2np}}{1+q^{(2n-1)p}}, \quad (11.57)$$

$$v_{\mu+2} = \sqrt{2}\left(\alpha^\mu q^{\frac{1}{p}}\right)^{\frac{1}{8}} \prod_{n=1}^{\infty} \frac{1+\left(\alpha^\mu q^{\frac{1}{p}}\right)^{2n}}{1+\left(\alpha^\mu q^{\frac{1}{p}}\right)^{2n-1}}, \quad \mu = 0, 1, \ldots, p-1. \quad (11.58)$$

An application of the Cotes formula gives us the relation

$$(-1)^{\frac{p^2-1}{8}} u^{p+1} = v_1 v_2 \cdots v_{p+1}. \quad (11.59)$$

The modular equation must have roots $v_1, v_2, \ldots, v_{p+1}$:

$$(v - v_1)(v - v_2) \cdots (v - v_{p+1}) = v^{p+1} + C_1 v^p + \cdots + C_{p+1} = 0, \quad (11.60)$$

where $C_1, \ldots, C_{p+1}$ denote the elementary symmetric functions of the roots. Note that by (11.59)

$$C_{p+1} = (-1)^{\frac{p^2-1}{8}} u^{p+1}, \quad (11.61)$$

since $p$ is an odd prime. To analyze the modular equation, note that $u$ has the power series:

$$u = \sqrt{2}\, q^{\frac{1}{8}}(1 - q + 2q^2 - 3q^3 + 4q^4 - 6q^5 + \cdots). \quad (11.62)$$

Observe also that since the modular equation remains unchanged by interchanging $k$ and $\lambda$, it follows that (11.60) is not altered when we write $v$ instead of $u$ and $(-1)^{\frac{p^2-1}{8}} u$ instead of $v$. Thus, since $v^{p+1}$ is the highest power of $v$ in the equation, $u^{p+1}$ must be the highest power of $u$ in the equation. Now by (11.62)

$$u^{p+1} = 2^{\frac{p+1}{2}} q^{\frac{p+1}{8}} (1 - q + 2q^2 - 3q^3 + \cdots)^{p+1}. \quad (11.63)$$

Again by the second equation in (11.57) we get

$$(-1)^{\frac{p^2-1}{8}} v_1 = \sqrt{2}\, q^{\frac{p}{8}}(1 - q^p + 2q^{2p} - \cdots) \quad (11.64)$$

and since $p + 1$ is even we have

$$v_1^{p+1} = 2^{\frac{p+1}{2}} q^{\frac{p(p+1)}{8}} (1 - q^p + 2q^{2p} - \cdots)^{p+1}. \quad (11.65)$$

Since the exponent of the lowest power of $q$ in (11.65) is

$$\frac{p(p+1)}{8} = \frac{p^2-1}{8} + \frac{p+1}{8} \quad (11.66)$$

---

[28] See also Enneper (1890) pp. 452–469 for a summary of Sohnke's two methods.

and since $\frac{p^2-1}{8}$ is an integer for odd $p$, we may conclude that the exponent of the lowest power of $q$ in $v_1^{p+1}$ leaves the same remainder as the exponent of the lowest power of $q$ in $u^{p+1}$ given in (11.63). Next, let $p + 1 = 8\mu + t$, where $t$ is the remaidner when $p + 1$ is divided by 8. Thus, $v^{p+1}$ and $u^{p+1}$ each contain the factor $q^{\frac{t}{8}}$; hence, each term $v^a u^b$ in the modular equation must have $q^{\frac{t}{8}}$ as a factor. By (11.62) and (11.64) and since

$$pa + b = (pa + b - t) + t, \tag{11.67}$$

$pa + b - t$ must be divisible by 8. This means that $v^a$ must have a coefficient of the form

$$u^b(\beta + \gamma u^8 + \delta u^{16} + \cdots). \tag{11.68}$$

We may conclude from this analysis that the term $vu$, where $a = 1$, $b = 1$, is certainly possible, because the lowest power of $q$ for this case is $p + 1$. In fact, we may determine the value of $\beta$ in (11.68) for this case, since $\beta uv$ and $(-1)^{\frac{p^2-1}{8}} u^{p+1}$ have the same lowest power of $q$. In fact, equations (11.62), (11.64), and (11.65) imply that

$$\beta(\sqrt{2})^2 + (\sqrt{2})^{p+1} = 0 \quad \text{or} \quad \beta = -2^{\frac{p-1}{2}}. \tag{11.69}$$

These calculations allow us to express the modular equation of order $p$ in the form

$$v^{p+1} + \cdots - 2^{\frac{p-1}{2}} uv + (-1)^{\frac{p^2-1}{8}} u^{p+1} = 0. \tag{11.70}$$

In order to be able to determine more coefficients of the modular equation, Sohnke showed[29] that if we take $u = 1$ in the modular equation, then it reduces to $(v - 1)^{p+1}$ when $p = 8\mu \pm 1$, and to $(v - 1)(v + 1)^p = 0$ in the case $p = 8\mu \pm 3$. Sketching a proof of this result, we observe that by (3.68) we have

$$\mathrm{sn}(K - \omega, k) = i\mathrm{tn}\left(\frac{K - \omega}{i}, k'\right). \tag{11.71}$$

Now if $u = 1$, then $k = 1$ and $k' = 0$, so that (11.71) becomes

$$\begin{aligned}
\mathrm{sn}(K - \omega, 1) &= i\mathrm{tn}\left(\frac{K - \omega}{i}, 0\right) = i\tan\frac{K - \omega}{i} \\
&= \frac{e^{K-\omega} - e^{-(K-\omega)}}{e^{K-\omega} + e^{-(K-\omega)}} = \frac{1 - e^{-2(K-\omega)}}{1 + e^{-2(K-\omega)}}.
\end{aligned} \tag{11.72}$$

When $\omega = \frac{4r(mK + m'iK')}{p}$, we obtain

$$\mathrm{sn}\left(K - \frac{4r(mK + m'iK')}{p}, 1\right) = \frac{e^{\frac{-2(4mr-p)}{p}K} - e^{\frac{8rm'iK'}{p}}}{e^{\frac{-2(4mr-p)}{p}K} + e^{\frac{8rm'iK'}{p}}}. \tag{11.73}$$

Observe that taking $k = 1$ gives us $K = \infty$ and $K' = \frac{\pi}{2}$; by (11.73) we then have

$$\mathrm{sn}\left(K - \frac{4r(mK + m'iK')}{p}, 1\right) = \begin{cases} +1, & p - 4mr > 0, \\ -1, & p - 4mr < 0. \end{cases} \tag{11.74}$$

[29] Sohnke (1837) pp. 111–112.

Recall that

$$v = u^p \operatorname{sn}(K - 4\,\omega)\operatorname{sn}(K - 8\,\omega)\cdots\operatorname{sn}\big(K - 2(p-1)\,\omega\big). \tag{11.75}$$

For $m = 0$ and $u = 1$, (11.74) shows that every factor on the right-hand side of (11.75) reduces to 1. Thus, $v = 1$. For $m = 1$, by (11.74) we see that

$$\operatorname{sn}(K - 4r\,\omega) = \begin{cases} +1, & p > 4r, \\ -1, & p < 4r. \end{cases}$$

Now if $p = 8\mu + 1$, then $r = 1, 2, \ldots, 4\mu$. Thus, on the right-hand side of (11.75), $2\mu$ factors reduce to 1 and $2\mu$ factors reduce to $-1$, so that $v = 1$. Hence, in the case $p = 8\mu + 1$, all the $p + 1$ values of $v$ will be 1 and the modular equation will reduce to $(v - 1)^{p+1} = 0$. The cases $p = 8\mu - 1$ and $8\mu \pm 3$ can be similarly worked out to complete Sohnke's proof.

It may be instructive to study the methods for deriving particular examples of modular equations. In the case $p = 3$, so that the degree of the equation must be 4, the value of $t$ in (11.67) will be 4. Therefore, the terms that can come into this equation are $v^4$, $v^3 u^3$, $vu$, $u^4$ and by (11.70) it will take the form

$$v^4 + \beta u^3 v^3 - 2uv - u^4 = 0. \tag{11.76}$$

Since (11.76) must reduce to $(v - 1)(v + 1)^3$ when $u = 1$, we see that $\beta = 2$ so that we have obtained Jacobi's modular equation of order 3. Next, take $p = 5$ so that $t = 6$; the terms in this case must be $v^6$, $v^5 u^5$, $v^4 u^2$, $v^2 u^4$, $vu$, $u^6$ and the modular equation takes the form

$$v^6 + \beta u^5 v^5 + \gamma u^2 v^4 + \delta u^4 v^2 - 4uv - u^6 = 0.$$

To determine $\beta, \gamma, \delta$, note that when $u = 1$, the modular equation reduces to $(v - 1)(v + 1)^5 = (v^2 - 1)(v + 1)^4 = 0$, yielding $\beta = 4$, $\gamma = 5$, $\delta = -5$. The modular equation of order 5 is thus

$$v^6 + 4u^5 v^5 + 5u^2 v^4 - 5u^4 v^2 - 4uv - u^6 = 0;$$

indeed, this is Jacobi's modular equation of order 5.

Note that $u^5 v^5$ and $uv$ have the same coefficient except for the sign and the same is true for $u^2 v^4$ and $u^4 v^2$. A similar result holds for the modular equation of order 3. Sohnke gave an explanation for this phenomenon:[30] Recall that Jacobi had shown that the modular equation would remain unaltered by an interchange of $k$ and $\frac{1}{k}$ and of $\lambda$ and $\frac{1}{\lambda}$. Setting $u = \frac{1}{u}$ and $v = \frac{1}{v}$, Sohnke argued that $v^a u^b$ and $v^{p+1-a} u^{p+1-b}$ would have the same factors, except in the case when $p = 8\mu \pm 3$ when the factors would would have different signs. Moreover, since the modular equation would remain unchanged when $k$ and $\lambda$ were interchanged, $v^a u^b$ and $v^b u^a$ must have the same coefficients, except for the change in signs when $p = 8\mu \pm 3$ with $a$, $b$ even.

The techniques for finding modular equations of orders 3 and 5 work equally well for orders 7 and 11. But for obtaining modular equations of orders 13 and higher, Sohnke required the expansions of $u, u^2, \ldots$ and $v, v^2, \ldots$ in powers of $q$. Taking

---

[30] Ibid. p. 111.

$u = \sqrt{2}\, q^{\frac{1}{8}} f(q)$, he calculated:[31] $f(q)$ up to the 26th power of $q$; $f^2(q)$ and $f^3(q)$ up to the 21st power; $f^4(q)$ through $f^{18}(q)$ up to the 18th power; $f^{19}(q)$ and $f^{20}(q)$ up to the 10th power.

To see how Sohnke utilized these expansions, take the case $p = 13$, in which the modular equation has the form:

$$v^{14} + v^{13}u^5(\alpha_1 + \alpha_2 u^8) + v^{12}u^2(\beta_1 + \beta_2 u^8) + \gamma v^{11}u^7 + v^{10}u^4(\delta_1 + \delta_2 u^8)$$
$$+ v^9 u(\epsilon_1 + \epsilon_2 u^8) + \zeta v^8 u^6 + v^7 u^3(\eta_1 + \eta_2 u^8) + \theta v^6 u^8 + v^5 u^5(\iota_1 + \iota_2 u^8)$$
$$+ v^4 u^2(\chi_1 + \chi_2 u^8) + \lambda v^3 u^7 + v^2 u^4(\mu_1 + \mu_2 u^8) + vu(\nu_1 + \nu_2 u^8) - u^{14} = 0.$$
$$(11.77)$$

Considering the fact that $u$ and $v$ can be interchanged, while keeping in mind the cases in which there is a change in sign, we see that

$$\iota_2 = \alpha_1, \quad \mu_2 = -\beta_1, \quad \delta_2 = -\beta_2, \quad \nu_2 = \gamma, \quad \chi_2 = -\delta_1,$$
$$\nu_2 = \epsilon_1, \quad \theta = -\zeta, \quad \lambda = \nu_1, \quad \mu_1 = \chi_1.$$

Then, since the modular equation remains unchanged when $u$ and $v$ are replaced by $\frac{1}{u}$ and $\frac{1}{v}$, respectively,

$$\nu_2 = -\alpha_1, \quad \nu_1 = -\alpha_2, \quad \mu_2 = -\beta_1, \quad \mu_1 = -\beta_2, \quad \lambda = -\gamma, \quad \chi_2 = -\delta_1,$$
$$\chi_1 = -\delta_2, \quad \iota_2 = -\epsilon_1, \quad \iota_1 = -\epsilon_2, \quad \theta = -\zeta, \quad \eta_2 = -\eta_1.$$

To obtain the values of the integers $\alpha_1, \alpha_2, \ldots$, etc, we observe that when $u = 1$ in the modular equation, it reduces to $(v + 1)^{13}(v - 1) = 0$ or

$$v^{14} + 12v^{13} + 65v^{12} + 208v^{11} + 429v^{10} + 572v^9 + 429v^8$$
$$- 429v^6 - 572v^5 - 429v^4 - 208v^3 - 65v^2 - 12v - 1 = 0. \qquad (11.78)$$

Set $u = 1$ in (11.77) and equate the coefficients of the powers of $v$ with those in (11.78) to arrive at

$$\alpha_1 + \alpha_2 = 12, \quad \beta_1 + \beta_2 = 65, \quad \gamma = 208, \quad \delta_1 + \delta_2 = 429,$$
$$\epsilon_1 + \epsilon_2 = 572, \quad \zeta = 429, \quad \eta_1 + \eta_2 = 0.$$

Moreover, (11.69) shows that $\alpha_2 = 64$. Thus, we have determined all the coefficients except one, and we can choose that to be $\beta_2$. The modular equation can now be expressed:

$$v^{14} + v^{13}u^5(-52 + 64u^8) + v^{12}u^2\big((65 - \beta_2) + \beta_2 u^8\big) + 208v^{11}u^7$$
$$+ v^{10}u^4\big((429 + \beta_2) - \beta_2 u^8\big) + v^9 u(52 + 520u^8) + 429v^8 u^6$$
$$- v^7 u^3(208 - 208u^8) - 429v^6 u^8 - v^5 u^5(520 + 52u^8) + v^4 u^2\big(\beta_2 - (429 + \beta_2)u^8\big)$$
$$- 208v^3 u^7 - v^2 u^4\big(\beta_2 + (65 - \beta_2)u^8\big) - vu(64 - 52u^8) - u^{14} = 0. \qquad (11.79)$$

---

[31] Ibid. pp. 113–116.

The lowest power of $v$ that contains $\beta_2$ in its coefficient is $v^2$. Now

$$v^2 = 2q^3 \, q^{\frac{2}{8}} \big( f(q^{13}) \big)$$
$$u^4 = 4q^{\frac{4}{8}}(1 - 3q + \cdots),$$

so that

$$-\beta_2 v^2 u^4 = -8\beta_2 \, q^{\frac{6}{8}}(q^3 + \cdots).$$

Other terms that have $q^{\frac{6}{8}} \cdot q^3$ in them are:

$$u^{14} = 128 q^{\frac{6}{8}}(-q + 14q^2 - 119q^3 + \cdots)$$
$$52vu^9 = -52(32) \, q^{\frac{6}{8}}(q^2 - 9q^3 + \cdots)$$
$$-64vu = 128 q^{\frac{6}{8}}(q - q^2 + 2q^3 + \cdots).$$

Equating the coefficients of $q^{\frac{6}{8}} \cdot q^3$ gives us

$$-119 \cdot 128 + 9 \cdot 32 \cdot 52 + 256 - 8\beta_2 = 0$$

Therefore, $\beta_2 = 0$, so that we have determined the modular equation of order 13.

Sohnke's second method[32] was to to employ the sums of powers of the roots of the modular equation

$$S_m = v_1^m + v_2^m + \cdots + v_{p+1}^m$$

to find the coefficients $C_1, C_2, \ldots, C_{p+1}$ in (11.60) by means of Newton's formulas:[33]

$$S_1 + C_1 = 0, \quad S_2 + C_1 S_1 + 2C_2 = 0, \quad S_3 + C_1 S_2 + C_2 S_1 + 3C_3 = 0, \ldots.$$

Beginning a very brief sketch of this method, we write equation (11.62) as

$$u = \sqrt{2} \, q^{\frac{1}{8}} \, f(q) = \sqrt{2} q^{\frac{1}{8}}(1 + A_1 q + A_2 q^2 + A_3 q^3 + \cdots), \qquad (11.80)$$

that is, putting $f$ in terms of the general symbols $A_1, A_2, A_3, \cdots$. We may similarly rewrite

$$v_1 = (-1)^{\frac{p^2-1}{8}} \sqrt{2} q^{\frac{p}{8}} \, f(q^p), \quad v_2 = \sqrt{2} q^{\frac{1}{8p}} \, f(q^{\frac{1}{p}}), \ldots,$$
$$v_{p+1} = \sqrt{2}\big(\alpha^{p-1} q^{\frac{1}{p}}\big)^{\frac{1}{8}} f\big(\alpha^{p-1} q^{\frac{1}{p}}\big)$$

in terms of those general symbols. Observe that the coefficient of $A_r$ in

$$2^{-\frac{1}{2}}(v_2 + v_3 + \cdots + v_{p+1})$$

is

$$\sum_{\mu=0}^{p-1} \big(\alpha^{\mu} q^{\frac{1}{p}}\big)^{\frac{8r+1}{8}}. \qquad (11.81)$$

---

[32] Ibid. pp. 118–130.

[33] Discovered by Newton in 1665–1666. See Newton (1967–1981) vol. 1, p. 519. This result was first published in Newton's 1707 *Arithmetica Universalis*.

Since $\alpha^p = 1$, the sum in (11.81) is zero, providing $8r + 1$ is not a multiple of $p$. When $8r + 1 = lp$, where $l$ is an integer, then the sum is $pq^{\frac{1}{8}}$. Recall that this is an example of Simpson's dissection of a series. Hence

$$S_1 = (-1)^{\frac{p^2-1}{8}} q^{\frac{p}{8}} (1 + A_1 q^p + A_2 q^{2p} + \cdots) + p\sqrt{2} q^{\frac{1}{8}} (A_r + A_{r+p}q + A_{r+2p}q^2 + \cdots).$$

Similarly, if we set

$$f^2(q) = 1 + B_1 q + B_2 q^2 + \cdots$$
$$f^3(q) = 1 + D_1 q + D_2 q^2 + \cdots,$$

then we can write

$$S_2 = 2q^{\frac{p}{4}} (1 + B_1 q^p + B_2 q^{2p} + \cdots) + 2pq^{\frac{l_1}{4}} (B_r + B_{r+p}q + \cdots) \tag{11.82}$$

$$S_3 = 2\sqrt{2} q^{\frac{3p}{8}} (1 + D_1 q^p + D_2 q^{2p} + \cdots) + 2\sqrt{2} pq^{\frac{l_2}{8}} (D_r + D_{r+p}q + \cdots), \tag{11.83}$$

where $8r + 2 = 2l_1 p$ and $8r + 3 = l_2 p$.

We illustrate the usefulness of these relations by applying $S_1$ and $S_2$ to determine the value of $\beta_2$ in (11.79). First observe that

$$-C_1 = S_1 = -52u^5 + 64u^{13}, \quad C_2 = ((65 - \beta_2)u^2 + \beta_2 u^{10}. \tag{11.84}$$

When $p = 13$, it can be shown from (11.82) that

$$S_2 = 26(-10 + 3348q + \cdots)q^{\frac{1}{4}}. \tag{11.85}$$

Now the factor $q^{\frac{1}{4}}$ can appear only in the terms $u^2, u^{10}, u^{18}, u^{26}$; we stop at $u^{26}$ because that is the highest power of $u$ that can appear in $S_2$. Thus, as can also be seen from the fact that $S_2 = S_1^2 - 2C_2$,

$$S_2 = mu^2 + m'u^{10} + m''u^{18} + m'''u^{26}.$$

Next, since

$$u^2 = 2q^{\frac{1}{4}} (1 - 2q + \cdots)$$
$$u^{10} = 32q^{\frac{1}{4}} (q - \cdots),$$

and since the smallest powers of $q$ in $u^{18}$ and $u^{26}$ are $q^{\frac{1}{4}+2}$ and $q^{\frac{1}{4}+3}$, it follows from (11.85) that

$$2m = -260, \quad -4m + 32m' = 26 \cdot 3348.$$

Thus, $m = -130$ and $m' = 2704$ and hence $\beta_2 = 0$. Sohnke also applied this method to determine modular equations of orders 17 and 19.[34]

Sohnke's method can also be employed to determine the $p + 1$ different values of the multiplier $M$ corresponding to the $p + 1$ different values of $\lambda$ in (11.1). It is possible that Brioschi used this method to calculate $M$ and attributed it to Sohnke. We need only

---

[34] Sohnke (1837) pp. 123–130.

compute

$$\prod_{s=1}^{\frac{p-1}{2}} \text{sn}^2 4s\,\omega$$

for the $p+1$ different values of $\omega$ given in (11.45). For example, when $\omega = \frac{K}{p}$, the product formula (3.157) for $\text{sn}\frac{2Kx}{\pi}$ with $x = \frac{2s\pi}{p}$ yields

$$\prod_{s=1}^{\frac{p-1}{2}} \text{sn}^2 \frac{4sK}{p} = \frac{2^{p-1}q^{\frac{p-1}{4}}}{k^{\frac{p-1}{2}}} \prod_{s=1}^{\frac{p-1}{2}} \sin^2 \frac{2s\pi}{p} \prod_{n=1}^{\infty} \prod_{s=1}^{\frac{p-1}{2}} \left( \frac{1 - 2q^{2n}\cos\frac{4s\pi}{p} + q^{4n}}{1 - 2q^{2n-1}\cos\frac{4s\pi}{p} + q^{4n-2}} \right)^2.$$

(11.86)

Again, by the de Moivre-Cotes formula,

$$\prod_{s=1}^{\frac{p-1}{2}} \left( 1 - 2y\cos\frac{4s\pi}{p} + y^2 \right) = \frac{1 - y^p}{1 - y};$$

letting $y \to 1$, we then obtain

$$\prod_{s=1}^{\frac{p-1}{2}} \sin^2 \frac{2s\pi}{p} = \frac{p}{2^{p-1}}.$$

Hence we can rewrite (11.86) as

$$\prod_{s=1}^{\frac{p-1}{2}} \text{sn}^2 \frac{4sK}{p} = \frac{pq^{\frac{p-1}{4}}}{k^{\frac{p-1}{2}}} \prod_{n=1}^{\infty} \left( \frac{1 - q^{2pn}}{1 - q^{2n}} \cdot \frac{1 - q^{2n-1}}{1 - q^{p(2n-1)}} \right)^2.$$

We have already given Sohnke's method for calculating

$$\prod_{s=1}^{\frac{p-1}{2}} \text{sn}^2 \left( K - \frac{4sK}{p} \right);$$

thus, by (3.167), (11.44), and (11.51), it follows that

$$\frac{1}{M} = (-1)^{\frac{p-1}{2}} p \prod_{n=1}^{\infty} \left( \frac{1 - (q^p)^{2n}}{1 - (q^p)^{2n-1}} \cdot \frac{1 + (q^p)^{2n-1}}{1 + (q^p)^{2n}} \right)^2 \left( \frac{1 - q^{2n-1}}{1 - q^{2n}} \cdot \frac{1 + q^{2n}}{1 + q^{2n-1}} \right)^2$$

$$= (-1)^{\frac{p-1}{2}} p \frac{K(q^p)}{K(q)}$$

$$= (-1)^{\frac{p-1}{2}} p \frac{\Lambda}{K},$$

where $\Lambda$ is given by

$$\Lambda = \int_0^1 \frac{dy}{\sqrt{(1 - y^2)(1 - \lambda^2 y^2)}},$$

as discussed in Section 3.4, and with $\lambda = k(q^p)$. Using (3.167) and (3.168) and taking $x = 0$ in (3.189), we see that

$$\frac{1}{M} = (-1)^{\frac{p-1}{2}} p \frac{\theta_3^2 (q^p)}{\theta_3^2(q)}. \tag{11.87}$$

We next compute the value of $\frac{1}{M}$ when $\omega = \frac{iK'}{p}$, with the remaining cases left to the reader as an exercise. To compute the value

$$\text{sn}^2 \frac{4siK'}{p}, \quad \text{we take} \quad x = \frac{2si\pi K'}{pK}$$

in the infinite product (3.157) for $\text{sn}\frac{2Kx}{\pi}$. The result is

$$\prod_{s=1}^{\frac{p-1}{2}} \text{sn}^2 \frac{4siK'}{p} = \frac{2^{p-1} q^{\frac{p-1}{4}}}{k^{\frac{p-1}{2}}}$$

$$\times \prod_{s=1}^{\frac{p-1}{2}} \left\{ \sin \frac{2si\pi K'}{pK} \prod_{n=1}^{\infty} \frac{\left(1 - e^{-\frac{2n\pi K'}{K} - \frac{4s\pi K'}{pK}}\right)\left(1 - e^{-\frac{2n\pi K'}{K} + \frac{4s\pi K'}{pK}}\right)}{\left(1 - e^{-\frac{(2n-1)\pi K'}{K} - \frac{4s\pi K'}{pk}}\right)\left(1 - e^{-\frac{(2n-1)\pi K'}{K} + \frac{4s\pi K'}{pk}}\right)} \right\}.$$

After simplifying and using the calculation for $\text{sn}^2 \left(K - \frac{4siK'}{p}\right)$, it emerges that

$$\frac{1}{M} = \frac{K(q^{\frac{1}{p}})}{K(q)} = \frac{\theta_3^2 (q^{\frac{1}{p}})}{\theta_3^2 (q)}. \tag{11.88}$$

## 11.4   Brioschi on Jacobi's Multiplier Equation

In a paper of 1858,[35] Brioschi proved Jacobi's formulas (11.42). Brioschi observed that the $p + 1$ values of the multiplier $\frac{1}{M}$ were given by (11.20):

$$z_\infty = (-1)^{\frac{p-1}{2}} \frac{p\Lambda_1}{K}, \quad z_0 = \frac{\Lambda_2}{K}, \quad \cdots, \quad z_{p-1} = \frac{\Lambda_p}{K}. \tag{11.89}$$

This notation for the subscripts of $z$ is due to Galois; it was later employed by Klein. Brioschi denoted these values by $z_1, z_2, \ldots, z_{p+1}$. He applied Jacobi's formula

$$\sqrt{\frac{2K}{\pi}} = \sum_{n=-\infty}^{\infty} q^{n^2}, \quad q = e^{-\frac{\pi K'}{K}}$$

and (11.22) to find that

$$\sqrt{z_\infty} = \sqrt{(-1)^{\frac{p-1}{2}} p} \frac{\sum_{n=-\infty}^{\infty} q^{pn^2}}{\sum_{n=-\infty}^{\infty} q^{n^2}} \tag{11.90}$$

---

[35] Brioschi (1858) pp. 275–277.

and

$$\sqrt{z_k} = \frac{\sum_{n=-\infty}^{\infty} p_k^{n^2}}{\sum_{n=-\infty}^{\infty} q^{n^2}}, \qquad (11.91)$$

where $p_k = e^{\frac{2\pi k i}{p}} q^{\frac{1}{p}}, k = 0, \ldots, p - 1$. Brioschi next noted that

$$\sum_{n=-\infty}^{\infty} q^{\frac{n^2}{p}} = \sum_{k=0}^{\frac{p-1}{2}} \sum_{n=-\infty}^{\infty} q^{\frac{(np+k)^2}{p}},$$

allowing him to set

$$A_0 = \frac{\sum_{n=-\infty}^{\infty} q^{pn^2}}{\sum_{n=-\infty}^{\infty} q^{n^2}}, \qquad A_k = \frac{2q^{\frac{k^2}{p}} \sum_{n=-\infty}^{\infty} q^{pn^2+2kn}}{\sum_{n=-\infty}^{\infty} q^{n^2}}, \qquad (11.92)$$

where $k = 1, \ldots, \frac{p-1}{2}$. Thus, he had Jacobi's relations (11.42); with $\epsilon$ a complex $p$th root of unity,

$$\sqrt{z_\infty} = \sqrt{(-1)^{\frac{p-1}{2}} p} \, A_0, \qquad (11.93)$$

$$\sqrt{z_k} = \sum_{j=0}^{\frac{p-1}{2}} \epsilon^{j^2 k} A_j, \qquad k = 0, 1, \ldots, p - 1. \qquad (11.94)$$

Brioschi then obtained relations satisfied by $A_0, A_1, \ldots$ for $p = 3$ and $p = 5$ by using the multiplier equations of orders 3 and 5. When $p = 3$, the multiplier equation was given by (11.18):

$$x^4 - 6x^2 + 8(1 - 2k^2)x - 3 = 0.$$

For $p = 3$ then

$$A_0 = \frac{\sum_{n=-\infty}^{\infty} q^{3n^2}}{\sum_{n=-\infty}^{\infty} q^{n^2}}, \qquad A_1 = \frac{2q^{\frac{1}{3}} \sum_{n=-\infty}^{\infty} q^{3n^2+2n}}{\sum_{n=-\infty}^{\infty} q^{n^2}},$$

and the roots were given by (11.93) and (11.94). Hence

$$z_\infty = -3A_0^2, \quad z_1 = (A_0 + A_1)^2, \quad z_2 = (A_0 + \epsilon A_1)^2, \quad z_3 = (A_0 + \epsilon^2 A_1)^2,$$

so that the elementary symmetric functions of these roots could be expressed as

$$z_\infty + z_1 + z_2 + z_3 = 0, \qquad (11.95)$$

$$z_\infty(z_1 + z_2 + z_3) + z_1(z_2 + z_3) + z_2 z_3 = 6, \qquad (11.96)$$

$$z_\infty(z_1 z_2 + z_1 z_3 + z_2 z_3) + z_1 z_2 z_3 = -8(1 - 2k^2), \qquad (11.97)$$

$$z_\infty z_1 z_2 z_3 = -3. \qquad (11.98)$$

Clearly, (11.95) was identically satisfied; (11.96) and (11.97) gave Brioschi the relations

$$A_0(A_0^3 + A_1^3) = 1, \quad 8A_0^6 - 20A_0^3 A_1^3 - A_1^6 = 8(1 - 2k^2). \qquad (11.99)$$

Here note that the first relation in (11.99) is equivalent to (11.98), so that (11.98) does not produce a new relation.

Jacobi[36] had predicted that the relations (11.93) and (11.94) would play an important role in the algebraic theory of transformations. Brioschi regarded his own application of these relations to the solution of a fifth-degree equation as an exemplification of Jacobi's remark.

The multiplier equation of order 5 was computed by Brioschi to be

$$z^6 - 10z^5 + 35z^4 - 60z^3 + 55z^2 - 2(13 - 2^7k^2k'^2)z + 5 = 0. \qquad (11.100)$$

As we show in the next section, (11.100) may be derived directly or may be obtained from Jacobi's equation (11.19).

For $p = 5$, the values of $A_0$, $A_1$, and $A_2$ were:

$$A_0 = \frac{\sum_{n=-\infty}^{\infty} q^{5n^2}}{\sum_{n=-\infty}^{\infty} q^{n^2}}, \quad A_1 = \frac{2q^{\frac{1}{5}} \sum_{n=-\infty}^{\infty} q^{5n^2+2n}}{\sum_{n=-\infty}^{\infty} q^{n^2}}, \quad A_2 = \frac{2q^{\frac{4}{5}} \sum_{n=-\infty}^{\infty} q^{5n^2+4n}}{\sum_{n=-\infty}^{\infty} q^{n^2}};$$

with $\epsilon = e^{\frac{2\pi i}{5}}$,

$$z_\infty = 5A_0^2, \qquad\qquad z_1 = (A_0 + A_1 + A_2)^2,$$
$$z_2 = (A_0 + \epsilon A_1 + \epsilon^4 A_2)^2, \qquad z_3 = (A_0 + \epsilon^2 A_1 + \epsilon^3 A_2)^2,$$
$$z_4 = (A_0 + \epsilon^3 A_1 + \epsilon^2 A_2)^2, \qquad z_5 = (A_0 + \epsilon^4 A_1 + \epsilon A_2)^2 \ .$$

Brioschi set[37]

$$A = A_0^2 + A_1 A_2$$
$$B = 8A_0^4 A_1 A_2 - 2A_0^2 A_1^2 A_2^2 + A_1^3 A_2^3 - A_0(A_1^5 + A_2^5)$$
$$C = 320A_0^6 A_1^2 A_2^2 - 160A_0^4 A_1^3 A_2^3 + 20A_0^2 A_1^4 A_2^4 + 6A_1^5 A_2^5$$
$$\quad - 4A_0(32A_0^4 - 20A_0^2 A_1 A_2 + 5A_1^2 A_2^2)(A_1^5 + A_2^5) + A_1^{10} + A_2^{10}$$

and then computed the elementary symmetric functions of $z_\infty, z_1, z_2, z_3, z_4, z_5$ to be

$$S_1 = 10A, \qquad\qquad S_2 = 35A^2,$$
$$S_3 = 60A^3 - 10B, \qquad\qquad S_4 = 55A^4 - 30AB,$$
$$S_5 = 26A^5 - 30A^2 B + C, \qquad S_6 = 5(A^3 - B)^2.$$

For the multiplier equation (11.100), the values of $S_i$ could then be written:

$$S_1 = 10, \quad S_2 = 35, \quad S_3 = 60,$$
$$S_4 = 55, \quad S_5 = 2(13 - 2^7k^2k'^2), \quad S_6 = 5.$$

Brioschi finally obtained $A = 1$, $B = 0$, and $C = 2^8k^2k'^2$. He also got the relations for $p = 5$ corresponding to the relations for $p = 3$ as given in (11.99); he employed these relations to show that the multiplier equation (11.100) could be reduced to a fifth-order equation. Following Hermite, he took the resolvent to be

$$x = (z_\infty - z_1)(z_5 - z_2)(z_3 - z_4)$$

[36] Jacobi (1969) vol. 1, p. 261.      [37] Brioschi (1901) vol. 1, pp. 335–338.

and demonstrated that $x$ satisfied the fifth-order equation

$$x(x^2 + 5^2 2^8 k^2 k'^2)^2 = 5\sqrt{5}\, 2^{23}\, k^4 k'^4 (1 - 4k^2 k'^2).  \qquad (11.101)$$

Setting

$$f(\omega) = \left( \sum_{n=-\infty}^{\infty} e^{i\pi n^2 \omega} \right)^2$$

and

$$F(\omega) = \left( 5f(5\,\omega) - f\left(\frac{\omega}{5}\right) \right) \left( f\left(\frac{\omega+8}{5}\right) - f\left(\frac{\omega+2}{5}\right) \right)$$
$$\cdot \left( f\left(\frac{\omega+4}{5}\right) - f\left(\frac{\omega+6}{5}\right) \right),$$

he arrived at the five roots of (11.101):

$$x_i = \frac{F(\omega + (i-1)2)}{f^3(\omega)}, \quad i = 1, \ldots, 5.$$

Brioschi also showed[38] that (11.101) could be converted to Hermite's form (5.18)

$$X^5 - X - \frac{2}{\sqrt[4]{5^5}} \frac{1 + k^2}{k' \sqrt{k}} = 0$$

by means of the substitution

$$\sqrt{x} = 4\sqrt[4]{-25k^2 k'^2}\, X.$$

## 11.5 Joubert on the Multiplier Equation

Father Charles Joubert obtained results for the multiplier equation by methods analogous to Sohnke's for the modular equation. Joubert graduated first in his class from the École Normal and taught mathematics for many years at the Collège Rollin in Paris during the second half of the nineteenth century. He made notable contributions to the application of elliptic functions to algebra and number theory. In addition to several other papers, in 1860 he published a thirty-page pamphlet, *Sur la théorie des fonctions elliptiques et son application à la théorie des nombres*. His 1875 short book, *Sur les équations qui se recontrent dans la théorie de la transformation des fonctions elliptiques*, contains his results with proofs on the multiplier equation, although he obtained and published most of these results, though mainly without proofs, almost two decades earlier.

Joubert employed two theorems[39] in his construction of multiplier equations:

**Theorem I.** Let $F(k, x) = 0$ denote the multiplier equation with $x = \frac{1}{M}$. Then the roots of the equation $F(\frac{1}{k}, x) = 0$ are given by the various values of $\frac{\lambda}{kM}$.

---

[38] Brioschi (1901) vol. 1, pp. 339–341.  [39] Joubert (1858) p. 341.

**Theorem II.** If one changes $k$ to $k'$ in $F(k, x) = 0$, the roots remain the same when the order of the transformation $p$ is of the form $4m + 1$; they change only in sign when $p$ is of the form $4m + 3$.

The first theorem was his own, and Joubert noted that the second theorem was due to Jacobi.[40] Joubert proved these theorems in his doctoral thesis and published it as his 1875 book. We will very briefly present Joubert's derivation of the multiplier equation. To prove his first theorem,[41] he first observed the fact that $k\left(\frac{\omega}{1+\omega}\right) = \frac{1}{k(\omega)}$, next expressed $M$ in terms of theta functions, and then changed $\omega$ to $\frac{\omega}{1+\omega}$ to see, by use of Hermite's transformation formula (5.7), that $M$ was changed to $\frac{Mk}{\lambda}$. For the second theorem,[42] he substituted $-\frac{1}{\omega}$ for $\omega$ to check that $M$ then changed to $(-1)^{\frac{p-1}{2}}M$.

Following Sohnke, Joubert set $k = 0$ in $F(k, x)$ to find the form of $F(0, x)$. For $\frac{1}{M}$ he took what would be the $p + 1$ values of $\frac{(-1)^{\frac{p-1}{2}}}{M}$, as given in (11.20), and expressed them as ratios of theta functions as in (11.87) and (11.88). He observed that as $k \to 0, q \to 0$, as one may understand from (3.161). Note that $\theta_3(q) \to 1$ as $q \to 0$. So when $k = 0$, one of the values of $x = \frac{1}{M}$ (our $\frac{(-1)^{p-1}}{M}$) would be $p$ and the rest of the values would be $(-1)^{\frac{p-1}{2}} \cdot 1$. Note that the positive sign would be chosen in the case $p = 4m + 1$ and the negative sign when $p = 4m + 3$. Thus, Joubert obtained the result

$$F(x, 0) = (x - 1)^p(x - p) \quad \text{for} \quad p = 4m + 1$$
$$F(x, 0) = (x + 1)^p(x - p) \quad \text{for} \quad p = 4m + 3. \tag{11.102}$$

Now $F(x, k^2)$ is a monic polynomial of degree $p + 1$, in which the coefficients are polynomials $\phi(k^2)$. The second theorem implies that $\phi(k^2) = \pm\phi(k'^2)$. With the upper sign in force, Joubert noted without proof that there existed a polynomial $\psi$ such that

$$\phi(k^2) = \phi(k'^2) = \psi(k^2k'^2) \tag{11.103}$$

and if the lower sign were in force, then there existed a polynomial $\psi_1$ such that

$$\phi(k^2) = -\phi(k'^2) = (k^2 - k'^2)\psi_1(k^2k'^2). \tag{11.104}$$

Note that (11.103) and (11.104) follow from the observation that with $f$ a polynomial, $a$ a constant, and $f(x) = \pm f(a - x)$, then there exist two polynomials $g_1$ and $g_2$ such that for the upper sign, $f(x) = g_1(x(a - x))$ and for the lower sign $f(x) = (a - 2x) g_2(x(a - x))$. Thus, for $p = 4m + 1$, (11.102) implied that the multiplier equation would take the form

$$F(x, k^2) = (x - 1)^p(x - p) + k^2k'^2xG(k^2k'^2, x) = 0, \tag{11.105}$$

[40] Jacobi (1969) p. 108.    [41] Joubert (1875) pp. 65–66.    [42] Ibid. pp. 67–68.

where $G$ was a polynomial of degree $p - 1$. On the other hand, if $p = 4m + 3$, then the coefficients of the odd powers of $x$ would contain $k^2 - k'^2$ as a factor, so that in that case Joubert had

$$F(x, k^2) = h(x^2) + (k^2 - k'^2)xh_1(x^2) + k^2k'^2x^2G_1(k^2k'^2, x^2)$$
$$+ (k^2 - k'^2)k^2k'^2x G_2(k^2k'^2, x^2) = 0, \tag{11.106}$$

where $h, h_1, G_1, G_2$ were polynomials. Observe that the first two terms in (11.106) could be rewritten as

$$k^2(h(x^2) + xh_1(x^2)) + k'^2(h(x^2) - xh_1(x^2))$$

and hence Joubert obtained

$$F(x, 0) = h(x^2) - xh_1(x^2) \quad \text{and} \quad F(-x, 0) = h(x^2) + xh_1(x^2).$$

With $f(x) = (x + 1)^p(x - p)$, he could recast equation (11.106) as

$$k^2 f(x) + k'^2 f(-x) + k^2k'^2x^2 G_1(k^2k'^2, x^2) + (k^2 - k'^2)k^2k'^2x G_2(k^2k'^2, x^2) = 0. \tag{11.107}$$

To further analyze equation (11.105), he changed $k$ to $\frac{1}{k}$ and took $f(x) = (x - 1)^p(x - p)$ so that (11.105) became

$$f\left(\frac{x}{k}\right) + \frac{k^2 - 1}{k^4}\frac{x}{k}G\left(\frac{k^2 - 1}{k^4}, \frac{x}{k}\right) = 0. \tag{11.108}$$

Next, Joubert observed that by theorem 1, the roots of (11.108) were the various values of $\frac{\lambda}{M}$. If $k = 0$, all these values would be zero. Thus, if the equation were multiplied by $k^{p+1}$, it would become monic and all terms except the first would vanish when $k = 0$. This in turn implied that if $G(k^2k'^2, x)$ contained the term

$$(A_0 + A_1k^2k'^2 + A_2k^4k'^4 + \cdots + A_hk^{2h}k'^{2h})x^g,$$

then (11.108) must contain the term

$$\frac{k^2 - 1}{k^{5+g}}\left(A_0 + A_1\frac{k^2 - 1}{k^4} + \cdots + A_h\frac{(k^2 - 1)^h}{k^{4h}}\right)x^{g+1},$$

which when multiplied by $k^{p+1}$ must be zero for $k = 0$. In this way, Joubert arrived at[43]

$$4h + g + 5 < p + 1$$

or

$$h < \frac{p - g - 4}{4}. \tag{11.109}$$

---

[43] Joubert (1875) pp. 73–76, especially p. 74.

Since (11.109) implies that $g \leq p - 4$, we see that the term of highest possible degree in $G(k^2 k'^2, x)$ is $p - 5$. We have thus verified (11.24), which Joubert applied to determine the multiplier equation of order 5. He first noted that such an equation must have the form

$$(x - 1)^5(x - 5) + Ak^2 k'^2 x = 0. \tag{11.110}$$

Changing $k$ to $\frac{1}{k}$ and $x$ to $\frac{x}{k}$, so that the equation

$$(x - k)^5(x - 5k) + A(k^2 - 1)kx = 0 \tag{11.111}$$

would have roots $\frac{\lambda}{M}$, Joubert then found $A = 256$ by determining the first term of the series in $k$ for $\frac{\lambda}{M}$. Thus, the multiplier equation of order 5 emerged as

$$(x - 1)^5(x - 5) + 256k^2 k'^2 x = 0. \tag{11.112}$$

Observe that (11.111), with $A = 256$, is the same as Jacobi's (11.19). A similar analysis of equation (11.107) leads to

$$G_1(k^2 k'^2, x^2) = Ax^{p-7} + Bx^{p-9}$$
$$+ (A' + A_1' k^2 k'^2)x^{p-11} + (B' + B_1' k^2 k'^2)x^{p-13} + \cdots$$

and

$$G_2(k^2 k'^2, x^2) = Cx^{p-7} + Dx^{p-9} + \cdots,$$

which were used by Joubert to determine the multiplier equations of orders 3 and 7:

$$k'^2(x + 1)^3(x - 3) + k^2(x - 1)^3(x + 3) = 0 \tag{11.113}$$

and

$$k'^2(x + 1)^7(x - 7) + k^2(x - 1)^7(x + 7) + Ak^2 k'^2 x^2 - 2^{11}(k^2 - k'^2)k^2 k'^2 x = 0. \tag{11.114}$$

Joubert determined the value of $A$, in his multiplier equation of order 7, to be $-21 \cdot 256$ by finding the sums of powers of the roots of the multiplier equation. This approach was similar to that used in Sohnke's second method for constructing the modular equation.

### 11.6  Kiepert and Klein on the Multiplier Equation

Ludwig Kiepert and Felix Klein worked independently on the multiplier equation, obtaining similar results with somewhat different approaches. Kiepert's work was based on a result of Felix Müller,[44] who proved the following theorem in his 1867 Ph.D. dissertation, supervised by Weierstrass:

---

[44]  Müller (1867); see also Enneper (1890) pp. 492–497, summarizing Müller's work.

Let $\omega_1$ and $\omega_2$ be periods of the Weierstrass elliptic function, $\wp$ and let $p$ be a prime. Then any symmetric polynomial in

$$\wp\left(\frac{\omega_2}{p}\right),\ \wp\left(\frac{2\omega_2}{p}\right),\ \ldots,\ \wp\left(\frac{(p-1)\omega_2}{2p}\right)$$

can be expressed as a polynomial in $g_2$ and $g_3$. Recall that Weierstrass and his followers took the periods to be $2\,\omega_1$ and $2\,\omega_2$.

Kiepert's motivation was to show that there existed many simple function values that satisfied algebraic equations of degree $p+1$ with coefficients that were rational functions in $g_2$ and $g_3$. As a first result he used (4.120) to show that the function $g(\omega)$ in (11.27), that we here denote by $h(\omega)$, and with $p = 6n \pm 1$, would satisfy the relation[45]

$$h^{-2}(\omega_2) = (-1)^n \prod_{\alpha=1}^{\frac{p-1}{2}} \left(\wp\left(\frac{\alpha\,\omega_2}{p}\right) - \wp\left(\frac{2\alpha\,\omega_2}{p}\right)\right). \tag{11.115}$$

By Müller's result, $h^{-2}(\omega_2)$ was a root of a monic equation of degree $p+1$ with coefficients that were polynomials in $g_2$ and $g_3$. Kiepert then employed the infinite product for the Weierstrass $\sigma$ function, obtained by substituting in (4.139) the value of $\phi(x)$ given in (4.8). He could thus show that

$$h^2(\omega_2) = \left(\frac{D}{\Delta^p}\right)^{\frac{1}{12}}, \tag{11.116}$$

where, with $q = e^{i\pi\frac{\omega_1}{\omega_2}}$,

$$D^{\frac{1}{12}} = \frac{2p\pi}{\omega_2} q^{\frac{p}{6}} \prod_{m=1}^{\infty} (1 - q^{2pm})^2. \tag{11.117}$$

He gave the other roots of the equation satisfied by $h^2$ as[46]

$$h_j^2(\omega_2) = (-1)^{\frac{p-1}{2}} \left(\frac{D_j}{\Delta^p}\right)^{\frac{1}{12}}, \quad j = 0, 1, \ldots, p-1, \tag{11.118}$$

where, with $\alpha = e^{\frac{4\pi i}{p}}$,

$$D_j^{\frac{1}{12}} = \frac{2\pi}{\omega_2} \alpha^j q^{\frac{1}{6p}} \prod_{m=1}^{\infty} (1 - \alpha^{12jm} q^{\frac{2m}{p}})^2. \tag{11.119}$$

He determined[47] that the product of the $p+1$ roots in (11.116) and (11.118) would be

$$\frac{(-1)^{\frac{p-1}{2}} p}{\Delta^{\frac{p^2-1}{12}}}. \tag{11.120}$$

---

[45] Kiepert (1879b) p. 202.   [46] Kiepert (1879b) pp. 203–204.   [47] Ibid. pp. 207–208.

Note that $\frac{p^2-1}{12}$ must be an integer when $p = 6n \pm 1$. Since $h^{-2}$ satisfied a monic polynomial of degree $p + 1$ with coefficients that were polynomials in $g_2$ and $g_3$, Kiepert could conclude from (11.120) that $h^2$ satisfied a monic polynomial, with coefficients that were polynomials in $g_2$ and $g_3$, divided by an integer power of $\Delta$. He also showed that the roots $h^2, h_0^2, h_1^2, \ldots, h_{p-1}^2$ satisfied a Jacobi-type property exemplified by (11.23). Kiepert also presented a method for determining the monic polynomial equation satisfied by $h^2$, an equation sometimes called the transformation equation by nineteenth-century mathematicians.

Denote the coefficients of this polynomial by $C_1, C_2, \ldots, C_{p+1}$. Kiepert noted that $C_{p+1}$ would be given by (11.120) and that the denominator of $C_\alpha$ had an integer exponent of $\Delta$ between $\frac{\alpha(p-1)}{12}$ and $\frac{\alpha p}{12}$. Thus, if the interval $\left[\frac{\alpha(p-1)}{12}, \frac{\alpha p}{12}\right]$ did not contain an integer, then $C_\alpha$ had to equal 0. For $p = 5$, he observed that $C_1 = 0$, $C_2 = 0$, and $C_4 = 0$, because between 4 and 5, 8 and 10, 16 and 20 there was no multiple of 12. Hence, the transformation equation of order 5 took the form

$$y^6 + \frac{a}{\Delta}y^3 + \frac{b}{\Delta^2}y + \frac{5}{\Delta^2} = 0, \tag{11.121}$$

where $a, b$ were polynomials in $g_2, g_3$. Similarly, he had for $n = 7$ and $n = 11$, respectively, the equations

$$y^8 + \frac{a}{\Delta}y^6 + \frac{b}{\Delta^2}y^4 + \frac{c}{\Delta^3}y^2 + \frac{d}{\Delta^4}y - \frac{7}{\Delta^4} = 0 \tag{11.122}$$

$$y^{12} + \frac{a}{\Delta^5}y^6 + \frac{b}{\Delta^6}y^5 + \frac{c}{\Delta^7}y^4 + \frac{d}{\Delta^8}y^3 + \frac{e}{\Delta^9}y^2 + \frac{f}{\Delta^{10}}y - \frac{11}{\Delta^{10}} = 0, \tag{11.123}$$

with $a, b, c, d, e, f$ some polynomials in $g_2, g_3$.

In order to determine $a, b, c, \ldots$, Kiepert assigned the dimensions 2, 3, 6 to $g_2, g_3$, $\Delta = g_2^3 - 27g_3^2$, respectively. It followed from (11.116) and (11.117) that the dimension of a root $y$ of the transformation equation would be $-\frac{p-1}{2}$ and that the dimension of $C_\alpha$ was $-\frac{\alpha(p-1)}{2}$. Because the dimension of the denominator of $C_\alpha$ lay between $\frac{\alpha(p-1)}{2}$ and $\frac{\alpha p}{2}$, the numerator had to have (with $m$ an integer)

dimension 0 if $\alpha(p-1)$ was a multiple of 12 $\left(C_\alpha = \dfrac{m}{\Delta^{\frac{\alpha(p-1)}{12}}}\right)$

dimension 1 if $\alpha(p-1)+2$ was a multiple of 12 $(C_\alpha = 0)$

dimension 2 if $\alpha(p-1)+4$ was a multiple of 12 $\left(C_\alpha = \dfrac{mg_2}{\Delta^{\frac{\alpha(p-1)+4}{12}}}\right)$

dimension 3 if $\alpha(p-1)+6$ was a multiple of 12 $\left(C_\alpha = \dfrac{mg_3}{\Delta^{\frac{\alpha(p-1)+6}{12}}}\right)$

....

Kiepert noted that since any polynomial in $g_2, g_3$ would necessarily have dimension of at least 2, it followed that if the dimension were one, then $C_\alpha = 0$. Thus, the transformation equations of orders 5, 7, 11, respectively, would take the form

$$y^6 + \frac{m}{\Delta}y^3 + \frac{m_1 g_2}{\Delta^2}y + \frac{5}{\Delta^2} = 0, \tag{11.124}$$

$$y^8 + \frac{m}{\Delta}y^6 + \frac{m_1}{\Delta^2}y^4 + \frac{m_2}{\Delta^3}y^2 + \frac{m_3 g_3}{\Delta^4}y - \frac{7}{\Delta^4} = 0, \tag{11.125}$$

$$y^{12} + \frac{m}{\Delta^5}y^6 + \frac{m_1 g_2}{\Delta^7}y^4 + \frac{m_2 g_3}{\Delta^6}y^3 + \frac{m_3 g_2^2}{\Delta^9}y^2 + \frac{m_4 g_2 g_3}{\Delta^{10}}y - \frac{11}{\Delta^{10}} = 0, \tag{11.126}$$

where $m, m_1, m_2, \ldots$ denoted integers. Kiepert then noted that this form of the transformation equation could be determined almost without calculation. Only the integer coefficients were left to be figured. He found these values by calculating the sums of powers of the roots

$$S_k = h^{2k} + \sum_{j=0}^{p-1} h_j^{2k}. \tag{11.127}$$

For example, by applying Newton's recurrence formula (6.37) to (11.124), he saw that if $p = 5$, then

$$S_3 = \frac{-3m}{\Delta}, \quad S_5 = \frac{-5m_1 g_2}{\Delta^2}. \tag{11.128}$$

He noted that by (11.117) and (11.119)

$$h^6 = \frac{5^3 q^{\frac{5}{2}} \prod_{k=1}^\infty (1 - q^{10k})^6}{\Delta q^{\frac{1}{2}} \prod_{k=1}^\infty (1 - q^{2k})^6}, \tag{11.129}$$

$$h_j^6 = \frac{\alpha^{3j} q^{\frac{1}{10}} \prod_{k=1}^\infty \left(1 - \alpha^{12jk} q^{\frac{2k}{5}}\right)^6}{\Delta q^{\frac{1}{2}} \prod_{k=1}^\infty (1 - q^{2k})^6}, \quad \alpha = e^{\frac{4\pi i}{5}}. \tag{11.130}$$

He also presented the series expansions

$$g_2 = \left(\frac{\pi}{\omega_2}\right)^4 \left(\frac{1}{12} + 20(q^2 + 9q^4 + 28q^6 + \cdots)\right), \tag{11.131}$$

$$g_3 = \left(\frac{\pi}{\omega_2}\right)^6 \left(\frac{1}{216} + \frac{7}{3}(-q^2 - 33q^4 - 244q^6 - \cdots)\right), \tag{11.132}$$

$$\Delta^{\frac{1}{12}} = \frac{\pi}{\omega_2} q^{\frac{1}{6}} \prod_{k=1}^\infty (1 - q^{2k})^2 = \frac{\pi}{\omega} q^{\frac{1}{6}}(1 - q^2 - q^4 + \cdots)^2. \tag{11.133}$$

Kiepert next observed that (11.128), (11.129), and (11.130) implied that the numerator in the sum in (11.127) with $k = 3$ must be divisible by

$$q^{\frac{1}{2}} \prod (1 - q^{2k})^6.$$

Moreover,

$$\sum_{j=0}^{4} \alpha^{3j} = 0 \quad \text{and} \quad \sum_{j=0}^{4} \alpha^{15j} = 5;$$

hence, the term with the lowest power of $q$ in the numerator was the $q^{\frac{1}{2}}$ term and its coefficient was $5(-6) = -30$. Therefore

$$-3m = -30 \quad \text{so that} \quad m = 10.$$

In a similar manner, Kiepert noted that the sum of the numerators of $h^{10}, h_j^{10}, j = 0, 1, 2, 3, 4$ had to be divisible by $q^{\frac{1}{6}} \prod(1 - q^{2k})^2$. The series expansion of the numerator began with $5h^{\frac{1}{2}}$, so the expansion of $-5m_1 g_2$ must begin with

$$-5m_1 \left(\frac{\pi}{\omega_2}\right)^4 \frac{1}{12}.$$

From this he could conclude that the second equation in (11.128) would produce $m_1 = -12$ and the fifth-order transformation equation (11.124) was obtained:

$$y^6 + \frac{10}{\Delta} y^3 - \frac{12g_3}{\Delta^2} y + \frac{5}{\Delta^2} = 0.$$

Employing the same reasoning but more calculation, he derived the equations of orders 7 and 11, respectively:

$$y^8 + \frac{14}{\Delta} y^6 + \frac{63}{\Delta^2} y^4 + \frac{70}{\Delta^3} y^2 + \frac{216g_3}{\Delta^4} y - \frac{7}{\Delta^4}, \tag{11.134}$$

$$y^{12} + 11 \left( -\frac{90}{\Delta^5} y^6 + \frac{40(12g_2)}{\Delta^7} y^4 - \frac{15(216g_3)}{\Delta^8} y^3 + \frac{2(12g_2)^2}{\Delta^9} y_3 \right)$$
$$- \frac{12g_2(216g_3)}{\Delta^{10}} y - \frac{11}{\Delta^{10}} = 0. \tag{11.135}$$

Without giving details, Kiepert had the coefficients of $y^3$ and $y$ in (11.135) as positive, so that he arrived at the equation obtained when $y$ was changed to $-y$.

Klein had simultaneously obtained Kiepert's results, but by means of modular forms methods. He found the form of the equation satisfied by $\Delta^{\frac{1}{12}}(p\omega_1, \omega_2)$, where $p$ was taken to be a prime. Recall that Klein had shown that $\Delta^{\frac{1}{12}}(\omega_1, \omega_2)$ could be regarded as a multiplier when an elliptic integral was transformed. In his 1879 paper, "Über Multiplicatorgleichungen,"[48] he let

$$\Delta'(\omega_1, \omega_2) = \Delta(a\omega_1 + b\omega_2, c\omega_1 + d\omega_2),$$

where $ad - bc = p$, $p = 6m \pm 1$, denote a transformed value of $\Delta(\omega_1, \omega_2)$, so that $\Delta'$ took $p + 1$ different values. Thus, $\sqrt[12]{\Delta'}$ had $12(p+1)$ distinct values for a given pair $\omega_1, \omega_2$. Klein wrote that it could be shown that a homogeneous modular form of integer weight, that reproduced itself except for a twelfth root of unity under the action of the modular group, was a polynomial in $g_2(\omega_1, \omega_2)$, $g_3(\omega_1, \omega_2)$, and $\sqrt[12]{\Delta(\omega_1, \omega_2)}$. By the

---

[48] Klein (1921–1923) vol. 3, pp. 137–139.

phrase, "reproduced itself except for a twelfth root of unity," Klein meant that the modular form had a multiplier system involving twelfth roots of unity. He then noted that there had to be $p + 1$ different values of $\sqrt[12]{\Delta'}$ with appropriately chosen twelfth roots of unity, such that symmetric functions of these $p + 1$ values would be modular forms with multiplier systems. The method for choosing these roots of unity was later given by Hurwitz.

We can thus see that these $p + 1$ values were roots of an equation of degree $p + 1$, called the multiplier equation, whose coefficients were polynomials in $g_2$, $g_3$, and $\sqrt[12]{\Delta}$. Klein denoted any of these $p + 1$ values by $z$ and observed that since $z = \sqrt[12]{\Delta'}$ had weight 1, the coefficient of $z^{p+1-m}$ in the multiplier equation would have weight $m$ and would be expressible as a polynomial in $g_2$, $g_3$ and $\sqrt[12]{\Delta}$.

He next considered a particular value of $z$, that is,

$$z\left(\frac{\omega_2}{2\pi}\right) = \left(\frac{\omega_2}{2\pi}\right) \sqrt[12]{\Delta'} = pq^{\frac{p}{6}} \prod_{n=1}^{\infty} (1 - q^{2pn})^2, \quad q = e^{i\pi\omega}, \tag{11.136}$$

where $\omega = \frac{\omega_1}{\omega_2}$, Im $\omega > 0$. He observed that the change $\omega \to \omega + 1$ caused $z$ to change by a factor $e^{\frac{p\pi i}{6}}$ and $\sqrt[12]{\Delta}$ to change by a factor of $e^{\frac{i\pi}{6}}$. He then argued that the equation satisfied by $z$ should remain unchanged under $\omega \to \omega + 1$, and thus each of the terms of that equation must contain the same root of unity. With this, Klein was able to state the theorem:

The coefficient of $z^{p+1-m}$ has the form $\Delta^{\frac{\lambda}{12}} G$, where $\lambda$ denotes the smallest nonnegative integer that is congruent to $pm$ modulo 12, and where $G$ is a polynomial in $g_2$ and $g_3$ only. In particular, if $\lambda > m$, then the coefficient vanishes.

Using his theorem, Klein was able to determine the form of a multiplier equation of an arbitrary prime order $p > 3$. In particular, when $p = 11$, the multiplier equation took the form

$$z^{12} + a\Delta^{\frac{6}{12}} z^6 + b\Delta^{\frac{4}{12}} g_2 z^4 + c\Delta^{\frac{3}{12}} g_3 z^3 + d\Delta^{\frac{2}{12}} g_2^2 z^2 + e\Delta^{\frac{1}{12}} g_2 g_3 z + f g_2^3 + g g_3^2 = 0, \tag{11.137}$$

where $a, b, c \ldots$ were constants. As Kiepert had done, Klein used the series expansions for $g_2$, $g_3$, $\Delta^{\frac{1}{2}}$ to determine the constants:

$$a = -90 \cdot 11, \quad b = 40 \cdot 11 \cdot 12, \quad c = -15 \cdot 11 \cdot 216, \quad d = 2 \cdot 11 \cdot 144,$$
$$f = -11, \quad g = -11 \cdot 27.$$

Note that $-11g_2^3 - 11 \cdot 27g_3^2 = -11\Delta$. The multiplier equation for $p = 11$ was therefore given by

$$z^{12} - 90 \cdot 11 \sqrt[2]{\Delta} z^6 + 40 \cdot 11 \cdot 12 g_2 \sqrt[3]{\Delta} z^4 - 15 \cdot 11 \cdot 216 g_3 \sqrt[4]{\Delta} z^3$$
$$+ 2 \cdot 11 \cdot (12g_2)^2 \sqrt[6]{\Delta} z^2 - 12g_2 \cdot 216g_3 \sqrt[12]{\Delta} z - 11 \cdot \Delta = 0. \tag{11.138}$$

Note that we can obtain Klein's multiplier equation of order 11 (11.138) from Kiepert's transformation equation of order 11 (11.135) by setting $z = \Delta^{\frac{11}{12}} y$.

Klein's work implies that there exist 12th roots of unity $\xi, \xi_0, \xi_1, \ldots, \xi_{p-1}$, with $p = 6m \pm 1$, such that

$$\xi \sqrt[12]{\Delta(p\,\omega_1, \omega_2)} + \sum_{j=0}^{p-1} \xi_j \sqrt[12]{\Delta(\omega_1 + j\,\omega_2, p\,\omega_2)} = \text{constant} \cdot \sqrt[12]{\Delta(\omega_1, \omega_2)}.$$

$$(11.139)$$

The values of these roots of unity $\xi, \xi_0, \xi_1, \ldots, \xi_{p-1}$ were determined by Hurwitz and these values, combined with the equation (11.139), formed the basis of Mordell's work on Ramanujan's conjectures.

Klein defined the multiplier equation only for the modular form $\sqrt[12]{\Delta}$ of level 1, though he was clearly aware that the concept could be defined for modular forms of arbitrary levels. In 1887, Paul Biedermann, Klein's student, wrote that the problem of defining the $n$th-order transformation consisted in determining the algebraic relationship between a modular form $F(\omega_1, \omega_2)$ belonging to the $s$th level and the entire set of transformed forms $F\left(\omega_1', \frac{\omega_2'}{n}\right)$, where

$$\omega_1' = (sa+1)\omega_1 + sb\,\omega_2, \quad \omega_2' = sc\,\omega_1 + (sd+1)\omega_2,$$

$$(sa+1)(sd+1) - sb \cdot sc = 1.$$

## 11.7 Hurwitz: Roots of the Multiplier Equation

Hurwitz's doctoral dissertation of 1881[49], written under the supervision of Klein, had two parts. The first part contained a proof, using only the theory of modular forms, that Dedekind's $\eta$ function was expressible as the infinite product (8.96). Recall that Dedekind had defined his $\eta$ function by (8.5); thus for an appropriately chosen constant $C$

$$\eta(\omega) = CJ^{-\frac{1}{16}}(1 - J^{-\frac{1}{8}}) \left(\frac{dJ}{d\omega}\right)^{\frac{1}{4}},$$

where $J$ denoted that Dedekind-Klein modular invariant. Hurwtiz showed that $\eta^4(\omega) = (2\pi)^{-4}\Delta^{\frac{1}{6}}(\omega)$, and he applied modular function theory to prove that

$$\Delta(\omega) = g_2^3(\omega) - 27g_3^2(\omega) = (2\pi)^{12} q^2 \prod_{n=1}^{\infty} (1 - q^{2n})^{24}, \quad q = e^{i\pi\omega}.$$

The topic of the second part of the thesis was the multiplier equation. Hurwitz explicity gave roots of the multiplier equation when the multiplier was taken as

$$\frac{\sqrt[12]{\Delta\left(\frac{\omega_1}{n}, \omega_2\right)}}{\sqrt[12]{\Delta(\omega_1, \omega_2)}}.$$

We briefly outline his derivation of the expression for the roots, first noting that it is clear that for any linear transformation

$$\omega_1' = a\,\omega_1 + b\,\omega_2, \quad \omega_2' = c\,\omega_1 + d\,\omega_2, \quad \Delta(\omega_1', \omega_2') = \Delta(\omega_1, \omega_2), \quad (11.140)$$

[49] Hurwitz (1962) vol. 1, pp. 1–66.

for $ad - bc = 1$. It follows that

$$\sqrt[12]{\Delta(\omega_1', \omega_2')} = \epsilon(a, b, c, d) \sqrt[12]{\Delta(\omega_1, \omega_2)}, \tag{11.141}$$

where $\epsilon$ is an appropriate 12th root of unity. Hurwitz determined that[50]

$$\epsilon(a, b, c, d) = e^{((1-c^2)(db+3d(c-1)+c+3)+c(d+a-3))\frac{\pi i}{6}}. \tag{11.142}$$

We mention that Hurwitz wrote $\begin{bmatrix} a & b \\ c & d \end{bmatrix}$ instead of $\epsilon(a, b, c, d)$. In a footnote, he stated that (11.142) was not given in the literature up to that time. Now observe that

$$\epsilon(1, 1, 0, 1) = e^{\frac{\pi i}{6}} \tag{11.143}$$

and

$$\epsilon(0, -1, 1, 0) = e^{\frac{3\pi i}{2}}. \tag{11.144}$$

We can see that (11.143) follows immediately from

$$\sqrt[12]{\Delta(\omega_1, \omega_2)} = \frac{2\pi}{\omega_2} q^{\frac{1}{6}} \prod_{n=1}^{\infty} (1 - q^{2n})^2, \quad q = e^{\frac{\omega_1}{\omega_2} \pi i}, \tag{11.145}$$

while (11.144) is obtainable from (11.141) and (11.145) by taking $\frac{\omega_1}{\omega_2} = i$. Since a general transformation $\begin{pmatrix} a & b \\ c & d \end{pmatrix}$ is generated by $\begin{pmatrix} 1 & 1 \\ 0 & 1 \end{pmatrix}$ and $\begin{pmatrix} 0 & -1 \\ 1 & 0 \end{pmatrix}$, the value of $\epsilon(a, b, c, d)$ is obtained in principle. Deriving (11.142) requires a little more work; for that, the reader may consult Hurwitz or attempt it as an exercise. Note that $\epsilon(a, b, c, d)$ is a multiplier system of weight 1 for $\Gamma(1)$.

It is interesting to note that, though Hurwitz considered in detail only the modular form $\sqrt[12]{\Delta\left(\frac{\omega_1}{n}, \omega_2\right)}$, he stated the "transformation problem" in some generality: To study the connection between the values of a modular form for $\omega_1, \omega_2$ and those for $\omega_1', \omega_2'$, where $\omega_1' = \frac{\omega_1}{n}, \omega_2' = \omega_2$, with $n$ a positive integer. Hurwitz clearly wished to determine an algebraic relation between a modular form and the modular form obtained after an $n$th-order transformation was applied to it. Hurwitz stated that in order to resolve the transformation problem, it was necessary to answer the question: How many inequivalent number pairs $\omega_1' = \frac{\omega_1}{n}, \omega_2' = \omega_2$ arise, if one replaces $\omega_1, \omega_2$ by all possible equivalent number pairs $\alpha \omega_1 + \beta \omega_2, \gamma \omega_1 + \delta \omega_2$?

Hurwitz's answer to this question was that the number of inequivalent pairs would be equal to the index of the subgroup $\Gamma^0(n)$ in the modular group $\Gamma = \Gamma(1)$. Recall that

$$\Gamma^0(n) = \left\{ \begin{pmatrix} \alpha & \beta \\ \gamma & \delta \end{pmatrix} : \quad \beta \equiv 0, \quad \alpha\delta - \beta\gamma = 1 \right\}$$

$$\Gamma_0(n) = \left\{ \begin{pmatrix} \alpha & \beta \\ \gamma & \delta \end{pmatrix} : \quad \gamma \equiv 0, \quad \alpha\delta - \beta\gamma = 1 \right\},$$

where $\alpha, \beta, \gamma, \delta$ are integers.

[50] Hurwitz (1962) vol. 1, p. 40.

To prove this result, Hurwitz began with the observation that the number pair $\left(\frac{\omega_1}{n}, \omega_2\right)$ was equivalent to the pair

$$\left(\frac{\alpha\,\omega_1 + \beta\,\omega_2}{n},\, \gamma\,\omega_1 + \delta\,\omega_2\right) \equiv \left(\alpha\frac{\omega_1}{n} + \frac{\beta}{n}\omega_2,\, n\gamma\frac{\omega_1}{n} + \delta\,\omega_2\right)$$

if and only if $\beta \equiv 0 \pmod{n}$, that is, if and only if

$$\begin{pmatrix} \alpha & \beta \\ \gamma & \delta \end{pmatrix} \in \Gamma^0(n).$$

Now for a modular transformation

$$P = \begin{pmatrix} a & b \\ c & d \end{pmatrix} \in \Gamma, \tag{11.146}$$

Hurwitz defined the number pair

$$P^{(n)} = \left(\frac{a\,\omega_1 + b\,\omega_2}{n},\, c\,\omega_1 + d\,\omega_2\right).$$

Hurwitz then reasoned that two number pairs $P^{(n)}$ and $Q^{(n)}$ were equivalent if and only if

$$P = SQ,$$

where $S = \begin{pmatrix} \alpha & \beta \\ \gamma & \delta \end{pmatrix} \in \Gamma^0(n)$; moreover,

$$P^{(n)} = S'Q^{(n)},$$

where $S' = \begin{pmatrix} \alpha & \frac{\beta}{n} \\ n\gamma & \delta \end{pmatrix} \in \Gamma_0(n)$. Hurwitz computed the index $N$ of $\Gamma^0(n)$ in $\Gamma(1)$ to be

$$N = n \prod_{p|n} \left(1 + \frac{1}{p}\right). \tag{11.147}$$

This completed Hurwitz's answer to the question of the number of inequivalent pairs. Recall that Dedekind had found that $N$ in (11.147) also represented the number of distinct matrices $\lambda$ in (8.110). This can be proved directly and we leave it as an exercise. Now by choosing a number pair $T_k^{(n)}$ from each of the $N$ classes of equivalent number pairs, Hurwitz had a complete system of representatives $T_1^{(n)}, T_2^{(n)}, \ldots, T_N^{(n)}$ such that each number pair $P^{(n)}$ was equivalent to one and only one of them.

Next, Hurwitz let

$$T_i^{(n)} = \left(\frac{A_i\,\omega_1 + B_i\,\omega_2}{n},\, C_i\,\omega_1 + D_i\,\omega_2\right), \quad i = 1, 2, \ldots, N, \tag{11.148}$$

where $A_i, B_i, C_i, D_i$ were integers such that $A_iD_i - B_iC_i = 1$. He observed that if $n$ was relatively prime to 12, then $A_i, B_i, C_i, D_i$ could be chosen such that

$$\begin{pmatrix} A_i & B_i \\ C_i & D_i \end{pmatrix} = \begin{pmatrix} 1 & 0 \\ 0 & 1 \end{pmatrix} \pmod{12}. \tag{11.149}$$

Hurwitz was then in a position to state his fundamental theorem:

The symmetric functions of $N$ quantities $z_i = \sqrt[12]{\Delta T_i^{(n)}}$, where $T_i^{(n)}$ are given by (11.148) and (11.149), change by the same 12th root of unity as $\sqrt[12]{\Delta(\omega_1, \omega_2)}$ when the transformation (11.140) is applied to $\omega_1, \omega_2$.

To verify this proposition, first let $P \in SL_2(\mathbb{Z})$ be given by (11.146). Then $z_i = \sqrt[12]{\Delta T_i^{(n)}}$ changes to $\sqrt[12]{\Delta(T_i P)^{(n)}}$. Since the $T_i^{(n)}$ constitute a full representative system, there exists an $S_i \in \Gamma^0(n)$ and a corresponding $S_i' \in \Gamma_0(n)$ such that for some $j$

$$T_i P = S_i T_j \tag{11.150}$$

and

$$(T_i P)^{(n)} = S_i'(T_j^{(n)}). \tag{11.151}$$

So $z_i$ changes to $z_j$ multiplied by that 12th root of unity by which $\sqrt[12]{\Delta}$ changes upon the application of $S_i'$. However, since by (11.149)

$$T_i \equiv T_j \equiv \begin{pmatrix} 1 & 0 \\ 0 & 1 \end{pmatrix} \pmod{12},$$

(11.150) implies that

$$S_i \equiv P \pmod{12}.$$

Thus, all $S_j$ and therefore all $S_i'$ are congruent modulo 12, so that an application of $P$ to $\omega_1, \omega_2$ produces the same change, by a 12th root of unity, in all the $z_i$, $i = 1, 2, \ldots, N$, as is produced in $\sqrt[12]{\Delta(\omega_1, \omega_2)}$. This proves Hurwitz's fundamental theorem.

In particular, this theorem implies that the elementary symmetric function

$$\sum_{i=1}^{N} z_i = \sum_{i=1}^{N} \sqrt[12]{\Delta T_i^{(n)}} \tag{11.152}$$

changes by the same 12th root of unity as $\sqrt[12]{\Delta}$ under the action of any element of the modular group. This means that

$$\frac{\sum_{i=1}^{N} z_i}{\sqrt[12]{\Delta}} = \frac{\sum_{i=1}^{N} \sqrt[12]{\Delta T_i^{(n)}}}{\sqrt[12]{\Delta(\omega_1, \omega_2)}} \tag{11.153}$$

is a modular function. Hurwitz noted that it was possible to give an explicit formula for the

$$z_i = \sqrt[12]{\Delta T_i^{(n)}} = \sqrt[12]{\Delta\left(\frac{A_i \omega_1 + B_i \omega_2}{n}, C_i \omega_1 + D_i \omega_2\right)}$$

$$= n \sqrt[12]{\Delta(A_i \omega_1 + B_i \omega_2, n(C_i \omega_1 + D_i \omega_2))} \tag{11.154}$$

where

$$\begin{pmatrix} A_i & B_i \\ C_i & D_i \end{pmatrix} \equiv \begin{pmatrix} 1 & 0 \\ 0 & 1 \end{pmatrix} \pmod{12}. \tag{11.155}$$

To obtain this explicit expression, he started with the observation that every number pair

$$A_i\,\omega_1 + B_i\,\omega_2,\ n(C_i\,\omega_1 + D_i\,\omega_2) \tag{11.156}$$

was equivalent to exactly one number pair

$$a\,\omega_1 + b\,\omega_2,\ d\,\omega_2 \tag{11.157}$$

where $a, b, d$ have no common factor greater than 1 and where

$$ad = n,\quad 0 \le b < d. \tag{11.158}$$

To verify Hurwitz's observation, we must show that there exist integers $\alpha', \beta', \gamma', \delta'$, with $\alpha'\delta' - \beta'\gamma' = 1$, such that either both relations

$$A_i\,\omega_1 + B_i\,\omega_2 = \alpha'(a\,\omega_1 + b\,\omega_2) + \beta'd\,\omega_2$$
$$nC_i\,\omega_1 + nD_i\,\omega_2 = \gamma'(a\,\omega_1 + b\,\omega_2) + \delta'd\,\omega_2,$$

hold, or the relations

$$A_i = \alpha'a,\quad B_i = \alpha'b + \beta'd, \tag{11.159}$$

$$nC_i = \gamma'a,\quad nD_i = \gamma'b + \delta'd \tag{11.160}$$

all hold. The existence of such numbers $a, b, c, d$ and $\alpha', \beta', \gamma', \delta'$ as satisfy the necessary conditions is not difficult to show and is left to the reader. See section 8.13 for similar calculations in the work of Dedekind. The converse of the theorem can also be shown: that for every number pair (11.157) that satisfies (11.158), there exists a corresponding number pair (11.156) that satisfies (11.155).

Now note that $n$ is relatively prime to 12. Thus, $n$ is of the form $12m \pm 1$ or $12m \pm 5$ while $n^2$ takes the form $12m + 1$. A similar argument may be used for $a$ and $d$, since $ad = n$. Hence, $n^2 \equiv a^2 \equiv d^2 \equiv 1 \pmod{12}$. Using this fact and (11.155), (11.159), and (11.160), we may conclude that

$$\begin{pmatrix} \alpha' & \beta' \\ \gamma' & \delta' \end{pmatrix} \equiv \begin{pmatrix} a & -nb \\ 0 & a \end{pmatrix} \pmod{12}. \tag{11.161}$$

Thus, since the root $\frac{z_i}{n}$ in (11.154) emerges from $\sqrt[12]{\Delta(a\,\omega_1 + b\,\omega_2, d\,\omega_2)}$ by means of the transformation $\begin{pmatrix} \alpha' & \beta' \\ \gamma' & \delta' \end{pmatrix}$, Hurwitz could apply (11.161) and (11.142) to obtain the important formula

$$z_i = n e^{(-bd + 3(a-1))\frac{\pi i}{6}} \sqrt[12]{\Delta(a\,\omega_1 + b\,\omega_2, d\,\omega_2)}. \tag{11.162}$$

By applying (11.145) to (11.162), we arrive at Hurwitz's explicit formula for $z_i$:[51]

$$z_i(a, b, d) = \frac{2\pi}{\omega_2} a(-1)^{\frac{a-1}{2}} e^{ab(1-d^2)\frac{\pi i}{6}} q^{\frac{a}{6d}} \prod_{m=1}^{\infty} \left(1 - e^{\frac{2abm\pi i}{n}} q^{\frac{2an}{d}}\right)^2. \tag{11.163}$$

In particular, when $n = p$ is a prime and $p \nmid 6$, we can take, in Hurwitz's notation for $z_i$ (following Galois and Klein):

$$z_b = e^{-\frac{pb\pi i}{6}} \sqrt[12]{\Delta(\omega_1 + b\,\omega_2, \omega_2)},\quad 0 \le b < p \tag{11.164}$$

---

[51] Hurwitz (1962) vol. 1, p. 47.

and

$$z_\infty = (-1)^{\frac{p-1}{2}} \sqrt[12]{\Delta(p\omega_1, \omega_2)}. \tag{11.165}$$

However, if we choose $b = 12h$ with $0 \le h < p$, as was done by Kiepert, then the number pairs $(\omega_1 + 12h\omega_2, p\omega_2)$ will already be included among the $T_h^{(p)}$ given by (11.148) and (11.149). In that case, we can take

$$z_h = \sqrt[12]{\Delta(\omega_1 + 12h\omega_2, p\omega_2)}, \quad 0 \le h < p \tag{11.166}$$

and

$$z_\infty = (-1)^{\frac{p-1}{2}} \sqrt[12]{\Delta(p\omega_1, \omega_2)}. \tag{11.167}$$

An application of (11.153) to (11.164) and (11.165) shows us that

$$\frac{\sum_{b=0}^{p-1} e^{-\frac{pb\pi i}{6}} \sqrt[12]{\Delta(\omega_1 + b\omega_2, p\omega_2)} + (-1)^{\frac{p-1}{2}} \sqrt[12]{\Delta(p\omega_1, \omega_2)}}{\sqrt[12]{\Delta(\omega_1, \omega_2)}} \tag{11.168}$$

is a modular function belonging to $\Gamma = \Gamma(1)$. In fact, Hurwitz observed that the series expansions at $i\infty$ of the functions in the numerator of (11.168) showed that expression to be a constant. Mordell used this result to prove Ramanujan's conjecture on the coefficients of the series expansion at $i\infty$ of $\Delta(\omega_1, \omega_2)$. As we shall see, it is possible that Mordell thought that Hurwitz's argument required a little more work. He showed that the zero of the numerator at $i\infty$ was of at least as high an order as the zero of the denominator. Therefore, the expression in (11.168) was a bounded modular function and thus a constant. A similar application of (11.153) to (11.166) and (11.167) reveals that

$$\frac{\sum_{h=0}^{p-1} \sqrt[12]{\Delta(\omega_1 + 12h\omega_2, p\omega_2)} + (-1)^{\frac{p-1}{2}} \sqrt[12]{\Delta(p\omega_1, \omega_2)}}{\sqrt[12]{(\omega_1, \omega_2)}} \tag{11.169}$$

is a constant.

Now let $e$ be any divisor of 12 and $ef = 12$. With the notation of (11.164) and (11.165), we can show that

$$\frac{\sum_{b=0}^{p-1} z_b^f + z_\infty^f}{\sqrt[e]{\Delta(\omega_1, \omega_2)}} \tag{11.170}$$

is a constant. Note that the expression in the numerator of (11.170) is the sum of the $f$th powers of the roots of the multiplier equation. By Newton's theorem, this expression can be written as a polynomial in the elementary symmetric functions of the roots; hence, by Hurwitz's result, under the action of $\Gamma(1)$, it changes by the same root of unity as the denominator. It then follows that (11.170) is a modular function. As we shall see in Section 13.6, Mordell proved that (11.170) was bounded and thus must be a constant. Similarly, using (11.166) and (11.167), we find that

$$\frac{\sum_{h=0}^{p-1} z_h^f + z_\infty^f}{\sqrt[e]{\Delta(\omega_1, \omega_2)}} \tag{11.171}$$

is a constant.

We observe that there exist multiplier equations of order 3 for $\Delta^{\frac{1}{4}}$ and $\Delta^{\frac{1}{2}}$; their roots are, respectively,

$$\Delta^{\frac{1}{4}}(3\omega_1, \omega_2), \quad \Delta^{\frac{1}{4}}(\omega_1 + h\omega_2, 3\omega_2), \quad h = 0, 1, 2,$$

and

$$\Delta^{\frac{1}{2}}(3\omega_1, \omega_2), \quad \Delta^{\frac{1}{2}}(\omega_1 + h\omega_2, 3\omega_2), \quad h = 0, 1, 2.$$

It may be shown in a manner similar to the derivation of (11.168) or (11.170) that

$$\Delta^{\frac{1}{2}}(3\omega_1, \omega_2) + \sum_{h=0}^{2} \Delta^{\frac{1}{2}}(\omega_1 + h\omega_2, 3\omega_2) = c\Delta^{\frac{1}{2}}(\omega_1, \omega_2). \tag{11.172}$$

For $\Delta^{\frac{1}{3}}$, there exists a multiplier equation of order 2 with roots

$$\Delta^{\frac{1}{3}}(2\omega_1, \omega_2), \quad \Delta^{\frac{1}{3}}(\omega_1 + h\omega_2, 2\omega_2), \quad h = 0, 1;$$

moreover,

$$\Delta^{\frac{1}{3}}(2\omega_1, \omega_2) + \sum_{h=0}^{1} \Delta^{\frac{1}{3}}(\omega_1 + h\omega_2, 2\omega_2) = c\Delta^{\frac{1}{3}}(\omega_1, \omega_2). \tag{11.173}$$

One can also verify that for all primes $p$

$$\Delta(p\omega_1, \omega_2) + \sum_{h=0}^{p-1} \Delta(\omega_1 + h\omega_2, p\omega_2) = c\Delta(\omega_1, \omega_2). \tag{11.174}$$

## 11.8  Exercises

1. Let

$$r = e^{2pi\,\omega i},$$

$$\Delta(\omega) = r \prod_{n=1}^{\infty}(1 - r^n)^{24} = \sum_{n=1}^{\infty} \tau(n) r^n,$$

$$F(s) = \sum_{n=1}^{\infty} \frac{\tau(n)}{n^s}.$$

Prove that for $t$ real and for sufficiently large Re $s$,

$$(2\pi)^{-s}\,\Gamma(s)F(s) = \int_{0}^{\infty} t^{s-1}\,\Delta(it)dt.$$

Prove that the integral is convergent for all $s$ and hence that $F(s)$ is an entire function. Next, prove that $(2\pi)^{-s}\,\Gamma(s)F(s)$ is invariant under $s \to 12 - s$. This result is due to the Australian mathematician J. R. Wilton (1929).

2. Prove the result of Jacobi and Sohnke that the list (11.46) gives all the $p + 1$ transformed values of $k(q)$.

3. Use (11.74) to complete the proof given in the text of Sohnke's result that when $u = 1$, the modular equation reduces to $(v - 1)^{p+1}$ for $p = 8\mu \pm 1$ and to $(v - 1)(v + 1)^p = 0$ for $p = 8\mu \pm 3$. See Sohnke (1837) p. 112 or Enneper (1890) pp. 460–461.

4. Derive by Sohnke's method a modular equation of order 7:

$$(1 - u^8)(1 - v^8) = (1 - uv)^8.$$

See Sohnke (1837) p. 120. This result is due to Guetzlaff (1834).

5. Read the proofs of the eleven modular equations of order 7 given by Ramanujan in Berndt (1985–1998) vol. 3, pp. 324–324. These proofs are of interest because they most probably resemble the proofs Ramanujan had in mind.

6. Suppose $f$ is a polynomial and $a$ is a constant; also suppose that $f(x) = \pm f(a - x)$. Show that there exist polynomials $g_1$ and $g_2$ such that when the upper sign holds, then $f(x) = g_1(x(a - x))$ and when the lower sign holds, then $f(x) = (a - 2x)g_2(x(a - x))$.

7. Derive the multiplier equations of orders 3 and 7 given by equations (11.113) and (11.114) due to Joubert.

8. Read the proofs of (11.36) and (11.37) on pages 397–400 and 363–372 respectively of Berndt (1985–1998) vol. 3.

9. Prove that when $\lambda = \lambda_1, \lambda_2, \ldots, \lambda_{p+1}$, the corresponding $p + 1$ values of $\frac{1}{M}$ are given by (11.20). Use the same method to show (11.87) and (11.88).

10. Fill in the details of the proofs of Joubert's two theorems given in Section 11.5. See Joubert (1875) pp. 65–68.

11. Show that the multiplier equation $F(x, k^2) = 0$ is irreducible. See Joubert (1875) pp. 68–70.

# 12

---

# The Theory of Modular Forms
## as Reworked by Hurwitz

### 12.1  Preliminary Remarks

In the first half of his doctoral dissertation of 1881, discussed in our earlier chapters, Hurwitz gave a new foundation to the theory of elliptic modular forms, based on the work of Dedekind and Eisenstien. In 1904, Hurwitz published a paper[1] in which he simplified the proofs of his basic theorems and cast them in a more accessible form, so they could be readily employed in a text or course on modular functions.

Recall that Dedekind had defined the invariant function $J(\omega)$ as a conformal mapping of the fundamental domain of $\Gamma(1)$ onto the complex plane. He then defined his $\eta$ function as

$$\eta(\omega) = \text{constant } J^{-\frac{1}{6}} (1 - J)^{-\frac{1}{8}} \left( \frac{dJ}{d\omega} \right)^{\frac{1}{4}}, \quad \text{Im } \omega > 0;$$

This led to his complicated proof of the result

$$\eta(\omega) = q^{\frac{1}{12}} \prod_{n=1}^{\infty} (1 - q^{2n}), \quad q = e^{i\pi \omega}.$$

In his thesis, Hurwitz adopted Dedekind's definitions of $J$ and $\eta$; then in his 1904 paper, Hurwitz proved that the function defined by the infinite product

$$(2\pi)^{12} q^2 \prod_{n=1}^{\infty} (1 - q^{2n})^{24} \tag{12.1}$$

was a cusp form of weight 12 for $\Gamma(1)$. He deduced this fact from Eisenstein's result that, with $(m_1, m_2) \neq (0, 0)$,

$$\sum_{m_1} \sum_{m_2} \frac{1}{(m_1 \omega_1 + m_2 \omega_2)^2} - \sum_{m_2} \sum_{m_1} \frac{1}{(m_1 \omega_1 + m_2 \omega_2)^2} = \pm \frac{2i\pi}{\omega_1 \omega_2}, \tag{12.2}$$

---

[1]  Hurwitz (1962) vol. 1, pp. 574–595.

giving a very elegant proof based on an idea of Eisenstein. Hurwitz next denoted the infinite product (12.1) by $\Delta$ and defined the function $J(\omega)$ by the equation

$$J(\omega) = \frac{g_2^3(\omega)}{\Delta(\omega)}. \tag{12.3}$$

Integrating $\frac{J'(\omega)}{J(\omega)}$ along the boundary of the fundamental domain, he was able to show that $J(\omega)$ took each complex value exactly once in the fundamental domain. He then used this result to prove that

$$\Delta(\omega) = g_2^3(\omega) - 27g_3^2(\omega). \tag{12.4}$$

Hurwitz concluded his paper by proving that any equation

$$y^2 = 4x^3 - c_2 x - c_3, \quad c_2^3 - 27c_3^2 \neq 0, \tag{12.5}$$

could be parametrized by means of elliptic functions. For this reason, equations reducible to the form (12.5) are called elliptic curves.

Hurwtiz's 1881 paper and its 1904 update were regarded during the early twentieth century as providing fundamental foundations for the theory of modular functions. In his 1917 paper on Ramanujan's conjectures, Mordell wrote in a footnote, concerning Hurwitz's original 1881 paper, "For an elementary introduction to the modular functions, see Hurwitz . . . ." Again, perhaps because of J. P. Serre's 1957 lecture on modular forms,[2] employing several key ideas from Hurwitz's 1904 paper, Hurwitz's proofs of basic results on modular forms have become well known and commonly used. For these reasons, we discuss Hurwitz's paper in its entirety and provide a translation into English as an appendix.

## 12.2   The Fundamental Domain

Hurwitz's proof that the fundamental domain of $\Gamma = SL_2(\mathbb{Z})$ is given by (8.23) was based on his definition of height; take the height of point $\omega \in \mathbb{H}$ to be

$$H(\omega) = 2i\frac{\omega\overline{\omega} + 1}{\omega - \overline{\omega}} = \frac{x^2 + y^2 + 1}{y}, \quad \omega = x + iy. \tag{12.6}$$

One may check that when $\omega_1$ is given by $\omega_1 = \frac{a\omega+b}{c\omega+d}$, then

$$\frac{1}{2i}(\omega_1 - \overline{\omega}_1) = \frac{1}{2i}(\omega - \overline{\omega})\frac{1}{(c\omega+d)(c\overline{\omega}+d)} \tag{12.7}$$

and

$$H(\omega_1) = 2i\frac{(a\omega+b)(a\overline{\omega}+b) + (c\omega+d)(c\overline{\omega}+d)}{\omega - \overline{\omega}}. \tag{12.8}$$

Taking $a = d = 0$ and $b = -c = 1$ in (12.8), we see that

$$H\left(-\frac{1}{\omega}\right) = H(\omega). \tag{12.9}$$

---

[2] Serre (1966).

Since (12.8) is a positive definite quadratic form with coefficients $a, b, c, d$, it follows that in the equivalence class given by $\Gamma \omega = \{g\omega | g \in \Gamma\}$, there are one or more elements with the smallest height $H(\omega_1)$. By (12.9) we know that an element $\omega$ with the smallest height in an equivalence class can be chosen to satisfy $|\omega| \geq 1$. Taking this $\omega$ and applying (12.8) with $\omega_1$ equivalent to $\omega$ (that is, $\omega_1 = \frac{a\omega+b}{c\omega+d}$), we may write

$$\frac{H(\omega_1)}{H(\omega)} = \frac{(a\omega+b)(a\overline{\omega}+b) + (c\omega+d)(c\overline{\omega}+d)}{\omega\overline{\omega}+1} \geq 1. \tag{12.10}$$

Taking $a = b = d = 1$ and $c = 0$ in (12.10), we obtain

$$|\omega|^2 + \omega + \overline{\omega} + 2 \geq |\omega|^2 + 1$$

or

$$\omega + \overline{\omega} \geq -1 \quad \text{or} \quad \text{Re}\,\omega \geq -\frac{1}{2}. \tag{12.11}$$

Similarly, setting $a = d = 1$, $b = -1$, and $d = 0$ yields

$$\omega + \overline{\omega} \leq 1 \quad \text{or} \quad \text{Re}\,\omega \leq \frac{1}{2}. \tag{12.12}$$

Taking (12.10) and (12.11) in combination with $|\omega| \geq 1$ will then produce the same fundamental region obtained by Dedekind.

### 12.3  An Infinite Product as a Modular Form

Hurwitz showed that the infinite product (12.1) was a modular form of weight 12. To prove this, first note that the modular group is generated by the two elements

$$S = \begin{pmatrix} 1 & 1 \\ 0 & 1 \end{pmatrix} \quad \text{and} \quad T = \begin{pmatrix} 0 & -1 \\ 1 & 0 \end{pmatrix};$$

hence, it is sufficient to show that $\Delta(\omega_1, \omega_2)$, defined by the expression in (12.1), satisfies

$$\Delta(\omega_1 + \omega_2, \omega_2) = \Delta(\omega_1, \omega_2) \tag{12.13}$$

$$\Delta(-\omega_2, \omega_1) = \Delta(\omega_1, \omega_2). \tag{12.14}$$

Recall that (12.13) is a consequence of the definition of $\Delta$ because $e^{2\pi i} = 1$ and, as discussed in Section 4.11, (12.14) follows from (12.2). To prove (12.2), start with the partial fractions expansion of $\cot \pi a$ in terms of Eisenstein summation notation:

$$\sum_{m=-\infty}^{\infty} \frac{1}{a+m} = \lim_{n\to\infty} \left( \frac{1}{a} + \sum_{k=1}^{n} \left( \frac{1}{a+k} + \frac{1}{a-k} \right) \right)$$

$$= \pi \cot a\pi = i\pi \frac{e^{i\pi a} + e^{-i\pi a}}{e^{i\pi a} - e^{-i\pi a}}. \tag{12.15}$$

Next consider the sum

$$S = \sum_{m_1} \sum_{m_2} \frac{1}{(m_1 - n_1)\omega_1 + (m_2 - n_2)\omega_2}, \tag{12.16}$$

where $n_1$, $n_2$ denotes any given pair of integers, and $m_1$, $m_2$ stands for all possible pairs of integers except $(m_1, m_2) = (n_1, n_2)$. We wish to show that

$$S = \pm \frac{2\pi i n_1}{\omega_2}, \qquad (12.17)$$

in which the positive sign is chosen when $\text{Im}\left(\frac{\omega_1}{\omega_2}\right)$ is positive and the negative sign is chosen otherwise. To confirm (12.17), observe that with $\omega = \frac{\omega_1}{\omega_2}$, we have by (12.15)

$$\sum_{m_2} \frac{1}{(m_1 - n_1)\omega_1 + (m_2 - n_2)\omega_2} = \frac{1}{\omega_2} \sum_{m_2} \frac{1}{(m_1 - n_1)\omega - n_2 + m_2}$$

$$= \frac{\pi}{\omega_2} \cot\left[(m_1 - n_1)\omega - n_2\right]\pi$$

$$= \frac{\pi}{\omega_2} \cot(m_1 - n_1)\omega\pi, \qquad (12.18)$$

when $m_1 \neq n_1$; when $m_1 = n_1$, the sum (12.18) is given by

$$\sum_{m_2} \frac{1}{(m_2 - n_2)\omega_2} = \lim_{n \to \infty} \frac{1}{\omega_2} \sum_{m_2=-n}^{n}{}' \frac{1}{m_2 - n_2} = 0,$$

where the prime symbol in the summation indicates that $m_2 \neq n_2$. Thus, the sum (12.15) is reduced to

$$S = \frac{\pi}{\omega_2} \sum_{m_1} \cot(m_1 - n_1)\omega\pi, \qquad (12.19)$$

where $m_1 \neq n_1$. Now observe that

$$\sum_{m_1=-\lambda}^{\lambda}{}' \cot(m_1 - n_1)\omega\pi = \sum_{m=-\lambda-n_1}^{\lambda-n_1}{}' \cot(m\omega\pi)$$

$$= \sum_{m=-\lambda-n_1}^{\lambda+n_1}{}' \cot(m\omega\pi) - \sum_{m=\lambda-n_1+1}^{\lambda+n_1} \cot(m\omega\pi). \qquad (12.20)$$

Since $\cot x$ is an odd function, we know that

$$\sum_{m=-\lambda-n_1}^{\lambda+n_1}{}' \cot(m\omega\pi) = 0,$$

so that we obtain

$$S = -\frac{\pi}{\omega_2} \lim_{\lambda \to \infty} \sum_{m=\lambda-n_1+1}^{\lambda+n_1} \cot(m\omega\pi). \qquad (12.21)$$

Based on

$$\cot(m\omega\pi) = i \frac{e^{im\omega\pi} + e^{-im\omega\pi}}{e^{im\omega\pi} - e^{-im\omega\pi}},$$

we see that as $\lambda \to \infty$ each of the $2n_1$ terms

$$\cot(m\omega\pi) \quad \text{with} \quad m = \lambda - n_1 + 1, \ \lambda - n_1 + 2, \dots, \lambda + n_1$$

contributes to the limit (12.21), as $\lambda \to \infty$, either $-i$ or $i$, depending on whether the imaginary part of $\omega$ is positive or negative. We have thus established (12.17) and this in turn implies that when the order of summation $S$ is changed, the result is

$$\sum_{m_2} \sum_{m_1} \frac{1}{(m_1 - n_1)\omega_1 + (m_2 - n_2)\omega_2} = \mp \frac{2\pi i n_2}{\omega_1},$$

where the plus sign is chosen when $\operatorname{Im} \frac{\omega_2}{\omega_1}$ is positive and the negative sign is chosen otherwise. This puts us in a position to determine the difference in the two sums:

$$G_2' = \sum_{m_1} {\sum_{m_2}}' \frac{1}{(m_1 \omega_1 + m_2 \omega_2)^2} \tag{12.22}$$

$$G_2'' = \sum_{m_2} {\sum_{m_1}}' \frac{1}{(m_1 \omega_1 + m_2 \omega_2)^2}, \tag{12.23}$$

where the $'$ indicates that the summation is over all $(m_1, m_2) \neq (0, 0)$. Now observe the absolute convergence of the sum

$$s = \sum_{m_2,m_1} \left[ \frac{1}{(m_1 - 1)\omega_1 + (m_2 - 1)\omega_2} - \frac{1}{m_1 \omega_1 + m_2 \omega_2} - \frac{\omega_1 + \omega_2}{(m_1 \omega_1 + m_2 \omega_2)^2} \right]$$

$$= (\omega_1 + \omega_2)^2 \sum_{m_2,m_1} \frac{1}{(m_1 \omega_1 + m_2 \omega_2)^2 ((m_1 - 1)\omega_1 + (m_2 - 1)\omega_2)}, \tag{12.24}$$

where the summation is performed over all pairs of numbers $m_1, m_2$ except $m_1 = 1$, $m_2 = 1$ and $m_1 = 0$, $m_2 = 0$. By summing the series $s$ first with respect to $m_2$ and then with respect to $m_1$, we obtain

$$s = \frac{3}{\omega_1 + \omega_2} \pm \frac{2i\pi}{\omega_2} - (\omega_1 + \omega_2) \sum_{m_1} {\sum_{m_2}}' \frac{1}{(m_1 \omega_1 + m_2 \omega_2)^2}. \tag{12.25}$$

Note that the term $\frac{3}{m_1+m_2}$ occurs because the term $-\frac{1}{\omega_1+\omega_2}$ is missing from the three terms within the square brackets in (12.24). We may similarly sum the series $s$ first with respect to $m_1$ and then with respect to $m_2$:

$$s = \frac{3}{\omega_1 + \omega_2} \mp \frac{2i\pi}{\omega_1} - (\omega_1 + \omega_2) \sum_{m_2} {\sum_{m_1}}' \frac{1}{(m_1 \omega_1 + m_2 \omega_2)^2}. \tag{12.26}$$

Subtracting (12.25) from (12.26) completes a streamlined proof of Eisenstein's result that

$$G_2' - G_2'' = \pm \frac{2i\pi}{\omega_1 \omega_2}, \tag{12.27}$$

where $G_2'$ and $G_2''$ are as given in (12.22) and (12.23) and where the positive sign is chosen when $\operatorname{Im} \frac{\omega_1}{\omega_2}$ is positive, the negative sign being chosen otherwise. From (12.27), Hurwitz deduced the fact that $\Delta(\omega)$ was a modular form of weight 12, as shown in our Section 4.11.

## 12.4 The *J*-Function

Hurwitz defined the *J*-function by equation (12.3), from which it easily follows that $J(\omega') = J(\omega)$ when $\omega'$ is equivalent to $\omega$ under the action of $\Gamma$. Observe that $g_2(\omega)$ has the Fourier series expansion

$$g_2(\omega) = (2\pi)^4 \left( \frac{1}{12} + 20 \sum_{n=1}^{\infty} \sigma_3(n) q^{2n} \right), \quad q = e^{\pi i \omega}, \tag{12.28}$$

where $\sigma_3(n) = \sum_{d|n} d^3$, the sum of the cubes of all divisors of $n$. Hence, $J(\omega)$ has the Fourier series expansion

$$J(\omega) = \frac{1}{12^3 q^2} (1 + c_1 q^2 + c_2 q^4 + \cdots), \tag{12.29}$$

where $c_1, c_2, \ldots$ are integers. Therefore, $J(\omega)$ has a pole of order 1 at $\omega = i\infty$.

Hurwitz proved that $J(\omega)$ maps the fundamental region one-to-one and onto the whole complex plane $\mathbb{C}$. Begin by observing that the equation $J(\omega) = a$ has at most a finite number of solutions in the fundamental domain. If it had an infinite number of roots, these roots would have an accumulation point at $\infty$, so that $\infty$ would be an essential singularity. However, $\infty$ is a pole of order 1. Note that the accumulation point cannot occur at a finite point because in that case $J(\omega)$ would be identically equal to the constant $a$. Next, to determine the number of solutions $N$ of $J(\omega) = a$ lying inside the fundamental domain $D$, draw a line $CC'$ parallel to the $x$-axis in order to cut off from $D$ the finite domain $D' = CABA'C'$. Make this domain sufficiently large to contain all the solutions of of $J(\omega) - a = 0$ contained in the fundamental domain. Also assume that none of the solutions lie on the boundary of this domain. It is a consequence of Cauchy's residue theorem that

$$N = \frac{1}{2\pi i} \int_{\partial D'} d \log(J(\omega) - a),$$

where the integral is taken over the boundary of $D'$, denoted by $\partial D'$ in the positive sense. See Figure 12.1.

This integral can be broken up into five parts:

$$\int_A^B - \int_{A'}^B + \int_{A'}^{C'} - \int_A^C + \int_{C'}^C d \log (J(\omega) - a).$$

The first two integrals are over the arcs $AB$ and $A'B'$; the remaining three integrals are over straight lines. The substitution $\omega = -\frac{1}{\omega'}$ in the first integral shows that

$$\int_A^B d \log (J(\omega) - a) = \int_{A'}^B d \log (J(\omega) - a).$$

We see that the first two integrals cancel and the similar substitution $\omega = \omega' + 1$ shows that the third and fourth integrals also cancel one another. We are therefore left with

$$N = - \int_C^{C'} d \log (J(\omega) - a).$$

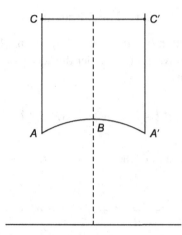

Figure 12.1. Path of Integration Along $\partial D'$.

Let the straight line $CC'$ be defined by $-\frac{1}{2} \leq \operatorname{Re}\omega \leq \frac{1}{2}$, $\operatorname{Im}\omega = y_0$. As $\omega$ runs through the straight line $CC'$, $q = e^{\pi i \omega}$ describes a circle of radius $e^{-\pi y_0}$. By (12.29)

$$J(\omega) - a = \frac{1}{12^3 q^2}\left(1 + (c_1 - 12^3 a)q^2 + c_2 q^4 + \cdots\right).$$

Therefore

$$\int_C^{C'} d\log(J(\omega) - a) = -1 \quad \text{and} \quad N = 1.$$

We note that a modification and reinterpretation are required if the point on the boundary is the vertex $A = \rho = -\frac{1}{2} + i\frac{\sqrt{3}}{2}$ or the vertex $B = i$. We will demonstrate that $g_2(\rho) = 0$ and that, since

$$J(\rho) = \frac{g_2^3(\rho)}{\Delta(\rho)},$$

$J$ has a zero of order 3 at $\omega = \rho$. However, the vertex $A$ is the meeting point of the three fundamental domains formed by $D$, $ST(D)$, and $(ST)^2(D)$, where $D$ is the principal fundamental domain. Therefore, $J$ can be taken to have a zero at of order 1 in each of the three fundamental domains. Similarly, at $\omega = i$, we have $J(i) = 1$ and for $J(\omega) - 1 = 0$, $i$ is a zero of order 2. But because $\omega = i$ is the meeting point of the two fundamental domains $D$ and $S(D)$, the contribution to the order of the zero of $J(\omega) - 1 = 0$ is 1 in each of the domains.

We must now show that $J(\omega)$ is real in $D$ if and only if $\omega$ lies on the boundary of $D$ or on the imaginary axis. Suppose $\omega = x + iy$ and $\omega^* = -x + iy$ denote the reflection points with respect to the imaginary axis. It follows that

$$q = e^{\pi i \omega} = e^{\pi i x - \pi y} \quad \text{and} \quad \bar{q} = e^{\pi i \omega^*} = e^{-\pi i x - \pi y}$$

are conjugates. Using (12.26) we then have $J(\omega^*) = \overline{J(\omega)}$ so that $J(\omega)$ is real if and only if $J(\omega) = J(\omega^*)$. For points in $D$ or on its boundary, this implies that either (a) $\omega = \omega^*$ or $\omega$ lies on the imaginary axis or (b) $\omega$ is equivalent to $\omega^*$. Since two points in the closure of $D$, denoted as $\overline{D}$, are equivalent if and only if they lie on the boundary of $D$, the result is verified.

Next, we begin the proof that $J(\rho) = 0$, $\rho = -\frac{1}{2} + i\frac{\sqrt{3}}{2}$ by recalling that

$$g_2(\omega) = 60 \sum_{m_1} \sum_{m_2} \frac{1}{(m_1\omega + m_2)^4}. \tag{12.30}$$

Because $1 + \rho + \rho^2 = 0$, we see that

$$\frac{1}{(m_1\rho + m_2)^4} + \frac{1}{((m_1 - m_2)\rho - m_1)^4} + \frac{1}{(-m_2\rho + m_1 - m_2)^4}$$
$$= \frac{1}{(m_1\rho + m_2)^4} \left(1 + \frac{1}{\rho^4} + \frac{1}{\rho^8}\right) = 0.$$

Since the double sum in (12.30) is absolutely convergent, its terms may be grouped in triples and we may write $g_2(\rho) = 0$. This implies that

$$J(\rho) = \frac{g_2^3(\rho)}{\Delta(\rho)} = 0.$$

We proceed to prove that

$$\Delta(\omega) = g_2^3(\omega) - 27g_3^2(\omega)$$

where $g_3(\omega)$ is given by

$$g_3(\omega) = 140 \sum_{m_1} \sum_{m_2}' \frac{1}{(m_1\omega + m_2)^6}.$$

We already know that

$$g_3(\omega) = (2\pi)^6 \left(\frac{1}{216} - \frac{7}{3} \sum_{n=1}^{\infty} \sigma_5(n)q^{2n}\right), \tag{12.31}$$

where $q = e^{\pi i \omega}$ and $\sigma_5(n) = \sum_{d|n} d^5$. Now define a new function

$$J_1(\omega) = \frac{g_3^2(\omega)}{\Delta(\omega)}.$$

By an argument similar to the one used to prove that $J(\omega) = J(\omega')$, one can show that $J_1(\omega') = J_1(\omega)$ when $\omega'$ is equivalent to $\omega$. Next observe that, because $g_3(i\infty) \neq 0$ and since $\Delta(\omega)$ has a zero of order 1 at $\omega = i\infty$, the function $J_1(\omega)$ has a pole of order 1. Hence, $J(\omega)$ and $J_1(\omega)$ are linear functions of one another:

$$\frac{g_2^3(\omega)}{\Delta(\omega)} = J(\omega) = aJ_1(\omega) + b = a\frac{g_3^2(\omega)}{\Delta(\omega)} + b,$$

or, by (12.1), (12.28), and (12.31),

$$\left(\frac{1}{12} + 20\sum_{n=1}^{\infty} \sigma_3(n)q^{2n}\right)^3 = a\left(\frac{1}{216} - \frac{7}{3}\sum_{n=1}^{\infty} \sigma_5(n)q^{2n}\right)^2 + bq^2\prod_{n=1}^{\infty}(1 - q^{2n})^{24}.$$

Equating constant terms, we obtain

$$a = \frac{216^2}{12^3} = 27$$

and by equating the coefficient of $q^2$ on each side, we get

$$\frac{3(20)}{12^2} = -\frac{2a(7)}{216(3)} + b, \quad b = \frac{5}{12} + \frac{7}{12} = 1.$$

Thus,

$$\frac{g_2^3}{\Delta} = J = \frac{27 g_3^2}{\Delta} + 1 \tag{12.32}$$

or

$$\Delta = g_2^3 - 27 g_3^2. \tag{12.33}$$

Clearly, $g_3(i) = 0$ and therefore $J(i) = 1$.

## 12.5   An Application to the Theory of Elliptic Functions

We have seen, for example in the work of Eisenstein, that the function

$$\wp(u) = \frac{1}{u^2} + \sum_{m_1} \sum_{m_2}{}' \left( \frac{1}{(u - m_1 \omega_1 - m_2 \omega_2)^2} - \frac{1}{(m_1 \omega_1 + m_2 \omega_2)^2} \right)$$

satisfies the differential equation

$$\wp'^2(u) = 4\wp^3(u) - g_2 \wp(u) - g_3$$
$$= 4 \left( \wp(z) - \wp\left(\frac{\omega_1}{2}\right) \right) \left( \wp(z) - \wp\left(\frac{\omega_2}{2}\right) \right) \left( \wp(z) - \wp\left(\frac{\omega_1 + \omega_2}{2}\right) \right)$$
$$\equiv 4(\wp(z) - e_1)(\wp(z) - e_2)(\wp(z) - e_3), \tag{12.34}$$

where

$$g_2 = 60 \sum_{m_1} \sum_{m_2}{}' \frac{1}{(m_1 \omega_1 + m_2 \omega_2)^4}, \quad g_3 = 140 \sum_{m_1} \sum_{m_2}{}' \frac{1}{(m_1 \omega_1 + m_2 \omega_2)^6}. \tag{12.35}$$

From this it is clear that

$$\Delta(\omega) = g_2^3 - 27 g_3^2 = (e_1 - e_2)^2 (e_2 - e_3)^2 (e_1 - e_3)^2$$

is the discriminant of the cubic on the right-hand side of (12.34). Now suppose we have the cubic

$$y^2 = 4x^3 - c_2 x - c_3, \quad \text{where} \quad c_2^3 - 27 c_3^2 \neq 0.$$

Must there always exist periods $\omega_1$ and $\omega_2$ such that

$$g_2(\omega_1, \omega_2) = c_2, \quad g_3(\omega_1, \omega_2) = c_3,$$
$$x = \wp(u; \omega_1, \omega_2), \quad y = \wp'(u; \omega_1, \omega_2)?$$

The theory of the $J$-function answers this question in the affirmative. Observe that there is exactly one value $\omega$ in the fundamental domain for which

$$J(\omega) = \frac{c_2^3}{c_2^3 - 27\, c_3^2}.$$

Now choose $\omega_2$ in such a manner that

$$\omega_2^2 = \frac{c_2}{c_3}\frac{g_3(\omega, 1)}{g_2(\omega, 1)} \quad \text{and} \quad \omega_1 = \omega\,\omega_2.$$

It is easy to check that these values of $\omega_1$ and $\omega_2$ will produce $g_2(\omega_1, \omega_2) = c_2$ and $g_3(\omega_1, \omega_2) = c_3$.

# Ramanujan's Euler Products and Modular Forms

## 13.1 Preliminary Remarks

When Srinivasa Ramanujan (1887–1920) was thirteen or fourteen years old, with only rudimentary training in mathematics, he rediscovered Euler's formulas for $\sin x$ and $\cos x$:[1]

$$\sin x = \frac{e^{xi} - e^{-xi}}{2i}, \quad \cos x = \frac{e^{xi} + e^{-xi}}{2}.$$

He was later disappointed to learn that these results were already known! With access to almost no other mathematics books, in 1903, at the age of fifteen, he started reading Carr's two-volume *Synopsis of Elementary Results in Pure and Applied Mathematics*, a compendium of thousands of theorems from various branches of mathematics, presented with virtually no proofs.[2] James Newman comments on Ramanujan's resourceful use of this text:

> In general, the mathematical knowledge contained in Carr's book went no further than the 1860s. Yet in areas that interested him, Ramanjuan was abreast, and often ahead, of contemporary mathematical knowledge when he arrived in England [in 1914]. Thus in a mighty sweep he had succeeded in re-creating in his field through his own unaided powers, a rich half-century of European mathematics. One may doubt that so prodigious a feat had ever before been accomplished in the history of thought.[3]

Although he entered college, Ramanjan's excessive devotion to mathematics prevented his overall success, so that he could not continue. However, in 1916 he was awarded a bachelors degree from Cambridge for his research work. He might remind us of some eighteenth-century British mathematicians who also lacked university training. However, those British mathematicians had access to up-to-date books and also had the opportunity to consult with other like-minded people. It appears, by contrast, that Ramanujan worked in mathematical isolation until he was almost twenty-three, when he met Ramaswami Aiyar, the founder of the Indian Mathematical Society, who put

---

[1] See Ramanujan (2000) p. xii.    [2] Carr (2013).    [3] Newman (1956) vol. 1, p. 366.

him into contact with other mathematicians. In this context, Ramanujan began a correspondence with G. H. Hardy, requesting Hardy's help in getting his theorems published and asking for guidance. In his first letter, written on January 16, 1913, Ramanujan enclosed eleven pages of theorems and formulas on the distribution of primes, infinite series, integrals, and continued fractions. We present, as samples, some of his results related to theta functions:

i

$$\int_0^\infty \prod_{n=0}^\infty \left(1 + r^{2n}x^2\right)^{-1} dx = \frac{\pi}{2\sum_{n=0}^\infty r^{\frac{n(n+1)}{2}}}.$$

ii  The coefficient of $x^n$ in $(1 - 2x + 2x^4 - 2x^9 + 2x^{16} - \cdots )^{-1}$ is equal to the integer nearest to

$$\frac{1}{4n}\left(\cosh(\pi\sqrt{n}) - \frac{\sinh(\pi\sqrt{n})}{\pi\sqrt{n}}\right).$$

iii  If

$$u = \frac{x}{1+} \frac{x^5}{1+} \frac{x^{10}}{1+} \frac{x^{15}}{1+} \frac{x^{20}}{1+} \cdots$$

and

$$v = \frac{\sqrt[5]{x}}{1+} \frac{x}{1+} \frac{x^2}{1+} \frac{x^3}{1+} \frac{x^4}{1+} \cdots$$

then

$$v^5 = u \cdot \frac{1 - 2u + 4u^2 - 3u^3 + u^4}{1 + 3u + 4u^2 + 2u^3 + u^4}.$$

iv

$$\frac{1}{1+} \frac{e^{-2\pi}}{1+} \frac{e^{-4\pi}}{1+} \frac{e^{-6\pi}}{1+} \cdots = \left(\sqrt{\frac{5+\sqrt5}{2}} - \frac{\sqrt5+1}{2}\right) e^{\frac{2\pi}{5}}.$$

v

$$\frac{1}{1-} \frac{e^{-\pi}}{1+} \frac{e^{-2\pi}}{1-} \frac{e^{-3\pi}}{1+} \cdots = \left(\sqrt{\frac{5-\sqrt5}{2}} - \frac{\sqrt5-1}{2}\right) e^{\frac{\pi}{5}}.$$

Hardy replied to Ramanujan in a letter of February 8, 1913[4] that some of the results were known and that (i) through (v) were probably difficult to prove, so that he would like to see Ramanujan's proofs. Concerning (iii) through (v), Hardy wrote in 1940: "I had never seen anything in the least like them before. A single look at them is enough to show they could only be written down by a mathematician of the highest class. They must be true because no one would have the imagination to invent them."[5] On theorem (ii), he wrote in his letter that he and Littlewood thought it could not be true. In his later

---

[4] For these letters, see Berndt and Rankin (1995).      [5] Hardy (1978) p. 9.

lectures on Ramanujan, Hardy said of (ii): "Ramanujan's false statement was one of the most fruitful he ever made, since it ended by leading us to all our joint work on partitions."[6] In one of their joint papers, Hardy and Ramanujan gave a corrected version of (ii),[7] whose proof employed the Hardy-Littlewood circle method. Berndt and Rankin commented on this method:[8] "Hardy and Ramanujan introduced the famous 'circle method', which since then has been one of the most powerful tools in additive number theory." Berndt and Rankin explain that "The crux of the 'circle method' depends upon precisely determining the behavior of the generating function about selected singularities." and that using several singularities was an idea Ramanujan had employed elsewhere. Thus, they conclude, "...he should be given more credit for the invention of the circle method."

In 1914, Ramanujan traveled to Cambridge to collaborate with Hardy; because of his complete lack of resources and formal training, Ramanujan's mathematical knowledge had gaping holes. Fortunately for Ramanujan and mathematics, Hardy possessed the wisdom and perspicacity necessary to guide Ramanujan's unique talent. In his obituary notice of Ramanujan Hardy wrote:[9]

> There was one great puzzle. What was to be done in the way of teaching him modern mathematics? The limitations of his knowledge were as startling as its profundity. Here was a man who could work out modular equations, and theorems of complex multiplication, to orders unheard of, whose mastery of continued fractions was, on the formal side at any rate, beyond that of any mathematician in the world, who had found for himself the functional equation of the Zeta-function, and the dominant terms of many of the most famous problems in the analytic theory of numbers; and he had never heard of a doubly periodic function or of Cauchy's theorem, and had indeed but the vaguest idea of what a function of a complex variable was. His ideas as to what constituted a mathematical proof were of the most shadowy description. All his results, new or old, right or wrong, had been arrived at by a process of mingled argument, intuition, and induction, of which he was entirely unable to give any coherent account.
>
> It was impossible to ask such a man to submit to systematic instruction, to try to learn mathematics from the beginning once more. I was afraid too that, if I insisted unduly on matters which Ramanujan found irksome, I might destroy his confidence or break the spell of his inspiration. On the other hand there were things of which it was impossible that he should remain in ignorance. Some of his results were wrong, and in particular those which concerned the distribution of primes, to which he attached the greatest importance. It was impossible to allow him to go through life supposing that all the zeros of the Zeta-function were real. So I had to try to teach him, and in a measure I succeeded, though obviously I learnt from him much more than he learnt from me. In a few years' time he had a very tolerable knowledge of the theory of functions and the analytic theory of numbers. He was never a mathematician of the modern school, and it was hardly desirable that he should become one; but he knew when he had proved a theorem and when he had not. And his flow of original ideas showed no symptoms of abatement.

The Hardy–Ramanujan collaboration was very fruitful, not only for the important work they did together, but also for the new research problems it suggested to each of them. For example, it was desirable to test their results in the process of developing the

---

[6] Hardy (1978) p. 9.    [7] Ramanujan (2000) p. 304.    [8] Berndt and Rankin (1995) p. 42.
[9] Ramanujan (2000) pp. xxi–xxxvi, especially p. xxx.

asymptotic formula for the number of partitions $p(n)$ of an integer $n$. With the help of the great calculator Major Percy MacMahon, Ramanujan constructed a table for $p(n)$ up to $n = 200$. Interestingly, Hardy wrote that though Ramanujan was a very fast calculator, MacMahon was even faster and more accurate.[10] However, Ramanujan noticed something in the table, not observed by anyone else at the time: $p(4)$, $p(9)$, $p(14)$... were all divisible by 5; $p(5)$, $p(12)$, $p(19)$... were all divisible by 7. He conjectured and proved that

$$p(5n + 4) \equiv 0 \ (\text{mod } 5), \quad \text{and} \quad p(7n + 5) \equiv 0 \ (\text{mod } 7).$$

On the basis of numerical evidence, he proceeded to make a more general conjecture:

$$\text{If} \quad \delta = 5^a 7^b 11^c \quad \text{and} \quad 24\lambda \equiv 1 \ (\text{mod } \delta),$$

$$\text{then} \quad p(\delta n + \lambda) \equiv 0 \ (\text{mod } \delta).$$

As it turns out, this conjecture is not quite correct. Hansraj Gupta extended MacMahon and Ramanujan's table[11] and on that basis, S. Chowla in 1934 perceived that $p(243)$ was not divisible by $7^3$, though $24 \cdot 243 \equiv 1 \bmod 7^3$. The conjecture was then salvageable because $p(243) \equiv 1 \bmod 7^2$. The corrected version of the conjecture, due to Chowla and eventually proved by Atkin, states:[12]

$$\text{Let} \quad \delta = 5^a 7^b 11^c, \quad \delta' = 5^a 7^{b'} 11^c,$$

$$\text{where} \quad b' = b \ \text{if} \ b = 0, 1, 2 \ \text{and} \ b' = \left[\frac{b+1}{2}\right] \ \text{if} \ b > 2.$$

$$\text{If} \quad 24\lambda \equiv 1 \ (\text{mod } \delta), \quad \text{then} \quad p(\delta n + \lambda) \equiv 0 \ (\text{mod } \delta'), \quad n = 0, 1, 2, \ldots.$$

Note that $p(n)$ is the coefficient in the expansion of a modular form

$$\frac{1}{\prod_{n=1}^{\infty} (1 - q^n)} = 1 + \sum_{n=1}^{\infty} p(n)q^n.[13]$$

As he investigated the properties of $p(n)$, Ramanujan also explored the arithmetical properties of coefficients of some other modular and theta functions. For example, he defined $\tau(n)$ as the coefficient in the series expansion of the discriminant function:

$$(2\pi)^{-12} \Delta(x) = x \prod_{n=1}^{\infty} (1 - x^n)^{24} = \sum_{n=1}^{\infty} \tau(n)x^n, \quad x = e^{2\pi i \omega}, \ \text{Im } \omega > 0. \quad (13.1)$$

He conjectured that the Dirichlet series $\sum_{n=1}^{\infty} \frac{\tau(n)}{n^s}$ corresponding to the modular form (13.1) had an Euler product expansion given by

$$\sum_{n=1}^{\infty} \frac{\tau(n)}{n^s} = \prod_p \left(1 - \tau(p) p^{-s} + p^{11-2s}\right)^{-1}.$$

---

[10] Ramanujan (2000) p. xxxv.    [11] Gupta (1935).    [12] Chowla (1934); Atkin (1967).
[13] Euler proved this result in 1740. See Eu.I-2, pp. 163–193.

It follows, in particular, that if $m$ and $n$ are relatively prime integers, then

$$\tau(m)\tau(n) = \tau(mn).$$

The 1916 paper, "On certain arithmetical functions,"[14] contains this striking result and offers an exposition of Ramanujan's theory of modular forms and, in particular, his derivation of the product formula for $\Delta(x)$. Later in this chapter, we discuss aspects of this 1916 paper, in particular on Euler products associated with some modular forms.

Note that because of his premature death, only a small portion of Ramanujan's work appeared in print during his lifetime. He recorded hundreds of results without proof in two notebooks and in the so-called lost notebook. The monumental task of editing these notebooks and verifying their contents is almost complete, thanks to the prodigious efforts of Bruce Berndt, on the two notebooks,[15] and George Andrews and Berndt on the lost notebook and other unpublished papers.[16] Insightful assessments of Ramanujan's legacy have been offered by Hardy and by M. Ram Murty and V. Kumar Murty.[17]

### 13.2   Ramanujan's $\tau$ Function

In his remarkable modular forms paper of 1916, Ramanujan derived the product formula (13.1) for the discriminant function $\Delta(x)$ by employing the differential equation satisfied by the Eisenstein series used in the definition of $\Delta$. The basic identity used in this process was the formula

$$\left( \frac{1}{4}\cot\frac{\theta}{2} + \frac{x\sin\theta}{1-x} + \frac{x^2\sin 2\theta}{1-x^2} + \frac{x^3\sin 3\theta}{1-x^3} + \cdots \right)^2$$
$$= \left( \frac{1}{4}\cot\frac{\theta}{2} \right)^2 + \frac{x\cos\theta}{(1-x)^2} + \frac{x^2\cos 2\theta}{(1-x^2)^2} + \frac{x^3\cos 3\theta}{(1-x^3)^2} + \cdots$$
$$+ \frac{1}{2}\left( \frac{x}{1-x}(1-\cos\theta) + \frac{2x^2}{1-x^2}(1-\cos 2\theta) + \frac{3x^3}{1-x^3}(1-\cos 3\theta) + \cdots \right).$$
$$(13.2)$$

This formula was not a mere technical device used in the derivation of the differential equation; Ramanujan found that it applied to other situations. For instance, he noted in passing that when $\theta = \frac{2\pi}{3}$ and $\theta = \frac{\pi}{2}$, then (13.2) reduced, respectively, to

$$\left( \frac{1}{6} + \frac{x}{1-x} - \frac{x^2}{1-x^2} + \frac{x^4}{1-x^4} - \frac{x^5}{1-x^5} + \cdots \right)^2$$
$$= \frac{1}{36} + \frac{1}{3}\left( \frac{x}{1-x} + \frac{2x^2}{1-x^2} + \frac{4x^4}{1-x^4} + \frac{5x^5}{1-x^5} + \cdots \right), \qquad (13.3)$$

---

[14] Ramanujan (2000) pp. 136–162.   [15] Berndt (1985–1998).
[16] Andrews and Berndt (2005–).   [17] Ram Murty and Kumar Murty (2013).

where $1, 2, 4, 5, \ldots$ were the natural numbers without the multiples of three; and

$$\left( \frac{1}{4} + \frac{x}{1-x} - \frac{x^3}{1-x^3} + \frac{x^5}{1-x^5} - \frac{x^7}{1-x^7} + \cdots \right)^2$$

$$= \frac{1}{16} + \frac{1}{2} \left( \frac{x}{1-x} + \frac{2x^2}{1-x^2} + \frac{3x^3}{1-x^3} + \frac{5x^5}{1-x^5} + \cdots \right), \qquad (13.4)$$

where $1, 2, 3, 5, \ldots$ were the natural numbers without the multiples of four.

In fact, Ramanujan's identity (13.4) gives an easy way to prove the four squares theorem from Fermat's theorem on the sums of two squares. Now the two squares theorem may be written as the identity

$$(1 + 2x + 2x^4 + 2x^9 + \cdots)^2 = 1 + 4 \left( \frac{x}{1-x} - \frac{x^3}{1-x^3} + \frac{x^5}{1-x^5} - \frac{x^7}{1-x^7} + \cdots \right);$$

$$(13.5)$$

this is equivalent to the result that

$$r_2(n) = 4 \left( d_1(n) - d_3(n) \right),$$

where $r_k(n)$ is the number of representations of $n$ as a sum of $k$ squares and $d_j(n)$ is the number of divisors of $n$ of the form $4m + j$. This is clear when one equates the coefficients of $x^n$ on each side of (13.5). In a similar way, the four squares theorem may be expressed as the identity

$$(1 + 2x + 2x^4 + 2x^9 + \cdots)^4 = 1 + 8 \left( \frac{x}{1-x} + \frac{2x^2}{1+x^2} + \frac{3x^3}{1-x^3} + \frac{4x^4}{1+x^4} + \cdots \right).$$

$$(13.6)$$

Identities (13.5) and (13.6) were presented by Jacobi; direct proofs of these results appear in our Chapter 15. It is easy to check that

$$\frac{x}{1-x} + \frac{2x^2}{1+x^2} + \frac{3x^3}{1-x^3} + \frac{4x^4}{1+x^4} + \cdots$$

$$= \frac{x}{1-x} + \frac{2x^2}{1-x^2} + \frac{3x^3}{1-x^3} + \frac{5x^5}{1-x^5} + \cdots.$$

For example, in order to understand what happened to $\frac{4x^4}{1+x^4}$, we note that

$$\sum_{n=1}^{m} \frac{2^n x^{2^n}}{1 + x^{2^n}} = \sum_{n=1}^{m} \left( \frac{2^n x^{2^n}}{1 - x^{2^n}} - \frac{2^{n+1} x^{2^{n+1}}}{1 - x^{2^{n+1}}} \right)$$

$$= \frac{2x^2}{1 - x^2} - \frac{2^{m+1} x^{2^{m+1}}}{1 - x^{2^{m+1}}}.$$

Hence

$$\sum_{n=1}^{\infty} \frac{2^n x^{2^n}}{1 + x^{2^n}} = \frac{2x^2}{1 - x^2}.$$

Similarly, it can be shown that

$$\sum_{n=1}^{\infty} \frac{2^n(2m+1)x^{2^n(2m+1)}}{1+x^{2^n(2m+1)}} = \frac{2(2m+1)x^{2(2m+1)}}{1-x^{2(2m+1)}}.$$

Thus, Ramanujan's identity (13.4) shows the square of the right-hand side of (13.5) to be the right-hand side of (13.6). The four squares theorem then follows from the two squares theorem.

## 13.3  Ramanujan: Product Formula for $\Delta$

To find the Eisenstein series associated with (13.2), Ramanujan expanded the trigono-metric functions as power series in $\theta$. We start with Euler's generating function for the Bernoulli numbers:

$$\frac{x}{e^x-1} = \sum_{k=0}^{\infty} B_k \frac{x^k}{k!} = 1 - \frac{1}{2}x + \sum_{k=1}^{\infty} B_{2k} \frac{x^{2k}}{(2k)!}.$$

This implies that

$$\begin{aligned}
x\cot x &= xi\,\frac{\left(e^{ix}+e^{-ix}\right)}{e^{ix}-e^{-ix}} \\
&= ix + \frac{2ix}{e^{2ix}-1} \\
&= 1 - B_2 \frac{2^2 x^2}{2!} + B_4 \frac{2^4 x^4}{4!} - \cdots .
\end{aligned} \qquad (13.7)$$

Using the fact that $\zeta(-2m+1) = (-1)^m \frac{B_{2m}}{2m}$,[18] where $m = 1, 2, 3, \ldots$, rewrite (13.7) as

$$\frac{1}{4}\cot\frac{\theta}{2} = \frac{1}{2}\left(\frac{1}{\theta} + \frac{\zeta(-1)}{1!}\theta + \frac{\zeta(-3)}{3!}\theta^3 + \frac{\zeta(-5)}{5!}\theta^5 + \cdots\right),$$

where $\zeta$ denotes the Riemann zeta function.

From the expansion

$$\sin k\theta = \frac{k\theta}{1!} - \frac{k^3\theta^3}{3!} + \frac{k^5\theta^5}{5!} - \cdots$$

we arrive at Ramanujan's result

$$\begin{aligned}
\frac{1}{4}\cot\frac{\theta}{2} &+ \frac{x\sin\theta}{1-x} + \frac{x^2\sin 2\theta}{1-x^2} + \frac{x^3\sin 3\theta}{1-x^3} - \cdots \\
&= \frac{1}{2\theta} + \frac{\theta}{1!}S_1 - \frac{\theta^3}{3!}S_3 + \frac{\theta^5}{5!}S_5 - \cdots ,
\end{aligned} \qquad (13.8)$$

---

[18]  For example, see Titchmarsh and Heath-Brown (1986) p. 19. For an account of this and related results in the context of their development, see Roy (2011) Chapter 26.

where
$$S_k = \frac{1}{2}\zeta(-k) + \sum_{n=1}^{\infty} \frac{n^k x^n}{1 - x^n}.$$

Note that $S_k$, except for a constant factor, is the Eisenstein series $G_{2k}$. A similar expansion for $\cos\theta$ applied to (13.2) then gave Ramanujan

$$\left(\frac{1}{2\theta} + \frac{\theta}{1!}S_1 - \frac{\theta^3}{3!}S_3 + \frac{\theta^5}{5!}S_5 - \cdots\right)^2$$

$$= \frac{1}{4\theta^2} + S_1 - \frac{\theta^2}{2!}\Phi_{1,2}(x) + \frac{\theta^4}{4!}\Phi_{1,4}(x) - \frac{\theta^6}{6!}\Phi_{1,6}(x) + \cdots$$

$$+ \frac{1}{2}\left(\frac{\theta^2}{2!}S_3 - \frac{\theta^4}{4!}S_5 + \frac{\theta^6}{6!}S_7 - \cdots\right), \tag{13.9}$$

where
$$\Phi_{1,s}(x) = \sum_{n=1}^{\infty} \frac{n^s x^n}{(1 - x^n)^2}.$$

Note that in Ramanujan's notation we would write

$$\Phi_{0,s} = \sum_{n=1}^{\infty} \frac{n^s x^n}{1 - x^n} = S_s - \frac{1}{2}\zeta(-s).$$

Equating the coefficients of $\theta^n$ in (13.9), where $n$ is even, we get Ramanujan's formula

$$\frac{n+3}{2(n+1)}S_{n+1} - \Phi_{1,n} = \sum_{l=0}^{\frac{n}{2}-1} \binom{n}{2l+1} S_{2l+1} S_{n-2l-1}. \tag{13.10}$$

In fact, as Ramanujan and others showed, $S_{2k+1}$ for $k \geq 1$ is a polynomial in $S_3$ and $S_5$ and hence he set

$$P = -24S_1 \quad = 1 - 24\sum_{n=1}^{\infty} \frac{nx^n}{1 - x^n} \quad = 1 - 24\,\Phi_{0,1}(x) \tag{13.11}$$

$$Q = 240S_3 \quad = 1 + 240\sum_{n=1}^{\infty} \frac{n^3 x^n}{1 - x^n} = 1 + 240\,\Phi_{0,3}(x) \tag{13.12}$$

$$R = -504S_5 = 1 - 504\sum_{n=1}^{\infty} \frac{n^5 x^n}{1 - x^n} = 1 - 504\,\Phi_{0,5}(x). \tag{13.13}$$

If we take $n = 2$ in (13.10), the result is

$$\frac{5}{6}S_3 - \Phi_{1,2}(x) = 2S_1^2.$$

This implies that

$$288\,\Phi_{1,2}(x) = 240\,S_3 - (-24\,S_1)^2$$
$$= Q - P^2. \tag{13.14}$$

The case $n = 4$ gives us $\frac{7}{10} S_5 - \Phi_{1,4}(x) = 4S_1 S_3 + 4S_3 S_1$. Hence

$$720\,\Phi_{1,4} = -5760\,S_1 S_3 + 504\,S_5$$
$$= PQ - R. \tag{13.15}$$

Now the case $n = 6$ yields $\frac{9}{14} S_7 - \Phi_{1,6} = 6S_1 S_5 + 20 S_3 S_3 + 6 S_1 S_5$. Thus,

$$1008\,\Phi_{1,6}(x) = 648\,S_7 - 12096\,S_1 S_5 - 20160\,S_3^2. \tag{13.16}$$

To compute $S_7$, Ramanujan deduced another identity, proved in the next section, similar to (13.10):

$$\frac{(n-2)(n+5)}{12\,(n+1)(n+2)}\,S_{n+3} = \sum_{k=1}^{\frac{n}{2}-1} \binom{n}{2k} S_{2k+1} S_{n-2k+1}, \tag{13.17}$$

where $n$ is an even integer. If we take $n = 4$, we obtain

$$\frac{1}{20} S_7 = 6 S_3^2 \quad \text{or} \quad \Phi_{0,7}(x) + \frac{1}{2}\zeta(-7) = 120 S_3^2 \quad \text{or} \quad 1 + 480\,\Phi_{0,7}(x) = Q^2.$$

As Ramanujan remarked, (13.17) may be used to show that the Eisenstein series $G_{2k}$, $k \geq 1$, is a polynomial in $G_4$ and $G_6$. Note that we can rewrite (13.16) as

$$1008\,\Phi_{1,6}(x) = 57600\,S_3^2 - 12096\,S_1 S_5$$
$$= Q^2 - PR. \tag{13.18}$$

Making use of (13.14), (13.15), and (13.18), Ramanujan obtained his basic differential equations

$$x\frac{dP}{dx} = -24\,\Phi_{1,2}(x) = \frac{P^2 - Q}{12} \tag{13.19}$$

$$x\frac{dQ}{dx} = 240\,\Phi_{1,4}(x) = \frac{PQ - R}{3} \tag{13.20}$$

$$x\frac{dR}{dx} = -504\,\Phi_{1,6}(x) = \frac{PR - Q^2}{2}. \tag{13.21}$$

We note that the discriminant function is given by $1728(2\pi)^{-12}\Delta = Q^3 - R^2$; then from (13.19), (13.20), and (13.21) we have Ramanujan's proof that

$$x\frac{d}{dx}\log\Delta = \frac{x}{Q^3 - R^2}\frac{d}{dx}(Q^3 - R^2) = \frac{\left[3xQ^2\frac{dQ}{dx} - 2xR\frac{dR}{dx}\right]}{Q^3 - R^2}$$
$$= \frac{[Q^2(PQ - R) - R(PR - Q^2)]}{Q^3 - R^2}$$
$$= P.$$

From the definition of $P$ it is clear that

$$P = x\frac{d}{dx}\log[x(1 - x^2)(1 - x^3)\cdots]^{24}$$

and that the coefficient of $x$ in $Q^3 - R^2$ is 1728. Hence

$$1728(2\pi)^{-12}\Delta = Q^3 - R^2 = 1728\,x\,[(1-x)(1-x^2)(1-x^3)\cdots]^{24}.$$

This is Ramanujan's very elegant proof of the product formula for $\Delta$.

## 13.4  Proof of Identity (13.2)

Ramanujan began his proof of (13.2) with the observation that

$$\cot\frac{1}{2}\theta\sin n\theta = 1 + 2\cos\theta + 2\cos 2\theta + \cdots + 2\cos(n-1)\theta + \cos n\theta. \quad (13.22)$$

This follows by an application of the formula

$$2\sin\frac{\theta}{2}\cos k\theta = \sin\left(k\theta + \frac{\theta}{2}\right) - \sin\left(k\theta - \frac{\theta}{2}\right).$$

Now, following Hardy's presentation of the proof,[19] rewrite the left-hand side of (13.2), setting $u_m = \frac{x^m}{1-x^m}$:

$$S^2 = \left(\frac{1}{4}\cot\frac{\theta}{2} + \sum_{m=1}^{\infty} u_m\sin m\theta\right)^2$$

$$= \left(\frac{1}{4}\cot\frac{\theta}{2}\right)^2 + \frac{1}{2}\cot\frac{\theta}{2}\sum_{m=1}^{\infty} u_m\sin m\theta + \left(\sum_{m=1}^{\infty} u_m\sin m\theta\right)^2,$$

or, in short,  $= \left(\frac{1}{4}\cot\frac{\theta}{2}\right)^2 + S_1 + S_2.$

Thus, using (13.22) we can provide expressions for $S_1$ and $S_2$:

$$S_1 = \sum_{m=1}^{\infty}\frac{1}{2}(1 + 2\cos\theta + 2\cos 2\theta + \cdots + 2\cos(m-1)\theta + \cos m\theta)\,u_m$$

$$S_2 = \sum_{m=1}^{\infty}\sum_{n=1}^{\infty}(u_m u_n\sin m\theta\sin n\theta)$$

$$= \frac{1}{2}\sum_{m=1}^{\infty}\sum_{n=1}^{\infty}[\cos(m-n)\theta - \cos(m+n)\theta\,]u_m u_n.$$

A rearrangement of $S_1 + S_2$ then yields

$$S_1 + S_2 = \sum_{k=0}^{\infty} C_k\cos k\theta. \quad (13.23)$$

---

[19]  See Hardy (1978) pp. 134–135.

Next, note that the contribution of $S_1$ to $C_0$ (the term independent of $\theta$) is $\frac{1}{2}\sum_{k=1}^{\infty} u_m$ and the contribution of $S_2$ to $C_0$ is $\frac{1}{2}\sum_{m=1}^{\infty} u_m^2$. Thus we obtain

$$C_0 = \frac{1}{2}\sum_{m=1}^{\infty} u_m(1+u_m) = \frac{1}{2}\sum_{m=1}^{\infty} \frac{x^m}{(1-x^m)^2}$$

$$= \frac{1}{2}\sum_{m=1}^{\infty}\sum_{n=1}^{\infty} nx^{mn} = \frac{1}{2}\sum_{n=1}^{\infty} nu_n.$$

Now when $k \geq 1$, the contribution of $S_1$ to $C_k$ is

$$\frac{1}{2}u_k + \sum_{l=0}^{\infty} u_{k+l}$$

and the contribution of $S_2$ to $C_k$ is

$$\frac{1}{2}\sum_{m-n=k} u_m u_n + \frac{1}{2}\sum_{n-m=k} u_m u_n - \frac{1}{2}\sum_{m+n=k} u_m u_n = \sum_{l=1}^{\infty} u_l u_{l+k} - \frac{1}{2}\sum_{l=1}^{k-1} u_l u_{k-l};$$

thus,

$$C_k = \frac{1}{2}u_k + \sum_{l=1}^{\infty} u_{k+l} + \sum_{l=1}^{\infty} u_l u_{k+l} - \frac{1}{2}\sum_{l=1}^{k-1} u_l u_{k-l}.$$

Observe that since

$$u_{k+l}(1+u_l) = \frac{x^{k+l}}{1-x^{k+l}}\left(1 + \frac{x^l}{1-x^l}\right)$$

$$= \frac{x^k}{1-x^k}\left(\frac{x^l}{1-x^l} - \frac{x^{k+l}}{1-x^{k+l}}\right) = u_k(u_l - u_{l+k})$$

and

$$u_l u_{k-l} = \frac{x^l}{1-x^l} \cdot \frac{x^{k-l}}{1-x^{k-l}}$$

$$= \frac{x^k}{1-x^k}\left(1 + \frac{x^l}{1-x^l} + \frac{x^{k-l}}{1-x^{k-l}}\right) = u_k(1 + u_l + u_{k-l}),$$

we may write

$$C_k = u_k\left(\frac{1}{2} + \sum_{l=1}^{\infty}(u_l - u_{k+l}) - \frac{1}{2}\sum_{l=1}^{k-1}(1 + u_l + u_{k-l})\right)$$

$$= u_k\left(\frac{1}{2} + u_1 + u_2 + \cdots + u_k - \frac{1}{2}(k-1) - (u_1 + u_2 + \cdots + u_{k-1})\right)$$

$$= u_k\left(1 + u_k - \frac{1}{2}k\right).$$

We complete the proof of (13.2) by substituting these values of $C_k$ and $C_0$ in (13.23):

$$\left(\frac{1}{4}\cot\frac{\theta}{2}\right)^2 + S_1 + S_2$$

$$= \left(\frac{1}{4}\cot\frac{\theta}{2}\right)^2 + \frac{1}{2}\sum_{m=1}^{\infty} mu_m + \sum_{m=1}^{\infty} u_m\left(1 + u_m - \frac{m}{2}\right)\cos m\theta$$

$$= \left(\frac{1}{4}\cot\frac{\theta}{2}\right)^2 + \sum_{m=1}^{\infty} u_m(1 + u_m)\cos m\theta + \frac{1}{2}\sum_{m=1}^{\infty} mu_m(1 - \cos m\theta)$$

$$= \left(\frac{1}{4}\cot\frac{\theta}{2}\right)^2 + \sum_{m=1}^{\infty}\frac{x^m\cos m\theta}{1 - x^m} + \frac{1}{2}\sum_{m=1}^{\infty}\frac{mx^m}{1 - x^m}(1 - \cos m\theta).$$

In a similar way, Ramanujan proved, without giving details, that

$$\left(\frac{1}{8}\cot^2\frac{\theta}{2} + \frac{1}{12} + \sum_{m=1}^{\infty}\frac{mx^m}{1 - x^m}(1 - \cos m\theta)\right)^2$$

$$= \left(\frac{1}{8}\cot^2\frac{\theta}{2} + \frac{1}{12}\right)^2 + \frac{1}{12}\sum_{m=1}^{\infty}\frac{m^3x^m}{1 - x^m}(5 + \cos m\theta) \qquad (13.24)$$

by making use of the identity

$$\cot^2\frac{\theta}{2}(1 - \cos n\theta)$$

$$= (2n - 1) + 4(n - 1)\cos\theta + 4(n - 2)\cos 2\theta + \cdots + 4\cos(n - 1)\theta + \cos n\theta. \qquad (13.25)$$

By expanding the trigonometric functions as power series, he had

$$\left(\frac{1}{2\theta^2} + \sum_{m=1}^{\infty}(-1)^{m-1}\frac{\theta^{2m}}{(2m)!}S_{2m-1}\right)^2 = \frac{1}{4\theta^2} + \frac{1}{2}S_3 - \frac{1}{12}\sum_{m=1}^{\infty}(-1)^{m-1}\frac{\theta^{2m}}{(2m)!}S_{2m+3},$$

where $S_m$ was defined by (13.8). Note that by equating the coefficient of $\theta^n$ on both sides, one obtains (13.17).

Ramanujan's formula (13.2) may appear to us unmotivated. On page 135 of his book on Ramanujan, Hardy gives an equivalent formula in terms of Weierstrass's $\wp$ and $\zeta$ functions and in Hardy's form, the formula may appear to have even less motivation. However, in a paper of 1993,[20] Li-Chien Shen cast Ramanujan's (13.2) in a slightly

---

[20] Shen (1993).

different form, by first setting

$$S = \cot x + 4 \sum_{n=1}^{\infty} \frac{q^{2n}}{1 - q^{2n}} \sin 2nx;$$

$$T_1 = -1 + 16 \sum_{n=1}^{\infty} \frac{q^{2n}}{(1 - q^{2n})^2} \cos 2nx + 8 \sum_{n=1}^{\infty} \frac{nq^{2n}}{1 - q^{2n}};$$

$$T_2 = 1 + \cot^2 x - 8 \sum_{n=1}^{\infty} \frac{nq^{2n}}{1 - q^{2n}} \cos 2nx.$$

Thus, Shen puts Ramanujan's identity in the form

$$S^2 = T_1 + T_2. \tag{13.26}$$

Shen proved that if we take the derivative of the logarithmic derivative of $\theta_1$, given by (3.214), then we get the identity

$$\left( \frac{\theta_1'}{\theta_1} \right)^2 = \frac{\theta_1''}{\theta_1} - \left( \frac{\theta_1'}{\theta_1} \right)'$$

and

$$\frac{\theta_1'}{\theta_1} = S; \quad \frac{\theta_1''}{\theta_1} = T_1; \quad \left( \frac{\theta_1'}{\theta_1} \right)' = -T_2. \tag{13.27}$$

### 13.5 The Arithmetic Function $\tau(n)$

In his 1916 paper, "On certain arithmetical functions," Ramanujan also investigated arithmetical properties of the Fourier coefficients of some modular forms. Recall that in his 1881 paper on modular functions and multiplier equations, Hurwitz had done this for the modular forms defined by Eisenstein series. Hurwitz's results were stated for the Eisenstein series

$$G_{2k}(\omega_1, \omega_2) = \sum_{m,n}' \frac{1}{(m\omega_1 + n\omega_1)^{2k}}, \tag{13.28}$$

where Im $\frac{\omega_1}{\omega_2} > 0$ and the sum was taken over all integers $m$ and $n$ except $(m, n) = (0, 0)$. In modern form, the Eisenstein series may be stated as

$$G_{2k}(\omega) = \sum_{m,n}' \frac{1}{(m\omega + n)^{2k}}, \tag{13.29}$$

where Im $\omega > 0$. The Fourier expansion of $G_{2k}$ found by Hurwitz may now be written as

$$G_{2k}(\omega) = \frac{(2\pi)^{2k}}{(2k)!} \left( B_{2k} + (-1)^k 2k \sum_{n=1}^{\infty} \sigma_{2k-1}(n)q^{2n} \right), \tag{13.30}$$

$$\equiv \frac{(2\pi)^{2k} B_{2k}}{(2k)!} E_{2k}(\omega), \tag{13.31}$$

where $q = e^{i\pi\omega}$ and

$$\sigma_{2k-1}(n) = \sum_{d|n} d^{2k-1}. \tag{13.32}$$

Note that Ramanujan's functions $P, Q, R$ are $E_2, E_4, E_6$, respectively. Thus, $\sigma_{2k-1}(n)$ is essentially the $n$th Fourier coefficient of the modular form, $G_{2k}(\omega)$ or $E_{2k}(\omega)$. Hurwitz observed[21] that it was a multiplicative arithmetic function; that is, for $m$ and $n$ relatively prime

$$\sigma_{2k-1}(mn) = \sigma_{2k-1}(m)\,\sigma_{2k-1}(n). \tag{13.33}$$

This multiplicativity can also be understood by considering the fact that, with $a \geq 0$, the Dirichlet series

$$\sum_{n=1}^{\infty} \frac{\sigma_a(n)}{n^s} \tag{13.34}$$

has an Euler product

$$\sum_{n=1}^{\infty} \frac{\sigma_a(n)}{n^s} = \zeta(s)\,\zeta(s-a) = \prod_p \left(1 - \frac{1}{p^s}\right)^{-1}\left(1 - \frac{p^a}{p^s}\right)^{-1}$$

$$= \prod_p \left(1 - \frac{1+p^a}{p^s} + \frac{1}{p^{2s-a}}\right)^{-1}$$

$$= \prod_p \left(1 - \frac{\sigma_a(p)}{p^s} + \frac{1}{p^{2s-a}}\right)^{-1}. \tag{13.35}$$

Note that the series (13.34) converges for Re $s > a + 1$ and formula (13.35) holds for these values of $s$.

Ramanujan defined the $\tau(n)$ function as the coefficient of $x^n$ in the power series expansion of $\Delta(\omega)$:

$$(2\pi)^{-12}\,\Delta(\omega) = x\left\{\prod_{n=1}^{\infty}(1-x^n)\right\}^{24} = \sum_{n=1}^{\infty} \tau(n)x^n, \quad x = e^{2\pi i\omega}. \tag{13.36}$$

He then gave a method for numerically computing $\tau(n)$: Differentiate (13.36) logarithmically with respect to $x$ to obtain

$$\sum_{n=1}^{\infty} n\tau(n)x^n = \left(1 - 24\sum_{n=1}^{\infty} \frac{nx^n}{1-x^n}\right)\sum_{n=1}^{\infty} \tau(n)x^n$$

$$= \left(1 - 24\sum_{n=1}^{\infty} \sigma_1(n)x^n\right)\sum_{n=1}^{\infty} \tau(n)x^n.$$

---

[21] Hurwitz (1962) p. 31.

Equating the coefficient of $x^n$ on both sides, one has

$$n\tau(n) = \tau(n) - 24 \sum_{k=1}^{n-1} \sigma_1(k)\,\tau(n-k)$$

or

$$\tau(n) = \frac{24}{1-n} \sum_{k=1}^{n-1} \sigma_1(k)\,\tau(n-k).$$

To obtain another interesting formula, Ramanujan employed Jacobi's relation (3.213), here restated:

$$\prod_{n=1}^{\infty} (1-x^n)^3 = \sum_{n=0}^{\infty} (-1)^n\,(2n+1)x^{\frac{n(n+1)}{2}}, \tag{13.37}$$

so that he could write

$$\sum_{n=1}^{\infty} \tau(n)x^n = x\left(\sum_{n=0}^{\infty} (-1)^n\,(2n+1)x^{\frac{n(n+1)}{2}}\right)^8.$$

Differentiating each side produces

$$\left(\sum_{n=0}^{\infty} (-1)^n\,(2n+1)x^{\frac{n(n+1)}{2}}\right) \sum_{n=1}^{\infty} (n-1)\tau(n)x^n$$

$$= \left(\sum_{n=1}^{\infty} \tau(n)x^n\right) 8x\left(-3 + 5\cdot 3x^2 - 7\cdot 6x^5 + \cdots\right).$$

Equating the coefficient of $x^n$ yields the recurrence relation

$$(n-1)\,\tau(n) - 3(n-10)\,\tau(n-1)$$
$$+ 5(n-28)\,\tau(n-3) - 7(n-55)\,\tau(n-6) + \cdots = 0,$$

where the $k$th term of the sequence $0, 1, 3, 6, \ldots$ is $\frac{k(k-1)}{2}$ and the $k$th term of the sequence $1, 10, 28, 55, \ldots$ is $1 + \frac{9}{2}k(k-1)$. Ramanujan made a table using these recurrence relations, giving values of $\tau(n)$ for $n = 1, 2, \ldots, 30$. It was apparently this set of numerical computations he used to make his conjecture concerning the Dirichlet series:

$$\sum_{n=1}^{\infty} \frac{\tau(n)}{n^s} = \prod_p (1 - \tau(p)p^{-s} + p^{11-2s})^{-1}. \tag{13.38}$$

This amounts to the relations

$$\tau(mn) = \tau(m)\,\tau(n), \quad \text{if} \quad (m, n) = 1 \tag{13.39}$$

and, if $p$ is prime,

$$\tau(p^{k+1}) = \tau(p^k)\,\tau(p) - p^{11}\,\tau(p^{k-1}) \quad \text{for} \quad k \geq 1. \tag{13.40}$$

Ramanujan presented these relations in the form:

If $n = p_1^{a_1} p_2^{a_2} \cdots p_k^{a_k}$, where $p_1$, $p_2$, ..., $p_k$ are the prime divisors of $n$, then

$$n^{-\frac{11}{2}} \tau(n) = \frac{\sin(1+a_1)\theta_{p_1}}{\sin \theta_{p_1}} \cdot \frac{\sin(1+a_2)\theta_{p_2}}{\sin \theta_{p_2}} \cdots \frac{\sin(1+a_k)\theta_{p_k}}{\sin \theta_{p_k}}, \tag{13.41}$$

$$\text{where} \quad \cos\theta_p = \frac{1}{2} p^{-\frac{11}{2}} \tau(p). \tag{13.42}$$

Ramanujan stated that (13.41) implied (13.39). He then made his famous conjecture[22] that $\theta_p$ was real, so that (13.42) entailed

$$\left(\frac{1}{2}\tau(p)\right)^2 \leq p^{11}. \tag{13.43}$$

He further observed that then (13.41) implied that

$$n^{-\frac{11}{2}} |\tau(n)| \leq (1+a_1)(1+a_2)\cdots(1+a_k),$$

so that Ramanunan conjectured

$$|\tau(n)| \leq n^{\frac{11}{2}} d(n), \tag{13.44}$$

where $d(n)$ was the number of divisors of $n$.

Next, to show that (13.38) is equivalent to the relation (13.41), first observe that

$$1 - \tau(p) p^{-s} + p^{11-2s} = \left(1 - e^{i\theta_p} p^{\frac{11}{2}-s}\right)\left(1 - e^{-i\theta_p} p^{\frac{11}{2}-s}\right),$$

$$\text{where} \quad 2\cos\theta_p = p^{-\frac{11}{2}} \tau(p).$$

Now for sufficiently large $s$, take

$$\left(1 - e^{i\theta_p} p^{\frac{11}{2}-s}\right)^{-1} = 1 + e^{i\theta_p} p^{\frac{11}{2}-s} + e^{2i\theta_p} p^{2(\frac{11}{2}-s)} + \cdots \tag{13.45}$$

$$\left(1 - e^{-i\theta_p} p^{\frac{11}{2}-s}\right)^{-1} = 1 + e^{-i\theta_p} p^{\frac{11}{2}-s} + e^{-2i\theta_p} p^{2(\frac{11}{2}-s)} + \cdots . \tag{13.46}$$

The coefficient of $p^{a(\frac{11}{2}-s)}$ in the product of (13.45) and (13.46) can be written as

$$e^{ai\theta_p} + e^{(a-2)i\theta_p} + e^{(a-4)i\theta_p} + \cdots + e^{-ai\theta_p}$$

$$= e^{-ai\theta_p} \frac{e^{(2a+2)i\theta_p} - 1}{e^{2i\theta} - 1} = \frac{e^{(a+1)i\theta_p} - e^{-(a+1)i\theta_p}}{e^{i\theta_p} - e^{-i\theta_p}} = \frac{\sin(a+1)\theta_p}{\sin\theta_p},$$

and, thus, the result is proved.

As a consequence of (13.41), Ramanujan noted that for $n = p^a$,

$$n^{-\frac{11}{2}} \tau(n) = \frac{\sin(1+a)\theta_p}{\sin\theta_p}. \tag{13.47}$$

Note that $a$ may be chosen as large as we wish and such that

$$\left| \frac{\sin(1+a)\theta_p}{\sin\theta} \right| \geq 1;$$

---

[22] Ramanujan (2000) p. 153.

Ramanujan concluded from (13.47) that $|\tau(n)| \geq n^{\frac{11}{2}}$ for an infinite number of values of $n$.

In order to check whether similar results might hold for other known situations, Ramanujan considered the expansion (with $x = q^2 = e^{2\pi i \omega}$, Im $\omega > 0$):

$$x\left\{\left(1 - x^{\frac{24}{\alpha}}\right)\left(1 - x^{\frac{48}{\alpha}}\right)\left(1 - x^{\frac{72}{\alpha}}\right) \cdots\right\}^{\alpha} = \sum_{n=1}^{\infty} \psi_{\alpha}(n) x^{n}, \tag{13.48}$$

where $\alpha$ is a divisor of 24. For example, when $\alpha = 1$, by application of Euler's pentagonal number theorem (2.43), Ramanujan had

$$\sum_{n=1}^{\infty} \psi_1(n) x^n = x - x^{5^2} - x^{7^2} + x^{11^2} + x^{13^2} - x^{17^2} - x^{19^2} + \cdots$$

$$= \sum_{n=0}^{\infty} (-1)^n x^{(6n\pm1)^2}.$$

Hence

$$\sum_{n=1}^{\infty} \frac{\psi_1(n)}{n^s} = \sum_{n=0}^{\infty} \frac{(-1)^n}{(6n \pm 1)^{2s}} = \prod (1 + q^{-2s})^{-1} (1 - p^{-2s})^{-1}, \tag{13.49}$$

where $q$ was a prime of the form $12n \pm 5$, and $p$ was a prime of the form $12n \pm 1$. It was then clear that $\psi_1(n)$ was multiplicative and

$$|\psi_1(n)| \leq 1 \quad \text{for all } n \text{ and } \quad |\psi_1(n)| = 1 \quad \text{for infinitely many } n.$$

Again, by taking $n = 3$ in (13.48), by Jacobi's identity (13.37) he obtained

$$\sum_{n=1}^{\infty} \psi_3(n) x^n = \sum_{n=0}^{\infty} (-1)^n (2n + 1) x^{(2n+1)^2}.$$

Hence

$$\sum_{n=1}^{\infty} \frac{\psi_3(n)}{n^s} = \sum_{n=0}^{\infty} \frac{(-1)^n}{(2n + 1)^{2s-1}} = \prod (1 + q^{1-2s})^{-1} (1 - p^{1-2s})^{-1}, \tag{13.50}$$

where $q$ was a prime of the form $4n + 3$ and $p$ was a prime of the form $4n + 1$. From this he could conclude that $\psi_3(n)$ was multiplicative and that

$$|\psi_3(n)| \leq \sqrt{n} \quad \text{for all } n \quad \text{and} \quad |\psi_3(n)| = \sqrt{n} \quad \text{for infinitely many } n.$$

For the cases $\alpha = 2, 4, 6, 8, 12$, Ramanujan conjectured results similar to his conjecture for the case $\alpha = 24$. Thus, he surmised

... that

$$\sum_{n=1}^{\infty} \frac{\psi_2(n)}{n^s} = \prod_1 \prod_2, \tag{13.51}$$

where

$$\prod_1 = \left( (1 + 5^{-2s})(1 - 7^{-2s})(1 - 11^{-2s})(1 + 17^{-2s}) \cdots \right)^{-1}, \tag{13.52}$$

$5, 7, 11, \ldots$ being the primes of the form $12n - 1$ and $12n \pm 5$, those of the form $12n + 5$ having the plus sign and the rest the minus sign; and

$$\prod_2 = \left( (1 + 13^{-s})^2 (1 - 37^{-s})^2 (1 - 61^{-s})^2 (1 + 73^{-s})^2 \cdots \right)^{-1}, \tag{13.53}$$

$13, 37, 61, \ldots$ being the primes of the form $12n + 1$, those of the form $m^2 + (6n - 3)^2$ having the plus sign and those of the form $m^2 + (6n)^2$ the minus sign.

This is equivalent to the assertion that if

$$n = \left( 5^{a_5} \cdot 7^{a_7} \cdot 11^{a_{11}} \cdot 17^{a_{17}} \cdots \right)^2 \cdot 13^{a_{13}} \cdot 37^{a_{37}} \cdot 61^{a_{61}} \cdot 73^{a_{73}} \cdots , \tag{13.54}$$

where $a_p$ is zero or a positive integer, then

$$\psi_2(n) = (-1)^{a_5 + a_{13} + a_{17} + a_{29} + a_{41} \cdots} (1 + a_{13})(1 + a_{37})(1 + a_{61}) \cdots , \tag{13.55}$$

where $5, 13, 17, 29, \ldots$ are the primes of the form $4n + 1$, excluding those of the form $m^2 + (6n)^2$; and that otherwise

$$\psi_2(n) = 0. \tag{13.56}$$

It follows that $|\psi_2(n)| \le d(n)$ for all values of $n$, and $|\psi_2(n)| \ge 1$ for an infinity of values of $n$. These results are easily proved to be actually true.[23]

Ramanujan's conjectures for $\alpha = 4$ and $6$ are given in the exercises at the end of this chapter. Note that when $\alpha = 12$, the product (13.48) can be written as

$$(2\pi)^{-6} \sqrt{\Delta(2\omega)}. \tag{13.57}$$

Ramanujan's also conjectured that

$$\sum_{n=1}^{\infty} \frac{\psi_{12}(n)}{n^s} = \prod_p \left( 1 - \psi_{12}(p) p^{-s} + p^{5-2s} \right)^{-1}, \tag{13.58}$$

where the product is over all odd primes and

$$\left( \psi_{12}(p) \right)^2 \le 4p^5. \tag{13.59}$$

Products such as those appearing in (13.35), (13.38), (13.49), (13.50), (13.51), (13.58) are called Euler products. All these conjectures of Ramanujan on the existence of Euler products were proved by Mordell in 1917.

In 1988, a fascinating volume, containing a facsimile of Ramanujan's lost notebook and other previously unpublished papers, was printed. This contains Ramanujan's statements of further results on Euler products associated with modular forms.[24] He gave Euler products for Dirichlet series corresponding to cusp forms of the type $\left( \eta(\omega)\eta(N\omega) \right)^k$, where $k \mid 24$ and $N + 1 \mid 24$, and where $\eta(\omega)$ denotes Dedekind's $\eta$ function. He considered five examples:

$$\eta(\omega)\eta(7\omega), \quad \eta(\omega)\eta(11\omega), \quad \eta(\omega)\eta(23\omega), \quad \left( \eta(\omega)\eta(11\omega) \right)^2, \quad \left( \eta(\omega)\eta(7\omega) \right)^3.$$

---

[23] Ramanujan (2000) p. 156.
[24] See Ramanujan (1988) pp. 146, 150, 160, 162, 163, 233–235, 247, 249, 328, 329, 403.

To see his results for the third and fourth cases, let $x = e^{2i\pi\,\omega}$. Now for the third case, if

$$\sum_{n=1}^{\infty} \phi(n)x^n = x \prod_{n=1}^{\infty} (1 - x^n)(1 - x^{23n}), \tag{13.60}$$

then

$$\sum_{n=1}^{\infty} \phi(n)n^{-s} = \prod_{p} (1 - p^{-2s})^{-1} \prod_{q} (1 + q^s + q^{2s})^{-1} \prod_{r} (1 - r^{-s})^{-2}, \tag{13.61}$$

where $p$ denotes a prime that is a quadratic nonresidue (mod 23), $q$ a quadratic residue (mod 23) of the form $23A^2 + B^2$, and $r$ a quadratic nonresidue not of the form $23A^2 + B^2$. Next, for the fourth case, if

$$\sum_{n=1}^{\infty} \lambda(n)x^n = x \prod_{n=1}^{\infty} (1 - x^n)^2(1 - x^{11n})^2, \tag{13.62}$$

then

$$\sum_{n=1}^{\infty} \lambda(n)n^{-s} = (1 - 11^{-s})^{-1} \prod_{p \neq 11} \left(1 - \lambda(p)p^{-s} + p^{1-2s}\right)^{-1}. \tag{13.63}$$

These examples were also considered by Hecke in his 1940 paper, "Analytische Arithmetik der Positiven Quadratischen Formen."[25] In 1975, B. J. Birch edited a list of even more of Ramanujan's formulas on Euler products.[26] The study of these formulas led Birch to remark, "Ramanujan realized that even when a modular form did not itself have an Euler product, one could regain the Euler product property by taking an appropriate linear combination of forms of the same level." Birch further concluded of these formulas, "They support the view that Ramanujan's insight into the arithmetic of modular forms was even greater than has been realized." In 1982, S. S. Rangachari published a paper presenting proofs, including some of his own, of all Ramanujan's theorems on Euler products.[27]

The proofs of inequalities such as (13.59) and (13.43) were given by Deligne in his famous 1974 paper on Weil's conjectures.

### 13.6 Mordell on Euler Products

Having perhaps learned of Ramanujan's conjectures from Hardy before their publication, Mordell, wrote a 1917 paper "On Mr. Ramanujan's Empirical Expansions of Modular Functions,"[28] in which he proved Ramanujan's conjectures on Euler products associated with $\left(\Delta(\frac{12}{a}\omega)\right)^{\frac{a}{12}}$ where $a \mid 12$. For this purpose, Mordell was able to employ results of Hurwitz on the multiplier equation for $\left(\Delta(\omega)\right)^{\frac{a}{12}}$. In effect, Mordell

---

[25] Hecke (1959) pp. 789–918, especially pp. 898–918.    [26] Birch (1975).
[27] Rangachari (1982).    [28] Mordell (1917a).

had defined for each prime $p$ an operator that could be identified with the Hecke operator $T_p$ on a space of cusp forms.

Louis J. Mordell (1888–1972) was born in Philadelphia, Pennsylvania, but after secondary school, he traveled to Cambridge, England, to study mathematics. He took the Mathematical Tripos examinations and went into research in number theory. In 1917, he published a paper[29] on the representations of numbers as sums of an even number of squares in which he utilized the theory of modular forms. According to J. W. S. Cassels, "Mordell was, apparently, the first to treat the representation of integers as the sum of a fixed number $n$ of squares by using the finite dimensionality of the space of modular forms of given dimensions to establish identities thereby unifying the existing mass of results for individual values of $n$."[30] This paper established Mordell as an expert in elliptic modular functions in Britain. When Hardy drew his attention to Ramanujan's conjectures, Mordell quickly found the solution within Hurwitz's work on the multiplier equation. Textbooks generally concentrate on Mordell's work on the Euler product of $\Delta(\omega_1, \omega_2)$, and it is not usually mentioned that Mordell also considered $\Delta^{\frac{a}{12}}(\omega_1, \omega_2)$, where $a$ is a divisor of 12. Mordell evidently did not attempt to generalize these results to other modular forms, and this was done almost two decades later by Hecke, then unaware of the earlier work on this topic of Ramanujan and Mordell.

Mordell also did fundamental work on the rational points on the elliptic curve

$$y^2 = 4x^3 - g_2 x - g_3,$$

where $g_2$ and $g_3$ are integers and $g_2^3 - 27g_3^2 \neq 0$. As shown in Section 12.6, such cubic curves are called elliptic curves because they can be parametrized by elliptic functions: $(x, y) = (\wp(z), \wp'(z))$ with appropriately chosen periods of $\wp(z)$. The rational points on such a curve can be given the structure of an Abelian group by means of the addition formula for $\wp(z)$. In 1922, Mordell proved[31] that this Abelian group is finitely generated, a result conjectured by Poincaré in 1901.

To set up Mordell's proof of Ramanujan's conjectures, recall that Hurwitz had proved that when $af = 12$ and $p \nmid 6$, with $p$ a prime, then

$$\frac{\sum_{h=0}^{p-1} \sqrt[f]{\Delta(\omega_1 + 12h\,\omega_2, p\,\omega_2)} + (-1)^{a\frac{p-1}{2}} \sqrt[f]{\Delta(p\,\omega_1, \omega_2)}}{\sqrt[f]{\Delta(\omega_1, \omega_2)}} \tag{13.64}$$

was a modular function for $\Gamma(1)$. Recall that Kiepert, Klein, and Hurwitz had shown that (13.64) was a constant for $f = 12$. However, Mordell deduced this fact by proving that (13.64) was bounded at the cusp at infinity; in that case, (13.64) would be a bounded modular function, and hence a constant. Following Mordell's argument, first note that

$$\left(\frac{\omega_2}{2\pi}\right)^a \Delta^{\frac{1}{f}}(\omega_1, \omega_2) = q^{\frac{a}{6}} \prod_{s=1}^{\infty} (1 - q^{2n})^{2a}, \quad q = e^{\pi i \omega}, \quad \omega = \frac{\omega_1}{\omega_2},$$

---

[29] Mordell (1917b).   [30] Cassels (1973) p. 499.   [31] Mordell (1923).

so that in order to obtain boundedness at infinity, it will be sufficient to show that the lowest power of $q$ in the numerator of (13.64) is $r$, where $r \geq \frac{a}{6}$. Now let

$$\left(\frac{\omega_2}{2\pi}\right)^a \Delta^{\frac{1}{7}}(\omega_1, \omega_2) = \sum_{s=0}^{\infty} B_s \, q^{\frac{a}{6}+2s}.$$

In that case

$$\left(\frac{\omega_2}{2\pi}\right)^a \sum_{h=0}^{p-1} \Delta^{\frac{1}{7}}(\omega_1 + 12h\,\omega_2, p\,\omega_2) = \sum_{h=0}^{p-1} \sum_{s=0}^{\infty} e^{\left(\frac{a}{6}+2s\right)12\frac{h\pi i}{p}} B_s \, q^{\frac{1}{p}\left(\frac{a}{6}+2s\right)}$$

$$= \sum_{s=0}^{\infty} \left( \sum_{h=0}^{p-1} e^{\frac{2\pi i h}{p}(a+12s)} \right) B_s \, q^{\frac{1}{p}\left(\frac{a}{6}+2s\right)}.$$

The sum in parentheses is identically zero, except when $a + 12s \equiv 0 \pmod{p}$. In the latter case, the sum in parentheses equals $p$. Thus, the lowest power of $q$ is the smallest value of $\frac{a+12s}{6p}$ when $a + 12s \equiv 0 \pmod{p}$. Since $a$ and $p$ are relatively prime,

$$a(1 + 12f) \equiv 0 \pmod{p} \text{ implies that } 1 + 12f \equiv 0 \pmod{p}.$$

Thus,

$$\frac{a + 12s}{6p} = \frac{a(1 + 12f)}{6p} \geq \frac{a}{6}$$

and there exists a constant $c$ such that

$$\sum_{h=0}^{p-1} \sqrt[7]{\Delta(\omega_1 + 12h\,\omega_2, p\,\omega_2)} + (-1)^a \frac{p-1}{2} \sqrt[7]{\Delta(p\,\omega_1, \omega_2)} = c\sqrt[7]{\Delta(\omega_1, \omega_2)}. \quad (13.65)$$

This proves (11.171); using a similar approach, we can prove (11.170), as well as (11.172) through (11.174).

To derive the Euler product for $\sqrt[7]{\Delta}$, change $\omega_1$ to $f\,\omega_1$ in (13.65). Next, supposing that

$$\sqrt[7]{\Delta(f\,\omega_1, \omega_2)} = \left(\frac{2\pi}{\omega_2}\right)^a \sum_{s=1}^{\infty} A_s \, e^{2\pi i s \omega}, \quad \omega = \frac{\omega_1}{\omega_2}, \quad A_1 = 1, \quad (13.66)$$

we may write

$$\sum_{h=0}^{p-1} \sqrt[7]{\Delta(f\,\omega_1 + 12h\,\omega_2, p\,\omega_2)} = \left(\frac{2\pi}{p\,\omega_2}\right)^a \sum_{s=1}^{\infty} \sum_{h=0}^{p-1} A_s \, e^{\frac{2\pi i s \omega}{p}} e^{\frac{24\pi i h s}{p}}. \quad (13.67)$$

Observe that

$$\sum_{h=0}^{p-1} e^{\frac{24\pi i h s}{p}} = \begin{cases} p, & \text{when } p \mid s, \\ 0, & \text{when } p \nmid s. \end{cases} \quad (13.68)$$

Substituting (13.68) in the right-hand side of (13.67), the latter becomes

$$\left(\frac{2\pi}{\omega_2}\right)^a p^{1-a} \sum_{s=1}^{\infty} A_{ps} \, e^{2\pi i s \omega}. \quad (13.69)$$

By applying (13.66) and (13.69) to (13.65), Mordell obtained the basic relation[32]

$$p^{1-a} \sum_{s=1}^{\infty} A_{ps}\, e^{2\pi i s \omega} + (-1)^{a\frac{p-1}{2}} \sum_{s=1}^{\infty} A_s\, e^{2\pi i s p\omega} = c \sum_{s=1}^{\infty} A_s\, e^{2\pi i s \omega}. \qquad (13.70)$$

Equating the coefficients of $e^{2\pi i s \omega}$ when $p \nmid s$, Mordell found that

$$A_{ps} = c\, p^{a-1}\, A_s. \qquad (13.71)$$

Since $A_1 = 1$, when $s = 1$, (13.71) reduces to

$$A_p = c\, p^{a-1}. \qquad (13.72)$$

Substituting (13.72) in (13.71) for $p \nmid s$ produces

$$A_{ps} = A_p A_s. \qquad (13.73)$$

By equating coefficients of $e^{2\pi i p s \omega}$, we find

$$p^{1-a} A_{p^2 s} + (-1)^{a\frac{p-1}{2}} A_s = c A_{ps};$$

by (13.72) this implies

$$A_{p^2 s} - A_p A_{ps} + (-1)^{a\frac{p-1}{2}} p^{a-1} A_s = 0. \qquad (13.74)$$

Mordell employed (13.73) and (13.74) to show that the Fourier coefficients $A_s$ were multiplicative and to show that the Dirichlet series $\sum_{n=1}^{\infty} A_n\, n^{-s}$ had an Euler product representation. He noted that to prove multiplicativity, it would be sufficient to prove that if $p \nmid s$, then

$$A_{sp^\lambda} = A_s A_{p^\lambda}. \qquad (13.75)$$

He gave an inductive proof of (13.75) by replacing $s$ in (13.74) by $p^\lambda s$, arriving at

$$A_{p^{\lambda+2}s} - A_p A_{p^{\lambda+1}s} + (-1)^{a\frac{p-1}{2}} p^{a-1} A_{p^\lambda s} = 0, \qquad (13.76)$$

and, in particular, on taking $s = 1$,

$$A_{p^{\lambda+2}} - A_p A_{p^{\lambda+1}} + (-1)^{a\frac{p-1}{2}} p^{a-1} A_{p^\lambda} = 0. \qquad (13.77)$$

He assumed (13.75) true for $\lambda$ and $\lambda + 1$ so that (13.76) produced

$$A_{p^{\lambda+2}s} - A_p A_{p^\lambda} A_s + (-1)^{a\frac{p-1}{2}} p^{a-1} A_{p^\lambda} A_s = 0;$$

by the application of (13.77), he could then conclude that

$$A_{p^{\lambda+2}s} = A_{p^{\lambda+2}} A_s, \qquad (13.78)$$

and that (13.75) held for $\lambda + 2$. Since (13.75) was true for $\lambda = 0$ and $\lambda = 1$, it followed that (13.75) was true for all $\lambda \geq 0$.

[32] Mordell (1917a) pp. 118–119.

Mordell next observed that (13.77) was a linear second-order difference equation with constant coefficients, so that de Moivre's method, described in (6.24) and (6.25) and applied in (6.40), yielded the relation

$$1 + A_p x + A_{p^2} x^2 + A_{p^3} x^3 + \cdots = \left(1 - A_p x + (-1)^{a\frac{p-1}{2}} p^{a-1} x^2\right)^{-1}, \qquad (13.79)$$

or, setting $x = p^{-s}$,

$$\sum_{k=0}^{\infty} A_{p^k} \, p^{-ks} = \left(1 - A_p \, p^{-s} + (-1)^{a\frac{p-1}{2}} p^{a-1} p^{-2s}\right)^{-1}. \qquad (13.80)$$

Next, Mordell noted[33] that from (13.75) it followed that if $x = e^{2\pi i \omega}$ and if

$$x \prod_{n=1}^{\infty} (1 - x^{fn})^{2a} = \sum_{s=1}^{\infty} A_{a,s} \, x^s, \qquad (13.81)$$

then $A_{a,l} A_{a,m} = A_{a,lm}$ for $l$ and $m$ relatively prime, and

$$\sum_{n=1}^{\infty} \frac{A_{a,n}}{n^s} = \prod_{p} \left(1 - A_{a,p} \, p^{-s} + (-1)^{a\frac{p-1}{2}} p^{a-1-2s}\right)^{-1}, \qquad (13.82)$$

where $p$ was taken to run over all primes excluding $p = 2$ except in the case $a = 4, 12$; and excluding $p = 3$ except in the case $a = 3, 6, 12$.

Mordell used (13.82) to prove all the cases of Euler products presented in Ramanujan's 1916 paper; we mention the two most important. When $a = 12$, Mordell had a proof of Ramanujan's famous result that

$$\sum_{n=1}^{\infty} \tau(n) n^{-s} = \prod_{p} \left(1 - \tau(p) p^{-s} + p^{11-2s}\right)^{-1}, \qquad (13.83)$$

where the product was over all primes. When $a = 6$, he had another of Ramanujan's results (13.58), with $\psi_{12}(n)$ replaced by Glaisher's notation $\Omega(n)$: If

$$x \prod_{n=1}^{\infty} (1 - x^{2n})^{12} = \sum_{n=1}^{\infty} \Omega(n) x^n,$$

then

$$\sum_{n=1}^{\infty} \Omega(n) n^{-s} = \prod_{p \neq 2} \left(1 - \Omega(p) \, p^{-s} + p^{5-2s}\right)^{-1}. \qquad (13.84)$$

where the product was taken over all primes except $p = 2$. Note that (13.84) implies that $\Omega(m)$ is multiplicative. This was first conjectured by Glaisher in his work on the number of representations of an integer as a sum of twelve squares.

These results of Ramanujan and Mordell were rediscovered independently by Erich Hecke in 1935.[34] He applied his theory of Hecke operators to obtain Euler products of modular forms belonging to congruence subgroups of $\Gamma(1)$.

---

[33] Mordell (1917a) p. 121.     [34] Hecke (1959) pp. 577–590.

Ramanujan's conjectures serve to exemplify his great gift of divining or foreseeing important directions in future mathematics. Commenting on this aspect of Ramanujan's work, Ken Ono has written, "... he should be remembered as the greatest anticipator of mathematics. Although he was recognized during his lifetime, his most important ideas, those that have powered mathematicians after his death, were largely viewed as insignificant while he was alive. Those ideas are typically found in his notebooks, letters, and papers as innocent-looking formulas and expressions. Those gems have offered visions of the future, hints of subjects conjured long after his death. ... He had a gift of imagination the like of which the world of mathematics had never seen before."[35]

## 13.7 Exercises

1. Let $x = e^{2\pi i \omega}$. Using the notation of (13.31), suppose

$$\sum_{n=1}^{\infty} \tau_2(n)x^n = (2\pi)^{-12}E_4(\omega)\Delta(\omega)$$

$$\sum_{n=1}^{\infty} \tau_3(n)x^n = (2\pi)^{-12}E_6(\omega)\Delta(\omega)$$

$$\sum_{n=1}^{\infty} \tau_4(n)x^n = (2\pi)^{-12}E_4^2(\omega)\Delta(\omega)$$

$$\sum_{n=1}^{\infty} \tau_5(n)x^n = (2\pi)^{-12}E_4(\omega)E_6(\omega)\Delta(\omega)$$

$$\sum_{n=1}^{\infty} \tau_7(n)x^n = (2\pi)^{-12}E_4^2(\omega)E_6(\omega)\Delta(\omega).$$

Then prove that

$$\sum_{n=1}^{\infty} \frac{\tau_2(n)}{n^s} = \prod_p \left(1 - \tau_2(p)\,p^{-s} + p^{15-2s}\right)^{-1}$$

$$\sum_{n=1}^{\infty} \frac{\tau_3(n)}{n^s} = \prod_p \left(1 - \tau_3(p)\,p^{-s} + p^{17-2s}\right)^{-1}$$

$$\sum_{n=1}^{\infty} \frac{\tau_4(n)}{n^s} = \prod_p \left(1 - \tau_4(p)\,p^{-s} + p^{19-2s}\right)^{-1}$$

$$\sum_{n=1}^{\infty} \frac{\tau_5(n)}{n^s} = \prod_p \left(1 - \tau_5(p)\,p^{-s} + p^{21-2s}\right)^{-1}$$

$$\sum_{n=1}^{\infty} \frac{\tau_7(n)}{n^s} = \prod_p \left(1 - \tau_7(p)\,p^{-s} + p^{25-2s}\right)^{-1}.$$

---

[35] Ono and Aczel (2016) pp. 225–226.

Ramanujan gave these in his unpublished papers; see Ramanujan (1988) pp. 159–160. He commented, "All these seem to be capable of proof as the case of $\tau(n)$ by Mr. Mordell's methods." Note that Ramanujan wrote $E_4(\omega)$ and $E_6(\omega)$ as $Q$ and $R$, respectively. Concerning this problem, see the last sentence in Section 16.5. See also Berndt and Ono (2001), especially page 102.

2. This three-part problem gives examples of the situations in which Ramanujan took linear combinations of modular forms to obtain Euler products.

   (a) Let $P, Q, R$ be defined by (13.11) through (13.13) and let

   $$f(x) = x^{\frac{1}{24}} \prod_{n=1}^{\infty} (1 - x^n).$$

   Note that $f(x)$ is identical with Dedekind's $\eta$ function. Suppose $A$ and $B$ are any two integers such that $A^2 + 2B^2 = p$ and $A \equiv 1 \pmod 3$, $p$ being a prime of the form $6k + 1$. Let

   $$\sum_{n=0}^{\infty} a_0(n)x^{\frac{n}{6}} = f^4, \qquad\qquad \sum_{n=0}^{\infty} a_1(n)x^{\frac{n}{6}} = f^4 p,$$

   $$\sum_{n=0}^{\infty} a_2(n)x^{\frac{n}{6}} = f^4 Q, \qquad\qquad \sum_{n=0}^{\infty} a_3(n)x^{\frac{n}{6}} = f^4 R,$$

   $$\sum_{n=0}^{\infty} a_4(n)x^{\frac{n}{6}} = f^4 Q^2 + 288\sqrt{7}\, f^{20}, \qquad\qquad \sum_{n=0}^{\infty} a_5(n)x^{\frac{n}{6}} = f^4 QR,$$

   $$\sum_{n=0}^{\infty} a_7(n)x^{\frac{n}{6}} = f^4 Q^2 R + 10080\sqrt{286}\, f^{20} R.$$

   In all these cases, prove that

   - $\sum_{n=1}^{\infty} a_k(n)n^{-s} = \prod_p \left(1 - a_k(p)p^{-s} + p^{2k+1-2s}\right)^{-1}$,
     where $p$ assumes all prime values greater than 3;
   - if $k = 0, 2, 3, 5$:
     $a_k(p) = 0$ for $p \equiv 5 \pmod 6$ and
     $a_k(p) = (A + \sqrt{-3}\,B)^{2k+1} + (A - \sqrt{-3}\,B)^{2k+1}$ for $p \equiv 1 \pmod 6$;
   - for all values of $n$, $a_1(n) = na_0(n)$.

   Ramanujan noted that $a_4(p)$ and $a_7(p)$ did not seem to follow such simple laws.

   (b) Suppose $P, Q, R, f, A, B$ are defined as in (a) and let

   $$\sum_{n=0}^{\infty} a_0(n)x^{\frac{n}{3}} = f^8,$$

   $$\sum_{n=0}^{\infty} a_1(n)x^{\frac{n}{3}} = f^8 P,$$

   $$\sum_{n=0}^{\infty} a_2(n)x^{\frac{n}{3}} = f^8 Q + 6\sqrt{10}\, f^{16},$$

$$\sum_{n=0}^{\infty} a_3(n)x^{\frac{n}{3}} = f^8 R,$$

$$\sum_{n=0}^{\infty} a_4(n)x^{\frac{n}{3}} = f^8 Q^2 + 6\sqrt{70}\, f^{16} Q,$$

$$\sum_{n=0}^{\infty} a_6(n)x^{\frac{n}{3}} = f^8 QR + 12\sqrt{55}\, f^{16} R,$$

$$\sum_{n=0}^{\infty} a_8(n)x^{\frac{n}{3}} = f^8 Q^2 R + 12\sqrt{910}\, f^{16} QR.$$

In all these cases, prove:

- $\sum_{n=1}^{\infty} a_k(n)n^{-s} = \prod_p \left(1 - a_k(p)p^{-s} + p^{2k+3-2s}\right)^{-1}$,
  where $p$ assumes all prime values except 3;

- if $k = 0$ or 3, then

  $$a_k(p) = 0 \quad \text{for} \quad p \equiv 2 \pmod 3$$

  and

  $$a_k(p) = (A + \sqrt{-3}B)^{2k+3} + (A - \sqrt{-3}B)^{2k+3} \quad \text{for} \quad p \equiv 1 \pmod 3;$$

- for all values of $n$, $a_1(n) = na_0(n)$.

(c) Let $P, Q, R, f$ be the same as in (b). Suppose $C$ and $D$ are integers such that $C^2 + 4D^2 = p$, where p is a prime of the form $4k + 1$. Let

$$\sum_{n=0}^{\infty} a_0(n)x^{\frac{n}{4}} = f^6, \qquad \sum_{n=0}^{\infty} a_1(n)x^{\frac{n}{4}} = f^6 P,$$

$$\sum_{n=0}^{\infty} a_2(n)x^{\frac{n}{4}} = f^6 Q, \qquad \sum_{n=0}^{\infty} a_3(n)x^{\frac{n}{4}} = f^6 R + 24\sqrt{-35}\, f^{18},$$

$$\sum_{n=0}^{\infty} a_4(n)x^{\frac{n}{4}} = f^6 Q^2, \qquad \sum_{n=0}^{\infty} a_5(n)x^{\frac{n}{4}} = f^6 QR + 24\sqrt{-1155}\, f^{18} Q,$$

$$\sum_{n=0}^{\infty} a_7(n)x^{\frac{n}{4}} = f^6 Q^2 R + 120\sqrt{-3003}\, f^{18} Q^2.$$

In all these cases prove that:

- $\sum_{n=1}^{\infty} a_k(n)n^{-s} = \prod_1 \prod_2$,

  where $\prod_1 = \prod_p \left(1 - a_k(p)p^{-s} - p^{2k+2-2s}\right)^{-1}$, with $p$ assuming the value of all primes of the form $4k - 1$,

$$\text{and where} \quad \prod_2 = \prod_p \left(1 - a_k(p)p^{-s} - p^{2k+2-2s}\right)^{-1},$$

with $p$ assuming the value of all primes of the form $4k + 1$;

- if $k = 0, 2,$ or $4$, then

  $a_k(p) = 0$ for $p \equiv 3 \pmod 4$

  $a_k(p) = (C + 2\sqrt{-1}\,D)^{2k+2} = (C - 2\sqrt{-1}\,D)^{2k+2}$ for $p \equiv 1 \pmod 4$;

- $a_1(n) = na_0(n)$.

  See Ramanujan (1988) pp. 233–235. The statements of the results follow Birch (1975), pp. 75–76. For proofs, see Rangachari (1982).

3. Prove (13.24).
4. Prove Shen's result (13.27).
5. Let

$$\frac{1}{16} \theta_0^4 \theta_2^4 \theta_3^4 \left(\theta_3^8 - \theta_0^4 \theta_2^4\right) = \sum_{n=1}^{\infty} \psi_1(n)q^n,$$

$$\frac{1}{16} \theta_0^4 \theta_2^4 \theta_3^4 \left(\theta_3^8 + 2\theta_0^4 \theta_2^4\right) = \sum_{n=1}^{\infty} \psi_2(n)q^n.$$

Prove that the functions $\psi_i$, $i = 1, 2$ are multiplicative, that is that $\psi_i(mn) = \psi_i(m)\psi_i(n)$, $i = 1, 2$, for $(m, n) = 1$. Prove also that, for $p$ an odd prime and $m$ a positive integer,

$$\psi_i(p^{m+1}) = \psi_i(p)\,\psi_i(p^m) - p^9\,\psi_i(p^{m-1}), \quad i = 1, 2.$$

For an elementary proof of these results, see Rankin (1946).

# 14

---

# Dirichlet Series and Modular Forms

## 14.1  Preliminary Remarks

The transformation connecting a modular form with a Dirichlet series is called a Mellin transform. The Mellin transform of a function $f(x)$ is defined by the integral

$$F(s) = \int_0^\infty x^{s-1} f(x)dx, \tag{14.1}$$

under the assumption that the integral exists. Perhaps the earliest example of a Mellin transform is Euler's formula

$$\Gamma(s) = \int_0^\infty x^{s-1} e^{-x} dx. \tag{14.2}$$

The integral in (14.2) converges for Re $s > 0$ and is often taken as the definition of a gamma function. For Euler, the integral in (14.2) was a theorem, since in 1729 he defined the gamma function as an infinite product[1]

$$\Gamma(s+1) = \prod_{n=1}^\infty \left(1 + \frac{1}{n}\right)^s \left(1 + \frac{s}{n}\right)^{-1}.$$

Initially, he wrote the integral in (14.2) as

$$\Gamma(s) = \int_0^1 \left(\log \frac{1}{t}\right)^{s-1} dt.$$

Note that Euler himself wrote $\Gamma(s+1)$ as $1.2\ldots s$ even if $s$ was not a positive integer; the $\Gamma$ notation is due to Legendre.[2]

Riemann was probably the first mathematician to connect modular forms with Dirichlet series. In his famous 1859 paper on prime numbers,[3] Riemann applied the

---

[1]  See Euler's first four letters to Goldbach in Fuss (1968). In these letters, he discussed the connection of the product with the integral. See also Eu.I-14, pp. 1–24, especially p. 23, where Euler mentioned how the gamma function could be employed to define fractional derivatives.

[2]  Legendre (1811–1817) vol. 2, p. 5.      [3]  Riemann (1990) pp. 177–185.

formula

$$\pi^{-\frac{s}{2}}\Gamma(s)\zeta(s) = \int_0^\infty x^{\frac{1}{2}s-1} \sum_{n=1}^\infty e^{-n^2\pi x}\, dx \qquad (14.3)$$

to prove that the quantity on the left-hand side of (14.3) was invariant under $s \to 1-s$. This gave him the functional equation for $\zeta(s)$. He made essential use of the transformation

$$1 + 2\sum_{n=1}^\infty e^{-n^2\pi x} = x^{-\frac{1}{2}}\left(1 + 2\sum_{n=1}^\infty e^{-\frac{n^2\pi}{x}}\right). \qquad (14.4)$$

In a paper of 1894, the French mathematician Eugène Cahen gave the Mellin inversion of (14.2):[4]

$$e^{-y} = \frac{1}{2\pi i}\int_{c-i\infty}^{c+i\infty} \Gamma(s)y^{-s}\, ds, \quad c > 0 \text{ and } \operatorname{Re} y > 0. \qquad (14.5)$$

He used this formula to establish transformation formulas for theta functions with twists. He showed for example, that where $\left(\frac{n}{p}\right)$ denotes the Legendre symbol, the twisted theta series

$$\phi_1(x) = \sum_{n=1}^\infty \left(\frac{n}{p}\right)e^{-\frac{n^2\pi x}{p}}, \quad p \equiv 1 \mod 4 \qquad (14.6)$$

and

$$\phi_2(x) = \sum_{n=1}^\infty n\left(\frac{n}{p}\right)e^{-\frac{n^2\pi x}{p}}, \quad p \equiv -1 \mod 4 \qquad (14.7)$$

satisfied the transformations

$$\sqrt{x}\phi_1(x) = \phi_1\left(\frac{1}{x}\right) \qquad (14.8)$$

and

$$x^{\frac{3}{2}}\phi_2(x) = \phi_2\left(\frac{1}{x}\right), \qquad (14.9)$$

respectively. For this purpose, he employed the functional equations, well known even in Cahen's time, of the $L$-functions

$$L(s) = \sum_{n=1}^\infty \frac{\left(\frac{n}{p}\right)}{n^s}, \quad \text{with } p \equiv 1 \pmod 4 \text{ or with } p \equiv -1 \pmod 4.$$

The functional equation for the $L$-function when $p \equiv 1 \pmod 4$ is given by the invariance under $s \to 1-s$ of $\left(\frac{\pi}{p}\right)^{-\frac{s}{2}}\Gamma\left(\frac{s}{2}\right)L(s)$; when $p \equiv -1 \pmod 4$ it is given by the invariance under $s \to 1-s$ of $\left(\frac{\pi}{p}\right)^{-\frac{s}{2}}\Gamma\left(\frac{1+s}{2}\right)L(s)$.

---

[4] See Cahen (1894) pp. 75–164, especially p. 155.

In his 1894 paper, Cahen also treated the more general Dirichlet series of type $\lambda_n$:

$$f(s) = \sum_{n=1}^{\infty} a_n e^{-\lambda_n s}, \tag{14.10}$$

where $\lambda_n$ is an increasing sequence of numbers whose limit is infinity. If $\lambda_n = n$, then $f(s)$ is a power series in $e^{-s}$ and if $\lambda_n = \log n$, then $f(s)$ is the usual Dirichlet series. According to Hardy and Riesz[5],

> ...the first attempt to construct a systematic theory of the function $f(s)$ was made by Cahen [Cahen (1894)] which, although much of the analysis which it contains is open to serious criticism, has served – and possibly just for that reason – as the starting point of most of the later researches in the subject.

Not too much is known about Cahen who was born in 1865, but he published a few papers and wrote two books on number theory. His 1891 paper[6] contains the constant named in his honor. He was surely alive in 1923 when he was awarded the Legion of Honor for his service to education in France, but the year of his death is not easy to determine.

In 1859, Riemann gave an example of an inverse Mellin transform, much earlier that Cahen's, but it was in 1896 that a general treatment of this topic was presented by the Finnish mathematician Mellin. Robert Hjalmar Mellin (1854–1933) studied under Mittag-Leffler in Stockholm and under Weierstrass in Berlin. His 1881 doctoral thesis on algebraic functions has been described as an exercise in Weierstrassian analysis.[7] From 1896 onward, he developed the Mellin transform and applied it to several problems, such as the asymptotic behavior of integrals and the solutions of difference equations.

The Mellin transform $F(s)$ of $f(x)$ in (14.1) is usually defined as an analytic function in a vertical strip $\alpha < \operatorname{Re} s < \beta$, where $\alpha$ and $\beta$ are determined by the behavior of $f(x)$ as $x \to \infty$ and as $x \to 0^+$. In 1935, Hecke applied the Mellin inversion formula to develop a general theory of the relation between an automorphic form, belonging to the group generated by $z \to z + \lambda$ and $z \to -\frac{1}{z}$, and the corresponding Dirichlet series. As we have seen, many examples of such a correspondence had been found before that time.

## 14.2 Functional Equations for Dirichlet Series

In his 1859 paper on the distribution of prime numbers, Riemann introduced the function $\zeta(s)$ such that $(s - 1)\zeta(s)$ was an entire function of order 1 and such that

$$\zeta(s) = \sum_{n=1}^{\infty} \frac{1}{n^s}, \quad \operatorname{Re} s > 1.$$

---

[5] Hardy and Riesz (1915) pp. 1–2.   [6] Cahen (1891).   [7] Elfving (1981).

The function $\zeta(s)$ satisfies a functional equation; Riemann formulated this as the invariance of

$$\Gamma\left(\frac{s}{2}\right)\pi^{-\frac{s}{2}}\zeta(s) \tag{14.11}$$

under the change $s \to 1 - s$, that is

$$\Gamma\left(\frac{s}{2}\right)\pi^{-\frac{s}{2}}\zeta(s) = \Gamma\left(\frac{1-s}{2}\right)\pi^{-\frac{1-s}{2}}\zeta(1-s). \tag{14.12}$$

He first proved this by means of Cauchy's residue theorem. However, formulation of the theorem in the form (14.12) suggested to Riemann a different proof. He began by noting that

$$n^{-s}\Gamma\left(\frac{s}{2}\right)\pi^{-\frac{s}{2}} = \int_0^\infty e^{-n^2\pi x}x^{\frac{s}{2}-1}\,dx \tag{14.13}$$

and hence

$$\Gamma\left(\frac{s}{2}\right)\pi^{-\frac{s}{2}}\zeta(s) = \int_0^\infty x^{\frac{s}{2}-1}\sum_{n=1}^\infty e^{-n^2\pi x}\,dx. \tag{14.14}$$

With this, the functional equation (14.12) is an easy consequence of the transformation formula for the theta function

$$\theta\left(-\frac{1}{\tau}\right) = \sqrt{\frac{\tau}{i}}\,\theta(\tau), \tag{14.15}$$

where

$$\theta(\tau) = 1 + 2\sum_{n=1}^\infty e^{in^2\pi\tau}, \quad \operatorname{Im}\tau > 0. \tag{14.16}$$

Taking $\tau = ix$ and setting

$$\psi(x) = \sum_{n=1}^\infty e^{-n^2\pi x},$$

Riemann wrote (14.16) in the form

$$2\psi(x) + 1 = x^{-\frac{1}{2}}\left(2\psi\left(\frac{1}{x}\right) + 1\right). \tag{14.17}$$

Moreover, by (14.14) and (14.17) he could then conclude

$$\begin{aligned}
\Gamma\left(\frac{s}{2}\right)\pi^{-\frac{s}{2}}\zeta(s) &= \int_1^\infty \psi(x)x^{\frac{s}{2}-1}\,dx + \int_0^1\left(\psi\left(\frac{1}{x}\right) + \frac{1}{2}\right)x^{\frac{s-3}{2}} - \frac{1}{2}x^{\frac{s}{2}-1}\,dx \\
&= \int_1^\infty \psi(x)x^{\frac{s}{2}-1}\,dx + \int_0^1 \psi\left(\frac{1}{x}\right)x^{\frac{s-3}{2}}\,dx + \frac{1}{s(s-1)} \\
&= \frac{1}{s(s-1)} + \int_1^\infty \psi(x)\left(x^{\frac{s}{2}-1} + x^{-\frac{s+1}{2}}\right)\,dx. \tag{14.18}
\end{aligned}$$

Note that the last integral in (14.18) is an entire function of $s$. Riemann observed that expression (14.18) would remain unchanged under $s \to 1 - s$.

In 1921, H. Hamburger proved that three properties uniquely determine $\zeta(s)$ up to a constant factor:

- $P(s)f(s)$ is an entire function of finite order for some polynomial $P$;
- $f(s)$ has an absolutely convergent Dirichlet series for Re $s > 1$;

$$f(s) = \sum_{n=1}^{\infty} \frac{a_n}{n^s}.$$

- $f$ satisfies the functional equation

$$f(s)\Gamma\left(\frac{s}{2}\right)\pi^{-\frac{s}{2}} = f(1-s)\Gamma\left(\frac{1-s}{2}\right)\pi^{-\frac{1-s}{2}}.$$

In other words, under these three conditions,

$$f(s) = a_1 \zeta(s). \tag{14.19}$$

This result is sometimes described in general terms by saying that the zeta function is essentially defined by its functional equation. In 1936, Hecke found a correspondence between zeta functions satisfying functional equations and certain types of automorphic forms. He was able to determine the dimensions of the spaces of these automorphic forms; this gave him another proof of Hamburger's theorem and a good deal more. In fact, Hamburger had also considered two functions, $f(s)$ and $g(s)$ with appropriate Dirichlet series that satisfied

$$f(s)\Gamma\left(\frac{s}{2}\right)\pi^{-\frac{s}{2}} = g(s)\Gamma\left(\frac{1-s}{2}\right)\pi^{-\frac{1-s}{2}},$$

and he showed that the same conclusion, (14.19), would follow. In his last published paper, that appeared in 1944,[8] Hecke explained how the correspondence he had found could in fact be extended to cover Hamburger's alternate situation. The reader may wish to read of Hecke's paper and its further developments in the work of Knopp and Berndt.[9]

To lay the groundwork for Hecke's correspondence theorem, suppose $\phi(s)$ can be expanded as a convergent Dirichlet series and suppose

$$R(s) = \left(\frac{2\pi}{\lambda}\right)^{-s}\Gamma(s)\phi(s). \tag{14.20}$$

Then $\phi(s)$ is said to have the signature $(\lambda, k, \gamma)$ if we first have (with $\lambda > 0$, $k > 0$, and $\gamma$ an arbitrary constant)

$$R(s) = \gamma R(k-s); \tag{14.21}$$

and secondly, $(s-k)\phi(s)$ is an entire function of finite order, that is,

$$\left|(s-k)\phi(s)\right| \le e^{|s|^M} \tag{14.22}$$

for some constant $M$ when $|s| \to \infty$.

---

[8] Hecke (1959) pp. 919–940.   [9] Knopp (2000) and Berndt and Knopp (2008).

Note that it follows by iteration from (14.21) that $\gamma^2 = 1$. Now suppose that $\phi(s)$ has a Dirichlet series expansion given by

$$\phi(s) = \sum_{n=1}^{\infty} \frac{a_n}{n^s},$$

and associate with $\phi(s)$ the series

$$f(\tau) = a_0 + \sum_{n=1}^{\infty} a_n e^{\frac{2\pi i n \tau}{\lambda}}. \tag{14.23}$$

Note that the constant $a_0$ is determined by the residue of $\phi(s)$ at $s = k$.

Hecke's fundamental theorem related the Dirichlet series with the (automorphic) function $f(\tau)$:[10]

Suppose $\phi(s)$ is a Dirichlet series with signature $(\lambda, k, \gamma)$ and suppose

$$f(\tau) = a_0 + \sum_{n=1}^{\infty} a_n e^{\frac{2\pi i n \tau}{\lambda}} \quad \text{with} \quad a_0 = \gamma \left( \frac{2\pi}{\lambda} \right)^{-k} \Gamma(k)\rho,$$

where $\rho$ is the residue of $\phi(s)$ at $s = k$. Then
1. $f(\tau + \lambda) = f(\tau)$ is analytic in the upper half-plane $\operatorname{Im} \tau > 0$;
2. $f(-\frac{1}{\tau}) = \gamma(-i\tau)^k f(\tau)$;
3. $f(x + iy) = O(y^{-c})$, $c > 0$, as $y \to 0+$, uniformly for all $x$.

Conversely, the statements (1), (2), and (3) imply that $\phi(s)$ is of signature $(\lambda, k, \gamma)$.

Note that $(-i\tau)^k = e^{k \log(-i\tau)}$ where $\log(-i\tau)$ is real for $\tau = iy, y > 0$.

We first prove the converse of this theorem. Begin by noting that the Fourier coefficient $a_n$ in (14.23) is given by

$$a_n = \frac{1}{\lambda} \int_{\tau_0}^{\tau_0 + \lambda} f(\tau) e^{-\frac{2\pi i n \tau}{\lambda}} d\tau = \frac{1}{\lambda} \int_0^{\lambda} f(x + iy_0) e^{\frac{2\pi n (y_0 - ix)}{\lambda}} dx.$$

From condition (3), taking $y_0 = \frac{1}{n}$, we obtain

$$a_n = \frac{1}{\lambda} \int_0^{\lambda} O(n^c) dx = O(n^c). \tag{14.24}$$

Therefore, the series $\sum_{n=1}^{\infty} a_n n^{-s}$ converges absolutely for $\operatorname{Re} s > c + 1$. Now from the $\Gamma$-integral (14.12) we get

$$\left( \frac{2\pi n}{\lambda} \right)^{-s} \Gamma(s) = \int_0^{\infty} e^{-\frac{2\pi n y}{\lambda}} y^{s-1} dy$$

and by (14.20) we have

$$R(s) = \int_0^{\infty} (f(iy) - a_0) y^{s-1} dy = \left( \int_0^1 + \int_1^{\infty} \right) (f(iy) - a_0) y^{s-1} dy. \tag{14.25}$$

---

[10] This theorem is also included in Hecke's lectures: Hecke (1983).

Next, we prove that $R(s)$ satisfies (14.21). Taking $\tau = iy$ in the relation $f(-\frac{1}{\tau}) = \gamma(-i\tau)^k f(\tau)$ gives us

$$f\left(\frac{i}{y}\right) = \gamma \, y^k f(iy). \tag{14.26}$$

Applying the change of variables $y \to \frac{1}{y}$ and then applying (14.26) to the first integral on the right-hand side of (14.25), that integral then becomes

$$\int_0^1 (f(iy) - a_0) y^{s-1} \, dy = \int_1^\infty f\left(\frac{i}{y}\right) y^{-s-1} \, dy - a_0 \int_0^1 y^{s-1} \, dy$$

$$= \gamma \int_1^\infty f(iy) y^{k-s-1} \, dy - \frac{a_0}{s}.$$

Thus, by (14.25), we have

$$R(s) = \int_1^\infty (f(iy) - a_0)\left(\gamma \, y^{k-s} + y^s\right) \frac{dy}{y} - \frac{a_0}{s} - \gamma \, \frac{a_0}{k-s},$$

a relation valid for all $s$, since the integral is an entire function of $s$. This implies that $R(s) = \gamma \, R(k - s)$.

Now we turn to the proof of (14.22). Let $s = \sigma + it$ and employ

$$f(iy) - a_0 = \sum_{n=1}^\infty a_n e^{-\frac{2n\pi y}{\lambda}} = O(e^{-\alpha y}), \quad \alpha = \frac{2\pi}{\lambda},$$

and Stirling's formula (6.94) to obtain

$$\left| \int_1^\infty (f(iy) - a_0) y^{s-1} \, dy \right| \le c \int_1^\infty e^{-\alpha y} y^{|\sigma|-1} \, dy$$

$$= c\alpha^{-|\sigma|} \Gamma(|\sigma|) = O\left(e^{\beta|\sigma| \log |\sigma|}\right) \quad \text{as } |\sigma| \to \infty.$$

Hence $(s - k)R(s)$ is an entire function of finite order. The relation

$$(s - k)\phi(s) = \frac{s-k}{\Gamma(s)} \left(\frac{2\pi}{\lambda}\right)^s R(s),$$

then implies that $(s - k)\phi(s)$ is of finite order; in fact, it has order 1.

To prove the direct implication of Hecke's theorem, the definition of $f(\tau)$ gives us

$$f(\tau + \lambda) = a_0 + \sum_{n=1}^\infty a_n e^{\frac{2\pi i n(\tau + \lambda)}{\lambda}} = a_0 + \sum_{n=1}^\infty a_n e^{\frac{2\pi i n \tau}{\lambda}} = f(\tau).$$

Now note that $\phi(s) = \sum_{n=1}^\infty \frac{a_n}{n^s}$ converges somewhere if and only if $a_n = O(n^c)$ as $n \to \infty$ with $c > 0$ a constant.

To prove (3), that $f(x + iy) = O(y^{-c})$ as $y \to 0^+$, observe that when we know that $|a_n| = O(n^c)$, then we have

$$|f(x + iy)| \leq \sum_{n=0}^{\infty} |a_n| e^{-\frac{2\pi n y}{\lambda}}$$

$$\leq K \sum_{n=1}^{\infty} n^c e^{-\frac{2\pi n y}{\lambda}}. \tag{14.27}$$

Observe that $t^c e^{-\frac{2\pi t y}{\lambda}}$ takes its maximum value when $t = \frac{c\lambda}{2\pi y}$. So (14.27) produces

$$|f(x + iy)| \leq K \left( \int_0^{\frac{c\lambda}{2\pi y}} t^c \, dt + \int_{\frac{c\lambda}{2\pi y}}^{\infty} t^c e^{-\frac{2\pi t y}{\lambda}} \, dt \right)$$

$$\leq \frac{K}{c+1} \left( \frac{c\lambda}{2\pi y} \right)^{c+1} + \int_0^{\infty} t^c e^{-\frac{2\pi t y}{\lambda}} \, dt$$

$$= O(y^{-c-1}) + \frac{\Gamma(c+1)\lambda^{c+1}}{(2\pi y)^{c+1}} = O(y^{-c-1}) \quad \text{as } y \to 0^+,$$

proving (3).

Since the Dirichlet series for $\phi(s)$ converges, it converges absolutely in a half-plane; take this half-plane to be $\operatorname{Re} s \geq \sigma_0 > c + 1 > k$. Clearly, $\phi(s)$ is bounded in this half-plane $\operatorname{Re} s \geq \sigma_0$ and so

$$\phi(\sigma_0 + it) = O(1). \tag{14.28}$$

By the functional equation (14.21), we have

$$\phi(s) = \phi(\sigma + it) = O\left( \frac{\Gamma(k-s)}{\Gamma(s)} \phi(k-s) \right) \tag{14.29}$$

for $\sigma \leq k - \sigma_0$, $|t| \to \infty$. Since $\operatorname{Re}(k - s) \geq \sigma_0$, we know that $\phi(k - s)$ is bounded for these values, and hence by the asymptotic formula for $\Gamma(s)$ in (6.93), we have from (14.28) that

$$\phi(-\sigma_0 + it) = O\left( \frac{\Gamma(k-s)}{\Gamma(s)} \right) = O\left( |t|^{k-2\sigma_0} \right). \tag{14.30}$$

Because $(s - k)\phi(s)$ is an entire function of finite order, the condition (6.92) holds and by the Phragmén-Lindelöf theorem, it follows that

$$\phi(\sigma + it) = O(|t|^A), \quad A > 0 \tag{14.31}$$

uniformly in the strip $-\sigma_0 \leq \sigma \leq \sigma_0$.

To prove (2), we start with Cahen's formula

$$e^{-y} = \frac{1}{2\pi i} \int_{c-i\infty}^{c+i\infty} \Gamma(s) y^{-s} \, ds, \quad c > 0.$$

Then

$$f(iy) - a_0 = \sum_{n=1}^{\infty} a_n\, e^{-\frac{2\pi n y}{\lambda}}$$

$$= \frac{1}{2\pi i} \int_{\sigma_0 - i\infty}^{\sigma_0 + i\infty} \Gamma(s) \left(\frac{2\pi}{\lambda}\right)^{-s} y^{-s} \phi(s)\, ds$$

$$= \frac{1}{2\pi i} \int_{\sigma_0 - i\infty}^{\sigma_0 + i\infty} R(s)\, y^{-s}\, ds.$$

Since $(s - k)\,\phi(s)$ is given as an entire function, $\phi(s)$ has a pole of order at most 1 at $s = k$. Now push the line of integration to the left, so that we pick up the residue $\gamma\, a_0\, y^{-k}$ at $s = k$:

$$\frac{1}{2\pi i} \int_{\sigma_0} R(s)\, y^{-s}\, ds = \frac{1}{2\pi i} \int_{-\sigma_0} R(s)\, y^{-s}\, ds + \gamma\, a_0\, y^{-k}.$$

It is possible to carry out this procedure because, for $s$ inside the strip $-\sigma_0 \le \sigma \le \sigma_0$, and for a fixed $y$, we have by (14.31)

$$\Gamma(s)\,\phi(s)\, y^{-s} = O\left(e^{\frac{-\pi}{2}|t|}|t|^{\text{constant}}\right), \quad s = \sigma + it.$$

Thus we finally have

$$G(y) \equiv f(iy) - a_0 \left(1 + \gamma y^{-k}\right)$$

$$= \frac{1}{2\pi i} \int_{\frac{k}{2} - i\infty}^{\frac{k}{2} + i\infty} R(s)\, y^{-s}\, ds$$

$$= \frac{\gamma}{2\pi i} \int_{\frac{k}{2} - i\infty}^{\frac{k}{2} + i\infty} R(s)\, y^{s-k}\, ds$$

$$= \gamma y^{-k} G\left(\frac{1}{y}\right)$$

$$= \gamma\, y^{-k} \left\{ f\left(\frac{i}{y}\right) - a_0(1 + \gamma\, y^k) \right\},$$

and hence

$$f\left(\frac{i}{y}\right) = \gamma\, y^k\, f(iy)$$

or, by analytic continuation for $\mathrm{Im}\, \tau > 0$,

$$f\left(-\frac{1}{\tau}\right) = \gamma(-i\tau)^k\, f(\tau).$$

We refer to the functions corresponding to Dirichlet series with signature $(\lambda, k, \gamma)$ as functions with signature $(\lambda, k, \gamma)$. Since the sum of two functions with signature $(\lambda, k, \gamma)$ is a function with the same signature, it follows that the set of functions of signature $(\lambda, k, \gamma)$ is a vector space. Hecke determined the dimensions of these spaces.

He found that he needed to consider three separate cases: $\lambda > 2, \lambda = 2$, and $0 < \lambda < 2$. He gave the results:[11]

**Theorem 1.** If $\lambda > 2$, then there are infinitely many linearly independent functions of signature $(\lambda, k, \gamma)$.

**Theorem 2.** When $\lambda = 2$, the number of linearly independent functions with signature $(2, k, \gamma)$ is finite: for $\gamma = 1$ exactly $\lfloor \frac{k}{4} \rfloor + 1$ and for $\gamma = -1$, if $k \geq 2$, exactly $\lfloor \frac{k-2}{4} \rfloor$, otherwise zero. [We will utilize this theorem in our Chapter 15 on sums of squares.]

**Theorem 3.** For $0 < \lambda < 2$, there exist functions of signature $(\lambda, k, \gamma)$ only if $\lambda, k$ are of the form

$$\lambda = 2\cos\frac{\pi}{q}, \quad k = \frac{4K}{q-2} + 1 - \gamma$$

with positive integers $K, q \geq 3$. The number of linearly independent Dirichlet series $\phi(s)$ is then

$$\leq \left\lfloor \frac{K + \frac{\gamma-1}{2}}{q} \right\rfloor + 1,$$

and the number of $\phi(s)$ without a singularity at $s = k$ is

$$\left\lfloor \frac{K + \frac{\gamma-1}{2}}{q} \right\rfloor.$$

For $0 < \lambda < 1$, there is therefore no function of signature $(\lambda, k, \gamma)$.

If $\lambda = 1$, then $q = 3$ and the group generated by $\omega \to \omega + 1, \omega \to -\frac{1}{\omega}$ is the full modular group. For the signature $(1, k, \gamma)$, there is no solution except when $k$ is an even integer and $\gamma = (-1)^{\frac{k}{2}}$. In that case, the number of independent functions of signature $(1, k, \gamma)$ is

$$\left\lfloor \frac{k}{12} \right\rfloor \quad \text{if } k \equiv 2 \ (\text{mod } 12) \tag{14.32}$$

$$1 + \left\lfloor \frac{k}{12} \right\rfloor \quad \text{if } k \neq 2 \ (\text{mod } 12). \tag{14.33}$$

Note that the space of modular forms of weight $k$ for the full modular group is identical with the space of functions of signature $(1, k, (-1)^{\frac{k}{2}})$, where $k$ is an even integer.

### 14.3  Theta Series in Two Variables

Based on Hecke's methods, we prove that if

$$\phi(s) = \sum_{m,n}' \frac{(m+in)^4}{(m^2+n^2)^s}, \quad R(s) = \pi^{-s}\Gamma(s)\phi(s), \tag{14.34}$$

then

$$R(s) = R(5-s). \tag{14.35}$$

---

[11] For proofs, see Hecke (1959) pp. 593–616, Hecke (1983), or Berndt and Knopp (2008). The latter book gives an excellent treatment of Hecke's work on Dirichlet series and modular forms and later contributions by Salomon Bochner and others.

Hecke needed this theorem to prove a result on the number of representations of an integer as a sum of ten squares, and we make use of it in our Chapter 15. First, apply the Poisson summation in two variables to the function

$$f(x_1, x_2) = \left( \frac{\partial}{\partial x_1} + i \frac{\partial}{\partial x_2} \right)^4 e^{-\pi t(x_1^2 + x_2^2)} = (2\pi)^4 t^4 (x_1 + ix_2)^4 e^{-\pi t(x_1^2 + x_2^2)}. \quad (14.36)$$

Now it is easy to derive the Poisson formula for a class of functions that contain (14.36), since this function is smooth and vanishes rapidly as $x_1^2 + x_2^2 \to \infty$, as do all its derivatives.

As an example of this, in 1929 L. J. Mordell[12] presented a form of the Poisson summation formula whose proof required only integration by parts: For $a, b = 0, 1, 2$ excluding $a = b = 2$, if $\frac{\partial^{a+b} f}{\partial x_1^a \partial x_2^b}$ are all continuous functions of $x_1, x_2$ for every real value $x_1, x_2$, and all these continuous functions tend to zero when either variable tends to $\pm\infty$, regardless of the value of the other variable, and if $f$ and its derivatives are such that

$$\int_{-\infty}^{\infty} \int_{-\infty}^{\infty} \left( |f| + \left| \frac{\partial^2 f}{\partial x_1^2} \right| + \left| \frac{\partial^2 f}{\partial x_2^2} \right| + \left| \frac{\partial^4 f}{\partial x_1^2 \partial x_2^2} \right| \right) dx_1 dx_2 < \infty,$$

then

$$\sum_{m=-\infty}^{\infty} \sum_{n=-\infty}^{\infty} f(m, n) = \sum_{m=-\infty}^{\infty} \sum_{n=-\infty}^{\infty} \int_{-\infty}^{\infty} \int_{-\infty}^{\infty} e^{-2m\pi\, i x_1 - 2n\pi\, i x_2} f(x_1, x_2) dx_1 dx_2.$$

$$(14.37)$$

Observe that formula (14.37) is the Poisson summation formula in two variables; also note that

$$\hat{f}(y_1, y_2) = \int_{-\infty}^{\infty} \int_{-\infty}^{\infty} f(x_1, x_2) e^{-2\pi i(y_1 x_1 + y_2 x_2)} dx_1 dx_2$$

is the Fourier transform of $f(x_1, x_2)$, so that (14.37) can be written as

$$\sum_{m=-\infty}^{\infty} \sum_{n=-\infty}^{\infty} f(m, n) = \sum_{m=-\infty}^{\infty} \sum_{n=-\infty}^{\infty} \hat{f}(m, n). \quad (14.38)$$

The reader may wish to attempt a proof of this theorem of Mordell as an exercise.

Next, see that the Fourier transform of the function $f(x_1, x_2)$, defined by (14.36), is

$$\int_{-\infty}^{\infty} \int_{-\infty}^{\infty} \left( \frac{\partial}{\partial x_1} + i \frac{\partial}{\partial x_2} \right)^4 e^{-\pi t(x_1^2 + x_2^2)} e^{-2\pi i(x_1 y_1 + x_2 y_2)} dx_1 dx_2. \quad (14.39)$$

Integration by parts shows that (14.39) is equal to

$$\hat{f}(y_1, y_2) = (-2\pi i)^4 (y_1 + iy_2)^4 \int_{-\infty}^{\infty} \int_{-\infty}^{\infty} e^{-\pi t(x_1^2 + x_2^2)} e^{-2\pi i(x_1 y_1 + x_2 y_2)} dx_1 dx_2$$

$$= (-2\pi i)^4 (y_1 + iy_2)^4 t^{-1} e^{-\frac{\pi}{t}(y_1^2 + y_2^2)}. \quad (14.40)$$

---

[12] Mordell (1929).

When the value of $f(x_1, x_2)$ from (14.36) and the value of $\hat{f}(x_1, x_2)$ from (14.40) are substituted in (14.38), we obtain the theta transformation

$$\sum_{m,n} (m + in)^4 \, e^{-(m^2 + n^2)\pi t} = t^{-5} \sum_{m,n} (m + in)^4 \, e^{-(m^2 + n^2)\frac{\pi}{t}}. \qquad (14.41)$$

For Re $s > 3$, we can take the Mellin transform of the left-hand side of (14.41) to obtain

$$R(s) = \pi^{-s}\Gamma(s) \sum_{m,n}{}' \frac{(m + in)^4}{(m^2 + n^2)^s} = \int_0^\infty t^s \sum_{m,n} (m + in)^4 \, e^{-(m^2 + n^2)\pi t} \frac{dt}{t}. \qquad (14.42)$$

Since the constant term in the theta series in (14.41) is zero, we can apply the method used to derive (14.18) to see that $R(s)$ must be entire. Making the change $t \to \frac{1}{t}$ in the integral in (14.42), we see that

$$R(s) = \int_0^\infty t^{5-s} \sum_{m,n} (m + in)^4 e^{-(m^2 + n^2)\frac{\pi}{t}} \frac{dt}{t}.$$

When Re $s < 2$, we can integrate the series term-by-term to obtain

$$\sum_{m,n} \frac{(m + in)^4 \, \Gamma(5 - s)\pi^{5-s}}{(m^2 + n^2)^{5-s}};$$

therefore, we conclude that $R(s) = R(5 - s)$. It can be proved in a similar way that if

$$\psi(s) = \sum_{m,n}{}' \frac{(m + in)^8}{(m^2 + n^2)^s}, \qquad R(s) = \pi^{-s}\Gamma(s)\psi(s),$$

then

$$R(s) = R(9 - s). \qquad (14.43)$$

## 14.4  Exercises

1. Prove Cahen's formula (14.5).
2. Prove transformation formulas (14.8) and (14.9).
3. Show that the series $\sum_{n=1}^\infty a_n e^{-\lambda_n s}$ may be convergent for all $s$, or for none, or for only some $s$. If the last case holds, show that there exists a number $\sigma_0$ such that the series is convergent for Re $s > \sigma_0$ and not convergent for Re $s < \sigma_0$. The number $\sigma_0$ is called the abscissa of convergence. See Hardy and Riesz (1915) p. 4.
4. Suppose that for the series $\sum_{n=1}^\infty a_n e^{-\lambda_n s}$ the abscissa of convergence $\sigma_0$ is positive. In that case, show that

$$\sigma_0 = \varlimsup_{n \to \infty} \frac{\log \left| \sum_{k=1}^n a_k \right|}{\lambda_n}.$$

   See Cahen (1894) p. 89.
5. Give a proof of a theorem stated by Cahen, later rigorously demonstrated by Oskar Perron: If the series $\sum_{n=1}^\infty a_n e^{-\lambda_n s}$ is convergent for $s = \beta + i\gamma$, and

$c > 0, c > \beta, \lambda_n < \omega < \lambda_{n+1}$, then

$$\frac{1}{2\pi i} \int_{c-i\infty}^{c+i\infty} f(s)e^{\omega s} \frac{ds}{s} = \sum_{k=1}^{n} a_k,$$

with the path of integration being the line $\sigma = c$. See Hardy and Riesz (1915) p. 12.

6. Prove (14.43).

7. Prove that a Dirichlet series $\phi(s)$ with real coefficients and with signature $(1, k, (-1)^{\frac{k}{2}})$ has an infinite number of zeros on Re $s = \frac{k}{2}$ if and only if $\phi(s)$ is analytic at $s = k$. Recall that $\phi(s)$ is analytic at $s = k$ if and only if the associated modular form is a cusp form. See Hecke (1959) pp. 714–716.

8. Prove that if $f(\omega)$ is a cusp form of weight $k$, and if $f(\omega) = \sum_{n=1}^{\infty} a_n e^{2\pi i n \omega}$, then $a_n = O(n^{\frac{k}{2}})$. See Hecke (1959) p. 484.

# 15

## Sums of Squares

### 15.1 Preliminary Remarks

The efforts to solve the problem of expressing an integer as a sum of two, three, or four squares date back at least to Diophantus (c. 200–c. 284). Much of his work is now lost, so that it is difficult to say how much he could prove, and how much he conjectured from experimental evidence. In 1621, Claude Gaspar Bachet (1581–1638) published a Latin translation of the available portions of Diophantus's *Arithmetica* with extensive commentary. Bachet pointed out that in two of his problems, Diophantus had assumed that every integer could be expressed as a sum of four squares. In connection with problems discussed by Diophantus, Bachet raised the question of the conditions under which an integer could be expressed as a sum of two or three squares. Fermat read Bachet's work and around 1636 he began thinking about these questions. In 1640, he wrote to Roberval that he was able to prove that if an integer $r$ was of the form $m^2 n$, with $n$ containing a factor of the form $4k - 1$ but no square factor, then $r$ could not be expressed as a sum of two squares.

In letters written in the 1650s, Fermat wrote that he could prove that every prime of the form $4n + 1$ could be expressed as a sum of two squares and that every positive integer could be expressed as a sum of four squares. He never gave any written details of his proofs. In 1747, Euler solved the two squares problem and a year later he reduced the four squares problem to that of primes by proving that the product of two integers expressible as a sum of four squares was another sum of four squares. The full four squares theorem was proved by Lagrange in 1770 and Euler simplified this proof two years later. In a letter of 1750 to his friend Goldbach, Euler remarked that the best way to prove the four squares theorem might be to show that the coefficient of $x^n$ in $(1 + x + x^4 + x^9 + \cdots)^4$ must be positive for every $n \geq 1$. It was Jacobi who demonstrated in a paper of 1828 how this idea could be executed.[1]

The number of representations of a positive integer $n$ as a sum of $s$ squares is the number of $s$-tuples, $(x_1, x_2, \ldots, x_s)$, of integers $x_1, x_2, \ldots, x_s$ such that

$$n = x_1^2 + x_2^2 + \cdots + x_s^2. \tag{15.1}$$

---

[1] For a discussion of and references to the results mentioned in this and the previous paragraphs, see Weil (1984).

384

Note that the integers $x_1, x_2, \ldots, x_s$ can be negative, positive, or zero. A representation is called primitive if $x_1, x_2, \ldots, x_s$ have no common factor other than 1.

In this chapter, we discuss a few methods that employ elliptic and modular functions for finding the number of representations of an integer as a sum of squares. In particular, we explain in some detail a fascinating method due to Hecke that has been virtually ignored in the literature on sums of squares. Hecke made brief mention of this method in his lectures of 1938, delivered at the University of Michigan and at Princeton.

In 1828, Jacobi published a number of groundbreaking identities,[2] derived from the theory of elliptic functions, that could be interpreted in terms of the number of representations of an integer as a sum of squares. He was the first mathematician to publish such identities, including:

$$\left( \sum_{n=-\infty}^{\infty} q^{n^2} \right)^4 = 1 + 8 \sum_{n=1}^{\infty} \frac{n q^n}{1 + (-q)^n} \tag{15.2}$$

$$= 1 + 8 \sum_{n=1}^{\infty} \frac{q^n}{(1 + (-q)^n)^2} \tag{15.3}$$

$$= 1 + 8 \sum_{n=1}^{\infty} \sigma_1(2n+1)$$
$$\times \left( q^{2n+1} + 3 q^{2(2n+1)} + 3 q^{4(2n+1)} + 3 q^{8(4n+1)} + \cdots \right), \tag{15.4}$$

where $\sigma_1(m) = \sum_{d|m} d$. Since

$$\left( \sum_{n=-\infty}^{\infty} q^{n^2} \right)^s = 1 + \sum_{n=1}^{\infty} r_s(n) q^n, \tag{15.5}$$

where $r_s(n)$ denotes the number of representations of $n$ as a sum of $s$ squares, Jacobi's result implies that

$$r_4(2n+1) = 8\sigma_1(2n+1), \tag{15.6}$$
$$r_4\left(2^m(2n+1)\right) = 24\sigma_1(2n+1), \quad \text{where } m \geq 1. \tag{15.7}$$

Jacobi explicitly gave an interpretation of (15.4) for sums of squares in his *Fundamenta Nova* of 1829, in which he presented formulas that imply results on $r_{2s}(n)$ for $1 \leq s \leq 4$ and gave clear interpretations for $s = 1$ and $s = 2$. For $2s = 6$ and $2s = 8$ he simply derived the identities:

$$\left( \sum_{n=-\infty}^{\infty} q^{n^2} \right)^6 = 1 + 16 \sum_{n=1}^{\infty} \frac{n^2 q^n}{1 + q^{2n}} + 4 \sum_{n=1}^{\infty} \frac{(-1)^n (2n-1)^2 q^{2n-1}}{1 - q^{2n-1}} \tag{15.8}$$

$$\left( \sum_{n=-\infty}^{\infty} q^{n^2} \right)^8 = 1 + 16 \sum_{n=1}^{\infty} \frac{n^3 q^n}{1 + (-1)^{n-1} q^n}. \tag{15.9}$$

[2] Jacobi (1969) vol. 1, p. 262.

*Sums of Squares*

In Section 127 of the 1865 part VI of his famous *Report on the Theory of Numbers*, H. J. S. Smith offered interpretations of (15.8) and (15.9), respectively:[3]

> The number of representations of any number $N$ as a sum of six squares is $4\sum(-1)^{\frac{\delta-1}{2}}(4\delta'^2 - \delta^2)$, $\delta$ denoting any uneven divisor of $n$, $\delta'$ its conjugate divisor [meaning $\delta\delta' = N$]. In particular if $N \equiv 1$, mod 4, the number of representations is $12\sum(-1)^{\frac{\delta-1}{2}}\delta^2$; if $N \equiv -1$, mod 4, it is $20\sum(-1)^{\frac{\delta-1}{2}}\delta^2$.
>
> The number of representations of any uneven number as a sum of eight squares is sixteen times the sum of the cubes of its divisors; for an even number it is sixteen times the excess of the cubes of the even divisors above the cubes of the uneven divisors.

Note that Gauss independently discovered (15.2) and (15.3); they were written without proof, and without explicit application to the sums of squares, in an undated manuscript published after his death.[4] The same manuscript contains the formulas

$$\left(\sum_{n=-\infty}^{\infty} q^{(n+\frac{1}{2})^2}\right)^4 = 16\sum_{n=1}^{\infty} \frac{(2n-1)q^{2n-1}}{1-q^{4n-2}} \tag{15.10}$$

$$= 16\sum_{n=1}^{\infty} \frac{q^{2n-1}(1+q^{4n-2})}{(1-q^{4n-2})^2}. \tag{15.11}$$

These formulas were also given by Jacobi.[5] Again, Smith presented an interpretation of them:[6]

> The number of compositions of the quadruple of any uneven number $N$ by the addition of four uneven squares is equal to the sum of the divisors of $N$.

We note that Smith gave his theorem in terms of the number of compositions, rather than the number of representations, because the "four uneven squares" are of positive integers.

In all these results of Jacobi, the number of squares is even. But Dirichlet proved that the number of primitive representations of an odd number as a sum of three squares was

$$24\sum_{k<\frac{n}{4}}\left(\frac{k}{n}\right) \quad (n \equiv 1 \text{ mod } 4); \quad 8\sum_{k<\frac{n}{4}}\left(\frac{k}{n}\right) \quad (n \equiv 3 \text{ mod } 8), \tag{15.12}$$

where $\left(\frac{k}{n}\right)$ was the Jacobi symbol. Eisenstein stated Dirichlet's result in the form (15.12) in a one-page paper of 1847 where he also stated without proof a theorem on the number of representations of a number, that is a product of distinct odd primes, as a sum of 5 squares.[7] For example, he let $\psi(m)$ denote the number of representations of $m$ as a sum

---

[3] Smith (1965) vol. 1, p. 307. Also see Smith (1865) p. 336. (Note that in the reprint of the *Report*, $\delta^2$ was omitted from the end of the equation for the number of representations for $N \equiv 1$ mod 4.)

[4] Gauss (1863–1927) vol. 3, p. 445.     [5] Jacobi (1969) vol. 1, p. 160.

[6] Smith (1965) vol. 1, p. 307.     [7] Eisenstein (1969) vol. 1, p. 160.

of 5 squares and then wrote that if $m \equiv 1 \pmod 8$, then

$$\psi(m) = -80 \sum_{k=1}^{\frac{1}{2}(m-1)} \left(\frac{k}{m}\right) k,$$

except when $m = 1$. He also stated this theorem for $m \equiv 3, 5, 7 \pmod 8$. Then in 1850, again without proof, Eisenstein stated a similar theorem for 7 squares.[8] In 1867, Smith generalized Eisenstein's theorems to the number of primitive representations of a number as a sum of 5 and 7 squares. Smith did not present a proof at that time, but fifteen years later, the French Academy proposed the proofs of these theorems as prize problems. Smith and Hermann Minkowski won this prize with independent proofs, both utilizing the arithmetical theory of quadratic forms.

Observe that the statements of the results for $r_s(n)$, $2 \le s \le 8$, are for the case in which $s$ is even, in terms of the powers of divisors of $n$. On the other hand, when $s$ is odd, the results are not for $r_s(n)$, but for the number of primitive representations of $n$, and the results employ the Jacobi symbol. Recently, Goro Shimura published a paper in which he stated and proved results for $r_s(n)$, $2 \le s \le 8$, whether $s$ is even or odd, that is, giving the results, when $s$ is odd, for primitive as well as nonprimitive representations.[9] The reader may profitably consult Shimura, whose terminology and framework greatly facilitated his purpose.

Starting in 1856, Joseph Liouville published a number of papers on the expressions of integers as sums of squares; he wished to obtain the results of Jacobi and others by means of simple identities and elementary methods, without the use of the machinery of elliptic functions or of quadratic forms. These papers contained a large number of examples and identities that he obtained through elementary methods, but he gave no detailed proofs. In 1865, he wrote a paper on the number of representations of an integer as a sum of 10 squares; this paper used divisors of $n$, $a$ and $b$ being integers, of the form $a + ib$; these are the Gaussian integers. It appears that this approach was rediscovered by Glaisher and perhaps by Hecke. Before moving on to discuss Glaisher's papers on sums of squares, we mention that in the 1920s, J. V. Uspensky developed elementary methods for sums of squares problems, similar to those of Liouville.[10] Treatments of the Liouville-Uspensky method can be found in the books of Moreno and Wagstaff and of Williams.[11]

James Whitbread Lee Glaisher (1848–1928) was for almost forty years the sole editor of two journals, *The Messenger of Mathematics* and *The Quarterly Journal of Mathematics*. In this connection, Hardy wrote, "Glaisher was, I suppose, the last of the old school of mathematical editors, the men who, like Liouville, contrived to run mathematical journals practically unaided. At any rate it is safe to say that no one will again attempt to run two simultaneously, as Glaisher did."[12] Glaisher was also a prolific writer on mathematical topics. He published more than 400 papers on a wide range of subjects. In the same article on Glaisher, Hardy comments:

[8] Ibid. p. 620.   [9] Shimura (2002).   [10] Uspensky (1928).
[11] Moreno and Wagstaff, Jr. (2006) and Williams (2011).   [12] Hardy (1966–1979) vol. 7, p. 731.

I should like to end with a few words about Glaisher's position as a mathematician, because I think that he has generally been underestimated. He wrote a great deal, of very uneven quality, and he was 'old-fashioned' in a sense which is most unusual now; but the best of his work is really good. This work is almost all arithmetical, but it belongs to a peculiar region of the theory of numbers; neither to the 'classical' theory on the one hand nor to the full-blooded 'analytic' theory on the other, but to the semianalytic theory of Kronecker, Liouville, Ramanujan, Mordell, or Bell, in which we apply to arithmetic not general principles of function theory but special properties of particular functions such as the elliptic modular functions.

The standard problem of this theory is that of the representation of numbers by sums of squares. Glaisher studied this problem in a series of elaborate papers in Vols. 36–39 of the *Quarterly Journal*, in which he considers not merely representations by any even number of squares from 2 to 18 but also 'classified' representations in which stated numbers of the squares are odd or even. Two of his most important results, concerning representations of numbers $4k + 3$ by 10 squares, and even numbers by 12, had been anticipated, as he points out himself, by Eisenstein and Liouville respectively; but he proves many other very interesting theorems and there are still unsolved problems about the new arithmetical functions which he introduced. This was probably Glaisher's most important work, but he did much more, and anyone who will take the trouble to work through the index of Dickson's *History* will probably be surprised at the number of striking theorems associated with Glaisher's name.

Glaisher published a paper in 1907[13] in which he gave a table of formulas for $r_{2s}(n)$ for $1 \leq s \leq 9$. These formulas up to $s = 5$ were given in terms of divisors of $n$ and the $n = 5$ case employed complex divisors of $n$. He gave the value of $r_{10}(n)$ in terms of the fourth powers of the real divisors of $n$ as well as the fourth powers of the complex divisors of $n$. Glaisher denoted the latter quantity $\chi_4(n)$ and defined it by the equation

$$\chi_4(n) = \frac{1}{4} \sum (a + ib)^4, \qquad (15.13)$$

where $a + ib$ was a Gaussian integer dividing $n$, and the sum was taken over all such divisors of $n$; note that a Gaussian integer is a complex number $a + ib$, $a$ and $b$ taken to be integers. Initially, Glaisher had obtained $\chi_4(n)$ as a coefficient in the series:

$$k^2 k'^2 \rho^5 = \theta_0^4 \theta_2^4 \theta_3^2 \equiv 16 \sum_{n=1}^{\infty} \chi_4(n) q^n, \qquad (15.14)$$

where $\rho = \frac{2K}{\pi}$. While investigating the arithmetic properties of $\chi_4(n)$, defined as the coefficient of $q^n$ in (15.14), he discovered (15.13). In particular, (15.13) implies that if $m$ and $n$ are relatively prime, then

$$\chi_4(mn) = \chi_4(m)\chi_4(n). \qquad (15.15)$$

In the case $6 \leq s \leq 9$, Glaisher found that he had to employ coefficients that appeared in the expansions of certain elliptic modular functions. For example, he expressed $r_{12}(n)$ in terms of $\Omega(n)$, defined as

$$k^2 k'^2 \left( \frac{2K}{\pi} \right)^6 = 16 \sum_{n-1}^{\infty} \Omega(n) q^n. \qquad (15.16)$$

---

[13] Glaisher (1907a) p. 480.

Note that the left-hand side of (15.16) can be written as $16q \prod_{n=1}^{\infty} (1 - q^n)^{12}$. Glaisher conjectured that $\Omega(n)$ was a multiplicative arithmetic function, that is, for relatively prime positive odd integers $m$ and $n$,

$$\Omega(mn) = \Omega(m)\Omega(n). \tag{15.17}$$

Glaisher's proof of (15.17) had a gap, as he himself noted.[14] However, Mordell gave a nice proof utilizing Hurwitz's theory of the multiplier equation, presented in our Chapter 11.

Glaisher used coefficients of other modular forms to determine the values of $r_{14}(n)$, $r_{16}(n)$, $r_{18}(n)$. He defined $W(n)$, $\Theta(n)$, and $G(n)$ by

$$k^2 k'^2 \rho^7 = 16 \sum_{n=1}^{\infty} W(n)q^n,$$

$$k^2 k'^4 \rho^8 = 16 \sum_{n=1}^{\infty} \Theta(n)q^n,$$

$$k^2 k'^2 (k'^2 - k^2)\rho^9 = 16 \sum_{n=1}^{\infty} G(n)q^n.$$

Using empirical calculations, at which he was adept, he conjectured the multiplicative properties of these coefficients: For $W(n)$, $\Theta(n)$, and $G(n)$, with $m$ and $n$ relatively prime and odd, and where $f$ denoted any of the three functions, he found $f(mn) = f(m)f(n)$, except when $m$ and $n$ were both of the form $4k + 3$. When $m$ and $n$ were both of the form $4k + 3$, he had

$$W(mn) = -15W(m)W(n) \text{ and } G(mn) = -\frac{39}{25}G(m)G(n).$$

In his 1907 paper, Glaisher made some instructive remarks on his empirical discovery of these multiplicative properties:[15]

> The most interesting of the results obtained relate to 10 and 12 squares. So long ago as 1884 I had obtained the formulae for the numbers of representations of any even and uneven number as a sum of 12 squares, but I deferred the publication of the results in hopes of proving the relation $\Omega(m_1 m_2) = \Omega(m_1)\Omega(m_2)$, $m_1$ and $m_2$ being prime to one another, which I had observed to be satisfied by the values of $\Omega(m)$ that I had calculated. .... It was in 1905 that I worked for 10 squares, and was surprised to find that the number of representations depended upon the sum of the fourth powers of complex divisors. .... The principle difficulty in the investigations was the selection of the new functions to be introduced. It is obvious that the number of representations in any given case could be expressed by means of one or more new functions defined as coefficients and introduced for the purpose. It also appeared on investigation that the number of new functions which it was absolutely necessary to introduce was smaller than might have seemed likely. But my object was not to find the fewest functions which would suffice to express the number of representations, but to select suitable functions which had an arithmetical significance. The functions from which the choice had to be made were given in the first instance as coefficients, and this kind of definition throws no light whatever upon the arithmetical properties of the function (except that it might indicate classes of numbers for which the function vanished.) The only method available to me

---

[14] Glaisher (1907a) p. 489.    [15] Glaisher (1907a) pp. 487–488.

for determining the arithmetical nature of a function was to calculate its numerical values and so find what relations it seemed to satisfy. For the calculation of the functions it was necessary to investigate suitable recurring formulae; such formulae affording the only convenient method of obtaining the numerical values of a function defined as a coefficient.

Clearly, Glaisher wanted to employ coefficients that had arithmetic properties, particularly of multiplicativity. He was able to prove that the coefficients had this property in some cases and in other cases he formulated conjectures. These conjectures were proved by Rankin,[16] who employed Hecke operators on spaces of modular forms belonging to subgroups of the modular group.

Glaisher found another arithmetic property of $\Omega(n)$ and $\Theta(n)$: Let $a^2 + b^2 + c^2 + d^2$ be any representations of $n$ as a sum of four squares. Then

$$\Omega(n) = \frac{1}{8} \sum^{(r)} \left( [a^4] - 2[a^2b^2] \right),$$

$$\Theta(n) = e \sum^{(r)} \left( [a^6] - 5[a^4b^2] + 30[a^2b^2c^2] \right),$$

where $e$ is $\frac{1}{8}$ or $\frac{1}{24}$, according as $n$ is odd or even. The square brackets indicate that the sum is taken over all terms of that form. Thus, $[a^2b^2]$ denotes a sum of six terms; also, note that $\sum^{(r)}$ means that the sum is taken over all representations of $n$ as a sum of four squares.

In his 1916 paper, "On certain arithmetical functions," Ramanujan also employed elliptic functions to find a formula for which he provided a table of twelve particular cases, including all the instances $4 \le 2s \le 18$ considered by Glaisher. Ramanujan first defined an arithmetic function $\delta_{2s}(n)$ depending on the divisors of $n$ and on the residue class $s \bmod 4$. For example, if $s \equiv 4 \pmod 4$ and $s > 2$, then

$$(2^s - 1)B_s\, \delta_{2s}(n) = s\left( \sigma_{s-1}(n) - 2^s \sigma_{s-1}\left(\frac{n}{4}\right) \right), \tag{15.18}$$

where $\sigma_{s-1}(x) = 0$ for noninteger values of $x$, and $B_s$ was the Bernoulli number defined by the expansion

$$\frac{x}{e^x - 1} = \sum_{k=0}^{\infty} B_k \frac{x^k}{k!}. \tag{15.19}$$

Note that as Ramanujan defined it, $r_s(n)$ had half the value it was given in (15.5); he set

$$\left( \sum_{n=-\infty}^{\infty} q^{n^2} \right)^s = 1 + 2 \sum_{n=1}^{\infty} r_s(n)\, q^n.$$

He let

$$f_{2s}(q) = \sum_{n=1}^{\infty} \left( r_{2s}(n) - \delta_{2s}(n) \right) q^n \equiv \sum_{n=1}^{\infty} e_{2s}(n) q^n.$$

---

[16] Rankin (1977) Chapter 10.

Thus, he wrote his comprehensive formula as

$$f_{2s}(q) = \frac{f^{4s}(-q)}{f^{2s}(q^2)} \sum_{1 \le n \le \frac{s-1}{4}} K_n \frac{f^{24n}(q^2)}{f^{24n}(-q)}, \tag{15.20}$$

where

$$f(q) = q^{\frac{1}{24}} \prod_{n-1}^{\infty} (1 - q^n).$$

Ramanujan also showed that $\delta_{2s}(n)$ was the dominating term in the expression for $r_{2s}(n)$ and that $e_{2s}(n) = 0$ for $1 \le s \le 4$. He gave no derivation for (15.20), though he wrote that it could be demonstrated by the theory of elliptic functions. In a 1917 paper,[17] Mordell supplied a proof of (15.20), using the finite dimensionality of the space of holomorphic modular forms for the subgroup $\Gamma_3$ of $SL_2(\mathbb{Z})$ generated by the transformations $\omega \to \omega + 2$ and $\omega \to -\frac{1}{\omega}$. He observed that

$$\left( 1 + \sum_{n=1}^{\infty} \delta_{2s}(n) q^n \right) \theta_3^{-2s}$$

was invariant under the action of $\Gamma_3$; this fact was important in his proof.

Initially, Mordell was unable to find a function corresponding to

$$\chi = 1 + \sum_{n=1}^{\infty} \delta_{2s}(n) q^n$$

when the number of squares was odd. At about the same time, G. H. Hardy applied the circle method to solve the problem. This method can be applied to functions having a singularity at each point of the circle $|q| = 1$. For example:

$$\theta_3(\omega) = 1 + 2q + 2q^4 + 2q^9 + \cdots .$$

The function $\theta_3$ is approximated by simpler functions at each of the 'rational points' on the circle, that is, points $q = e^{\frac{2\pi h i}{k}}$, where $h$ and $k$ are integers. The sum $g(q)$ of these simpler functions over all distinct rational numbers then gives a good approximation of $\theta_3(\omega)$. Now the coefficient of $q^n$ in $\theta_3^s(\omega)$ provides us with the number of representations $r_s(n)$; hence, the coefficient, $\rho_s(n)$, of $q^n$ in $g^s(q)$ should yield a good approximation for $r_s(n)$. Hardy was able to prove that $\rho_s(n) = r_s(n)$ for $5 \le s \le 8$ by showing that $f(\omega) = \frac{g^s(q)}{\theta_3^s(\omega)}$ was invariant under the action of $\Gamma_3$ and that $f(\omega)$ was bounded in the fundamental region of $\Gamma_3$. He called the infinite series for $\rho_s(n)$ the singular series.

Mordell learned of the series $g^s(q)$ for odd $s$ from Hardy and recognized that it allowed him to define his function $\chi$ for odd $s$. He was then able to publish his paper on the number of representations of an integer as a sum of an odd number of squares. In 1936, Hecke found yet another way of using modular forms to find expressions for

---

[17] Mordell (1917b).

$r_{2s}(n)$ by showing a connection between modular forms and Dirichlet series; we will examine his method later in this chapter.

Thus, while Jacobi and Glaisher employed methods taken from the theory of elliptic functions to derive results on $r_s(n)$, Hardy, Mordell, and Hecke instead applied the theory of modular forms. But recently S. C. Milne[18] has reintroduced elliptic functions methods in order to construct infinite families of identities, of which Jacobi's sums of squares identities were the simplest first cases.

In the past seventy-five years, results on sums of squares of integers have been derived by myriad methods and techniques. Emil Grosswald[19] gives a very interesting discussion of several such methods up to 1985 and he provides an almost exhaustive bibliography. We mention a few examples of the use of $q$-series methods, different from those involving modular or elliptic functions.

In the late 1940s, Nathan Fine developed his theory of basic or $q$-hypergeometric series, in the course of which he discovered and rediscovered a number of identities. In particular, he rediscovered Ramanujan's $_1\psi_1$ sum: For $|q| < 1$, $|ba^{-1}| < |x| < 1$,

$$\sum_{n=-\infty}^{\infty} \frac{(a;q)_n}{(b;q)_n} x^n = \frac{(ax;q)_\infty \left(\frac{q}{ax};q\right)_\infty (q;q)_\infty \left(\frac{b}{a};q\right)_\infty}{(x;q)_\infty \left(\frac{b}{ax};q\right)_\infty (b;q)_\infty \left(\frac{q}{a};q\right)_\infty}, \qquad (15.21)$$

where

$$(a;q)_n = (1-a)(1-aq)\cdots(1-aq^{n-1}), \quad n \text{ a positive integer}; \qquad (15.22)$$

$$(a;q)_{-n} = \frac{1}{(aq^{-n};q)_n}, \quad n \text{ a positive integer}; \qquad (15.23)$$

$$(a;q)_\infty = \prod_{n=0}^{\infty}(1-aq^n). \qquad (15.24)$$

Fine observed that the particular case $b = aq$ of (15.21)

$$\sum_{n=-\infty}^{\infty} \frac{x^n}{1-aq^n} = \frac{(ax;q)_\infty \left(\frac{q}{ax};q\right)_\infty (q;q)_\infty^2}{(x;q)_\infty \left(\frac{q}{x};q\right)_\infty (a;q)_\infty \left(\frac{q}{a};q\right)_\infty}, \quad |q| < |x| < 1, \qquad (15.25)$$

was very useful in its applications to the problem of finding the number of representations of an integer as a sum of squares. In his book[20], Fine gave formulas for the number of representations of integers as sums of two, four, and eight squares, as well as special types of sums of three, five, and six squares, including the case of three squares in which at least two are the same. Shaun Cooper[21] has greatly extended the scope of this method by employing an identity of K. Venkatachaliegar and Ramanujan

$$F(a,x)F(b,x) = x\frac{\partial}{\partial x}F(ab,x) + F(ab,x)\big(\rho_1(a) + \rho_1(b)\big), \qquad (15.26)$$

---

[18]  Milne (2002). Ono (2002) and Zagier (2000) have used modular function methods to prove some of Milne's results. See also Chan and Chua (2003).
[19]  Grosswald (1985).      [20]  Fine (1988) pp. 72–77.      [21]  Cooper (2001).

where $F(x, a)$ denotes the Lambert series on the left-hand side of (15.25), where the prime on the sum indicates that $j = 0$ is excluded, and

$$\rho_1(z) = \frac{1}{2} + \sideset{}{'}\sum_{j=-\infty}^{\infty} \frac{z^j}{1 - q^{2j}}. \tag{15.27}$$

Cooper used this result to greatly facilitate the computation of generating functions related to sums of squares.

In a well-known paper,[22] George Andrews showed that a particular type of $_6\psi_6$ summation formula would readily produce the results for two, four, six, and eight squares. In a different direction, Michael Hirschhorn has shown that ingenious applications of the triple product identity are sufficient to prove the two and four squares theorems. For these proofs and references, see Berndt.[23] The notes at the end of Berndt's chapter three also give references to recent work on sums of squares based on elementary methods and $q$-series. We mention that Cooper has used $q$-series and Hecke operators to obtain good results on sums of odd numbers of squares, some in collaboration with Hirschhorn.

## 15.2 Jacobi's Elliptic Functions Approach

The problem of representing an integer as a sum of squares is an old one. In his letters to his mathematical correspondents in the period 1635–1660, Fermat wrote that he could prove that prime numbers of the form $4n + 1$ could be expressed as sums of two squares and conversely. He could also prove that every integer was a sum of four squares. Fermat used the method of infinite descent and his methods were probably rediscovered by Euler and Lagrange. Euler was perhaps the first mathematician to think of the possibility of using series; in a letter of 1750 to Goldbach,[24] Euler discussed the use of infinite series to address this problem. He wrote that a natural way to prove Fermat's remarkable theorem that every integer could be expressed as a sum of three triangular numbers, four squares, five pentagonal numbers, and so on, would be to show every power of $x$ to be positive in the series

$$(1 + x + x^3 + x^6 + \cdots)^3, (1 + x + x^4 + x^9 + \cdots)^4, \ldots.$$

Responding to Euler's challenge, Jacobi published an 1828 note[25] on elliptic functions containing the formula

$$(1 + 2q + 2q^4 + 2q^9 + \cdots)^4$$

$$= 1 + 8 \sum_{n=1}^{\infty} \frac{nq^n}{1 + (-q)^n}$$

$$= 1 + 8 \sum_p \phi(p)(q^p + 3q^{2p} + 3q^{4p} + 3q^{8p} + 3q^{16p} + \cdots), \tag{15.28}$$

---

[22] Andrews (1974).     [23] Berndt (2005) pp. 55–84.
[24] Fuss (1968) vol. 1, pp. 530–533.     [25] Jacobi (1969) vol. 1, p. 262.

where $p$ denoted an arbitrary odd integer, and $\phi(p)$ the sum of the divisors of $p$. This formula in fact yielded the number of representations of a positive integer as a sum of four squares. Jacobi gave no proof of (15.28) in his 1828 note, but he presented a complete proof in his 1829 *Fundamenta Nova Theoriae Functionum Ellipticarum*. In that work, he also presented formulas, without explicitly working out the details, for the number of representations of an integer as a sum of two, six, and eight squares. In addition, Jacobi discovered and proved many relations expressing simple functions of $K, k, k'$ in terms of series in $q = e^{-\frac{K'}{K}}$. Among these were the interesting relations:[26]

$$\sqrt{\frac{2K}{\pi}} = 1 + 2q + 2q^4 + 2q^9 + 2q^{16} + \cdots \tag{15.29}$$

$$\sqrt{\frac{2kK}{\pi}} = 2q^{\frac{1}{4}} + 2q^{\frac{9}{4}} + 2q^{\frac{25}{4}} + 2q^{\frac{49}{4}} + \cdots \tag{15.30}$$

$$\frac{2K}{\pi} = 1 + \frac{4q}{1-q} - \frac{4q^3}{1-q^3} + \frac{4q^5}{1-q^5} - \cdots \tag{15.31}$$

$$\frac{2kK}{\pi} = \frac{4q^{\frac{1}{2}}}{1-q} - \frac{4q^{\frac{3}{2}}}{1-q^3} + \frac{4q^{\frac{5}{2}}}{1-q^5} - \cdots \tag{15.32}$$

$$\frac{4K^2}{\pi^2} = 1 + \frac{8q}{1-q} + \frac{16q^2}{1+q^2} + \frac{24q^3}{1-q^3} + \cdots \tag{15.33}$$

$$\frac{4k^2K^2}{\pi^2} = \frac{16q}{1-q^2} + \frac{48q^3}{1-q^6} + \frac{80q^5}{1-q^{10}} + \cdots \tag{15.34}$$

$$\frac{4k'^2K^2}{\pi^2} = 1 - \frac{8q}{1+q} + \frac{16q^2}{1+q^2} - \frac{24q^3}{1+q^3} + \cdots . \tag{15.35}$$

Note that (15.29) combined with (15.31) gives the well-known result on the number of representations of an integer as a sum of two squares; (15.29) and (15.33) together yield the four squares theorem.

## 15.3 Glaisher

J. W. L. Glaisher's many papers on elliptic functions greatly increased the number of formulas similar in form to Jacobi's (15.29) through (15.35). In the course of his researches, Glaisher found the number of representations of an integer as a sum of ten, twelve, fourteen, sixteen, and eighteen squares. To express these results on sums of squares in a congenial form, he defined a number of arithmetic functions, several of which were at that time already known:

Let $n$ be a given positive integer and let $d$ denote any divisor of $n$; let $d'$ be the conjugate of $d$, that is, $dd' = n$. Let $\delta$ denote any odd divisor of $n$ and let $\delta'$ be the

---

[26] Jacobi (1969) vol. 1, pp. 179–180, 235, 259.

conjugate of $\delta$. Then set

$$E_i(n) = \sum (-1)^{\frac{\delta-1}{2}} \delta^i, \qquad E'_i(n) = \sum (-1)^{\frac{\delta-1}{2}} \delta'^i \qquad (15.36)$$

$$\Delta_i(n) = \sum_{\delta|n} \delta^i, \qquad \Delta'_i(n) = \sum_{\delta|n} \delta'^i, \qquad (15.37)$$

where the sums are over all odd divisors of n. If $i = 1$, the subscript would be omitted. Glaisher also found formulas for the number of representations of $n$ as a sum of $p$ odd and $s$ even squares,

$$r_{p,s}(n). \qquad (15.38)$$

Denoting $\rho = \frac{2K}{\pi}$, Glaisher used (15.29) and (15.30) to consider series of the form

$$k^{\frac{p}{2}} \rho^{\frac{p+s}{2}} = \left( \sum_{m=-\infty}^{\infty} q^{\left( \frac{(2m+1)^2}{4} \right)} \right)^p \left( \sum_{n=-\infty}^{\infty} q^{\frac{(2m)^2}{4}} \right)^s. \qquad (15.39)$$

The series on the right-hand side of (15.39) can be expanded in powers of $q^{\frac{1}{4}}$; also, $r'_{p,s}(n)$, the coefficient of $q^{\frac{n}{4}}$, denotes the number of representations of $n$ as a sum of $p$ odd and $s$ even squares, where the $p$ odd squares are given first, and $r_{p,s}(n)$ designates the number of representations of $n$ as a sum of $p$ odd squares and $s$ even squares. It is easy to see that

$$r_{p,s}(n) = \binom{p+s}{p} r'_{p,s}(n) = \frac{(p+s)!}{p!s!} r'_{p,s}(n). \qquad (15.40)$$

For example, from

$$k^{\frac{1}{2}} \rho^2 = 2 \sum_{n=0}^{\infty} \Delta(4n+1) q^{\frac{4n+1}{4}},$$

and

$$k^2 \rho^2 = 16 \sum_{n=0}^{\infty} \Delta(2n+1) q^{2n+1}$$

Glaisher found, respectively, the values[27]

$$r_{1,3}(4n+1) = 8\Delta(4n+1), \quad r_{4,0}(8n+4) = 16\Delta(2n+1). \qquad (15.41)$$

Incidentally, the second formula in (15.41) was first published by Jacobi in a brief note of 1828;[28] that formula follows from (15.34). Glaisher also found a group of

---

[27] Glaisher (1907b) p. 8.    [28] Jacobi (1969) vol. 1, p. 247.

twenty-four $q$-series formulas involving the function $\rho^5$; we present three of these:[29]

$$(5 - 6k^2 + k^4)\rho^5 = 5 + 4\sum_{n=1}^{\infty} E_4(n)q^n \tag{15.42}$$

$$(4k^2 + k^4)\rho^5 = 64\sum_{n=1}^{\infty} E_4'(n)q^n \tag{15.43}$$

$$k^2 k'^2 \rho^5 = 16\sum_{n=1}^{\infty} \chi_4(n)q^n, \tag{15.44}$$

with

$$\chi_4(n) = \frac{1}{4}\sum (a + ib)^4,$$

where the summation is over all Gaussian integers $a + ib$ whose norm is $n$. By adding (15.42) and (15.43) to the double of (15.44), Glaisher obtained

$$5\rho^5 = 5 + 4\sum_{n=1}^{\infty}\left(E_4(n) + 16E_4'(n) + 8\chi_4(n)\right)q^n. \tag{15.45}$$

This result implied the formula for $r_{10}(n)$, that is, the number of representations of $n$ as a sum of ten squares:

$$r_{10}(n) = \frac{4}{5}\left(E_4(n) + 16E_4'(n) + 8\chi_4(n)\right). \tag{15.46}$$

Since $\chi_4(4n + 3) = 0$, from (15.46) it follows that

$$r_{10}(4n + 3) = -12E_4(4n + 3), \tag{15.47}$$

a result stated by Eisenstein in 1847. Eisenstein[30] wrote that he had obtained this equation by arithmetical methods but he did not provide details. He did not give the explicit formula (15.47), but rather stated that the number of representations of an integer of the form $4n + 3$ by means of ten squares would be twelve times the difference between the sum of the fourth powers of factors of the form $4n + 3$ and the sum of the fourth powers of factors of the form $4n + 1$. He also observed that there was no similar result for integers of the form $4n + 1$. Of course, it is easy to see from (15.46) that

$$r_{10}(4n + 1) = \frac{4}{5}\left(17E_4(4n + 1) + 8\chi_4(4n + 1)\right). \tag{15.48}$$

We note that formula (15.46) was apparently first stated in 1866 by Liouville, so that in fact Glaisher rediscovered it. Note that equations (15.42) through (15.44) also imply

---

[29] Glaisher (1907b) pp. 18–19.  [30] Eisenstein (1847) vol. 1, p. 501.

two other formulas discovered by Glaisher:[31]

$$5k^2\rho^5 = 16 \sum_{n=1}^{\infty} \left(4E_4'(n) + \chi_4(n)\right)q^n \tag{15.49}$$

$$5k^4\rho^5 = 64 \sum_{n=1}^{\infty} \left(4E_4'(n) + \chi_4(n)\right)q^n. \tag{15.50}$$

As he noted, these in turn imply

$$r_{4,6}(4n) = 672\left(4E_4'(n) + \chi_4(n)\right) \tag{15.51}$$
$$r_{8,2}(4n) = 576\left(E_4'(n) + \chi_4(n)\right). \tag{15.52}$$

Glaisher's 1907 paper for the *Proceedings of the London Mathematical Society*[32] summarized his results on the number of representations of an integer as a sum of an even number of up to eighteen squares.

## 15.4 Ramanjuan's Arithmetical Functions

In 1916, Ramanujan published his results on sums of squares in an article, "On certain arithmetical functions." Then in 1918, he published another paper on this topic, "On certain trigonometrical series." These two depended upon Ramanujan's theory of elliptic functions, developed in India before he traveled to Cambridge in 1914. Concerning the origin of the ideas in these two papers, Hardy wrote:

> It is always difficult to say how much Ramanujan owed to other writers, and the difficulty is at its maximum when he is developing work which he began before he came to England, and when he is concerned, as he is here, with some side of the theory of elliptic functions. There is no book on elliptic functions which he could have seen in India and which says anything about the arithmetical applications of the theory. I believe therefore that Ramanujan had rediscovered Jacobi's formulae, which certainly lay well within his powers.
>
> By the time he published these papers, Ramanujan had read a great deal more, and knew all about Jacobi's and much later work. In particular he had read Glaisher's papers, and treats their content as known, and a reader who takes his acknowledgements at their face value will be liable to underestimate his originality. The papers are highly original, whatever deductions we may make: they are characteristic of Ramanujan at his best. They contain many remarkable theorems which are undeniably new, and conjectures still more remarkable, which were confirmed later by Mordell; and the general level of analysis is astonishingly high.

In his papers, Ramanujan briefly indicated the application of modular forms to the representation of numbers as sums of squares. Using series to define an arithmetic function $\delta_{2s}(n)$ depending on the divisors of $n$, for $s = 1$ he got

$$\sum_{n=1}^{\infty} \delta_2(n)\, q^n = 2 \sum_{n=1}^{\infty} \frac{(-1)^{n-1}q^{2n-1}}{1 - q^{2n-1}};$$

[31] Glaisher (1907b) pp. 21–23.   [32] Glaisher (1907a).

when $s$ was a multiple of 4,

$$(2^s - 1)|B_s| \sum_{n=1}^{\infty} \delta_{2s}(n)q^n = s \sum_{n=1}^{\infty} \frac{n^{s-1}q^n}{1 - (-q)^n};$$

when $s + 2$ was a multiple of 4,

$$(2^s - 1)|B_s| \sum_{n=1}^{\infty} \delta_{2s}(n)q^n = s \sum_{n=1}^{\infty} \frac{n^{s-1}q^n}{1 + (-q)^n};$$

when $s - 1$ was a multiple of 4,

$$E_s^* \sum_{n=1}^{\infty} \delta_{2s}(n)\, q^n = 2^s \sum_{n=1}^{\infty} \frac{n^{s-1}q^n}{1 + q^{2n}} + 2 \sum_{n=1}^{\infty} (-1)^{n-1} \frac{(2n-1)^{s-1}q^{2n-1}}{1 - q^{2n-1}};$$

when $s + 1$ was a multiple of 4,

$$E_s^* \sum_{n=1}^{\infty} \delta_{2s}(n)\, q^n = 2^s \sum_{n=1}^{\infty} \frac{n^{s-1}\, q^n}{1 + q^{2n}} - 2 \sum_{n=1}^{\infty} (-1)^{n-1} \frac{(2n-1)^{s-1}q^{2n-1}}{1 - q^{2n-1}}.$$

Note here that $B_s$ denotes the absolute value of $s$th Bernoulli number and $E_s^*$ represents the $s$th Euler number defined by the series

$$\sec x = \sum_{n=0}^{\infty} E_{n+1}^* \frac{x^n}{n!}.$$

Thus, $E_1^* = 1$, $E_3^* = 1$, $E_5^* = 5$, $E_7^* = 61$, $E_9^* = 1385,\ldots$. Note that we have used the "$*$" to denote Ramanujan's notation; the usual notation for Euler numbers gives Ramanujan's $E_{n-1}^*$ as $E_n$.

The expressions for $\delta_{2s}(n)$ in terms of the divisor functions can be obtained from these series and Ramanujan gave one expression explicitly: When $s + 2$ is a multiple of 4, then

$$(2^s - 1)|B_s|\, \delta_{2s}(n) = s\left\{\sigma_{s-1}(n) - 2^s \sigma_{s-1}\left(\frac{n}{4}\right)\right\},$$

where $\sigma_s(x) = 0$ when $x$ is not an integer.

Ramanujan also observed that $\frac{1}{2}r_{2s}(n) = \delta_{2s}(n) + e_{2s}(n)$; he proved that $e_{2s}(n) = 0$ for all $n$ when $s = 1, 2, 3, 4$ and that

$$e_{2s}(n) = O\left(n^{s-1-\frac{1}{2}[\frac{2s}{3}]+\epsilon}\right). \tag{15.53}$$

He also noted that it could be shown, for $s \geq 3$, that there exist positive constants $H$ and $K$ such that

$$Hn^{s-1} < \delta_{2s}(n) < Kn^{s-1}$$

and that this implied, for all positive integer values of $s$, that

$$\frac{1}{2}r_{2s}(n) \sim \delta_{2s}(n).$$

Ramanujan then defined the series

$$f_{2s}(q) = \sum_{n=1}^{\infty} e_{2s}(n) q^n \equiv \sum_{n=1}^{\infty} \left( \frac{1}{2} r_{2s}(n) - \delta_{2s}(n) \right) q^n, \tag{15.54}$$

and noted that it could be shown from the theory of elliptic functions that there exist constants $K_n$ such that

$$f_{2s}(q) = \theta_3^{2s}(\omega) \sum_{1 \le n \le \frac{s-1}{4}} K_n(kk')^{2n}, \quad q = e^{i\pi \omega}, \tag{15.55}$$

or that

$$f_{2s} = \frac{f^{4s}(-q)}{f^{2s}(q^2)} \sum_{1 \le n \le \frac{s-1}{4}} K_n \frac{f^{24n}(q^2)}{f^{24n}(-q)}, \tag{15.56}$$

where

$$f(q) = q^{\frac{1}{24}} \prod_{n=1}^{\infty} (1 - q^n). \tag{15.57}$$

On the basis of (15.56), Ramanujan then gave a short table of formulas for $f_{2s}(q)$, when $s = 1, 2, \ldots, 12$. This yields alternate derivations for Glaisher's and Jacobi's formulas for $r_{2s}(n)$, $1 \le s \le 9$. As we have seen from Hardy's remarks on Ramanjuan's work on sums of squares, it appears very likely that these derivations were done independently. Ramanujan's table illustrates his deep fascination with numbers:[33]

1.

$$f_2(q) = 0, \quad f_4(q) = 0, \quad f_6(q) = 0, \quad f_8(q) = 0.$$

2.

$$5 f_{10}(q) = 16 \frac{f^{14}(q^2)}{f^4(-q)}, \quad f_{12}(q) = 8 f^{12}(q^2).$$

3.

$$61 f_{14}(q) = 728 f^4(-q) f^{10}(q^2), \quad 17 f_{16}(q) = 256 f^8(-q) f^8(q^2).$$

4.

$$1385 f_{18}(q) = 24416 f^{12}(-q) f^6(q^2) - 256 \frac{f^{30}(q^2)}{f^{12}(-q)}.$$

5.

$$31 f_{20}(q) = 616 f^{16}(-q) f^4(q^2) - 128 \frac{f^{28}(q^2)}{f^8(-q)}.$$

[33] Ramanujan (2000) p. 159. Also note that V. Ramamani has corrected formula 6: 1103272 and 821888 should be changed to 1109416 and 944768, respectively; see Berndt (2006) p. 84 for the reference.

6.

$$50521 f_{22}(q) = 1103272 f^{20}(-q) f^2(q^2) - 821888 \frac{f^{26}(q^2)}{f^4(-q)}.$$

7.

$$691 f_{24}(q) = 16576 f^{24}(-q) - 32768 f^{24}(q^2).$$

Now

$$f^{24}(q) = q \prod_{n=1}^{\infty} (1 - q^n)^{24} = \sum_{n=1}^{\infty} \tau(n) q^n;$$

and Ramanuman observed that the seventh line of his table implied that

$$\frac{691}{64} e_{24}(n) = (-1)^{n-1} 259 \tau(n) - 512 \tau\left(\frac{n}{2}\right), \tag{15.58}$$

where $\tau(x) = 0$, $x$ not an integer. This gave him his famous formula for the number of representations of an integer $n$ as a sum of twenty-four squares:

$$r_{24}(n) = \frac{16}{691} \sigma_{11}^*(n) + \frac{128}{691} \left[(-1)^{n-1} 259 \tau(n) - 512 \tau\left(\frac{n}{2}\right)\right], \tag{15.59}$$

where $\sigma_v^*(n) = \sigma_v$, $n$ odd; $\sigma_v^*(n) = \sigma_v^e(n) - \sigma_v^o(n)$, $n$ even; and

$$\sigma_v^e(n) = \sum_{d|n, 2|d} d^v \; ; \; \sigma_v^o(n) = \sum_{d|n, 2\nmid d} d^v.$$

Recall that he proved (15.53), and on empirical evidence, he thought it probable that

$$e_{2s}(n) = O\left(n^{\frac{1}{2}(s-1)+\epsilon}\right), \tag{15.60}$$

implying, in particular,

$$e_{24}(n) = O\left(n^{\frac{11}{2}+\epsilon}\right). \tag{15.61}$$

Of course, (15.61) follows when Ramanujan's famous conjecture

$$|\tau(n)| = O\left(n^{\frac{11}{2}+\epsilon}\right) \tag{15.62}$$

is applied to (15.58); note that (15.62) was proved by Pierre Deligne in 1974.[34] Ramanujan also conjectured that $\tau(n)$ was multiplicative; recall that this was proved by Mordell,[35] who was impressed that Ramanujan had made this conjecture on empirical grounds. In his 2001 paper on sums of an even number of squares, Cooper has extended Ramanujan's table of formulas for $f_{2s}(q)$ to $s = 1, \ldots, 25$.

## 15.5   Mordell: Spaces of Modular Forms

In both his papers on sums of squares, Ramanujan also stated a general formula from which one could obtain expressions for the number of representations of an integer as sums of $2r$ squares. This formula yielded "simple results for many values of $r$,"

---

[34] Deligne (1974).     [35] Mordell (1917a).

in the words of Mordell. Ramanujan wrote that his formula followed from the theory of elliptic functions, but he omitted the proof. This inspired Mordell to search for a proof. In his 1917 paper on the representation of integers as sums of $2r$ squares, he wrote: "The object of this paper is to point out that all the various results for different values of $r$ can be found in a uniform and direct manner by considering the subject from the point of view of the modular functions."[36] In this effort, Mordell employed the finite dimensionality of the space of modular forms of a given weight to obtain identities relating modular forms. This type of argument was foreshadowed in the work of Hurwitz; the reader may see Section 8.12.

Mordell considered the subgroup of the modular group generated by $S^2 z = z + 2$ and $Tz = -\frac{1}{z}$. In the notation of Klein and Fricke's 1890–92 book, this subgroup is denoted by $\Gamma_3$, while Hecke denoted it by $G(2)$. Up to 1910, Klein and Fricke's book and the third volume of H. Weber's *Lehrbuch der Algebra* together contained almost everything known on the topic of modular forms. Now from Klein and Fricke, Mordell utilized an essential theorem about $\Gamma_3$: If $s \equiv 0 \pmod 4$, then there exist $\frac{s}{4} + 1$ linearly independent modular forms of weight $s$ for $\Gamma_3$.[37] Recall here that

$$\theta_3(\omega) = 1 + 2q + 2q^4 + 2q^9 + \cdots, \quad q = e^{i\pi\omega}, \quad \operatorname{Im}\omega > 0$$

satisfies the relations

$$\theta_3(\omega + 2) = \theta_3(\omega) \quad \text{and} \quad \theta_3\left(-\frac{1}{\omega}\right) = \sqrt{\frac{\omega}{i}}\,\theta_3(\omega).$$

Hence, $\theta_3^{2r}(\omega)$ is not quite a modular form for $\Gamma_3$ unless $r \equiv 0 \bmod 4$. Now let $\beta$ be the least positive residue of $r \pmod 4$, but when $r \equiv 0 \pmod 4$, let $\beta = 4$; in that case, $\theta_3^{2(r-\beta)}(\omega)$ is a modular form for $\Gamma_3$, of weight $r - \beta$.

Mordell also employed three theta constants $\theta_0, \theta_2, \theta_3$ and their behavior under the action of $S^2$ and $T$. With $q = e^{\pi\omega i}$, $\operatorname{Im}\omega > 0$ he utilized the formulas:

$$\theta_0(\omega) = \sum_{n=-\infty}^{\infty} (-1)^n q^{n^2},$$

$$\theta_2(\omega) = \sum_{n=-\infty}^{\infty} q^{\left(n+\frac{1}{2}\right)^2},$$

$$\theta_3(\omega) = \sum_{n=-\infty}^{\infty} q^{n^2},$$

$$\theta_0(\omega + 2) = \theta_0(\omega), \qquad \theta_0\left(-\frac{1}{\omega}\right) = \sqrt{\frac{\omega}{i}}\,\theta_2(\omega), \qquad (15.63)$$

$$\theta_2(\omega + 2) = i\theta_2(\omega), \qquad \theta_2\left(-\frac{1}{\omega}\right) = \sqrt{\frac{\omega}{i}}\,\theta_0(\omega), \qquad (15.64)$$

$$\theta_3(\omega + 2) = \theta_3(\omega), \qquad \theta_3\left(-\frac{1}{\omega}\right) = \sqrt{\frac{\omega}{i}}\,\theta_3(\omega). \qquad (15.65)$$

---

[36] Mordell (1917b) p. 94.   [37] Klein and Fricke (1890–92) vol. 2, p. 365.

Recall from Chapter 5 that (15.63) through (15.65) are particular cases of formulas given by Jacobi in 1848,[38] except that he left undetermined a certain eighth root of unity; this root of unity was found by Hermite in 1858.[39] Also recall that (15.63) was earlier published by Cauchy and Poisson as (8.2).

Modular forms came into play in Mordell's theory of representations of integers as sums of an even number of squares because

$$\theta_3^{2s}(\omega) = \sum_{n=0}^{\infty} r_{2s}(n)\, q^n \qquad (15.66)$$

was a relative invariant of the subgroup $\Gamma_3$ (generated by $S^2$ and $T$) of the modular group

$$\theta_3^{2s}(\omega + 2) = \theta_3^{2s}(\omega), \quad \theta_3^{2s}\left(-\frac{1}{\omega}\right) = (-1)^{\frac{s}{2}}\, \omega^s\, \theta_3(\omega). \qquad (15.67)$$

Following Mordell, we call $\theta^{2s}(\omega)$ a relative invariant of $\Gamma_3$ because of the factor $(-1)^{\frac{s}{2}}$ in the second equation in (15.67). Note also that when $s \equiv 0 \mod 4$, then $\theta_3^{2s}(\omega)$ is an invariant and hence, as we shall see, a modular form of weight $s$ for $\Gamma_3$. One need only check the behavior of the function at the cusps.

Now consider the behavior of $\theta_3^{2s}$ at the two inequivalent cusps, 1 and $\infty$, of the fundamental region for $\Gamma_3$ and observe that $-1$ is a cusp but is equivalent to 1. It follows from the $q$-series for $\theta_3$ that $\theta_3^{2s}(\omega)$ is holomorphic at the cusp at $\infty$, but it does not vanish at $\infty$. To determine the behavior of $\theta_3^{2s}(\omega)$ at $\omega = 1$, set

$$\omega = 1 - \frac{1}{\tau} \quad \text{or} \quad \tau = \frac{1}{1-\omega},$$

so that when $\omega \to 1$ inside the cusp, we have $\tau \to \infty$ in the fundamental region for $\Gamma_3$. Now let $q_1 = e^{\pi \tau i}$ and observe that

$$\theta_3^{2s} = \theta_3^{2s}\left(1 - \frac{1}{\tau}\right) = \theta_0^{2s}\left(-\frac{1}{\tau}\right) = \tau^s\, \theta_2^{2s}(\tau)$$
$$= \tau^4 \left(2q_1^{\frac{1}{4}} + 2q_1^{\frac{9}{4}} + \cdots\right)^{2s} = 2^{2s}\tau^4\, q_1^{\frac{s}{2}}(1 + q_1^2 + \cdots). \qquad (15.68)$$

Thus, $\theta_3^{2s}(\omega)$ has a zero at the cusp at 1, since $\tau^4 e^{\frac{s\tau}{2}\pi i} \to 0$ as $\tau \to i\infty$. Now let $\beta$ be the least positive residue of $s \pmod 4$, except that when $s \equiv 0 \pmod 4$, take $\beta = 4$. Let $\chi_1, \chi_2, \ldots$, be relative invariants belonging to $\Gamma_3$ with the same behavior under $S^2$ and $T$ as $\theta_3^{2\beta}(\omega)$, with the condition that $\frac{\chi_i}{\theta_3^{2\beta}}$ is a modular form and hence is holomorphic at the cusps. By (15.68), this implies that the expansions at the cusp at 1 of $\chi_1, \chi_2, \ldots$, have first terms of the form $q_1^\lambda, \lambda \geq \frac{\beta}{2}$.

---

[38] Jacobi (1969) vol. 2, pp. 171–190.     [39] Hermite (1905–1917) vol. 1, pp. 487–496.

Now the theorem of Fricke and Klein on the number of linearly independent forms shows that we need at most $t = \frac{s-\beta}{4} + 1$ modular forms $\frac{\chi}{\theta_3^{2\beta}}$ such that

$$\theta_3^{2(s-\beta)} = \sum_{\lambda=0}^{t-1} C_\lambda \left( \chi_\lambda \, \theta_3^{-2\beta} \right)$$

or

$$\theta_3^{2s} = \sum_{\lambda=0}^{t} C_\lambda \chi_\lambda. \tag{15.69}$$

When $s$ was even, Mordell took $\chi_0 \equiv \chi$ to be defined by the Eisenstein series

$$\chi(\omega) = \sum_{a,b} (2a\omega + 2b + 1)^{-s} + (-1)^{\frac{s}{2}} \sum_{a,b} (2a+1)\omega + 2b)^{-s}, \tag{15.70}$$

where $a$ and $b$ ran through the full set of integers. Note that the series (15.70) is absolutely convergent for all $s \geq 2$, provided that when $s = 2$, the two sums are combined. Now it can be seen that

$$\chi(\omega + 2) = \chi(\omega) \quad \text{and} \quad \chi\left(-\frac{1}{\omega}\right) = (-1)^{\frac{s}{2}} \omega^s \chi(\omega),$$

that is, $\chi$ is a relative invariant for $\Gamma_3$. Using Hurwitz's method, Mordell showed that the Eisenstein series (15.70) had the $q$-series expansion

$$\chi(\omega) = (2^s - 1) B_s + 2s \sum_{k=1}^{\infty} \frac{k^{s-1} q^k}{1 - (-1)^{k+\frac{s}{2}} q^k}, \tag{15.71}$$

where $B_s$ denoted the $s$th Bernoulli number. Next note that there is a nonzero constant term $(2^s - 1)B_s$ in (15.71), arising from the constant term in the first sum in (15.70):

$$\sum_{b=-\infty}^{\infty} (2b+1)^{-s} = 2\left(1 + \frac{1}{3^s} + \frac{1}{5^s} + \cdots\right)$$

$$= \frac{(2\pi)^s}{s!}\left(1 - \frac{1}{2^s}\right) B_s.$$

It is also necessary to check that $\chi$, given by (15.70), has an expansion at the cusp at 1 whose first term is $q_1^\lambda$, $\lambda \geq \frac{\beta}{2}$. In particular, it is easy to see that this means that the constant term in $\chi\left(1 - \frac{1}{\tau}\right)$ is zero. To obtain the expansion of $\chi\left(1 - \frac{1}{\tau}\right)$ in powers of $q_1$, apply the same procedure used to obtain (15.71).

When $s$ was odd, Mordell defined $\chi$ by the equation

$$\chi(\omega) = \sum_{a,b} (2a\omega + 4b + 1)^{-s} + i^s \sum_{a,b} \left((4a+1)\omega + 2b\right)^{-s}. \tag{15.72}$$

Again, we see that $\chi(\omega + 2) = \chi(\omega)$ and $\chi\left(-\frac{1}{\omega}\right) = (-1)^{\frac{s}{2}} \omega^s \chi(\omega)$. Moreover, $\chi$ has the $q$-series expansion

$$\chi(\omega) = \frac{1}{2} E_{\frac{s-1}{2}} + \sum_{k=1}^{\infty} \frac{2^s k^{s-1} q^k}{1 + q^{2k}} + 2 \sum_{k=1}^{\infty} (-1)^{\frac{s-3}{2}+k} \frac{(2k-1)^{s-1} q^{2k-1}}{1 - q^{2k-1}}, \qquad (15.73)$$

where $E_1, E_2, E_3, \ldots$ are the Euler numbers defined by

$$\sec x = 1 + \sum_{n=1}^{\infty} E_n \frac{x^{2n}}{(2n)!}. \qquad (15.74)$$

In order to generate the full space of relative invariants, $\frac{1}{4}(s - \beta)$ additional relative invariants $\chi_\lambda$ are required. Mordell defined these as

$$\chi_\lambda = \theta_3^{2s} \left( \frac{\theta_0^2 \theta_2^2}{\theta_3^4} \right)^{2\lambda}, \qquad (15.75)$$

where $\lambda = 1, 2, \ldots, \lfloor \frac{s-1}{4} \rfloor$. That these $\chi_\lambda$ are relative invariants may be checked by using (15.63) through (15.65). It is sufficient here to verify that the factor

$$\left( \frac{\theta_0^2 \theta_2^2}{\theta_3^4} \right)^{2\lambda}$$

is invariant under $S^2$ and $T$ and we leave the details to the reader. The linear independence of $\chi_1, \chi_2, \ldots,$ can be shown by employing Jacobi's identity (3.197)

$$\theta_3^4 = \theta_0^4 + \theta_2^4;$$

we replace $\theta_3^4$ by $\theta_0^4 + \theta_2^4$, so that any linear relation among the terms in the series $\chi_1, \chi_2, \ldots,$ can be expressed using only $\theta_2$ and $\theta_3$. It is also clear that $\chi_1, \chi_2, \ldots,$ are all holomorphic and that their expansions in powers of $q$ has a zero constant term. It follows that

$$\theta_3^{2s} = \frac{1}{a} \chi + \sum_{k=1}^{\lfloor \frac{s-1}{4} \rfloor} A_k \theta_0^{4k} \theta_2^{4k} \theta_3^{2(s-2k)}, \qquad (15.76)$$

where

$$a = (2^s - 1) B_s \quad \text{when } s \text{ is even;}$$

$$= \frac{1}{2} E_{\frac{s-1}{2}} \quad \text{when } s \text{ is odd.}$$

Note that the right-hand side of (15.76) consists of a suitable Eisenstein series $\chi$ and a sum of terms that vanish at the cusps. Rankin[40] proved that for $s > 4$ this sum cannot be identically equal to zero.

Recall that Ramanujan gave (15.76) without proof in the form (15.55) or (15.56). Also note that (15.76) contains Jacobi's formulas for the number of representations of

---

[40] Rankin (1965).

an integer as s sum of $2s$ squares for $s = 2, 3, 4$. When $s = 6$, we have

$$\theta_3^{12} = 1 + 8 \sum_{k=1}^{\infty} \frac{k^5 q^k}{1 + (-1)^k q^k} + \theta_0^4 \theta_2^4 \theta_3^4.$$ (15.77)

Observe that

$$\theta_0^4 \theta_2^4 \theta_3^4 = q \prod_{n=1}^{\infty} (1 - q^{2n})^{12} = \sqrt{\Delta(\omega)},$$

so that we obtain Glaisher's formula for the number of representations of $n$ as a sum of twelve squares. The reader may wish to read Mordell's alternative method[41] for selecting the invariants $\chi_1, \chi_2, \dots$, such that the number of representations of an integer as a sum of squares could be more easily expressible in terms of arithmetical functions.

The problem of determining the number of representations of an integer as a sum of an odd number of squares is more difficult than the case of an even number of squares. Mordell's problem was to find a function corresponding to $\chi$, given by (15.70) and (15.72), for an odd number of squares. At around the same time, Hardy was tackling the same problem along very different lines. His approach was to use the circle method that he and Ramanujan had discovered and then applied to the problem of finding an asymptotic formula for $p(n)$, the number of partitions of an integer $n$.[42] The circle method produced a series, called the singular series, that yielded an asymptotic formula for the number of representations of an integer as a sum of squares. When Mordell saw Hardy's result, he realized how he could define $\chi$ in the case of an odd number of squares. He was then able to write up and publish his paper on the topic.[43]

## 15.6 Hardy's Singular Series

The idea behind Hardy's construction of the singular series for the number of representations of an integer is to first determine the behavior of the function

$$f(s) = (1 + 2q + 2q^4 + \cdots)^s$$

as $q$ tends radially to the point $e^{\frac{2h\pi i}{k}}$ on the unit circle. Clearly, all cases will be covered if we take $h = 0$, $k = 1$, and when $k > 1$ we take $h$ and $k$ to be relatively prime and $0 < h < k$. If we set $q = re^{\frac{2h\pi i}{k}}$ and let $r \to 1$, then

$$\theta_3 = 1 + 2 \sum_{n=1}^{\infty} r^{n^2} e^{\frac{2n^2 h\pi i}{k}}$$

$$= 1 + 2 \sum_{j=1}^{k} \sum_{l=0}^{\infty} r^{(lk+j)^2} e^{\frac{2(lk+j)^2 h\pi i}{k}}$$

$$= 1 + 2 \sum_{j=1}^{k} e^{\frac{2j^2 h\pi i}{k}} \sum_{l=0}^{\infty} r^{(lk+j)^2}.$$

[41] Mordell (1917b).    [42] Ramanujan (2000) pp. 276–309.    [43] Mordell (1919).

We next approximate the infinite series $\sum_{l=0}^{\infty} r^{(lk+j)^2}$ by an integral so that we can replace the infinite series by a simpler function whose behavior is similar to that of the series as $r \to 1$. If we set $r = e^{-\delta}$ and let $\delta \to 0$, then

$$\sum_{l=0}^{\infty} r^{(lk+j)^2} = \sum_{l=0}^{\infty} e^{-\delta(lk+j)^2} \sim \int_0^{\infty} e^{-\delta(xk+j)^2} \, dx$$

$$\sim \int_0^{\infty} e^{-\delta x^2 k^2} \, dx$$

$$= \frac{1}{2k} \sqrt{\frac{\pi}{\delta}} = \frac{\sqrt{\pi}}{2k} \left( \log \frac{1}{r} \right)^{\frac{1}{2}}. \tag{15.78}$$

Thus, the behavior of $1 + 2q + 2q^9 + \cdots$, $q = re^{\frac{2h\pi i}{k}}$ as $r \to 1$ is similar to that of

$$\frac{\sqrt{\pi}}{k} S_{h,k} \left( \log \frac{1}{r} \right)^{-\frac{1}{2}}, \tag{15.79}$$

where $S_{h,k}$ is the quadratic Gauss sum

$$S_{h,k} = \sum_{j=1}^{k} e^{\frac{2j^2 h \pi i}{k}}. \tag{15.80}$$

Gauss was the first mathematician to calculate the value of the sum (15.80), given in a paper of 1811:[44]

$$S_{h,k} = \left( \frac{h}{k} \right) \frac{1 + i^{-k}}{1 + i^{-1}} \sqrt{k}. \tag{15.81}$$

Gauss's formula shows that when $k$ is of the form $4m + 2$, then $S_{h,k} = 0$. In this case, (15.79) can be interpreted as $o \left( \log \frac{1}{r} \right)^{-\frac{1}{2}}$. Thus,

$$\theta_3^s \sim \pi^{\frac{s}{2}} \left( \frac{S_{h,k}}{k} \right)^s \left( \log \frac{1}{r} \right)^{-\frac{s}{2}}.$$

Hardy next found an alternate infinite series in $q = re^{\frac{2h\pi i}{k}}$ whose behavior was similar to that of $\left( \log \frac{1}{r} \right)^{-\frac{s}{2}}$ as $r \to 1$. For this purpose, he used a formula of Mellin:[45]

$$F_s(q) \equiv \sum_{n=1}^{\infty} n^{\frac{s}{2}-1} q^n = \Gamma \left( \frac{s}{2} \right) \left( \log \frac{1}{r} \right)^{-\frac{s}{2}} + \sum_{k=0}^{\infty} \frac{\zeta(1 - \frac{s}{2} - k)(\log r)^k}{k!}, \tag{15.82}$$

where $|\log r| < 2\pi$. This implies that

$$F_s(q) - \Gamma \left( \frac{s}{2} \right) \left( \log \frac{1}{r} \right)^{-\frac{s}{2}}$$

---

[44] Gauss (1811). See also Roy (2011) pp. 611–615 and 422–424.
[45] See Lindelöf (1905) p. 139, where the formula is proved with a reference to the original paper of Mellin.

is analytic at $r = 1$. So if we set

$$f_{h,k}(q) = \frac{\pi^{\frac{s}{2}}}{\Gamma\left(\frac{s}{2}\right)} \left(\frac{S_{h,k}}{h}\right)^s F_s\left(qe^{-\frac{2h\pi i}{k}}\right), \tag{15.83}$$

then the behavior of $f_{h,k}(q)$ is similar to that of $\theta_3^s$ as $q \to e^{\frac{2h\pi i}{k}}$. Therefore, the series

$$\Theta_s(q) = 1 + \sum_{h,k} f_{h,k}(q) \tag{15.84}$$

behaves in a manner similar to $\theta_3^s$ near all points $e^{\frac{2h\pi i}{k}}$ on the unit circle. We may be justified therefore in expecting $\Theta_s(q)$ to behave so similarly to $\theta_3^s$ that the coefficient of $q^n$ in $\Theta_s(q)$ gives us a close approximation of $r_s(n)$, the number of representations of $n$ as a sum of $s$ squares. To determine the coefficient of $q^n$ in $\Theta_s(q)$, note that

$$\Theta_s(q) = 1 + \frac{\pi^{\frac{s}{2}}}{\Gamma\left(\frac{s}{2}\right)} \sum_{h,k} \left(\frac{S_{h,k}}{k}\right)^s \sum_{n=1}^{\infty} n^{\frac{s}{2}-1} e^{-\frac{2nh\pi i}{k}} q^n \tag{15.85}$$

$$= 1 + \sum_{n=1}^{\infty} \rho_s(n) q^n, \tag{15.86}$$

where the coefficient of $\rho_s(n)$ of $q^n$ can be written as

$$\rho_s(n) = \frac{\pi^{\frac{s}{2}}}{\Gamma\left(\frac{s}{2}\right)} n^{\frac{s}{2}-1} \sum_{k=1}^{\infty} A_k(n), \tag{15.87}$$

where $A_1(n) = 1$ and

$$A_k(n) = k^{-s} \sum_{h=1}^{k-1} S_{h,k}^s e^{-\frac{2nh\pi i}{k}}, \quad k > 1, \ (h,k) = 1. \tag{15.88}$$

Hardy named the series defined by (15.87) and (15.88) the singular series. In a paper published in 1920,[46] he showed that $r_s(n) = \rho_s(n)$ for $5 \le s \le 8$. We note that P. T. Bateman[47] proved in 1951 that this equation holds also when $s = 3$ or $s = 4$. We give a brief indication of Hardy's proof for the case $s = 5$, since it was after seeing this proof that Mordell was able to construct his function $\chi$ for the case in which $s$ was an odd integer.

Hardy employed a special case of a formula of Lipschitz to expand $\Theta_5(q)$ in partial fractions so that he could show that $\frac{\Theta_5(q)}{\theta_3^5(\omega)}$ was a modular function for $\Gamma_3$ that was bounded in the fundamental region. This implied that $\Theta_5(q) = \theta_3^5(\omega)$ and hence that $r_5(n) = \rho_5(n)$.

Lipschitz's formula: Suppose $\omega \in \mathbb{H}$, the upper half-plane, and $\sigma > 1$. Then

$$\frac{\pi^{\frac{s}{2}}}{\Gamma\left(\frac{s}{2}\right)} \sum_{m=1}^{\infty} m^{\frac{s}{2}-1} e^{m\pi\omega i} = \sum_{n=-\infty}^{\infty} \frac{1}{\left((2n-\omega)i\right)^{\frac{s}{2}}}, \tag{15.89}$$

---

[46] Hardy (1966–1979) vol. 1, pp. 345–374.   [47] Bateman (1951).

where

$$((2n - \omega)i)^{\frac{s}{2}} = \exp\left(\frac{s}{2} \log |(2n - \omega)i| + \frac{s}{2}\phi i\right), \quad -\frac{\pi}{2} < \phi < \frac{\pi}{2}.$$

Note that the left-hand side of (15.89) is

$$\frac{\pi^{\frac{s}{2}}}{\Gamma\left(\frac{s}{2}\right)} F_s(q).$$

Now by (15.81) and with $S_{h,k}^4 = \eta_k k^2$, where $\eta_k$ is $1, 0,$ or $-4$ according as $k$ is odd, oddly even (two times an odd number), or evenly even (two times an even number) and taking n = 5 in (15.85), it follows that

$$\Theta_5(q) = 1 + \frac{4\pi^2}{3} \sum \left(\frac{S_{h,k}}{k}\right)^5 F_5\left(qe^{-\frac{2h\pi i}{k}}\right)$$

$$= 1 + \frac{4\pi^2}{3} \sum \frac{\eta_k}{k^3} S_{h,k} F_5\left(qe^{-\frac{2h\pi i}{k}}\right).$$

Taking $s = 5$ in (15.89), we obtain after simplification

$$\Theta_5(q) = 1 + \sum \frac{\eta_k}{\sqrt{k}} \frac{S_{h,k}}{((2h - k\omega)i)^{\frac{5}{2}}}, \tag{15.90}$$

where the sum is taken over all $k = 1, 2, 3, \ldots$ and $h$ takes all values that are relatively prime to $k$. Now set

$$T_{h,k} = \sum_{j=0}^{k-1} e^{\frac{j^2 h\pi i}{k}}, \quad (h, k) = 1.$$

A calculation shows that (15.90) can be written as

$$\Theta_5(q) = 1 + \sum \frac{(-1)^h}{\sqrt{k}} \frac{T_{h,k}}{((h - k\omega)i)^{\frac{5}{2}}}, \tag{15.91}$$

where the sum is now over all $k = 1, 2, 3, \ldots$ and $h$ takes all values of opposite parity and prime to $k$. Now multiply both sides of (15.91) by

$$\frac{\pi^2}{8} = \frac{1}{1^2} + \frac{1}{3^2} + \frac{1}{5^2} + \cdots$$

to get

$$\chi(\omega) \equiv \frac{\pi^2}{8} \Theta_5(q) = \frac{\pi^2}{8} + \sum \frac{(-1)^h}{\sqrt{k}} \frac{T_{h,k}}{((h - k\omega)i)^{\frac{5}{2}}}, \tag{15.92}$$

where $h$ takes all values of opposite parity to $k$. Hardy then proved that $\chi(\omega)$, defined by (15.92), satisfied the relations

$$\chi(\omega + 2) = \chi(\omega) \quad \text{and} \quad \chi\left(-\frac{1}{\omega}\right) = \left(\frac{\omega}{i}\right)^{\frac{5}{2}} \chi(\omega). \tag{15.93}$$

Proving the second equation in (15.93) required a careful analysis because of the presence of the many-valued functions within the summation sign in (15.92). Hardy next showed that $\frac{\chi(\omega)}{\theta_3^5}$ was bounded in the fundamental region for $\Gamma_3$. This gave him the result that $r_5(n) = \rho_5(n)$. Hardy had to take one final step to reach his result: He had to show that his series for $\rho_5(n)$ was equal to the formula, (15.94), given by Smith and Eisenstein, for the number of primitive representations of $n$ as a sum of five squares. Hardy gave an original proof of this in about five pages of intricate calculations that did not employ the theory of quadratic forms.

Now in a one-page paper of 1847,[48] Eisenstein had stated without proof his results for the number of representations of a number, containing no square factor, as a sum of five squares. No doubt he had a proof by means of his theory of quadratic forms in many variables, but he did not find the time to publish it. Later, H. J. S. Smith gave a more general result on the number of primitive representations of any positive integer as a sum of five squares. Note that a representation

$$n = x_1^2 + x_2^2 + x_3^2 + x_4^2 + x_5^2$$

is designated primitive if $x_1, x_2, x_3, x_4, x_5$ do not have a common factor other than 1. In a paper of 1867,[49] Smith explicitly stated the results for the number of primitive representations of a number as a sum of five and seven squares. He applied his theory of quadratic forms in many indeterminates, but he did not provide detailed proofs. Then in 1882, much to Smith's annoyance, the French Academy announced as a prize problem the decomposition of a number as a sum of five squares. Smith's correspondence with Hermite shows that the Academy was unaware of Smith's 1867 work, but when Smith wrote up detailed demonstrations for his results,[50] the French Academy awarded him the prize, jointly with the eighteen-year-old Hermann Minkowski.[51]

The result proved by Smith and Minkowski was that the number of primitive representations of $n$ as a sum of five squares was given by the series

$$\frac{Cn^{\frac{3}{2}}}{\pi^2} \sum_{\substack{m \text{ odd} \\ (m,n)=1}} \left(\frac{n}{m}\right) \frac{1}{m^2}, \tag{15.94}$$

where

$$C = 80 \ (n \equiv 0, 1, 4), \quad C = 160 \ (n \equiv 2, 3, 6, 7), \quad C = 112 \ (n \equiv 5),$$

in which the congruences were to modulus 8.

Smith gave a theorem for seven squares for odd integers $n$ that are products of distinct primes. For such integers $n$, the number of primitive representations equals the total number of representations. Thus

$$r_7(n) = \frac{C_1 n^{\frac{5}{2}}}{\pi^3} \sum_{\substack{m \text{ odd} \\ (m,n)=1}} \left(-\frac{n}{m}\right) \frac{1}{m^2}, \tag{15.95}$$

[48] Eisenstein (1975) vol. 2, p. 505.    [49] Smith (1965) vol. 1, pp. 510–523, especially pp. 521–522.
[50] Ibid. vol. 2, pp. 623–680.    [51] Minkowski (1967) vol. 1, pp. 3–144.

where $C_1$ is valued at 448, 560, 448, or 592 according as $n$ is congruent to $1, 3, 5, 7$ (mod 8), respectively. The series (15.94) and (15.95) can be summed in finite terms by methods given by Dirichlet in 1839[52] and by Cauchy in 1840. Bachmann[53] provides details on these methods.

After studying Hardy's work, Mordell discovered the expression for $\chi$:

$$\chi = \sum \left( \frac{\theta_3(\omega)}{\theta_3\left(\frac{a\omega+b}{c\omega+d}\right)} \right)^s,$$

where the summation is over all transformations

$$\begin{pmatrix} a & b \\ c & d \end{pmatrix} \in \Gamma_3,$$

except that for a given pair of values for $c$ and $d$, we take only one set of values for $a$ and $b$, since it follows from Hermite's work on the transformation of theta functions that the general term of the series is independent of the values of $a$ and $b$. In addition, we need not consider negative values of $c$, and when $c = 0$, we take $a = d = 1$ so that each term of the series occurs only once. It can be seen that $\frac{\chi}{\theta_3^s}$ is invariant under the action of $\Gamma_3$. This need be checked for only the two transformations $\omega \to \omega + 2$ and $\omega \to \omega - \frac{1}{\omega}$. Hermite's formulas (5.7) through (5.9) imply that

$$\chi = 1 + \sum_{c,d} \left( \frac{e^{-\frac{\pi i c^2\, da}{4}} H_{c,d}}{\sqrt{-ic(c\omega + d)}} \right)^s,$$

where

$$H_{c,d} = \sum_{k=0}^{c-1} e^{-\pi i \frac{d}{c}(k-\frac{1}{2}c)^2}.$$

The summation in $\chi$ is over all coprime $c$ and $d$ of opposite parity, except that $c$ is always positive. For details on the application of $\chi$ to the problem of finding the number of representations of an integer as a sum of $5, 7, 9, 11, 13$ squares, one might wish to study Mordell's paper.[54] To tackle these cases, Mordell proved the very interesting formula

$$\theta_3^s = \chi + C_s\, \theta_3^{s-8} \theta_0^4\, \theta_2^4,$$

where

$$C_5 = 0, \ C_7 = 0, \ C_9 = \frac{2}{17}, \ C_{11} = \frac{22}{31}, \ C_{13} = \frac{871}{691}, \ C_{15} = \frac{861}{308}.$$

## 15.7   Hecke's Solution to the Sums of Squares Problem

Hecke's results[55] on the number of representations of an integer as a sum of $2k$ squares were based on his discovery of the correspondence between Dirichlet series of signature

---

[52] Dirichlet (1969) vol. 1, pp. 411–497.   [53] Bachmann (1923) vol. 4, pp. 600–625.
[54] Mordell (1919).   [55] Hecke (1983) pp. 62–66.

$(2, k, \pm 2)$ and functions of signature $(2, k, \pm 2)$, that is, functions $f$ analytic in the upper half-plane and satisfying

$$1. \ f(\tau) = a_0 + \sum_{n=1}^{\infty} a_n e^{\pi i n \tau} \qquad 2. \ f\left(-\frac{1}{\tau}\right) = \gamma(-i\tau)^k f(\tau)$$

$$3. \ f(x + iy) = O(y^{-c}) \text{ as } y \to 0^+ \text{ uniformly for all real } x.$$

As we mentioned in Chapter 14, Hecke proved[56] that the number of linearly independent Dirichlet series with signature $(2, k, \gamma)$ was

$$\left\lfloor \frac{k}{4} \right\rfloor + 1 \qquad \text{if } \gamma = 1, \tag{15.96}$$

$$\left\lfloor \frac{k-1}{4} \right\rfloor + 1 \quad \text{if } \gamma = -1. \tag{15.97}$$

Note that by (15.65)

$$\theta_3^{2k}(\tau) = \sum_{n=0}^{\infty} r_{2k}(n) e^{\pi i n \tau}$$

is a function of signature $(2, k, 1)$ and the corresponding Dirichlet series is

$$\sum_{n=1}^{\infty} \frac{r_{2k}(n)}{n^s}.$$

Let $A = \left\lfloor \frac{k}{4} \right\rfloor + 1$ and suppose $\phi_1(s), \phi_2(s), \ldots, \phi_A(s)$ are independent Dirichlet series of signature $(2, k, 1)$. Then there exist constants $c_1, c_2, \ldots, c_A$ such that

$$\sum_{n=1}^{\infty} \frac{r_{2k}(n)}{n^s} = \sum_{j=1}^{A} c_j \phi_j(s). \tag{15.98}$$

If $2k = 2, 4$ or $6$, then $A = 1$ and we must find only one Dirichlet series for each of these values of $k$. If $2k = 8, 10, 12$ or $14$, then $A = 2$; and so on.

We now turn to an explanation of Hecke's method for finding independent Dirichlet series, since they are more convenient than modular functions for finding the number of representations of an integer as sums of squares. Of course, Glaisher would have required the coefficients of the Dirichlet series to be multiplicative. However, some of the Dirichlet series that we have to work with do not have multiplicative coefficients. Still, we would like to choose Dirichlet series whose properties are as arithmetical as possible.

---

[56] Hecke (1983) p. 62.

We will see in Chapter 16 that Hecke initially developed his theory of Euler products for the Dirichlet series of signature $(1, k, (-1)^{\frac{k}{2}})$, with $k \geq 4$ an even integer; these are the Dirichlet series corresponding to the modular forms of weight $k$ for the full modular group.

Now a modular form of weight $k \geq 3$ is nonzero if and only if $k$ is even, for when $k$ is odd, a modular form $f$ of weight $k$ satisfies

$$f(\omega) = f\left(\frac{-\omega}{-1}\right) = (-1)^k f(\omega) = -f(\omega)$$

and thus

$$f(\omega) \equiv 0;$$

when $k$ is even and greater than 2, there exists an Eisenstein series of weight $k$. Since these functions are of period 1, we get $\lambda = 1$. It follows from Hecke's theorem, stated in Chapter 14, that a Dirichlet series of signature $(1, k, \gamma)$ exists if and only if: $k$ is even, $k \geq 4$, and $\gamma = (-1)^{\frac{k}{2}}$.

Recall from (14.32) and (14.33) that the dimension of the space of modular forms of even weight $k > 2$ is

$$\left\lfloor \frac{k}{12} \right\rfloor + 1, \quad \text{if } k \not\equiv 2 \ (\text{mod } 12)$$

$$\left\lfloor \frac{k}{12} \right\rfloor, \qquad \text{if } k \equiv 2 \ (\text{mod } 12).$$

So for even $k > 2$, there exist $\left\lfloor \frac{k}{12} \right\rfloor + 1$ independent Dirichlet series of signature $(1, k, \gamma)$ when $k \not\equiv 2 \ (\text{mod } 12)$ and $\left\lfloor \frac{k}{12} \right\rfloor$ such series when $k \equiv 2 \ (\text{mod } 12)$. The Dirichlet series corresponding to the Eisenstein series $G_k(\omega)$, as in (4.173), is

$$\sum_{n=1}^{\infty} \frac{\sigma_{k-1}(n)}{n^s} = \sum_{n=1}^{\infty} \frac{\sum_{d|n} d^{k-1}}{n^s} = \sum_{n=1}^{\infty} \frac{1}{n^s} \sum_{de=n} d^{k-1}$$

$$= \sum_{e=1}^{\infty} \frac{1}{e^s} \sum_{d=1}^{\infty} \frac{d^{k-1}}{d^s} = \zeta(s)\zeta(s - k + 1), \quad \operatorname{Re} s > k.$$

Hence, $\phi(s) = \zeta(s)\zeta(s - k + 1)$ has signature $(1, k, (-1)^{\frac{k}{2}})$ when $k \geq 4$ is an even integer. Note that $\phi(s)$ has a pole of order 1 at $s = k$ and no other poles; the pole of $\zeta(s)$ at $s = 1$ cancels with $\zeta(2 - k) = 0$ because $k \geq 4$ is even. That $\phi(s)$ has signature $(1, k, (-1)^{\frac{k}{2}})$ can be proved directly from the functional equation of $\zeta(s)$:

$$\pi^{-\frac{s}{2}} \Gamma\left(\frac{s}{2}\right) \zeta(s) = \pi^{-\frac{1-s}{2}} \Gamma\left(\frac{1-s}{2}\right) \zeta(1 - s).$$

Moreover, $\phi(s)$ has an Euler product: For Re $s > k > 1$,

$$\sum_{n=1}^{\infty} \frac{\sigma_{k-1}(n)}{n^s} = \zeta(s)\zeta(s-k+1)$$

$$= \prod_{p}(1-p^{-s})^{-1}\prod_{p}(1-p^{-s+k-1})^{-1}$$

$$= \prod_{p}(1-(1+p^{k-1})p^{-s}+p^{k-1-2s})^{-1}$$

$$= \prod_{p}(1-\sigma_{k-1}(p)\,p^{-s}+p^{k+1-2s})^{-1}.$$

We note that, unaware of Glaisher's work, Hecke applied his theory to, in essence, derive Glaisher's results up to sums of 12 squares. Recall that Glaisher himself had worked out the additional results on sums of 14, 16, and 18 squares; we will present these, as well as Ramanujan's result for 24 squares, by applying Hecke's method.

The Dirichlet series $\sum \frac{r_4(n)}{n^s}$ has signature $(2, 2, 1)$, so we would like to find a more familiar series with the same signature. Recall that $\phi(s) = \zeta(s)\zeta(s-k+1)$ has signature $(1, k, (-1)^{\frac{k}{2}})$ when $k$ is even and $k \geq 4$. Now the series with $\phi(s) = 2^{-s}\zeta(s)\zeta(s-k+1)$ has signature $(2, k, (-1)^{\frac{k}{2}})$. We are unable to take $k = 2$ because $\zeta(s)\zeta(s-1)$ has a pole $s = 2$ as well as a pole $s = 1$, counter to our requirements. However, if we multiply $\zeta(s)\zeta(s-1)$ by $2^s - 2^{2-s}$, then the zero of this factor at $s = 1$ cancels the pole at $s = 1$. Moreover, $2^s - 2^{2-s}$ changes sign when $s$ is changed to $2 - s$; thus, by (14.12), the series

$$\phi(s) = 2^{-s}(2^s - 2^{2-s})\zeta(s)\zeta(s-1)$$

has signature $(2, 2, 1)$ and may be used for our purpose. Since a Dirichlet series of signature $(2, 2, 1)$ is unique up to a constant factor, we have

$$\sum_{n=1}^{\infty} \frac{r_4(n)}{n^s} = c\left(1 - \frac{4}{4^s}\right)\sum_{n=1}^{\infty}\frac{\sigma_1(n)}{n^s}$$

$$= c\sum_{n=1}^{\infty}\left(\frac{\sigma_1(n)}{n^s} - \frac{4\sigma_1(n)}{(4n)^s}\right).$$

Note that here $r_4(1) = 8$, since

$$1 = (\pm 1)^2 + 0^2 + 0^2 + 0^2 = 0^2 + (\pm 1)^2 + 0^2 + 0^2$$

$$= 0^2 + 0^2 + (\pm 1)^2 + 0^2 = 0^2 + 0^2 + 0^2 + (\pm 1)^2.$$

Thus, if $n_o$ denotes an odd integer, then

$$r_4(n_o) = 8\sigma_1(n_o), \tag{15.99}$$

$$r_4(2n_o) = 8\sigma_1(2n_o), \tag{15.100}$$

$$r_4(2^k n_o) = 8\big(\sigma_1(2^k n_o) - 4\sigma_1(2^{k-2}n_o)\big), \quad \text{for } k \geq 2. \tag{15.101}$$

Observe that

$$\sigma_1(2^k n_o) = (1 + 2 + 2^2 + \cdots + 2^k)\sigma_1(n_o)$$

and

$$4\sigma_1(2^{k-2} n_o) = (2^2 + \cdots + 2^k)\sigma_1(n_o).$$

Thus, with $\sigma_1^o(n)$ denoting the sum of the odd divisors of $n$, (15.99) through (15.101) may be written as

$$r_4(n) = 8\sigma_1(n) \quad \text{when } n \text{ is odd}$$

and

$$r_4(n) = 24\sigma_1^o(n) \quad \text{when } n \text{ is even.}$$

Hecke found Dirichlet series of signature $(2, 4, 1)$ in order to obtain the formula for $r_8(n)$. As (15.96) shows, there are two independent series with this signature. Clearly, the first can be taken to be $2^{-s}\zeta(s)\zeta(s-3)$. Next, if we multiply this function by $2^s + 2^{4-s}$, we get a second series with signature $(2, 4, 1)$. The two Dirichlet series are thus

$$2^{-s}\zeta(s)\zeta(s-3) \quad \text{and} \quad (1 + 4^{2-s})\zeta(s)\zeta(s-3).$$

The linear combination of these two series is then

$$\sum_{n=1}^{\infty} \frac{r_8(n)}{n^s} = \left(c_1 2^{-s} + c_2(1 + 4^{2-s})\right)\zeta(s)\zeta(s-3)$$

$$= \left(c_1 2^{-s} + c_2(1 + 4^{2-s})\right) \sum_{n=1}^{\infty} \frac{\sigma_3(n)}{n^s}.$$

Since $r_8(1) = 16$ and $r_8(2) = 112$, we have the equations

$$c_2 = r_8(1) = 16, \quad c_1 + \sigma_3(2)c_2 = r_8(2) \quad \text{or} \quad c_1 = -32.$$

Thus, we obtain the formula

$$\sum_{n=1}^{\infty} \frac{r_8(n)}{n^s} = 16 \sum_{n=1}^{\infty} \frac{\sigma_3(n)}{n^s} - 32 \sum_{n=1}^{\infty} \frac{\sigma_3(n)}{(2n)^s} + 256 \sum_{n=1}^{\infty} \frac{\sigma_3(n)}{(4n)^s}. \tag{15.102}$$

If we write $n = 2^k n_o$, with $n_o$ odd, then from (15.102) we get

$$r_8(2^k n_o) = 16\left(\sigma_3(2^k n_o) - 2\sigma_3(2^{k-1} n_o) + 16\sigma_3(2^{k-2} n_o)\right), \tag{15.103}$$

where $\sigma_3\left(\frac{n_o}{2}\right) \equiv 0$ and $\sigma_3\left(\frac{n_o}{4}\right) \equiv 0$. Formula (15.103) is easily seen to be equivalent to the statement:

$r_8(n)$ equals 16 times the sum of the cubes of the divisors of $n$, when $n$ is odd, and equals 16 times the amount by which the sum of the cubes of the even divisors of $n$ exceeds sum of the cubes of the odd divisors of $n$, when $n$ is even.

Formula (15.103) can also be written as

$$r_8(n) = (-1)^n 16 \sum_{d|n} (-1)^d d^3.$$

Again by (15.96), there are two independent Dirichlet series of signature $(2, 6, 1)$. Observe that $2^{-s}\zeta(s)\zeta(s-5)$ has signature $(2, 6, -1)$, but after multiplication by $2^s - 2^{6-s}$, yielding

$$(1 - 2^{6-2s})\zeta(s)\zeta(s-5),$$

its signature becomes $(2, 6, 1)$. As Hecke noted, the other series arises from the modular form

$$\sqrt{\Delta(\omega)} = (2\pi)^6 q \prod_{n=1}^{\infty} (1 - q^{2n})^{12}, \quad q = e^{i\pi\omega}.$$

Hecke verified that $\sqrt{\Delta(\omega)}$ was of type (or signature) $(2, 6, 1)$ by observing that

$$\frac{\sqrt{\Delta(-\frac{1}{\omega})}}{(-i\omega)^6} = \sqrt{\Delta(\omega)};$$

this follows from the fact that $\Delta(\omega)$ is a modular form of weight 12. So if we write

$$\sqrt{\Delta(\omega)} = (2\pi)^6 \sum_{n=1}^{\infty} \Omega(n)q^n, \tag{15.104}$$

then the second series we are seeking turns out to be

$$\sum_{n=1}^{\infty} \frac{\Omega(n)}{n^s}, \quad \text{with } \Omega(1) = 1.$$

Thus,

$$\sum_{n=1}^{\infty} \frac{r_{12}(n)}{n^s} = c_1(1 - 2^{6-2s}) \sum_{n-1}^{\infty} \frac{\sigma_5(n)}{n^s} + c_2 \sum_{n=1}^{\infty} \frac{\Omega(n)}{n^s}.$$

This leads to the two equations

$$r_{12}(1) = 24 = c_1 + c_2$$
$$r_{12}(2) = 264 = c_1\sigma_5(2) + c_2\Omega(2) = 33c_1, \quad \text{since } \Omega(2) = 0.$$

Thus, we obtain

$$\sum_{n=1}^{\infty} \frac{r_{12}(n)}{n^s} = 8\left(1 - \frac{64}{4^s}\right) \sum_{n=1}^{\infty} \frac{\sigma_5(n)}{n^s} + 16 \sum_{n=1}^{\infty} \frac{\Omega(n)}{n^s}. \tag{15.105}$$

Let $n_0$ denote an odd positive integer. Then (15.105) implies that

$$r_{12}(n_0) = 8\sigma_5(n_0) + 16\Omega(n_0),$$
$$r_{12}(2n_0) = 8\sigma_5(2n_0),$$
$$r_{12}(4^k n_0) = 8\sigma_5(4^k n_0) - 8^3\sigma_5(4^{k-1}n_0), \quad k \geq 1.$$

Note that Glaisher[57] wrote the last two formulas more succinctly as

$$r_{12}(2n) = 8 \sum (-1)^{d+d'-1} d^5,$$

where $n$ was any positive integer and $d$ and $d'$ were positive integers such that $dd' = 2n$. Glaisher concluded, though he commented that he found it remarkable, that the number of representations of an even integer as a sum of twelve squares could be expressed completely in terms of the real divisors of that integer.

Note that, while Hecke thought in terms of the modular form $\Delta(\omega)$, Glaisher worked in terms of the Jacobi theory and considered $k^2 k'^2 \theta_3^{12}$. Now taking Glaisher's starting point and applying Hecke's method, we note that since

$$k^2 \left( -\frac{1}{\omega} \right) = k'^2(\omega) \quad \text{and} \quad k^2(\omega + 2) = k^2(\omega),$$

it follows that

$$(k^2 k'^2) \left( -\frac{1}{\omega} \right) = (k^2 k'^2)(\omega) \quad \text{and} \quad (k^2 k'^2)(\omega + 2) = (k^2 k'^2)(\omega).$$

This tells us that $k^2 k'^2 \theta_3^{12}$ is of type (or signature) $(2, 6, 1)$, the weight of $k^2 k'^2$ being zero. In fact, it is easy to check that

$$k^2 k'^2 \theta_3^{12} = 16q \prod_{n=1}^{\infty} (1 - q^{2n})^{12}$$

$$= 16 \sum_{n=1}^{\infty} \Omega(n) q^n, \tag{15.106}$$

the same function used by Hecke (15.104). Recall that Glaisher sought to use series whose coefficients had multiplicative properties. He established[58] that for $m$ and $n$ odd and relatively prime

$$\Omega(mn) = \Omega(m)\Omega(n) \tag{15.107}$$

by showing that $\Omega(n)$ was equal to one-eighth of the sum of the fourth powers of the integer quaternion divisors of $n$. It can be seen that (15.106) implies that $\Omega(2n) = 0$. By contrast, Hecke's method was to prove (15.107) using Euler products. Recall that Mordell had already proved (15.107) by showing that (15.106) had an Euler product.

There are three independent Dirichlet series of signature $(2, 8, 1)$. Two of these series are given by

$$2^{-s} \zeta(s) \zeta(s - 7) \quad \text{and} \quad (1 + 2^{8-2s}) \zeta(s) \zeta(s - 7);$$

the third series can be taken as the Dirichlet series corresponding to the function

$$(2\pi)^{-6} \sqrt{\Delta(\omega)} \, \theta_3^4(\omega) = q \prod_{n=1}^{\infty} (1 - q^{2n})^{16} (1 + q^{2n-1})^8$$

$$= \sum_{n=1}^{\infty} (-1)^{n-1} \Theta(n) q^n. \tag{15.108}$$

---

[57] Glaisher (1907a) p. 484.      [58] Glaisher (1908) p. 299.

Note that that (15.108) is the same function as Glaisher's $\frac{1}{16}k^2k'^2\theta_3^{16}$ and is of type (2, 8, 1). Glaisher found by numerical calculation of particular cases that if $m$ and $n$ were relatively prime odd integers, then

$$\Theta(mn) = \Theta(m)\Theta(n).$$

Thus, the third Dirichlet series of signature (2, 8, 1) can be taken to be

$$\sum_{n=1}^{\infty} \frac{\Theta(n)}{n^s}.$$

Glaisher showed that

$$r_{16}(n) = (-1)^{n-1}\frac{32}{17}(\zeta_7(n) + 16\Theta(n)), \tag{15.109}$$

where $\zeta_7(n)$ denotes the sum of the seventh powers of the odd divisors of $n$ diminished by (minus) the sum of the seventh powers of the even divisors of $n$.

In order to derive formulas for $r_2(n)$, $r_6(n)$, and $r_{10}(n)$ by obtaining Dirichlet series of signatures (2, 1, 1), (2, 3, 1), and (2, 5, 1), Hecke needed an $L$-function. This $L$-function can be defined by the series

$$L(s) = 1 - \frac{1}{3^s} + \frac{1}{5^s} - \frac{1}{7^s} + \frac{1}{9^s} - \frac{1}{11^s} + \cdots = \sum_{n=1}^{\infty} \frac{\chi(n)}{n^s}, \tag{15.110}$$

where

$$\chi(n) = \begin{cases} (-1)^{\frac{n-1}{2}}, & n \text{ odd}, \\ 0 & , n \text{ even}. \end{cases}$$

Recall that $L(1) = \frac{\pi}{4}$ and that $L(s)$ satisfies the functional equation

$$\left(\frac{\pi}{4}\right)^{-\frac{s}{2}}\Gamma\left(\frac{s+1}{2}\right)L(s) = \left(\frac{\pi}{4}\right)^{-\frac{1-s}{2}}\Gamma\left(1 - \frac{s}{2}\right)L(1 - s). \tag{15.111}$$

Using (15.111), it can be checked that $\zeta(s)L(s)$ has signature (2, 1, 1). Hence

$$\sum_{n=1}^{\infty} \frac{r_2(n)}{n^s} = 4\zeta(s)L(s) = 4\prod_p \left(1 - \frac{1}{p^s}\right)^{-1} \prod_{p_1,p_2} \left(1 - \frac{1}{p_1^s}\right)^{-1} \left(1 + \frac{1}{p_3^s}\right)^{-1}, \tag{15.112}$$

where $p$ denotes any prime; $p_1$ and $p_3$ denote primes congruent to 1 and 3 modulo 4, respectively. Formula (15.112) can be rewritten as

$$\sum_{n=1}^{\infty} \frac{r_2(n)}{n^s}$$

$$= 4\left(1 - \frac{1}{2^s}\right)^{-1} \prod_{p_1} \left(1 - \frac{1}{p_1^s}\right)^{-2} \prod_{p_3} \left(1 - \frac{1}{p_3^{2s}}\right)^{-1}$$

$$= 4\left(\frac{1}{2^s} + \frac{1}{2^{2s}} + \cdots\right) \prod_{p_1} \left(1 + \frac{2}{p_1^s} + \cdots + \frac{k+1}{p_1^{ks}} + \cdots\right)$$

$$\cdot \prod_{p_3} \left(1 + \frac{1}{p_3^{2s}} + \frac{1}{p_3^{4s}} + \cdots\right). \tag{15.113}$$

Observe that the coefficient $k+1$ in $\frac{k+1}{p_1^{ks}}$ indicates the number of divisors of the $p_1^k$. Now let $n = 2^l n_1 n_3$, where $n_1$ is the product of primes of the form $4m+1$, while $n_3$ is the product of primes of the form $4m+3$. By (15.113), if $n_3$ is not a square, then $r_2(n) = 0$; otherwise, $r_2(n) = 4d(n_1)$, where $d(n_1)$ denotes the number of divisors of $n_1$. Note that (15.112) can be more clearly expressed as an Euler product: For Re $s > 1$,

$$\zeta(s)L(s) = \prod_{p} (1 - p^{-s})^{-1} (1 - \chi(p)p^{-s})^{-1}$$

$$= \prod_{p} \left(1 - (1 + \chi(p))p^{-s} + \chi(p)p^{-2s}\right)^{-1}$$

$$= \prod_{p} \left(1 - \frac{r_2(p)}{2}p^{-s} + \chi(p)p^{-2s}\right)^{-1}.$$

To find the unique Dirichlet series (up to a constant multiplier) of signature $(2, 3, 1)$, one considers

$$\phi(s) = c_1 \zeta(s)L(s-2) + c_2 \zeta(s-2)L(s).$$

Note that $c_1 c_2 \neq 0$, because of the manner in which $\zeta(s)L(s-2)$ and $\zeta(s-2)L(s)$ individually transform. However, the functional equations for $\zeta(s)$ and $L(s)$ produce

$$\pi^{-s} \Gamma(s) \zeta(s) L(s-2) = -4\pi^{-(3-s)} \Gamma(3-s) \zeta(1-s) L(3-s)$$

$$-4\pi^{-s} \Gamma(s) L(s) \zeta(s-2) = \pi^{-(3-s)} \Gamma(3-s) L(1-s) \zeta(3-s).$$

Hecke therefore took $c_1 = 1$ and $c_2 = -4$ to obtain

$$\phi(s) = \zeta(s)L(s-2) - 4\zeta(s-2)L(s)$$

and

$$\sum_{n-1}^{\infty} \frac{r_6(n)}{n^s} = 4\left(4 \sum_{n=1}^{\infty} \frac{n^2}{n^s} \sum_{m=1}^{\infty} \frac{\chi(m)}{m^s} - \sum_{n=1}^{\infty} \frac{1}{n^s} \sum_{m=1}^{\infty} \frac{m^2 \chi(m)}{m^s}\right)$$

$$= 4\left(4 \sum_{n=1}^{\infty} \frac{\sum_{dd'=n} d^2 \chi(d')}{n^s} - \sum_{n=1}^{\infty} \frac{\sum_{d|n} d^2 \chi(d)}{n^s}\right);$$

hence

$$r_6(n) = 4 \left( 4 \sum_{dd'=n} d'^2 \chi(d) - \sum_{d|n} d^2 \chi(d) \right). \tag{15.114}$$

Observe that since $\chi(d) = (-1)^{\frac{d-1}{2}}$ when $d$ is odd, (15.114) can also be expressed in the form given by H. J. S. Smith:

$$r_6(n) = 4 \sum_{\delta} (-1)^{\frac{\delta-1}{2}} (4\delta'^2 - \delta^2),$$

where $\delta$ is an odd divisor of $n$ and $\delta'$ is a conjugate divisor, that is, $\delta\delta' = n$.

Again from (15.96), we see that there are two independent Dirichlet series of signature $(2, 5, 1)$ and one of them is

$$\zeta(s)L(s-4) + 4^2\zeta(s-4)L(s). \tag{15.115}$$

Hecke took the second series to be

$$\sum_{\mu} \frac{\mu^4}{|\mu|^{2s}} = \sum_{n=1}^{\infty} \frac{\sum_{a^2+b^2=n}(a+bi)^4}{n^s}, \tag{15.116}$$

where $\mu$ is an integer $a + bi$ in $Q(i)$. Note that (15.116) was defined in (14.34) and was shown to satisfy (14.35), proving that it was a series of signature $(2, 5, 1)$. Therefore

$$\sum_{n=1}^{\infty} \frac{r_{10}(n)}{n^s} = c_1 \left( 16 \sum_{n=1}^{\infty} \frac{\sum_{dd'=n} d'^4 \chi(d)}{n^s} + \sum_{n=1}^{\infty} \frac{\sum_{d|n} d^4 \chi(d)}{n^s} \right)$$

$$+ c_2 \sum_{n=1}^{\infty} \frac{\sum_{a^2+b^2=n}(a+bi)^4}{n^s}.$$

Using the facts that $r_{10}(1) = 20$ and $r_{10}(2) = 180$, we can conclude that $c_1 = \frac{4}{5}$ and $c_2 = \frac{8}{5}$; hence

$$r_{10}(n) = \frac{4}{5} \left( 16 \sum_{dd'=n} d'^4 \chi(d) + \sum_{d|n} d^4 \chi(d) \right) + \frac{8}{5} \sum_{a^2+b^2=n} (a+bi)^4. \tag{15.117}$$

Observe that (15.117) is equivalent to the formula given by Glaisher:[59]

$$r_{10}(n) = \frac{4}{5} \left( \sum_{\delta} (-1)^{\frac{\delta-1}{2}}(\delta^4 + 16\delta'^4) + 2 \sum_{a^2+b^2=n} (a+bi)^4 \right),$$

where $\delta$ is an odd divisor of $n$ and $\delta'$ is a conjugate divisor. To find $r_{10}(n)$, Glaisher considered the function $k^2 k'^2 \theta_3^{10}$ of type $(2, 5, 1)$ and defined $\chi_4(n)$ by

$$(k^2 k'^2 \theta_3^{10})(\omega) = 16 \sum_{n=1}^{\infty} \chi_4(n)q^n.$$

[59] Glaisher (1907a) p. 482.

He had considerable difficulty in proving that $\chi_4(n)$ was multiplicative. He finally succeeded when he took complex divisors of $n$, that is, divisors $t$ of the form $a + ib$ with $a$ and $b$ integers such that $a^2 + b^2 = n$. He was then able to prove[60]

$$\chi_4(n) = \frac{1}{4} \sum t^4,$$

where the sum was taken over all such numbers $t$. Hecke also proved that

$$f(\omega) := \sum_\mu \mu^4 e^{2\pi i |\mu|^2 \omega} = \frac{8}{5} \frac{\sqrt{\Delta(\omega)}}{(2\pi)^6 \theta_3^2(\omega)}, \tag{15.118}$$

where

$$\theta_3(\omega) = \prod_{n=1}^{\infty} (1 - q^{2n})(1 + q^{2n-1})^2, \quad q = e^{\pi i \omega}.$$

Since $f(\omega)$, the function $\frac{\sqrt{\Delta(\omega)}}{\theta_3^2(\omega)}$, and the function corresponding to the Dirichlet series (15.115) each have signature $(2, 5, 1)$, it follows that $f(\omega)$ is a linear combination of the other two functions. Observe that, while $f(i\infty) = 0$, the function corresponding to (15.115) does not vanish at infinity because the series (15.115) has a pole at $s = 6$. Thus, we have confirmed (15.118). We here remark that the function $f(\omega)$, except for a constant factor, is the same as Glaisher's $k^2 k'^2 \theta_3^{10}$.

The case of fourteen squares is similar to the case of ten squares; the number of independent Dirichlet series of signature $(2, 7, 1)$ is again 2. It can be shown that these two independent series are

$$\zeta(s)L(s-6) - 64\zeta(s-6)L(s) \quad \text{and} \tag{15.119}$$

$$\sum_{n=1}^{\infty} \frac{W(n)}{n^s}, \quad \text{where } W(n) \text{ is defined by} \tag{15.120}$$

$$\sum_{n=1}^{\infty} W(n)q^n = \frac{\sqrt{\Delta(\omega)}}{(2\pi)^6} \theta_3^2(\omega) = q \prod_{n=1}^{\infty} (1 - q^{2n})^{14}(1 + q^{2n-1})^4. \tag{15.121}$$

Note that Glaisher wrote the function (15.121) as $k^2 k'^2 \theta_3^{14}$. From series (15.119) and (15.120), we can deduce Glaisher's formula that

$$r_{14}(n) = \frac{4}{61} \left( 64E_6'(n) - E_6(n) + 364W(n) \right), \tag{15.122}$$

where $E_6(n) = \sum_{d|n} d^6 \chi(d)$ and $E_6'(n) = \sum_{dd'=n} d^6 \chi(d')$.

Recall that Glaisher also derived a formula for $r_{18}(n)$;[61] this derivation requires us to find three independent series of signature $(2, 9, 1)$. For this, set

$$\chi_8(n) = \frac{1}{4} \sum_{a^2 + b^2 = n} (a + bi)^8,$$

where $a + bi$ denotes a Gaussian integer divisor of $n$. The three series can then be taken

---

[60] Glaisher (1908) pp. 270–271.  [61] Glaisher (1907a).

to be

$$\zeta(s)L(s-8) + 4^4\zeta(s-8)L(s), \qquad \sum_{n=1}^{\infty} \frac{\chi_8(n)}{n^s}, \qquad (15.123)$$

and the Dirichlet series corresponding to

$$(2\pi)^{-6}\sqrt{\Delta(\omega)}\,\theta_3^6(\omega) = \sum_{n=1}^{\infty} U(n)q^n,$$

that is,

$$\sum_{n=1}^{\infty} \frac{U(n)}{n^s}. \qquad (15.124)$$

On the basis of a conjecture, Glaisher showed[62] that

- if $m_1$ and $m_2$ were both of the form $4k+3$, then

$$U(m_1 m_2) = -\frac{5}{3}U(m_1)U(m_2);$$

- if $m_1$ was of the form $4k+3$ and $m_2$ was of the form $4k+1$ and divisible by a prime factor of the form $4k+3$ not raised to an even power, then

$$U(m_1 m_2) = \frac{13}{5}U(m_1)U(m_2).$$

Glaisher verified that there was no multiplication theorem for $U$ if if either $m_1$ or $m_2$ was of the form $4k+1$ and was not divisible by a factor of the form $4k+3$. This fact led him to consider other functions whose coefficients might have better multiplicative properties. He considered the series

$$(2\pi)^{-6}k^2(\omega)\sqrt{\Delta(\omega)}\,\theta_3^6(\omega) = 16\sum_{n=1}^{\infty} V(n)q^n,$$

and found that $V(n)$ had properties similar to those of $U(n)$. However, Glaisher's conjecture, based upon numerical evidence, was that

$$G(n) = U(n) - 32V(n)$$

had the properties

-

$$G(mn) = G(m)G(n),$$

when $m$ and $n$ were odd integers and were not both of the form $4k+3$

-

$$G(mn) = -\frac{39}{25}G(m)G(n),$$

when both $m$ and $n$ were of the form $4k+3$.

---

[62] Ibid. p. 336.

Thus, Glaisher preferred to state his result for $r_{18}(n)$ in terms of $G(n)$ instead of $U(n)$. To facilitate this, he found the relation between $U(n)$ and $G(n)$:[63]

$$26U(n) = 16\chi_8(n) - 10G(n) - G(2n). \tag{15.125}$$

Now within Hecke's theory, there exist constants $c_1, c_2, c_3$ such that

$$\sum_{n=1}^{\infty} \frac{r_{18}(n)}{n^s}$$

$$= c_1\left(\zeta(s)L(s-8) + 256\zeta(s-8)L(s)\right) + c_2 \sum_{n=1}^{\infty} \frac{\chi_8(n)}{n^s} + c_3 \sum_{n=1}^{\infty} \frac{U(n)}{n^s}$$

$$= c_1 \sum_{n=1}^{\infty} \frac{E_8(n) + 256E_8'(n)}{n^s} + c_2 \sum_{n=1}^{\infty} \frac{\chi_8(n)}{n^s} + c_3 \sum_{n=1}^{\infty} \frac{U(n)}{n^s}, \tag{15.126}$$

where $\quad E_8(n) = \sum_{d|n} d^8 \chi(d) \quad$ and $\quad E_8' = \sum_{dd'=n} d'^8 \chi(d).$

After determining $c_1, c_2, c_3$ by the method we have described, and substituting the value of $U(n)$ in (15.126), Glaisher's formula can be obtained:

$$r_{18}(n) = \frac{4}{18005}\left(13E_8(n) + 3328E_8'(n) + 97504\chi_8(n) - 61200G(n) - 6120G(2n)\right). \tag{15.127}$$

The multiplicative properties of $G(n)$, conjectured by Glaisher, were later confirmed by Rankin.[64] He also obtained a formula for $r_{20}(n)$ in terms of multiplicative functions. See exercise 8 at the end of this chapter.

We finally derive, by Hecke's method, Ramanujan's formula for the number of representations of a number as a sum of twenty-four squares:

$$r_{24}(n) = \frac{16}{691}\sigma_{11}^*(n) + \frac{128}{691}\left((-1)^{n-1}259\,\tau(n) - 512\,\tau\left(\frac{n}{2}\right)\right), \tag{15.128}$$

where

$$\sigma_{11}^*(n) = \sigma_{11}(n), \text{ for } n \text{ odd,}$$
$$\sigma_{11}^*(n) = \sigma_{11}^e(n) - \sigma_{11}^o(n), \text{ for } n \text{ even,}$$

$\sigma_{11}^e(n)$ and $\sigma_{11}^o(n)$ being the sum of the eleventh powers of, respectively, the even divisors of n and the odd divisors of $n$.

To derive Ramanujan's formula, we need four independent Dirichlet series of signature $(2, 12, 1)$. Two of these are clearly

$$2^{-s}\zeta(s)\zeta(s-11) \quad \text{and} \quad (1+4^{6-s})\zeta(s)\zeta(s-11).$$

The third Dirichlet series should correspond to the function

$$\theta_3^{12}(\omega)\frac{\sqrt{\Delta(\omega)}}{(2\pi)^6}. \tag{15.129}$$

[63] Ibid. p. 325.   [64] Rankin (1977) p. 357.

Note that

$$\frac{\sqrt{\Delta(\omega)}}{(2\pi)^6} = q \prod_{n=1}^{\infty} (1 - q^{2n})^{12}, \quad q = e^{i\pi\omega},$$

and

$$\theta_3^{12}(\omega) = \prod_{n=1}^{\infty} (1 - q^{2n})^{12}(1 + q^{2n-1})^{24}.$$

Hence, the function (15.129) can be rewritten as the infinite product

$$q \prod_{n=1}^{\infty} (1 - q^{2n})^{24}(1 + q^{2n-1})^{24}. \tag{15.130}$$

Using the infinite product for $\Delta(\omega)$, it is easy to see that (15.130) can be rendered as

$$-\Delta\left(\frac{\omega + 1}{2}\right) = \sum_{n=1}^{\infty} (-1)^{n-1} \tau(n) q^n.$$

Thus, corresponding to this series, the third Dirichlet series emerges as

$$\sum_{n=1}^{\infty} \frac{(-1)^{n-1} \tau(n)}{n^s}.$$

Since $\Delta(\omega)$ is a cusp form of weight 12, it can be seen to be a function of signature $(2, 12, 1)$. Also,

$$\Delta(\omega) = \sum_{n=1}^{\infty} \tau(n) q^{2n} = \sum_{n=1}^{\infty} \tau\left(\frac{n}{2}\right) q^n,$$

where $\tau(t) = 0$ when $t$ is not an integer. Thus, we have the fourth series:

$$\sum_{n=1}^{\infty} \frac{\tau\left(\frac{n}{2}\right)}{n^s}.$$

By Hecke's theorem it follows that

$$\sum_{n=1}^{\infty} \frac{r_{24}(n)}{n^s} = c_1 \sum_{n=1}^{\infty} \frac{\sigma_{11}(n)}{(2n)^s} + c_2 \left(\sum_{n=1}^{\infty} \frac{\sigma_{11}(n)}{n^s} + 4^6 \sum_{n=1}^{\infty} \frac{\sigma_{11}(n)}{(4n)^s}\right)$$

$$+ c_3 \sum_{n=1}^{\infty} \frac{(-1)^{n-1} \tau(n)}{n^s} + c_4 \sum_{n=1}^{\infty} \frac{\tau\left(\frac{n}{2}\right)}{n^s}.$$

The constants $c_1, c_2, c_3, c_4$ can be found from the four linear equations obtained by equating the coefficients of $\frac{1}{n^s}$ for $n = 1, 2, 3, 4$. Solving these equations gives the values

$$c_1 = -\frac{32}{691}, \quad c_2 = \frac{16}{691}, \quad c_3 = \frac{128 \times 259}{691}, \quad c_4 = -\frac{128 \times 512}{691}.$$

This means that if $n = 2^k n_0$, with $n_0$ odd, then

$$r_{24}(n) = \frac{16}{691}\left(\sigma_{11}(2^k n_0) - 2\sigma_{11}(2^{k-1} n_0) + 2^{12}\sigma_{11}(2^{k-2} n_0)\right)$$
$$+ \frac{128}{691}\left((-1)^{n-1} 259\,\tau(n) - 512\,\tau\left(\frac{n}{2}\right)\right). \tag{15.131}$$

As in the case of four and eight squares, we can write

$$\sigma_{11}(2^k n_0) - 2\sigma_{11}(2^{k-1} n_0) + 2^{12}\sigma_{11}(2^{k-2} n_0) = \sigma^*(n),$$

thus completing the proof of (15.128).

## 15.8 Exercises

1. Give a rigorous proof of (15.78), using the Euler-Maclaurin summation formula to show that

$$\sum_{l=0}^{\infty} e^{-\delta(lk+j)^2} = \int_0^{\infty} e^{-\delta(xk+j)^2}\,dx + O(1).$$

2. Complete the proof of Glaisher's formula (15.122):

$$r_{14}(n) = \frac{4}{61}\left(64E_6'(n) - E_6(n) + 364W(n)\right).$$

3. Let $f_{2s}(q)$ be defined by (15.54) and $f(q)$ by (15.57). Then prove Ramanjuan's formulas:

$$1385 f_{18}(q) = 24416 f^{12}(-q) f^6(q^2) - 256\frac{f^{30}(q^2)}{f^{12}(-q)};$$

$$31 f_{20}(q) = 616 f^{16}(-q) f^4(q^2) - 128\frac{f^{28}(q^2)}{f^8(-q)};$$

$$50521 f_{22}(q) = 1109416 f^{20}(-q) f^2(q^2) - 944768\frac{f^{20}(q^2)}{f^4(-q)}.$$

4. Prove Glaisher's formula for $r_{16}(n)$, given in (15.109), by Hecke's method.
5. Find the values of the constants $c_1, c_2, c_3$ in (15.126) to prove Glaisher's formula (15.127).
6. Prove the Venkatachaliengar-Ramanujan identity (15.26). See Cooper (2001).
7. Use (15.128) to show that, with $n$ odd,

$$\tau(n) \equiv \sigma_{11}(n) \pmod{691}.$$

Ramanujan proved this result for all positive integers $n$ by means of the identity

$$E_6^2(z) = E_{12}(z) - \frac{762048}{691}(2\pi)^{-12}\Delta(z).$$

Prove this identity. See Ramanujan (1988), p. 153.

8. Prove that, with $\psi_1(n)$, $\psi_2(n)$ as defined in exercise 5 of Chapter 13,

$$r_{20}(n) = \frac{8}{31}\left(\sigma_9(n) - 2^{10}\,\sigma_9\left(\frac{n}{4}\right)\right) + \frac{16}{93}\left(155\,\psi_1(n) + 76\,\psi_2(n)\right),$$

where $\sigma\left(\frac{n}{4}\right) = 0$ when 4 does not divide $n$. See Rankin (1962).

# 16

## The Hecke Operators

### 16.1  Preliminary Remarks

The problem of determining Dirichlet series of signature $(1, k, (-1)^{\frac{k}{2}})$, $k$ even, was touched upon in Chapter 15. Recall that the number of linearly independent Dirichlet series with signature $(1, k, (-1)^{\frac{k}{2}})$ is

$$\kappa = \left[\frac{k}{12}\right] + 1, \text{ if } k \not\equiv 2 \pmod{12}; \quad \kappa = \left[\frac{k}{12}\right], \text{ if } k \equiv 2 \pmod{12}. \tag{16.1}$$

Hecke considered whether the $\kappa$ independent Dirichlet series could be so chosen that the coefficients $a_n$ in the series $\sum \frac{a_n}{n^s}$ would have interesting arithmetical properties. For example, when $k = 12$, the two independent Dirichlet series of signature $(1, 12, 1)$ can be taken to be

$$\zeta(s)\zeta(11 - s) = \sum_{n=1}^{\infty} \frac{\sigma_{11}(n)}{n^s} = \prod_{p} \left(1 - \sigma_{11}(p)p^{-s} + p^{11-25}\right)^{-1} \tag{16.2}$$

$$\sum_{n=1}^{\infty} \frac{\tau(n)}{n^s} = \prod_{p} \left(1 - \tau(p)p^{-s} + p^{11-2s}\right)^{-1}, \tag{16.3}$$

where $\sigma_{11}(n) = \sum_{d|n} d^{11}$ and $\tau(n)$ is the Ramanujan function. In particular, (16.2) and (16.3) imply that $\sigma_{11}(n)$ and $\tau(n)$ are multiplicative arithmetic functions. Recall that (16.3) was conjectured by Ramanujan and proved by Mordell; it seems that Hecke was initially unaware of this work. In his paper of 1935,[1] summarizing his results on the connections between modular forms and the corresponding Dirichlet series, Hecke mentioned neither Ramanujan nor Mordell. However, Hecke's later papers on the same topic, of 1936 and 1937, grant them appropriate credit.

Hecke proved that when $4 \leq k \leq 34$, there existed $\kappa$ linearly independent Dirichlet series with signature $(1, k, (-1)^{\frac{k}{2}})$ that had Euler products. For this purpose, he defined

---

[1] Hecke (1959) pp. 577–590.

operators $T_n$ on the space of modular forms with signature $(1, k, (-1)^{\frac{k}{2}})$: For each positive integer $n$,

$$T_n(f(\omega)) = n^{k-1} \sum d^{-k} f\left(\frac{a\omega + b}{d}\right), \tag{16.4}$$

where the summation was taken over all integers $a, b, d$ such that $ad = n$, $0 \le b < d$. Thus, the number of terms in this sum came to $\sum_{d|n} d = \sigma_1(n)$.

Observe that Mordell had considered the operators $T_p$, where $p$ is prime. Hecke too remarked that the origin of the sum on the right-hand side of (16.4) lay in transformation theory (i.e., the modular and multiplier equations) and he mentioned Hurwitz's work.[2] In his lecture notes of 1938, Hecke wrote that the new point of view was to consider (16.4) as defining an operator, whereas the older view took the term on the right-hand side of (16.4) as a coefficient in the multiplier equation.

We have seen that a convergent Dirichlet series $\sum_{n=1}^{\infty} \frac{c_n}{n^s}$ has an Euler product

$$\prod_p (1 - c_p \, p^{-s} + p^{k-1-2s})^{-1} \tag{16.5}$$

if and only if, for any two positive integers $m$ and $n$ and with $c(n) \equiv c_n$,

$$c(m)c(n) = \sum_{d|(m,n)} d^{k-1} c\left(\frac{mn}{d^2}\right), \tag{16.6}$$

where $(m, n)$ denotes the gcd of $m$ and $n$. Hecke proved that the operators $T_n$ were commutative and, moreover, with $T(n) \equiv T_n$,

$$T(m)T(n) = \sum_{d|(m,n)} d^{k-1} T\left(\frac{mn}{d^2}\right). \tag{16.7}$$

He also showed that if a modular form of signature $(1, k, (-1)^{\frac{k}{2}})$ had a Fourier expansion

$$f(\omega) = \sum_{m=0}^{\infty} c(m)e^{2\pi i m \omega}, \tag{16.8}$$

then

$$T_n(f(\omega)) = \sum_{m=0}^{\infty} c(m)e^{2\pi i m \omega} \sum_{d|(m,n)} d^{k-1} c\left(\frac{mn}{d^2}\right) \tag{16.9}$$

would have the same signature. Note that the formulas related to (16.9) were given by Klein and Fricke.[3] Equations (16.6) through (16.9) imply that a Dirichlet series $\sum_{n=1}^{\infty} \frac{c_n}{n^s}$ of signature $(1, k, (-1)^{\frac{k}{2}})$ has an Euler product if and only if the corresponding modular form is an eigenfunction of all the Hecke operators $T_n$. One such Dirichlet

---

[2] Hecke (1959) p. 582.
[3] Klein and Fricke (1890–1892) vol. 2, pp. 600–607. Also see Hurwitz (1962) vol. 2, pp. 54–56.

series would be

$$\zeta(s)\zeta(s-k+1) = \sum_{n=1}^{\infty} \frac{\sigma_{k-1}(n)}{n^s},$$

which corresponds to the Eisenstein series $G_k(\omega)$. Recall that the Fourier series for $G_k(\omega)$ has a nonzero constant term. A function $f$ is a cusp form when $c(0) = 0$ in (16.8). From (16.9) we see that $T(n)$ maps cusp forms to cusp forms. This insight enabled Hecke to prove that if the space of modular forms of weight $k$ had dimension 2, then all $T(n)$ would have the same two independent eigenfunctions. Note that it follows from (16.1) that $1 \leq \kappa \leq 2$ for $4 \leq k \leq 22$ and for $k = 26$. By means of explicit calculations, Hecke showed that for the cases $k = 24$, $k = 38$, and $28 \leq k \leq 34$, there would be three independent eigenfunctions for all $T(n)$ and that two of these were cusp forms. If $c(1) = 1$ for these cusp forms that are also eigenfunctions, then we call them normalized eigenforms. Hecke showed that for $k = 24$, all the coefficients of the normalized eigenforms were in $Q(\sqrt{144169})$.

In 1939, Hecke's former student, Hans Petersson, defined an inner product on the space of cusp forms of weight $k$, enabling him to prove that the $T(n)$ were Hermitian operators. He could then conclude that the operators $T(n)$ would have $\kappa$ linearly independent eigenfunctions on the space of modular forms of dimension $\kappa$.

## 16.2  The Hecke Operators $T(n)$

Erich Hecke defined an infinite sequence of operators on the finite dimensional vector of homogeneous modular forms of weight $k$. He first worked out the theory for level 1 functions, that is, modular forms for the full modular group. This set of functions is equivalent to the set of functions of signature $(1, k, (-1)^{\frac{k}{2}})$. To develop the idea of the Hecke operator $T(n)$ for each positive integer $n$, we fix an integer $k \geq 4$, and let $M$ be a transformation of order $n$. Thus, $M$ is a $2 \times 2$ matrix with integer entries $\left(\begin{smallmatrix} a & b \\ c & d \end{smallmatrix}\right)$ such that $ad - bc = n$. Let $F$ be a homogeneous modular form of weight $k$ and level 1; this means that:

- $F(\omega_1, \omega_2)$ is a holomorphic function of $\frac{\omega_1}{\omega_2}$ in the upper half-plane, with $\omega_2 \neq 0$;
- $F(\lambda\omega_1, \lambda\omega_2) = \lambda^{-k} F(\omega_1, \omega_2)$;
- $F(\omega_1, \omega_2)$ is invariant under the action of the homogeneous modular group, that is, if $N = \left(\begin{smallmatrix} p & q \\ r & s \end{smallmatrix}\right) \in SL_2(\mathbb{Z})$, then

$$F(p\omega_1 + q\omega_2, r\omega_1 + s\omega_2) = F(\omega_1, \omega_2); \qquad (16.10)$$

- 

$$\omega_2^k F(\omega_1, \omega_2) = a_0 + \sum_{n=1}^{\infty} a_n e^{2\pi i \tau}, \qquad (16.11)$$

with $\tau = \frac{\omega_1}{\omega_2}$, that is, $\omega_2^k F(\omega_1, \omega_2)$ is a holomorphic function in $h = e^{2\pi i \tau}$ for $|h| < 1$.

Hecke's approach was to study the effect of a transformation of order $n$ on $F(\omega) \equiv F(\omega_1, \omega_2)$; thus, he wished to study $F(M(\omega)) \equiv F(a\omega_1 + b\omega_2, c\omega_1 + d\omega_2)$ with

$ad - bc = n$. Note that since $F(\omega_1, \omega_2)$ is invariant under the modular group, we obtain $F(M(\omega)) = F(NM(\omega))$ when $N \in SL_2(\mathbb{Z})$. Hecke therefore defined an equivalence relation among the transformations of order $n$:

$$M_1 \sim M_2 \text{ if } M_1 = NM_2, \quad N \in SL_2(\mathbb{Z}). \tag{16.12}$$

The simplest representations of the equivalence classes are of the form

$$\begin{pmatrix} a_1 & b_1 \\ 0 & d_1 \end{pmatrix}, \text{ with } a_1 d_1 = n, \; d_1 > 0, \; b_1 \bmod d_1. \tag{16.13}$$

Recall that our Chapter 8 gives Dedekind's proof of a similar result, with $a, b, d$ relatively prime.

It is clear that the number of representations can be given as $\sum_{d|n} d = \sigma_1(n)$. To show that the representations are of the form given by (16.13), let $M = \begin{pmatrix} a & b \\ c & d \end{pmatrix}$ be a transformation of order $n$, that is, $ad - bc = n$. Choose relatively prime integers $r$ and $s$ such that $ra + sc = 0$. Next, choose integers $p$ and $q$ in such a way that $ps - qr = 1$. Then

$$\begin{pmatrix} p & q \\ r & s \end{pmatrix} \begin{pmatrix} a & b \\ c & d \end{pmatrix} = \begin{pmatrix} a_1 & b_1 \\ 0 & d_1 \end{pmatrix}. \tag{16.14}$$

Note that, since $\det \begin{pmatrix} a_1 & b_1 \\ 0 & d_1 \end{pmatrix} = \det \begin{pmatrix} p & q \\ r & s \end{pmatrix} \det \begin{pmatrix} a & b \\ c & d \end{pmatrix}$, it follows that $a_1 d_1 = n$. Now consider the relation between two transformations of the form given on the right-hand side of (16.14), when they belong to the same class. In that case, we have

$$\begin{pmatrix} a_1 & b_1 \\ 0 & d_1 \end{pmatrix} = \begin{pmatrix} p & q \\ r & s \end{pmatrix} \begin{pmatrix} a & b \\ 0 & d \end{pmatrix} = \begin{pmatrix} pa & pb + qd \\ ra & rb + sd \end{pmatrix}, \tag{16.15}$$

for some $\begin{pmatrix} p & q \\ r & s \end{pmatrix} \in SL_2(\mathbb{Z})$. Thus, $ra = 0$ and, since $ad = n$, it follows that $r = 0$. By definition we know that $ps - qr = 1$ and hence we may write $ps = 1$. Thus, we know that either $p = s = 1$ or $p = s = -1$. In the latter case, $d$ and $d_1$ would have to have opposite signs and this is clearly impossible because $d > 0$ and $d_1 > 0$. Hence, $p = s = 1$ and (16.15) becomes

$$\begin{pmatrix} a_1 & b_1 \\ 0 & d_1 \end{pmatrix} = \begin{pmatrix} a & b + qd \\ 0 & d \end{pmatrix};$$

this implies that $a = a_1$, $d = d_1$, and $b_1 \equiv b \bmod d$. Thus, we have proved that the complete list of inequivalent matrices of order $n$ is given by the set

$$\left\{ \begin{pmatrix} a & b \\ 0 & d \end{pmatrix} \, \middle| \, ad = n, \; d > 0, \; b \bmod d \right\}. \tag{16.16}$$

We are now in a position to state Hecke's proposition that if $F(\omega_1, \omega_2)$ is a (homogeneous) modular form of weight $k$ for the full modular group $\Gamma(1)$, then for each $n$, the form

$$F_1 = \sum_{h=1}^{\sigma_1(n)} F(M_h(\omega_1, \omega_2)) \tag{16.17}$$

is also a (homogeneous) modular form of weight $k$ for the full modular group. Here $\{M_h | h = 1, 2, \ldots, \sigma_1(n)\}$ represents the full set of inequivalent matrices of order $n$, as given by (16.16). Recall that this proposition was known in exactly this form to Dedekind, Klein, and Hurwitz within their theory of transformation or multiplier equations. To prove the proposition, first note that if

$$N = \begin{pmatrix} a & b \\ c & d \end{pmatrix} \in SL_2(\mathbb{Z}), \quad \text{then} \quad F(M_h(a\omega_1 + b\omega_2, \ c\omega_1 + d\omega_2)) = F(M_h N(\omega_1, \ \omega_2)).$$

Since

$$\det(M_h N) = \det M_h \det N = n,$$

there exist $N_1 \in SL_2(\mathbb{Z})$ and $M_{h_1}$, in the list (16.16), such that $M_h N = N_1 M_{h_1}$. Thus, since $F$ is invariant under $\Gamma(1) = SL_2(\mathbb{Z})$, with $(\omega) \equiv (\omega_1, \omega_2)$, we have

$$F(M_h N(\omega)) = F(N_1 M_{h_1}(\omega)) = F(M_{h_1}(\omega)).$$

Note that changing $(\omega)$ to $N(\omega)$ merely permutes the terms contained in the sum and so we may write

$$\sum_{h=1}^{\sigma_1(n)} F(M_h N(\omega)) = \sum_{h_1=1}^{\sigma_1(n)} F(M_{h_1}(\omega)).$$

We must also check that $F_1$ in (16.17) satisfies (16.11). But each term in the sum for $F_1$ is a holomorphic function of $h^{\frac{1}{n}}$, so that it must also be a holomorphic function of $h = e^{2\pi i \tau}$, because $F_1$ is invariant under $\tau \to \tau + 1$. Thus, our condition is satisfied, completing our proof.

To define the operator $T(n)$, take

$$F(\tau) = \omega_2^k F(\omega_1, \omega_2), \quad \tau = \frac{\omega_1}{\omega_2}, \tag{16.18}$$

to be a modular form of weight $k$ belonging to $\Gamma(1)$::

$$F | T(n) = n^{k-1} \omega_2^k \sum_{ad=n} F(a\omega_1 + b\omega_2, \ d\omega_2)$$

$$= n^{k-1} \sum_{\substack{ad=n \\ b \bmod d, \, d > 0}} F\left(\frac{a\tau + b}{d}\right) d^{-k}. \tag{16.19}$$

It follows from Hecke's proposition that the operator $T(n)$ maps the vector space of holomorphic $k$-forms to itself. It is also clear that if $F$ is a cusp form, that is, if $a_0 = 0$ in (16.11), then $F | T(n)$ is a cusp form. Hecke defined the sum $T(n) + T(m)$ and the product $T(n) \cdot T(m)$ as

$$F \big| (T(n) + T(m)) = F \big| T(n) + F \big| T(m);$$

$$F \big| T(n) \cdot T(m) = \left(F \big| T(n)\right) \big| T(m).$$

Moreover, for any constant $a$,

$$aT(n) = T(n)a \quad \text{and} \quad F \big| aT(n) = aF \big| T(n).$$

Thus, the set of operators $T(n)$ forms an algebra, referred to by Hecke as a ring. Two operators in this algebra, $T_1$ and $T_2$, are equal if $F|T_1 = F|T_2$ for every $k$-form $F$. Hecke's main theorem on this algebra of operators may now be stated: The algebra of operators $T(n)$ is commutative and

$$T(n) \cdot T(m) = \sum_{0 < d | m, n} T\left(\frac{mn}{d^2}\right) d^{k-1}. \tag{16.20}$$

Hecke proved this formula in three steps.[4] First, he showed that if $(m, n) = 1$, that is, if $m$ and $n$ are relatively prime, then

$$T(n) \cdot T(m) = T(nm) = T(m) \cdot T(n). \tag{16.21}$$

Next, he showed that

$$T(p) \cdot T(p^r) = p^{k-1} T(p^{r-1}) + T(p^{r+1}). \tag{16.22}$$

This implies the particular result, provable by induction, that $T(p^r)$ is a polynomial in $T(p)$. Combined with (16.21), this result implies the commutativity of the algebra of operators $T(n)$. Finally, Hecke used an inductive argument to prove that

$$T(p^s) \cdot T(p^r) = \sum_{0 \le n \le s, r} T(p^{r+s-2n}) p^{n(k-1)}. \tag{16.23}$$

We consider the case $(m, n) = 1$ to give a proof of (16.21). Let

$$\phi = F\big|T(n) = n^{k-1} \sum_{\substack{a'd'=n \\ b' \bmod d'}} F\left(\frac{a'\tau + b'}{d'}\right) d'^{-k};$$

$$\phi\big|T(m) = m^{k-1} \sum_{\substack{ad=m \\ b \bmod d}} \phi\left(\frac{a\tau + b}{d}\right) d^{-k}.$$

Hence

$$F\big|T(n) \cdot T(m) = (nm)^{k-1} \sum_{\substack{ad=m \\ b \bmod d}} d^{-k} \left\{ \sum_{\substack{a'd'=n \\ b' \bmod d'}} d'^{-k} F\left(\frac{\frac{a'(a\tau+b)}{d} + b'}{d'}\right) \right\}$$

$$= (nm)^{k-1} \sum_{\substack{ad=n, \, b \bmod d \\ a'd'=m, \, b' \bmod d'}} (dd')^{-k} F\left(\frac{aa'\tau + b'd + a'b}{dd'}\right). \tag{16.24}$$

We need to know the possible values of the term $b'd + a'b$. Since $a'$ and $d$ are relatively prime integers, the values of $b'd + a'b$, for fixed $a'$ and $d$, run over the complete set of incongruent numbers mod $dd'$ as $b$ and $b'$ run over all integers mod $d$, $d'$ respectively. Moreover, if $b'd + a'b \equiv b'_1 d + a'b_1 \bmod dd'$, then $b' \equiv b'_1 \bmod d'$ and $b \equiv b_1 \bmod d$. Thus, it follows from (16.24) that

$$T(n) \cdot T(m) = T(nm) \quad \text{for} \quad (m, n) = 1.$$

---

[4] See Hecke (1959) pp. 656–658 or Hecke (1983) pp. 71–73.

For the proof of (16.22), note that the $p + 1$ inequivalent matrices defining $T(p)$ are given by $\begin{pmatrix} p & 0 \\ 0 & 1 \end{pmatrix}$ and the remaining $p$ matrices are given by $\begin{pmatrix} 1 & b \\ 0 & p \end{pmatrix}$, $b \bmod p$. Also, the inequivalent matrices defining $T(p^r)$ may be written as

$$\begin{pmatrix} a & b \\ 0 & d \end{pmatrix} = \begin{pmatrix} p^{r-s} & b_s \\ 0 & p^s \end{pmatrix}, \quad b_s \bmod p^s, \ s = 0, 1, \ldots, r.$$

Then

$$(F|T(p))T(p') = p^{(r+1)(k-1)} \sum_{s,\, b_s} F\left( \begin{pmatrix} p & 0 \\ 0 & 1 \end{pmatrix} \begin{pmatrix} p^{r-s} & b_s \\ 0 & p^s \end{pmatrix} (\omega) \right)$$

$$+ p^{(r+1)(k-1)} \sum_{\substack{s,\, b_s \\ b \bmod p}} F\left( \begin{pmatrix} 1 & b \\ 0 & p \end{pmatrix} \begin{pmatrix} p^{r-s} & b_s \\ 0 & p^s \end{pmatrix} (\omega) \right) \qquad (16.25)$$

$$= p^{(r+1)(k-1)} \left( \sum_{s,\, b_s} F\left( \begin{pmatrix} p^{r+1-s} & pb_s \\ 0 & p^s \end{pmatrix} (\omega) \right) \right.$$

$$\left. + \sum_{\substack{s,\, b_s \\ b \bmod p}} F\left( \begin{pmatrix} p^{r-s} & b_s + bp^s \\ 0 & p^{s+1} \end{pmatrix} (\omega) \right) \right). \qquad (16.26)$$

Note that, except when $s = 0$,

$$\begin{pmatrix} p^{r+1-s} & pb_s \\ 0 & p^s \end{pmatrix} \qquad (16.27)$$

has a common factor $p$, so that the homogeneity of $F(\omega_1, \omega_2)$ produces

$$F\left( \begin{pmatrix} p^{r+1-s} & pb_s \\ 0 & p^s \end{pmatrix} (\omega) \right) = p^{-k} F\left( \begin{pmatrix} p^{r-s} & b_s \\ 0 & p^{s-1} \end{pmatrix} (\omega) \right).$$

Next, observe that when $s = 0$ (16.27) becomes $\begin{pmatrix} p^{r+1} & 0 \\ 0 & 1 \end{pmatrix}$. Thus, the first sum in (16.26) is given by

$$p^{(r+1)(k-1)} \sum_{s,\, b_s} F\left( \begin{pmatrix} p^{r+1-s} & pb_s \\ 0 & p^s \end{pmatrix} (\omega) \right) \qquad (16.28)$$

$$= p^{(r+k)(k-1)} p^{-k} \sum_{\substack{s=1,\ldots,r \\ b_s \bmod p^s}} F\left( \begin{pmatrix} p^{r-s} & b_s \\ 0 & p^{s-1} \end{pmatrix} (\omega) \right)$$

$$+ p^{(r+1)(k-1)} F\left( \begin{pmatrix} p^{r+1} & 0 \\ 0 & 1 \end{pmatrix} (\omega) \right). \qquad (16.29)$$

In the expression within the summation sign in equation (16.29), set $b_s = b_{s-1} + p^{s-1}b_1$, where $b_{s-1}$ and $b_1$ run through all the integers mod $p^{s-1}$ and mod $p$ respectively. Then, because $\begin{pmatrix} 1 & b_1 \\ 0 & 1 \end{pmatrix} \in \Gamma(1)$, we may write

$$F\left(\begin{pmatrix} p^{r-s} & b_{s-1} + p^{s-1}b_1 \\ 0 & p^{s-1} \end{pmatrix}(\omega)\right) = F\left(\begin{pmatrix} 1 & b_1 \\ 0 & 1 \end{pmatrix}\begin{pmatrix} p^{r-s} & b_{s-1} \\ 0 & p^{s-1} \end{pmatrix}(\omega)\right)$$

$$= F\left(\begin{pmatrix} p^{r-s} & b_{s-1} \\ 0 & p^{s-1} \end{pmatrix}(\omega)\right).$$

Since there are $p$ different values of $b_1$ mod $p$, we see that expression (16.29) equals

$$p^{(r+1)(k-1)} p^{-k} p \sum_{\substack{s=1,\ldots,r \\ b_{s-1} \bmod p^{s-1}}} F\left(\begin{pmatrix} p^{r-s} & b_{s-1} \\ 0 & p^{s-1} \end{pmatrix}(\omega)\right) + p^{(r+1)(k-1)}F\left(\begin{pmatrix} p^{r+1} & 0 \\ 0 & 1 \end{pmatrix}(\omega)\right).$$

$$(16.30)$$

In (16.26), substitute $b_{s+1}$ for $b_s + bp^s$ within the second sum to obtain

$$p^{(r+1)(k-1)} \sum_{\substack{s=0,\ldots,r \\ b_{s+1} \bmod p^{s+1}}} F\left(\begin{pmatrix} p^{r-s} & b_{s+1} \\ 0 & p^{s+1} \end{pmatrix}(\omega)\right)$$

$$= p^{(r+1)(k-1)} \sum_{\substack{s=1,\ldots,r+1 \\ b_s \bmod p^s}} F\left(\begin{pmatrix} p^{r+1-s} & b_s \\ 0 & p^s \end{pmatrix}(\omega)\right). \qquad (16.31)$$

Observe that the expression (16.29) is the sum of the expressions (16.30) and (16.31). But observe further that the sum of (16.31) and the second summand in (16.30) turns out to be $F|T(p^{r+1})$; moreover, the first summand within (16.31) is $p^{k-1}F|T(p^{r-1})$. Thus, we have Hecke's proof that

$$T(p) \cdot T(p^r) = p^{k-1} T(p^{r-1}) + T(p^{r+1}). \qquad (16.32)$$

As already mentioned, this result and (16.21) together imply that the $T(n)$ are commutative.

Recall that the generalization of (16.32) is (16.23):

$$T(p^s) \cdot T(p^r) = \sum_{0 \leq n \leq s, r} T(p^{r+s-2n}) p^{n(k-1)};$$

this expression is symmetric in $r$ and $s$. We can therefore assume that $s \leq r$ and proceed to prove the result for a fixed $r$ by induction on $s$. Taking $T(1)$ to be the identity operator, the result is true for $s = 0$ and $s = 1$. We then assume the result true up to $s \leq r$ and show that it is true for $s + 1 \leq r$. Note that it follows from (16.32) that

$$T(p^{s+1}) T(p^r) = T(p) T(p^s) T(p^r) - p^{k-1} T(p^{s-1}) T(p^r)$$

$$= T(p) \sum_{0 \leq n \leq s} T(p^{r+s-2n}) p^{n(k-1)} - p^{k-1} \sum_{0 \leq n \leq s-1} T(p^{r+s-2n-1}) p^{n(k-1)}. \quad (16.33)$$

Again using (16.32), we see that

$$T(p)\,T(p^{r+s-2n}) = p^{k-1}\,T(p^{r+s-2n-1}) + T(p^{r+s+1-2n})$$

and by combining (16.32) and (16.33) we may conclude that

$$T(p^{s+1})\,T(p^r) = \sum_{0 \le n \le s} T(p^{r+s+1-2n})\,p^{n(k-1)} + p^{k-1}\,T(p^{r-s-1})\,p^{s(k-1)}$$

$$= \sum_{0 \le n \le s+1} T(p^{r+s+1-2n})\,p^{n(k-1)}.$$

This proves (16.23) and, following Hecke, with the use of (16.21) we are now able to show that

$$T(n)T(m) = \sum_{d \mid m,n} T\left(\frac{mn}{d^2}\right) d^{k-1}.$$

If $n = p_1^{n_1} p_2^{n_2} \cdots p_k^{n_k}$ and $m = p_1^{m_1} p_2^{m_2} \cdots p_k^{m_k}$ and since the operators commute, we obtain

$$
\begin{aligned}
T(n)T(m) &= T\big(p_1^{n_1}\big)T\big(p_2^{n_2}\big)\cdots T\big(p_k^{n_k}\big)T\big(p_1^{m_1}\big)T\big(p_2^{m_2}\big)\cdots T\big(p_k^{m_k}\big)\\
&= T\big(p_1^{n_1}\big)T\big(p_1^{m_1}\big)T\big(p_2^{n_2}\big)T\big(p_2^{m_2}\big)\cdots T\big(p_k^{n_k}\big)T\big(p_k^{m_k}\big)\\
&= \left(\sum_{0 \le l \le n_1, m_1} T\big(p_1^{n_1+m_1-2l}\big)p_1^{l(k-1)}\right) \cdots \left(\sum_{0 \le l \le n_k, m_k} T\big(p_k^{n_k+m_k-2l}\big)p_k^{l(k-1)}\right)\\
&= \sum_{d \mid n, m} T\left(\frac{mn}{d^2}\right) d^{k-1}.
\end{aligned}
$$

### 16.3  The Operators $T(n)$ in Terms of Matrices $\lambda(n)$

Next, let $M_k$ denote the finite dimensional vector space of all holomorphic, homogeneous $k$-forms, $k \ge 4$, and let $F^\rho(\omega_1, \omega_2)$, $\rho = 1, \ldots, \kappa$ be a basis for $M_k$. We employ $\rho, \sigma, \nu$ to run over the integers $1, 2, \ldots, \kappa$. The Hecke operator $T(n)$ is then a linear transformation from $M_k$ to $M_k$. We set $\lambda(n)$ as the corresponding $\kappa \times \kappa$ matrix, determined by the basis $F^\rho(\omega_1, \omega_2)$, $\rho = 1, 2, \ldots, \kappa$. With this in place, (16.20) implies the theorem:[5] The algebra of matrices $\Lambda$ generated by $\lambda(n)$ is commutative. Moreover,

$$\lambda(n) \cdot \lambda(m) = \sum_{d \mid m, n} \lambda\left(\frac{nm}{d^2}\right) d^{k-1}. \tag{16.34}$$

The entries of the matrix $\lambda(n)$ with respect to the basis $F^\rho$, $\rho = 1, \ldots, \kappa$, are found by expressing the functions $F^\rho | T(n)$, $\rho = 1, \ldots, \kappa$ in terms of $F^\rho$. This means that $\lambda(n) = (\lambda_{\rho\sigma}(n))$, where

$$F^\rho | T(n) = \sum_{\sigma=1}^{\kappa} \lambda_{\rho\sigma}(n)\,F^\sigma. \tag{16.35}$$

---

[5]  See Hecke (1959) pp. 658–660 or Hecke (1983) pp. 73–76.

We note that by (16.11) $F^\rho$ has a Fourier expansion given by

$$F^\rho(\tau) = \omega_2^k F^\rho(\omega_1, \omega_2) = \sum_{l=0}^{\infty} a^\rho(l) e^{2\pi i l \tau}, \quad \tau = \frac{\omega_1}{\omega_2}. \tag{16.36}$$

It is evident that the entries $\lambda_{\rho\sigma}(n)$ of the matrix $\lambda(n)$ depend upon the coefficients $a^\rho(l)$. To make this connection explicit, we apply the definition (16.19) of $T(n)$, (16.18) and (16.36) to obtain

$$F^\rho(\tau)|T(n) = n^{k-1} \sum_{\substack{ad=n \\ b \bmod d}} d^{-k} F^\rho \left( \frac{a\tau + b}{d} \right)$$

$$= n^{k-1} \sum_{\substack{ad=n \\ b \bmod d}} d^{-k} \sum_{l=0}^{\infty} a^\rho(l) e^{2\pi i l \frac{a\tau+b}{d}}$$

$$= n^{k-1} \sum_{l,a,b,d} a^\rho(l) e^{2\pi i \frac{la\tau}{d} + \frac{2\pi i l b}{d}} d^{-k}, \tag{16.37}$$

where the last sum is extended over all nonnegative integers $a$, $b$, $d$ with $ad = n$ and $b$ reduced mod $d$ and all $l \geq 0$. Next, note that since $F^\rho|T(n)$ is a holomorphic $k$-form, that is, a modular form of weight $k$, it has a Fourier expansion of the form

$$\sum_{N=0}^{\infty} b^\rho(N) e^{2\pi i N \tau},$$

for suitable $b^\rho(N)$. To express this $b^\rho(N)$ in terms of the coefficients $a^\rho(l)$ of the expansion of $F^\rho(\tau)$, we apply the formula

$$\sum_{b \bmod d} \left( e^{\frac{2\pi i l b}{d}} \right) = d \quad \text{if } d|l \quad \text{and } = 0 \text{ if } d \nmid l \tag{16.38}$$

to the sum in (16.37). Setting $l = md$, this sum reduces to

$$F^\rho(\tau)|T(n) = n^{k-1} \sum_{m, \, ad=n} a^\rho(md) e^{2\pi i a m \tau} d^{1-k}.$$

Since $\frac{n}{d} = a$, we get

$$F^\rho(\tau)|T(n) = \sum_{m, \, a|n} a^\rho \left( \frac{mn}{a} \right) e^{2\pi i a m \tau} a^{k-1}.$$

Comparing this with $\sum_{N=0}^{\infty} b^\rho(N) e^{2\pi i N \tau}$, we see that

$$b^\rho(N) = \sum_{\substack{a|n \\ am=N}} a^\rho \left( \frac{mn}{a} \right) d^{k-1} = \sum_{d|n, N} a^\rho \left( \frac{Nn}{d^2} \right) d^{k-1}, \tag{16.39}$$

where $d|n, N$ means: $d|n$ and $d|N$. Thus, by (16.35), the matrix entries $\lambda_{\rho\sigma}(n)$ satisfy the relation

$$\sum_{d|n, N} a^\rho \left( \frac{Nn}{d^2} \right) d^{k-1} = \sum_{\sigma} \lambda_{\rho\sigma}(n) \cdot a^\sigma(N), \quad N \geq 0, \ n \geq 1, \ \rho = 1, \dots, \kappa. \tag{16.40}$$

Now set $N = 1$ in (16.40) to obtain

$$a^\rho(n) = \sum_\sigma \lambda_{\rho\sigma}(n) \cdot a^\sigma(1), \quad n = 1, 2, 3, \ldots. \tag{16.41}$$

This implies that the functions $F^\rho(\tau) - a^\rho(0)$, $\rho = 1, \ldots, \kappa$, are linear combinations of the $\kappa^2$ functions

$$\tilde{f}_{\rho\sigma}(\tau) = \sum_{n=1}^\infty \lambda_{\rho\sigma}(n) e^{2\pi i n\tau}. \tag{16.42}$$

In fact, each function $F^\rho(\tau) - a^\rho(0)$ is a linear combination of $\kappa$ functions

$$\sum_{n=1}^\infty \lambda_{\rho\sigma}(n) e^{2\pi i n\tau}, \quad \sigma = 1, \ldots, \kappa.$$

Conversely, we can show that we can construct functions $f_{\rho\sigma}$ out of $\tilde{f}_{\rho\sigma}$ such that $f_{\rho\sigma}$ can be expressed as a linear combination of $F^\rho$, $\rho = 1, \ldots, \kappa$. Consider the $\kappa^2$ functions

$$f_{\rho\sigma}(\tau) = \lambda_{\rho\sigma}(0) + \sum_{n=1}^\infty \lambda_{\rho\sigma}(n) e^{2\pi i n\tau}, \tag{16.43}$$

where the matrix $\lambda(0) = (\lambda_{\rho\sigma}(0))$ is defined by analogy with (16.41) as

$$a^\rho(0) = \sum_\sigma \lambda_{\rho\sigma}(0) a^\sigma(1). \tag{16.44}$$

Then, taking (16.44) and combining it with (16.41), we see that

$$F^\rho(\tau) = \sum_\sigma f_{\rho\sigma}(\tau) a^\sigma(1), \quad \rho = 1, \ldots, \kappa, \tag{16.45}$$

that is, each $F_\rho(\tau)$ is a linear combination of the $\kappa$ functions $f_{\rho\sigma}(\tau)$, $\sigma = 1, 2, \ldots, \kappa$. To verify this converse, note that the function $F'^\rho(\tau) = dF^\rho/dt$ can be expressed as a linear combination of

$$f'_{\rho\sigma}(\tau) = 2\pi i \sum_{n=1}^\infty n\lambda_{\rho\sigma}(n) e^{2\pi i n\tau};$$

note that these functions do not involve the constants $\lambda_{\rho\sigma}(0)$. Next note that the derivatives of $F^\rho$, $\rho = 1, 2, \ldots, \kappa$, are linearly independent. To see this, observe that if they were dependent, then a nontrivial linear combination of $F^\rho$ would be a constant; this would imply that the weights $k$ of the functions would be zero. But this impossible because $k \geq 4$, proving that the $F'^\rho$ are independent. Thus, the matrix of the coefficients of $F'^\rho$,

$$na^\sigma(n), \quad \sigma = 1, \ldots, \kappa, \quad n = 0, 1, 2, \ldots,$$

must be of rank $\kappa$. This means that there exist $\kappa$ integers $n_1, n_2, \ldots, n_k$ such that the determinant of the $\kappa \times \kappa$ matrix $(n_\nu a^\sigma(n_\nu))$ is nonzero. Now from the symmetry of the left-hand side of (16.40) in $n$ and $N$ for $N \geq 1$, it follows that

$$\sum_\sigma \lambda_{\rho\sigma}(n) a^\sigma(N) = \sum_\sigma \lambda_{\rho\sigma}(N) a^\sigma(n), \quad N, n \geq 1. \tag{16.46}$$

If, for the cases $n = n_1, n_2, \ldots, n_\kappa$, we multiply the equations (16.46) by $n_\nu N$, we arrive at the equations

$$\sum_\sigma N \lambda_{\rho\sigma}(N) a^\sigma(n_\nu) n_\nu = \sum_\sigma n_\nu \lambda_{\rho\sigma}(n_\nu) a^\sigma(N) N, \quad \nu = 1, \ldots, \kappa.$$

It follows from the nonvanishing of the determinant $|n_\nu a^\sigma(n_\nu)|$ that we may solve for $N\lambda_{\rho\sigma}(N)$ in terms of $Na^\sigma(N)$; thus

$$N \lambda_{\rho\sigma}(N) = \sum_\nu b^\nu_{\rho\sigma} N a^\sigma(N), \quad N = 1, 2, 3, \ldots, \tag{16.47}$$

where the coefficients $b^\nu_{\rho\sigma}$ are independent of $N$. After canceling $N$ from both sides of (16.47), we see that (16.41) and (16.44) imply that

$$\lambda_{\rho\sigma}(0) = \sum_\nu b^\nu_{\rho\sigma} a^\sigma(0). \tag{16.48}$$

Using (16.47) and (16.48) along with the expansions of $f_{\rho\sigma}(\tau)$ and $F^\rho(\tau)$ from (16.43) and (16.36), respectively, we may conclude that

$$f_{\rho\sigma}(\tau) = \sum_\nu b^\nu_{\rho\sigma} F^\nu(\tau). \tag{16.49}$$

We are now able to state an important theorem of Hecke:[6]

> Given a basis $F^\rho(\tau)$ of $\kappa$ linearly independent modular $k$-forms, there exist $\kappa$ constant, commutative $\kappa \times \kappa$ matrices $B^\nu$ and these form a basis of the commutative algebra of matrices $\Lambda$ corresponding to the algebra of operators generated by $T(n)$. Moreover, we have the Fourier series
>
> $$B(\tau) = \sum_\nu F^\nu(\tau) B^\nu = \sum_{n=0}^\infty \lambda(n) e^{2\pi i n \tau}, \tag{16.50}$$
>
> where the $\lambda(n) \in \Lambda$ satisfy the relation
>
> $$\lambda(n) \cdot \lambda(m) = \sum_{d | n, m} \lambda\left(\frac{nm}{d^2}\right) d^{k-1}, \quad m, n \geq 1. \tag{16.51}$$

We observe that this theorem also holds for any subset of $F^\rho(\tau)$, $\rho = 1, 2, \ldots, \kappa$, that is invariant under $T(n)$.

To prove this, consider the fact that the matrices $B^\nu = (b^\nu_{\rho\sigma})$, $\nu = 1, \ldots, \kappa$, are defined by the coefficients in (16.49). Then from (16.43) it follows that

$$B(\tau) = (f_{\rho\sigma}(\tau)) = \sum_\nu F^\nu(\tau) B^\nu = \sum_{n=0}^\infty \lambda(n) e^{2\pi i n \tau}.$$

---

[6] Hecke (1959) pp. 663–667 or Hecke (1983) pp. 77–78.

Now (16.36) implies that

$$\lambda(n) = \sum_\nu a^\nu(n) B^\nu. \tag{16.52}$$

Here, if we employ the special values $n_1, n_2, \ldots, n_\kappa$ with determinant $|a^\nu(n_\rho)| \neq 0$, we have for some constants $x_{n_\rho}$

$$B^\nu = \sum_\rho \lambda(n_\rho) \cdot x_{n_\rho}. \tag{16.53}$$

Thus, the algebra of matrices generated by $B^\nu$ is the same as the algebra generated by the matrices $\lambda(n)$ and, moreover, the matrices $B^\nu$ are linearly independent. Note that (16.51) may be verified by (16.34).

Observe that Hecke added the constants $\lambda_{\rho\sigma}(0)$ to the functions $\tilde{f}_{\rho\sigma}(\tau)$ in (16.42); he could then obtain $f_{\rho\sigma}(\tau)$ in (16.43). This allowed the matrix $B(\tau)$ in (16.50) to be expressed in terms of the functions $F^\nu(\tau)$. As we shall see, this in turn enabled him to prove that the characteristic roots of $B(\tau)$ were linear combinations of the functions $F^\nu(\tau)$, $\nu = 1, 2, \ldots, \kappa$. Hecke then went on to show that the characteristic roots of $B(\tau)$ were eigenfunctions of all the operators $T(n)$.

## 16.4  Euler Products

We now consider the Dirichlet series

$$\phi_{\rho\sigma}(s) = \sum_{n=1}^\infty \lambda_{\rho\sigma}(n) n^{-s}$$

associated with the modular forms $f_{\rho\sigma}(\tau)$. Note that if

$$\phi^\nu(s) = \sum_{n=1}^\infty \frac{a^\nu(n)}{n^s}$$

is the Dirichlet series corresponding to the modular form $F^\nu(\tau)$, $\nu = 1, \ldots, \kappa$ with the matrix $B^\nu$ defined by (16.50), then

$$\Phi(s) = \sum_\nu \phi^\nu(s) B^\nu = \sum_{n=1}^\infty \lambda(n) n^{-s} = (\phi_{\rho\sigma}(s)). \tag{16.54}$$

We may now prove another significant theorem of Hecke: The matrix $\Phi(s)$, whose entry in row $\rho$ and column $\sigma$ is the Dirichlet series $\phi_{\rho\sigma}(s)$, has an Euler product

$$\Phi(s) = \prod_p (I - \lambda(p) p^{-s} + p^{k-1-2s} I)^{-1}$$

within the region of absolute convergence of the Dirichlet series, with $I$ the identity matrix.

To begin the proof, take $p$ to be a prime and let $n \geq 1$ be expressed as $n = p^m N$ where $p \nmid N$. Since $(p, N) = 1$, it follows from (16.34) that

$$\lambda(n) = \lambda(p^m N) = \lambda(p^m) \lambda(N).$$

Hence we have

$$\Phi(s) = \sum_{n=1}^{\infty} \lambda(n)\, n^{-s} = \sum_{\substack{m \geq 0 \\ (N,\,p)=1}} \lambda(N)\,\lambda(p^m)\, N^{-s}\, p^{-ms}$$

$$= \left( \sum_{(N,\,p)=1} \lambda(N)\, N^{-s} \right) \left( \sum_{m \geq 0} \lambda(p^m)\, p^{-ms} \right). \qquad (16.55)$$

To complete the proof, we must show that

$$\left( \sum_{m=0}^{\infty} \lambda(p^m)\, p^{-ms} \right) \left( I - \lambda(p)\, p^{-s} + p^{k-1-2s} I \right) = \lambda(1) = I. \qquad (16.56)$$

However, by (16.22) and (16.34), the left-hand side of (16.56) may be written as

$$\sum_{m=0}^{\infty} \lambda(p^m)\, p^{-ms} - \sum_{m=0}^{\infty} \lambda(p)\,\lambda(p^m)\, p^{-(m+1)s} + \sum_{m=0}^{\infty} p^{k-1}\, p^{-(m+2)s}\, \lambda(p^m)$$

$$= \sum_{m=0}^{\infty} \lambda(p^m) p^{-ms} - \sum_{m=1}^{\infty} \left( \lambda(p^{m+1}) + \lambda(p^{m-1})\, p^{k-1} \right) p^{-(m+1)s} - \lambda(p)\, p^{-s}$$

$$+ \sum_{m=0}^{\infty} \lambda(p^m)\, p^{k-1}\, p^{-(m+2)s} \;=\; \lambda(1) \;=\; I.$$

From (16.56) we have

$$\sum_{m=0}^{\infty} \lambda(p^m)\, p^{-ms} = \left( I - \lambda(p)\, p^{-s} + p^{k-1-2s} I \right)^{-1}$$

and this, combined with (16.55), proves the theorem.

## 16.5  Eigenfunctions of the Hecke Operators

The $\kappa$ matrices $B^{\nu} = (b^{\nu}_{\rho\sigma})$ play a central role in the study of Hecke operators. Since they are mutually commutative, there exists a matrix $A$ such that $A^{-1} B^{\nu} A$, $\nu = 1, \ldots, \kappa$, are all upper triangular, that is, all entries below the main diagonal are zero. This implies that the matrix

$$B^*(\tau) = A^{-1} B(\tau) A = \sum_{\nu} F^{\nu}(\tau) A^{-1} B^{\nu} A$$

is also upper triangular. Now the eigenvalues of $B(\tau)$ are the same as those for $B^*(\tau)$, but $B^*(\tau)$ is upper triangular, so that the diagonal elements of $B^*(\tau)$ are the eigenvalues of $B(\tau)$. Thus, the eigenvalues of $B(\tau)$ are linear combinations of the $F^{\rho}(\tau)$. With this in mind, we may state Hecke's theorem on eigenfunctions of the operators $T(n)$:[7]

For a $k$-form $F$, that is, a modular form of weight $k$, the following statements are equivalent:

---

[7] See Hecke (1983) p. 80 or Hecke (1959) p. 668.

1. $F(\tau)$ is an eigenfunction of the algebra of operators generated by $T(n)$; that is, with $c$ a constan

$$F(\tau)\big|T(n) = c\,F(\tau), \quad n \geq 1.$$

2. $F(\tau)$ has an Euler product, that is,

if     $$F(\tau) = \sum_{n=0}^{\infty} a(n)\,e^{2\pi i n \tau},$$

then     $$\sum_{n=1}^{\infty} a(n)\,n^{-s} = a(1) \prod_{p} \left(1 - a(p)\,p^{-s} - p^{k-1-2s}\right)^{-1}.$$

3. $F(\tau)$, except for a constant factor, is a characteristic root of the matrix $B(\tau)$.

To prove that statement 1 implies 3, let $F$ be an eigenfunction and let $F = F^1$, $F^2$, ..., $F^\kappa$ be a basis of the vector space of $k$-forms. Then the entries of the first row of $\lambda(n)$ are given by $\lambda_{11}(n), 0, 0, \ldots, 0$; therefore the function

$$f_{11}(\tau) = \sum_{n} \lambda_{11}(n)\,e^{2\pi i n \tau}$$

is a characteristic root of $B(\tau)$.

To prove that statement 3 implies 2, let $F(\tau)$ be a characteristic root of $B(\tau)$ and let $A$ be a matrix such that $A^{-1}B(\tau)A$ is an upper triangular form. Then $F(\tau)$ must be on the diagonal, say $F = f_{\rho\rho}$. We may observe that for each form on the diagonal

$$f_{\sigma\sigma}(\tau) = \sum_{n} \lambda_{\sigma\sigma}(n)\,e^{2\pi i n \tau},$$

$$\lambda_{\sigma\sigma}(n) \cdot \lambda_{\sigma\sigma}(m) = \sum_{d \mid n, m} \lambda_{\sigma\sigma}\left(\frac{nm}{d^2}\right) d^{k-1}, \quad \sigma = 1, 2, \ldots, \kappa. \tag{16.57}$$

But when $\sigma = \rho$, $\lambda_{\sigma\sigma}(n) = a(n)$; it then follows from (16.57) that

$$a(n) \cdot a(m) = \sum_{d \mid n, m} a\left(\frac{nm}{d^2}\right) d^{k-1}. \tag{16.58}$$

Thus, $F(\tau)$ has an Euler product.

To show that statement 2 implies 1, let $F(\tau)$ have an Euler product. This means that the coefficients $a(n)$ in the Fourier expansion of $F(\tau)$ satisfy (16.58). However, by (16.39)

$$F(\tau)\big|T(n) = \sum_{N} b(N)\,e^{2\pi i N \tau}, \quad b(N) = \sum_{d \mid N, n} a\left(\frac{Nn}{d^2}\right) d^{k-1}.$$

Compare this with (16.58) to obtain

$$b(N) = a(n) \cdot a(N).$$

This leads us to conclude that

$$F(\tau)\big|T(n) = a(n)F(\tau),$$

or that $F(\tau)$ is an eigenfunction of the operator $T(n)$ corresponding to the eigenvalue $a(n)$ and the proof is complete.

A holomorphic $k$-form, that is, a modular form of weight $k$, providing it does not vanish at $\tau = i\infty$, is a constant times the Eisenstein series $G_k(\tau)$ and thus it has an Euler product. Any holomorphic $k$-form $F(\tau)$ may be written as

$$F(\tau) = CG_k(\tau) + f(\tau),$$

where $f(\tau)$ is a cusp form, that is, $f(i\infty) = 0$. The vector subspace of cusp forms of weight $k$ is invariant under the action of the operators $T(n)$. Therefore, there exists at least one eigenfunction that is a cusp form if $\kappa > 1$. Moreover, if $\kappa = 2$, then each cusp form has an Euler product. For $\kappa > 2$, Hecke could not be sure that there were $\kappa$ independent eigenfunctions, since there was no guarantee that the eigenvalues of the matrix $B(\tau)$ were distinct.

Recall that the dimension $\kappa$ of the space of holomorphic $k$-forms, or the functions of signature $(1, k, (-1)^{\frac{k}{2}})$ is given by (16.1). For example, when $k = 12$, the dimension would be $\kappa = 2$, the two eigenfunctions would be $G_{12}(\tau)$ and the discriminant function

$$\Delta(\tau) = g_2^3 - 27g_3^2.$$

The dimension formulas (16.1) make it clear that $\kappa = 2$ when $k = 12, 16, 18, 20, 22, 26$ and that the corresponding cusp forms

$$\Delta, \ \Delta G_4, \ \Delta G_6, \ \Delta G_8, \ \Delta G_{10}, \ \Delta G_{14} \tag{16.59}$$

have Euler products. Concerning this fact, Hecke wrote, "This is a curious arithmetical property of the coefficients of $\Delta(\tau)$ and of the Eisenstein series, the sum $\sigma_r(n)$ of powers of divisors of $n$ for $r = 3, 5, 7, 9, 13$."[8] Hecke also noted that he had computed the $2 \times 2$ matrices $B(\tau)$, acting on two-dimensional spaces of cusp forms, that occurred when $24 \leq k \leq 38$, $k \neq 26, 36$. In each case he found that there existed two linearly independent cusp forms of weight $k$ with an Euler product. He wrote,[9]

> These forms are roots of a quadratic equation and this equation may be solved – in general – by adjunction of a square root $\sqrt{D}$. In all numerical cases I have found that $D > 0$. For example, consider the case $k = 24$, the system of cusp-forms being $\Delta^2$, $\Delta \cdot G_{12}$. Here $D = 144169$ (a prime number!). This means that the coefficients of the series with an Euler product are numbers of the field $K(\sqrt{D})$ [$= Q(\sqrt{D})$].

In 1939, Hecke's student, Hans Petersson (1902–1984), defined an inner product on the space of cusp forms of dimension $\kappa - 1$. The operators $T(n)$ are Hermitian with respect to this inner product, so that Petersson was able to show that there exist $\kappa - 1$ linearly independent eigenfunctions of $T(n)$. By combining this with $G_k(\tau)$, he showed that there were $\kappa$ independent functions with Euler products. Moreover, since the eigenvalues of Hermitian matrices are real, we can see why $D > 0$, as Hecke found.

Ramanujan maintained that the problem contained in the first exercise in Chapter 13 could be solved by Mordell's method. Observe that the solution to that

---

[8] Hecke (1983) p. 82.    [9] Ibid.

problem is equivalent to Hecke's statement that the last five functions in (16.59) have Euler products.

## 16.6   The Petersson Inner Product

Let $F$ denote a fundamental domain of the group $\Gamma(1)$. Next, set $f$ and $g$ as two modular $k$-forms, at least one of which is a cusp form. Suppose $\tau = x + iy$, $x$ real, and take $y > 0$ as a point in the upper half-plane $H$. Now observe that the measure

$$f(\tau)\overline{g(\tau)}\, y^{k-2} dx\, dy = \frac{i}{2} f(\tau)\overline{g(\tau)}\, (\mathrm{Im}\,\tau)^{k-2} d\tau \wedge d\overline{\tau}$$

is invariant under the action of $\Gamma(1)$. This is true because for $A \in \Gamma(1)$

$$
\begin{aligned}
f(A\tau)&\overline{g(a\tau)}\, (\mathrm{Im}\,A\tau)^{k-2}\, dA\tau \wedge d\overline{A\tau} \\
&= f(\tau)A'(\tau)^{-k}\overline{g(\tau)}A'(\tau)^{-k}\, (\mathrm{Im}\,\tau)^{k-2}\, |A'(\tau)|^{k-2}A'(\tau)\overline{A'(\tau)}\, d\tau \wedge d\overline{\tau} \\
&= f(\tau)\overline{g(\tau)}\, (\mathrm{Im}\,\tau)^{k-2}\, d\tau \wedge d\overline{\tau}.
\end{aligned}
\tag{16.60}
$$

Petersson defined the inner product of $f$ and $g$ by a double integral over $F$:

$$
\begin{aligned}
(f,\, g) &= \int\!\!\int_F f(\tau)\overline{g(\tau)}\, y^{k-2} dx\, dy \\
&= \int\!\!\int_{AF} f(\tau)\overline{g(\tau)}\, y^{k-2} dx\, dy,
\end{aligned}
\tag{16.61}
$$

where $A \in \Gamma(1)$; note that (16.61) follows by an application of (16.60). Thus, the inner product (16.60) is independent of the choice of the fundamental domain. The integrals in (16.61) converge when at least one of the functions $f$ and $g$ is a cusp form. Suppose $f$ is a cusp form. Then

$$f(\tau) = a_1 e^{2\pi i \tau} + a_2 e^{4\pi i \tau} + \cdots$$

$$f(x + iy) = O(e^{-2\pi y}) \quad \text{as } y \to \infty.
\tag{16.62}$$

Thus,

$$
\begin{aligned}
(f,\, g) &= \int_0^1 \int_{\sqrt{1-x^2}}^{\infty} f(x+iy)\overline{g(x+iy)}\, y^{k-2} dx\, dy \\
&= \int_0^1 \int_{\sqrt{1-x^2}}^{\infty} O(e^{-2\pi y})\, O(1)\, y^{k-2} dy\, dx \; < \infty.
\end{aligned}
$$

It is easy to see that $(f,\, g)$ defines an inner product on the space of cusp forms of weight $k$ because it satisfies the characteristic properties:

- $(f, g) = \overline{(g, f)}$
- $(a_1 f_1 + a_2 f_2, g) = a_1\,(f_1, g) + a_2\,(f_2, g)$
- $(f, f) \geq 0$ and $(f, f) = 0$ only when $f = 0$.

Petersson's proof that the operators $T(n)$ are Hermitian had three parts. In the first and main part of the proof, he showed that for cusp forms $f$ and $g$

$$(f(\tau)|T(p),\, g(\tau)) = \big(f(\tau),\, g(\tau)|T(p)\big), \quad p \text{ prime}.
\tag{16.63}$$

Secondly, repeated application of this result gave a proof for any positive integer power $k$ or $T(p)$. Since $T(n)$ was a polynomial

$$T(p_1), \ T(p_2), \ \ldots, \ T(p_r) \quad \text{where} \quad n = p_1^{l_1} p_2^{l_2} \cdots p_r^{l_r},$$

Petersson was finally able to deduce from (16.63) that for any positive integer $n$

$$\left( f(t) \big| T(n), \ g(\tau) \right) = \left( f(t), \ g(\tau) \big| T(n) \right). \tag{16.64}$$

Continue to denote by $\lambda(n)$ the $\kappa - 1 \times \kappa - 1$ matrix obtained by the action of $T(n)$ on the space of cusp forms of weight $k$ and with respect to a basis $F^\sigma, \sigma = 1, 2, \ldots, \kappa - 1$ of this space. Thus, taking $B^\nu, \nu = 1, 2, \ldots, \kappa - 1$, as $\kappa - 1 \times \kappa - 1$ matrices, (16.52) continues to hold. Moreover, the $B^\nu$ form a basis for the algebra generated by the commutative Hermitian matrices $\lambda(n)$. From this we may perceive that the $\kappa - 1$ matrices $B^\nu$ are commutative and Hermitian, so that there exists a unitary matrix $A = (a_{\nu\sigma})$ such that $A, B^\nu, A^{-1}$, with $\nu = 1, 2, \ldots, \kappa - 1$, are all diagonal matrices. Therefore all the matrices in $\Lambda$ and, in particular, all the $\lambda(n)$, are diagonalizable by $A$:

$$A \, \lambda(n) A^{-1} = \left( \Lambda_\nu(n) \, \delta_{\nu\sigma} \right),$$

where $\delta_{\nu\sigma}$ denotes the Kronecker $\delta$ function. Now define

$$H^\nu(\tau) = \sum_{\sigma=1}^{\kappa-1} a_{\nu\sigma} F^\sigma(\tau), \quad \nu = 1, 2, \ldots, \kappa - 1. \tag{16.65}$$

We show that (16.65) are the required $\kappa - 1$ eigenfunctions for all $T(n)$: If we take $F$ and $H$ as each denoting the $\kappa - 1$ dimensional vectors:

$$F = \begin{pmatrix} F^1 \\ F^2 \\ \cdot \\ \cdot \\ \cdot \\ F^{\kappa-1} \end{pmatrix} \quad \text{and} \quad H = \begin{pmatrix} H^1 \\ H^2 \\ \cdot \\ \cdot \\ \cdot \\ H^{\kappa-1} \end{pmatrix},$$

then

$$H \big| T(n) = AF \big| T(n) = A\lambda(n)F = A\lambda(n)A^{-1}H$$
$$= \left( \Lambda_\nu(n) \, \delta_{\nu\sigma} \right) H,$$

or

$$H^\nu(\tau) \big| T(n) = \Lambda_\nu(n) H^\nu(\tau), \quad \nu = 1, 2, \ldots, \kappa - 1.$$

The Dirichlet series corresponding to $H^\nu(\tau)$, $\nu = 1, 2, \ldots, \kappa - 1$, have Euler products as well as analytic continuations to entire functions. This means that there must be exactly $\kappa$ Dirichlet series of signature $\{1, k, (-1)^{\frac{k}{2}}\}$, with $k$ even and greater than 4. One of these series is $\zeta(s)\zeta(s - k + 1)$ and the others are entire functions.

Joseph Lehner's remarks on Petersson and his work are insightful:[10]

---

[10] Lehner (1964) p. 35.

The most important contributor to the theory of automorphic functions in recent times is H. Petersson, whose investigations begin about 1930. He was a student of Hecke and much of his work consists in extending to more general discontinuous groups what Hecke developed for congruence subgroups of the modular group. Petersson, like Klein and Hecke, is greatly interested in the correspondence between Riemann surface theory and automorphic function theory. Because he insists on considerable generality, his papers are hard to read, but it is also true that he returns again and again to the classical examples from which the general theory sprang, reinterpreting and deepening them with newer results.

## 16.7  Exercises

1. Show that $\Delta^2$ is a cusp form of weight 24 but that the corresponding Dirichlet series cannot have an Euler product. See Hecke (1959) p. 588.

2. Observe that $\Delta E_6^2$ is also a cusp form of weight 24, where (with $h = e^{2\pi i\tau}$)

$$E_6(\tau) = 1 - 504(h + 33h^2 + 244h^3 + 1057h^4 + \cdots)$$

   and

$$\Delta(\tau) = h - 24h^2 + 252h^3 - 1472h^4 + \cdots .$$

   Let $f_1 = \Delta E_6^2$ and $f_2 = \Delta^2$. Calculate $f_1$ and $f_2$ up to the fourth power of $h$. Use (16.39) to figure $f_1 | T(2)$ and $f_2 | T(2)$ up to the second power of $h$. Use these results to find $a, b, c, d$ such that

$$\begin{pmatrix} f_1 \\ f_2 \end{pmatrix} \bigg| T(2) = \begin{pmatrix} a & b \\ c & d \end{pmatrix} \begin{pmatrix} f_1 \\ f_2 \end{pmatrix}.$$

   Let $\lambda_1, \lambda_2$ be the eigenvalues of the $2 \times 2$ matrix. Prove that $cf_1 + (d - \lambda_1)f_2$ and $cf_1 + (d - \lambda_2)f_2$ are the simultaneous eigenforms of all the operators $T(n)$. Show also that the coefficients of the normalized eigenforms are in $Q(\sqrt{144169})$.

# *Appendix*
# *Translation of Hurwitz's Paper of 1904*

Über die Theorie der elliptischen Modulfunktionen
Von
A. Hurwitz in Zürich
(*Mathematische Annalen*, Bd. 58, 1904, S. 343–360.)
On the theory of elliptic modular functions
by Adolf Hurwitz

In my dissertation,[1] in connection with the work of R. Dedekind[2] and F. Klein,[3] I developed the foundations for a theory of elliptic modular functions. However, I have since found that my presentation can be considerably simplified on some points. Therefore, in the present work I return to the subject discussed in my dissertation. To avoid writing at too great length, I here limit myself to the first elements of the theory. However, I will deal with these in full detail in order, on the one hand, to permit a complete overview of the tools I employ to ground the theory and, on the other hand, to lay out the present work in such a manner that it is understandable without the aid of other treatises. That the reflections communicated here may also be applicable to the theory of automorphic functions, if one generalizes them appropriately, will not escape the informed reader.

## §1. Equivalent Quantities

That part of the plane of complex numbers representing numbers with positive imaginary component is to be designated the "positive half-plane," the part representing numbers with negative imaginary component the "negative half-plane." The common boundary of these half-planes is formed by the axis of the real numbers.

---

[1] Gundlagen einer independenten Theorie der elliptischen Modulfunktionen und Theorie der Multiplikatorgleichungen erster Stufe, Mathem. Annalen, Bd. 18 (1881), S. 528–592.

[2] Schreiben an Herrn Borchardt über die Theorie der elliptischen Modulfuntionen, Crelle's Journal, Bd. 83 (1877), S. 265–292.

[3] Über die Transformation der elliptischen Funktionen und die Auflösung der Gleichungen fünften Grades, Mathem. Annalen, Bd. 14 (1879/79), S. 111–172. F. Klein and R. Fricke gave a comprehensive presentation of the theory in their "Vorlesungen über die Theorie der elliptischen Modulfunktionen" (Bd. I, Leipzig 1890; Bd. II, Leipzig 1892).

445

Now if

$$\omega = x + iy, \quad (y > 0) \tag{1}$$

is a point in the positive half-plane, then

$$H(\omega) = \frac{x^2 + y^2 + 1}{y} \tag{2}$$

should be called the "height" of the point $\omega$.[4] The height $H(\omega)$ always has a positive finite value. It may also be expressed by $\omega$ and the conjugate of $\omega$,

$$\overline{\omega} = x - iy,$$

in the form

$$H(\omega) = 2i \cdot \frac{\omega \overline{\omega} - 1}{\omega - \overline{\omega}}. \tag{3}$$

Furthermore, if there exists an equation of the form

$$\omega' = \frac{\alpha \omega + \beta}{\gamma \omega + \delta} \tag{4}$$

between the two values $\omega$ and $\omega'$, in which $\alpha$, $\beta$, $\gamma$, $\delta$ denote whole numbers of determinant $\alpha\delta - \beta\gamma = 1$, then we designate the values $\omega$ and $\omega'$ as "equivalent," a term we also assign to the points in the complex plane representing these values. The equation following from (4)

$$\frac{1}{2i}(\omega' - \overline{\omega}) = \frac{1}{2i}(\omega - \overline{\omega}) \cdot \frac{1}{(\gamma \omega + \delta)(\gamma \overline{\omega} + \delta)} \tag{5}$$

shows us that the imaginary components of $\omega$ and $\omega'$ have the same sign. So if $\omega$ lies in the positive half-plane, then the same holds for every point $\omega'$ equivalent to $\omega$.

Because two quantities equivalent to a third are also equivalent to one another (a fact illustrated by the "group property" of substitution (4)), the points of the positive half-plane arrange themselves such that two points belonging to the same system are equivalent to one another and two points belonging to different systems are not equivalent. If $\omega$ is any point in such a system, then (4) produces all points $\omega'$ of this system, providing that $\alpha$, $\beta$, $\gamma$, $\delta$ run through all possible integers of determinant $\alpha\delta - \beta\gamma = 1$.

The height of the points $\omega'$ defined by equation (4) may now be expressed according to (3) and (5) in the form

$$H(\omega') = 2i \frac{(\alpha\omega + \beta)(\alpha\overline{\omega} + \beta) + (\gamma\omega + \delta)(\gamma\overline{\omega} + \delta)}{\omega - \overline{\omega}} \tag{6}$$

From this we extract two important inferences.

On the one hand, it is clear for the special case $\alpha = \delta = 0$, $\beta = -1$, $\gamma = 1$ that

$$H\left(-\frac{1}{\omega}\right) = H(\omega). \tag{7}$$

Thus, the two points $\omega$ and $-\frac{1}{\omega}$ always have the same height.

---

[4] The introduction of the notion of height is justified by considerations to be given later. But I would like to note here that I have been led to this notion by very general researches on automorphic functions of one and several variables.

On the other hand, we note that the numerator of expression (6) is a definite (quadratic) form of the substitution coefficients $\alpha, \beta, \gamma, \delta$. From this it follows that among all whole-number systems $\alpha, \beta, \gamma, \delta$ $(\alpha\delta - \beta\gamma = 1)$ there are one or more for which $H(\omega')$ turns out to be as small as possible. Thus, in a system of equivalent points there is always one or more of minimum height.

Now we consider an arbitrary system of equivalent points! Within this system, let $\omega$ be one (or one of those) of minimum height. According to equation (7) we may assume

$$|\omega| \geqq 1, \tag{8}$$

that is, the absolute value of $\omega$ is not smaller than 1. Otherwise, one could replace the point $\omega$ with the point $-\frac{1}{\omega}$, having the same minimum height.

For every arbitrary point

$$\omega' = \frac{\alpha\omega + \beta}{\gamma\omega + \delta}$$

in the system of points under consideration, we have

$$\frac{H(\omega')}{H(\omega)} = \frac{(\alpha\omega + \beta)(\alpha\overline{\omega} + \beta) + (\gamma\omega + \delta)(\gamma\overline{\omega} + \delta)}{\omega\overline{\omega} + 1} \geqq 1.$$

For the special cases,

$$\begin{pmatrix} \alpha & \beta \\ \gamma & \delta \end{pmatrix} = \begin{pmatrix} 1 & 1 \\ 0 & 1 \end{pmatrix}, \quad \begin{pmatrix} 1 & -1 \\ 0 & 1 \end{pmatrix}$$

this inequality yields

$$\omega + \overline{\omega} \geqq -1, \quad \omega + \overline{\omega} \leqq 1. \tag{9}$$

A point $\omega$, satisfying the inequalities (8) and (9) lies then on the inside or on the boundary of a region of the positive half-plane, such region extending to infinity and bordered by the circle with center 0 and radius 1 and by two straight lines running parallel to the imaginary axis in the interval $+\frac{1}{2}, -\frac{1}{2}$, respectively. This region will hereafter always be denoted by $G$.

We therefore have the following theorem:

"In a system of equivalent points there is always one which lies on the inside or on the boundary of the region $G$."

Let $AC$, $A'C'$ be the straight portions and $ABA'$ the curved portion of the boundary of $G$, in which $B$ denotes the center of the arc $ABA'$; this center lies on the imaginary axis. This is shown in Figure 1.

If the point $\omega$ traverses the line $AC$, then the equivalent point $\omega + 1$ determines the line $A'C'$, and if the point $\omega$ traverses the arc $AB$, then the equivalent point $-\frac{1}{\omega}$ determines the arc $A'B$. Thus, the boundary points are pairwise equivalent and this fact leads us to the following determination:

Of the points on the boundary of the region $G$, only those located on the lines $AC$ and $AB$ may be considered as part of the region $G$; that is, when one states that a point

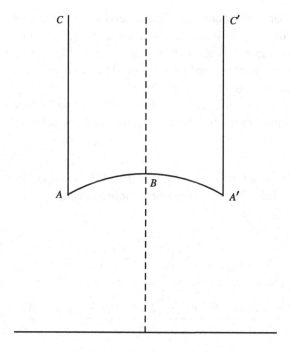

Figure 1. [Hurwitz's Figure 21.]

"lies in the region *G*," or "belongs to the region *G*," then we are to understand by this, that the point lies either inside *G* or on one of the boundary lines *AC* or *AB*.

From the above theorem it now follows immediately:

*In a system of equivalent points, there is always one belonging to the domain G.*

It would now be easy to also show that there is always only one single such point in a system of equivalent points.[5] But we need not stop to prove this fact, since in the course of our further researches, it will become apparent on its own.

## §2. The Modular Forms $G_n(\omega_1, \omega_2)$

Now let $\omega_1$ and $\omega_2$ be two complex variables, which are initially subject only to the condition that their quotient $\frac{\omega_1}{\omega_2}$ is finite and nonreal.

---

[5] It follows from this, that the inequality

$$\frac{H(\omega')}{H(\omega)} = \frac{(\alpha\omega + \beta)(\alpha\overline{\omega} + \beta) + (\gamma\omega + \delta)(\gamma\overline{\omega} + \delta)}{\omega\overline{\omega} + 1} > 1$$

is always satisfied if $\omega$ lies on the inside of *G*, apart from the case

$$\begin{pmatrix} \alpha & \beta \\ \gamma & \delta \end{pmatrix} = \begin{pmatrix} 1 & 0 \\ 0 & 1 \end{pmatrix}, \quad \begin{pmatrix} 0 & -1 \\ 1 & 0 \end{pmatrix};$$

it also follows that the analogous situation holds if $\omega$ lies on the boundary of *G*.

The sum

$$G_n \equiv G_n(\omega_1, \omega_2) = \sum \frac{1}{(m_1\,\omega_1 + m_2\,\omega_2)^n}, \tag{10}$$

extending over all pairs of whole numbers $m_1, m_2$ with the exception of the pair $m_1 = 0, m_2 = 0$, is thus always of finite value, independent of the order of summation, as long as $n$ is greater than 2.

From this it follows that $G_n(\omega_1, \omega_2)$ for $n > 2$ is a "modular form," by which we wish to imply nothing other than the fact that $G_n(\omega_1, \omega_2)$ represents a homogeneous function of $\omega_1, \omega_2$, which remains unchanged if $\omega_1, \omega_2$ are subjected to a linear whole-number substitution of determinant 1.

That is,

$$G_n\big(\omega_1', \omega_2'\big) = G_n(\omega_1, \omega_2), \quad (n > 2) \tag{11}$$

holds when

$$\omega_1' = \alpha\,\omega_1 + \beta\,\omega_2, \quad \omega_2' = \gamma\,\omega_1 + \delta\,\omega_2 \quad (\alpha\delta - \beta\gamma = 1), \tag{12}$$

with $\alpha, \beta, \gamma, \delta$ understood to be whole numbers.

Incidentally, of the sums $G_n$, only a few with even index $n$ are of interest, since for odd $n$ the terms in the sum (10) corresponding to number pairs $(m_1, m_2)$ and $(-m_1, -m_2)$ cancel one another, so that $G_n$ is identically zero.

As we shall soon see, the sums $G_1$ and $G_2$, corresponding to the indices $n = 1, n = 2$, are not independent of the order of summation. The considerations below yield further details on these sums.

Let $m$ denote the members of any given arbitrarily chosen system of whole numbers, all different from one another. Then, to simplify, the meaning of the notation

$$\sum_m \varphi(m) \tag{13}$$

will be assigned as follows. One understands

$$\sum_{-\lambda}^{+\lambda} \varphi(m)$$

to mean the sum of those values of $\varphi(m)$ which correspond to the values of $m$ satisfying the condition

$$-\lambda \leqq m \leqq +\lambda.$$

We then define the symbol (13) by the equation

$$\sum_m \varphi(m) = \lim_{\lambda=\infty} \sum_{-\lambda}^{+\lambda} \varphi(m). \tag{14}$$

Furthermore, if $m_1, m_2$ run through all the members of an infinite system of number pairs, all different from one another, then the symbol

$$\sum_{m_1} \sum_{m_2} \varphi(m_1, m_2) \tag{15}$$

should have the following meaning: We first construct for a fixed value of $m_1$,

$$\sum_{m_2} \varphi(m_1, m_2) = \psi(m_1);$$

this sum is a function of $m_1$ and as such it can be denoted as $\psi(m_1)$. Now we define:

$$\sum_{m_1} \sum_{m_2} \varphi(m_1, m_2) = \sum_{m_1} \psi(m_1). \tag{16}$$

Having stipulated this, one is able, for example, to distinguish between the two sums

$$\sum_{m_1} \sum_{m_2} \varphi(m_1, m_2) \quad \text{and} \quad \sum_{m_2} \sum_{m_1} \varphi(m_1, m_2).$$

These sums will, of course, represent the same value in the case of absolute convergence; they can have different values, however, in the case of conditional convergence. On these premises, we consider the equation, valid for every finite nonwhole number $a$

$$\sum_{m} \frac{1}{a+m} = \pi \cot(a\pi) = i\pi \frac{e^{ai\pi} + e^{-ai\pi}}{e^{ai\pi} - e^{-ai\pi}}, \tag{17}$$

in which $m$ runs through all whole-number values.

Using this equation, we easily derive the value of the sum

$$S = \sum_{m_1} \sum_{m_2} \frac{1}{(m_1 - n_1)\omega_1 + (m_2 - n_2)\omega_2} \tag{18}$$

in which $n_1$ and $n_2$ denote any two fixed whole numbers and $m_1, m_2$ run through all number pairs with the exception of the pairs $m_1 = n_1$ and $m_2 = n_2$. If we set

$$\frac{\omega_1}{\omega_2} = \omega$$

then we have, first,

$$\sum_{m_2} \frac{1}{(m_1 - n_1)\omega_1 + (m_2 - n_2)\omega_2} = \frac{1}{\omega_2} \sum_{m_2} \frac{1}{(m_1 - n_1)\omega - n_2 + m_2}$$

$$= \frac{\pi}{\omega_2} \cot \pi[(m_1 - n_1)\omega - n_2]$$

$$= \frac{\pi}{\omega_2} \cot(m_1 - n_1)\omega\pi,$$

if $m_1$ is different from $n_1$. For $m_1 = n_1$, however, the value of this sum

$$\left( \sum_{m_2} \frac{1}{(m_2 - n_2)\omega_2} \right),$$

as one may easily discern, is equal to zero. Therefore we have

$$S = \frac{\pi}{\omega_2} \sum_{m_1} \cot(m_1 - n_1)\,\omega\pi, \tag{19}$$

where $m_1$ runs through all whole numbers different from $n_1$. Furthermore,

$$\sum_{-\lambda}^{+\lambda} \cot(m_1 - n_1)\,\omega\pi = \sum_{m=-\lambda-n_1}^{m=\lambda-n_1} \cot(m\,\omega\pi)$$

$$= \sum_{-\lambda-n_1}^{+\lambda+n_1} \cot(m\,\omega\pi) - \sum_{m=\lambda-n_1+1}^{m=\lambda+n_1} \cot(m\,\omega\pi).$$

Because the cotangent is an odd function, each two terms in the penultimate sum cancel one another, so that

$$S = -\frac{\pi}{\omega_2} \cdot \lim_{\lambda=\infty} \sum_{m=\lambda-n_1+1}^{m=\lambda+n_1} \cot(m\,\omega\pi). \tag{20}$$

Each of the $2n_1$ terms

$$\cot(m\omega\pi) \qquad (m = \lambda - n_1 + 1, \lambda - n_1 + 2, \ldots, \lambda + n_1)$$

of the last sum approaches, as $\lambda$ increases to infinity, the limit $-i$ or $i$, depending on whether $\omega$ has a positive or negative imaginary component. This follows immediately from the equation

$$\cot(m\,\omega\pi) = i \cdot \frac{e^{im\,\omega\pi} + e^{-im\,\omega\pi}}{e^{im\,\omega\pi} - e^{-im\,\omega\pi}}.$$

Thus, we finally obtain

$$S = \sum_{m_1} \sum_{m_2} \frac{1}{(m_1 - n_1)\,\omega_1 + (m_2 - n_2)\,\omega_2} = \pm\frac{2i\pi}{\omega_2}\,n_1, \tag{21}$$

where the upper or lower sign is applied according as $\omega$ lies in the positive or the negative half-plane. Equation (21) clearly provides for the determination of the value of the sum $G_1(\omega_1, \omega_2)$ by a specific ordering of the terms of the sum.[6]

We may reduce the sum

$$\sum_{m_2} \sum_{m_1} \frac{1}{(m_1 - n_1)\,\omega_1 + (m_2 - n_2)\,\omega_2}$$

to the one we just considered by exchanging $\omega_1$ and $\omega_2$. Inasmuch as $\frac{\omega_2}{\omega_1}$ has a positive-or-negative imaginary component, according to whether $\frac{\omega_1}{\omega_2}$ has a negative-or-positive imaginary component, from (21) one immediately gleans:

$$\sum_{m_2} \sum_{m_1} \frac{1}{(m_1 - n_1)\,\omega_1 + (m_2 - n_2)\,\omega_2} = \mp\frac{2i\pi}{\omega_1}\,n_2, \tag{22}$$

---

[6] In the above proof of equation (21) it was assumed that $n_1 \geq 0$. But if one multiplies the equation by $-1$ and then replaces $m_1, m_2, n_2$ by $-m_1, -m_2, -n_2$, then one perceives that equation (21) also holds for $n_1 < 0$.

where, again, the upper or lower sign holds, according as $\frac{\omega_1}{\omega_2}$ lies in the positive or negative half-plane.

Equations (21) and (22) provide evidence of the conditional convergence of the sum $G_1(\omega_1, \omega_2)$.

Concerning the sum $G_2(\omega_1, \omega_2)$, it is sufficient for our purposes to compare the following two orderings:

$$G_2' = \sum_{m_1} \sum_{m_2} \frac{1}{(m_1 \omega_1 + m_2 \omega_2)^2}, \quad G_2'' = \sum_{m_2} \sum_{m_1} \frac{1}{(m_1 \omega_1 + m_2 \omega_2)^2}. \tag{23}$$

Finally, we note that the sum

$$s = \sum \left[ \frac{1}{(m_1 - 1)\omega_1 + (m_2 - 1)\omega_2} - \frac{1}{m_1\omega_1 + m_2\omega_2} - \frac{\omega_1 + \omega_2}{(m_1\omega_1 + m_2\omega_2)^2} \right]$$

$$= (\omega_1 + \omega_2)^2 \sum \frac{1}{(m_1\omega_1 + m_2\omega_2)^2((m_1 - 1)\omega_1 + (m_2 - 1)\omega_2)} \tag{24}$$

is absolutely convergent. The summation here should extend over all pairs of whole numbers $m_1, m_2$, with the exception of the two pairs $m_1 = 1, m_2 = 1$ and $m_1 = 0, m_2 = 0$. If one now sums first with respect to $m_2$, and then with respect to $m_1$, then by (21) we get

$$s = \frac{3}{\omega_1 + \omega_2} \pm \frac{2i\pi}{\omega_2} - (\omega_1 + \omega_2) \sum_{m_1} \sum_{m_2} \frac{1}{(m_1\omega_1 + m_2\omega_2)^2}. \tag{25}$$

However, if one sums first with respect to $m_1$ and then with respect to $m_2$, then the result is

$$s = \frac{3}{\omega_1 + \omega_2} \mp \frac{2i\pi}{\omega_1} - (\omega_1 + \omega_2) \sum_{m_2} \sum_{m_1} \frac{1}{(m_1\omega_1 + m_2\omega_2)^2}. \tag{26}$$

A comparison of these two representations of $s$ produces the relation

$$\sum_{m_1} \sum_{m_2} \frac{1}{(m_1\omega_1 + m_2\omega_2)^2} - \sum_{m_2} \sum_{m_1} \frac{1}{(m_1\omega_1 + m_2\omega_2)^2} = \pm \frac{2i\pi}{\omega_1\omega_2}, \tag{27}$$

where the upper or lower sign holds, according as $\frac{\omega_1}{\omega_2}$ lies in the positive or negative half-plane. That the sums (23) considered here have finite values is apparent from equations (25) and (26).

## §3.  The Representation of the Function $G_n$ by Power Series

From now on we assume that

$$\omega = \frac{\omega_1}{\omega_2} = x + iy$$

has a positive imaginary component.

The absolute value of the quantity

$$h = e^{2i\pi\omega} = e^{2i\pi x} \cdot e^{-2\pi y} \tag{28}$$

is thus less than 1.

In equation (17) $a$ will now be replaced by $m_1\omega_1$, where $m_1$ is understood to be a positive whole number, and then the right side will be expanded in terms of the power of $h$. We thus have

$$\sum_{m_2} \frac{1}{m_1\omega + m_2} = -i\pi - 2i\pi \sum_{r=1}^{\infty} h^{m_1 r}. \tag{29}$$

Differentiating $(n-1)$ times with respect to $\omega$ further yields

$$\sum_{m_2} \frac{1}{(m_1\omega + m_2)^n} = (-1)^n \cdot \frac{(2i\pi)^n}{(n-1)!} \sum_{r=1}^{\infty} r^{n-1} h^{m_1 r}, \quad (n \geqq 2). \tag{30}$$

If we sum over all positive whole numbers $m_1$, we obtain

$$\sum_{m_1=1}^{\infty} \sum_{m_2} \frac{1}{(m_1\omega + m_2)^n} = (-1)^n \cdot \frac{(2i\pi)^n}{(n-1)!} \sum_{r=1}^{\infty} r^{n-1} \frac{h^r}{1-h^r}, \quad (n \geqq 2). \tag{31}$$

Now here, when we replace $m_2$ by $-m_2$, as is clearly permitted, and then multiply both sides by $(-1)^n$, we obtain

$$\sum_{m_1=-1}^{-\infty} \sum_{m_2} \frac{1}{(m_1\omega + m_2)^n} = \frac{(2i\pi)^n}{(n-1)!} \sum_{r=1}^{\infty} r^{n-1} \frac{h^r}{1-h^r}, \quad (n \geqq 2). \tag{A.31a}$$

From (31), (A.31a), and the well-known equation

$$\sum_{m_1} \frac{1}{m_1^{2n}} = \frac{(2\pi)^{2n}}{(2n)!} B_n, \tag{32}$$

in which $B_n$ denotes the $n$th Bernoulli number, we finally arrive at

$$\sum_{m_1} \sum_{m_2} \frac{1}{(m_1\omega_1 + m_2\omega_2)^{2n}}$$
$$= \left(\frac{2\pi}{\omega_2}\right)^{2n} \cdot \frac{1}{(2n)!} \left[ B_n + (-1)^n 4n \sum_{r=1}^{\infty} r^{2n-1} \frac{h^r}{1-h^r} \right], \quad (n \geqq 1). \tag{33}$$

The modular forms $G_n$ may thus be represented by means of power series. If one arranges the sum occurring on the right side according to powers of $h$, then it takes the form

$$\sum_{r=1}^{\infty} \psi_{2n-1}(r)h^r = h + (1 + 2^{2n-1})h^2 + (1 + 3^{2n-1})h^3$$
$$+ (1 + 2^{2n-1} + 4^{2n-1})h^4 + \cdots, \tag{34}$$

where $\psi_{2n-1}(r)$ denotes the sum of the $(2n-1)$th powers of the divisors of the number $r$.

### §4. The Modular Form $\Delta(\omega_1, \omega_2)$

In the case $n = 1$ the equation (33) states

$$\sum_{m_1} \sum_{m_2} \frac{1}{(m_1\omega_1 + m_2\omega_2)^2} = \frac{1}{3}\left(\frac{\pi}{\omega_2}\right)^2\left[1 - 24\sum_{r=1}^{\infty} r\frac{h^r}{1 - h^r}\right]. \qquad (35)$$

Now here, if one replaces $\omega_1$ by $-\omega_2$ and $\omega_2$ by $\omega_1$, so that

$$h = e^{2i\pi\omega} \quad \text{is replaced by} \quad h' = e^{-\frac{2i\pi}{\omega}}, \qquad (36)$$

and if at the same time the summation indices $m_1, m_2$ are replaced by $-m_2$ and $m_1$, respectively, then

$$\sum_{m_2} \sum_{m_1} \frac{1}{(m_1\omega_1 + m_2\omega_2)^2} = \frac{1}{3}\left(\frac{\pi}{\omega_1}\right)^2\left[1 - 24\sum_{r=1}^{\infty} r\frac{h'^r}{1 - h'^r}\right]. \qquad (37)$$

Accordingly, equation (27) may be represented as:

$$\left(\frac{\pi}{\omega_2}\right)^2\left[1 - 24\sum_{r=1}^{\infty} \frac{rh^r}{1 - h^r}\right] - \left(\frac{\pi}{\omega_1}\right)^2\left[1 - 24\sum_{r=1}^{\infty} \frac{rh'^r}{1 - h'^r}\right] = \frac{6i\pi}{\omega_1\omega_2}$$

or

$$\log h\left[1 - 24\sum_{r=1}^{\infty} \frac{rh^r}{1 - h^r}\right] + \log h'\left[1 - 24\sum_{r=1}^{\infty} \frac{rh'^r}{1 - h'^r}\right] = -12. \qquad (38)$$

Next, from (36) follows $\log h \cdot \log h' = 4\pi^2$ and thence

$$\frac{d\log h}{\log h} = -\frac{d\log h'}{\log h'}.$$

If one multiplies (38) by $\frac{d\log h}{\log h}$, then one may write the result as

$$\frac{dh}{h}\left[1 - 24\sum_{r=1}^{\infty} \frac{rh^r}{1 - h^r}\right] - \frac{dh'}{h'}\left[1 - 24\sum_{r=1}^{\infty} \frac{rh'^r}{1 - h'^r}\right] = -12\frac{d\log h}{\log h}.$$

If you integrate this equation term-by-term and then change from the logarithm to the number, you obtain

$$(\log h)^{12}h\prod_{r=1}^{\infty}(1 - h^r) = C \cdot h'\prod_{r=1}^{\infty}(1 - h'^r)^{24}$$

or

$$\left(\frac{2\pi}{\omega_2}\right)^{12}h\prod_{r=1}^{\infty}(1 - h^r) = \left(\frac{2\pi}{\omega_1}\right)^{12}h'\prod_{r=1}^{\infty}(1 - h'^r)^{24}, \qquad (39)$$

in which the constant of integration $C$ is found by taking $\omega = 1$, for which $h = h' = e^{-2\pi}$.

The equation (39) shows that the function

$$\Delta \equiv \Delta(\omega_1, \omega_2) = \left(\frac{2\pi}{\omega_2}\right)^{12} h \prod_{r=1}^{\infty} (1 - h^r)^{24} \tag{40}$$

remains unchanged if one replaces $\omega_1, \omega_2$ by $-\omega_2, \omega_1$, respectively. Clearly, however, $\Delta$ also remains unchanged if one replaces $\omega_1, \omega_2$: either by $-\omega_1, -\omega_2$, respectively, or by $\omega_1 + \omega_2, \omega_2$, respectively. Now by the substitutions

$$\begin{array}{ll}
\omega_1' = -\omega_2, & \omega_2' = \omega_1; \\
\omega_1' = -\omega_1, & \omega_2' = -\omega_2; \\
\omega_1' = \omega_1 + \omega_2, & \omega_2' = \omega_2
\end{array}$$

all the homogeneous whole-number substitutions of determinant 1 can be compiled, so that it finally follows that

*The function $\Delta(\omega_1, \omega_2)$ remains unchanged if $\omega_1, \omega_2$ are subjected to arbitrary, linear, homogeneous whole-number substitutions of determinant 1.*

In this respect, this function has an especially simple analytic characterization as the product

$$h \prod_{r=1}^{\infty} (1 - h^r)^{24},$$

representing a function in $\omega$, analytic in the positive half-plane and vanishing nowhere.

## §5. The Modular Function $J(\omega)$

We now put into place the abbreviation

$$g_2 \equiv g_2(\omega_1, \omega_2) = 60 \sum_{m_1} \sum_{m_2} \frac{1}{(m_1\omega_1 + m_2\omega_2)^4}, \tag{41}$$

a function, which by (33) and (34) admits of the form

$$g_2 = \left(\frac{2\pi}{\omega_2}\right)^4 \left[\frac{1}{12} + 20 \sum_{r=1}^{\infty} \psi_3(r)h^r\right]. \tag{42}$$

Using $g_2$ and $\Delta$ we then construct the quotient

$$J(\omega) = \frac{g_2^3}{\Delta} = \frac{\left[\frac{1}{12} + 20 \sum_{r=1}^{\infty} \psi_3(r)h^r\right]^3}{h \prod_{r=1}^{\infty} (1 - h^r)^{24}}, \tag{43}$$

which remains dependent only on $\omega = \frac{\omega_1}{\omega_2}$.

Out of this equation defining the function $J(\omega)$, there emerges an expansion, valid for the whole positive half-plane, taking the form

$$J(\omega) = \frac{1}{12^3 h}[1 + c_1 h + c_2 h^2 + \cdots]. \tag{44}$$

Furthermore, in accordance with the properties of $g_2$ and $\Delta$, $J(\omega)$ satisfies the equation

$$J(\omega') = J(\omega), \tag{45}$$

in which $\omega$ and $\omega'$ denote any two equivalent points in the positive half-plane.

If the ordinate $y$ of $\omega = x + iy$ increases beyond all limit, then

$$h = e^{2i\pi\omega} = e^{2i\pi x} \cdot e^{-2\pi y}$$

tends to zero. Thus, by equation (44), $J(\omega)$ will then become infinitely large. Hence, this function has the singular point

$$\omega = \infty.$$

Consequently, the points equivalent to the point $\infty$, that is, the representations of the real rational numbers, are also singular points of $J(\omega)$. Since the latter are everywhere dense on the axis of the real numbers, the function $J(\omega)$ cannot be continued beyond the positive half-plane. According to (44), the function $J(\omega)$ is everywhere analytic on the inner half of the positive half-plane. —

We are now in a position to determine how often the function $J(\omega)$ takes on a given finite value $a$ in the region $G$.

In any case, the equation

$$J(\omega) = a \tag{46}$$

has only finitely many solutions in the region $G$. Infinitely many solutions would result in an accumulation point, which must necessarily be finite, since with an infinitely increasing ordinate of $\omega$, $|J(\omega)|$ also increases beyond limit, and in particular beyond $|a|$. This accumulation point, however, would be a singular point of $J(\omega)$, but no such point exists in the positive half-plane.

Now in order to determine the number of solutions $N$ of the equation (46) lying in the region $G$, we first cut off from $G$ the finite region

$$G' = CABA'C'$$

by means of a line $CC'$ parallel to the real axis, as shown in Figure 2.

The distance of the parallel $CC'$ from the real axis should subsequently increase to infinity; it was assumed from the outset to be of such a size, that the $N$ solutions of the equation (46) would lie in the region $G'$. As is well known, we have

$$N = \frac{1}{2\pi i} \int d\log[J(\omega) - a], \tag{47}$$

the integral in the positive sense stretching through the boundary of $G'$. (Here we have assumed that none of the $N$ solutions lie on the boundary of $G$. Should the latter be the case, one may circumvent the zero points lying on the boundary by integrating $J(\omega) - a$ using infinitesimal deviations.)

We decompose the integral according to the following scheme

$$\int_A^B - \int_{A'}^B + \int_{A'}^{C'} - \int_A^C + \int_{C'}^C.$$

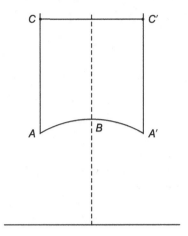

Figure 2. [Hurwitz's Figure 22.]

The first two integrals extend through the arcs $AB$ and $A'B$, respectively, and the rest of the integrals are rectilinear.

If one substitutes $-\frac{1}{\omega}$ for $\omega$ in the first integral, then it changes over to the second; analogously, by the substitution of $\omega + 1$ for $\omega$, the third changes over to the fourth.

These integrals thus cancel one another. This means that we have

$$N = \frac{1}{2\pi i} \int_{C'}^{C} d \log [J(\omega) - a]. \tag{48}$$

If, however, $\omega$ traverses the straight line $CC'$, then

$$h = e^{2i\pi\omega}$$

describes a circle around the zero point, whose radius decreases without limit as the distance of the straight line $CC'$ from the real axis increases without limit, and, indeed, $h$ describes this circle in a negative sense.

Seen as a function of $h$, the argument of the integral (48)

$$J(\omega) - a = \frac{1}{12^3 h}[1 + (c_1 - 12^3 a)h + c_2 h^2 + \cdots],$$

indicates of which order it will be infinite [indicates the order of the pole] at the point $h = 0$. This means that

$$N = 1.$$

With this, we have reached the fundamental theorem:

*In the region G, the function J(ω) assumes every finite value a once and only once.*

The same also holds for the value $\infty$, which $J(\omega)$ assumes only at the infinitely distant point of the region $G$.

From this theorem we can now draw a series of conclusions. First, we infer that in a system of equivalent points in the positive half-plane, there always exists only one belonging to the region $G$.

If there were two such points $\omega'$ and $\omega''$, the function $J(\omega)$ would take the value

$$a = J(\omega') = J(\omega'')$$

at the two points ($\omega'$ and $\omega''$) in the region $G$.

We infer further: If

$$J(\omega') = J(\omega''),$$

then $\omega'$ and $\omega''$ are equivalent points.

If $\omega_0$ is a point equivalent to $\omega$ in the region $G$, and if, likewise, $\omega_0'$ is a point equivalent to $\omega'$ in the region $G$, then it follows from

$$J(\omega_0') = J(\omega_0),$$

that $\omega_0'$ and $\omega_0$ coincide. Thus, $\omega$ and $\omega'$ are themselves equivalent to the point $\omega_0$ and to each other.

We now consider the two points

$$\omega = x + iy, \quad \omega^* = -x + iy,$$

which are reflection points with respect to the axis of the pure imaginary numbers!

The corresponding values

$$h = e^{2i\pi\omega} = e^{2i\pi x - 2\pi y}, \quad h^* = e^{2i\pi\omega^*} = e^{-2i\pi x - 2\pi y}$$

are conjugates. According to the defining equation (43), $J(\omega)$ and $J(\omega^*)$ are therefore also conjugates.

Now for a point $\omega$ in the region $G$, if the value of $J(\omega)$ is real, then we must have

$$J(\omega) = J(\omega^*)$$

and, consequently, the points $\omega$ and $\omega^*$ must be equivalent with each other. However, this is clearly the case only if $\omega$ lies on the boundary of $G$ or on the imaginary axis.

Hence, we have:

*The function $J(\omega)$ takes all real values (and each only once), if $\omega$ traverses the boundaries CA, AB of G and that portion, lying in G, of the axis of the pure imaginary numbers.*

Finally, we also determine the values that $J(\omega)$ takes at the boundary points $A$ and $B$ of the region $G$. The point $A$ is the representation of the third root of unity

$$\omega = -\frac{1}{2} + \frac{i}{2}\sqrt{3}.$$

Now if $\omega = \frac{\omega_1}{\omega_2}$ is an imaginary third root of unity, then in the sum

$$\frac{1}{60}g_2(\omega_1, \omega_2) = \sum_{m_1}\sum_{m_2} \frac{1}{(m_1\omega_1 + m_2\omega_2)^4} = \frac{1}{\omega_2^4}\sum_{m_1}\sum_{m_2}\frac{1}{(m_1\omega + m_2)^4}$$

the terms cancel each other by threes; specifically, we have $\omega^2 + \omega + 1 = 0$ and hence

$$\frac{1}{(m_1\omega + m_2)^4} + \frac{1}{((m_2 - m_1)\omega - m_1)^4} + \frac{1}{(-m_2\omega + m_1 - m_2)^4}$$
$$= \frac{1}{(m_1\omega + m_2)^4}\left[1 + \frac{1}{\omega^4} + \frac{1}{\omega^8}\right] = 0.$$

*At point A then we have*

$$J(\omega) = \frac{g_2^3(\omega_1, \omega_2)}{\Delta(\omega_1, \omega_2)} = 0. \tag{49}$$

The value of $J(\omega)$ at point $B$ emerges together with other noteworthy results in the following manner.

We set up the abbreviation

$$g_3 \equiv g_3(\omega_1, \omega_2) = 140 \sum_{m_1} \sum_{m_2} \frac{1}{(m_1\omega_1 + m_2\omega_2)^6} \tag{50}$$

and we then have, by (33) and (34)

$$g_3 = \left(\frac{2\pi}{\omega_2}\right)^6 \left[\frac{1}{216} - \frac{7}{3}\sum_{r=1}^{\infty} \psi_5(r)h^r\right]. \tag{51}$$

Now for the function

$$J_1(\omega) = \frac{g_3^2}{\Delta} = \frac{\left[\frac{1}{216} - \frac{7}{3}\sum_{r=1}^{\infty} \psi_5(r)h^r\right]^2}{h\prod_{r=1}^{\infty}(1 - h^r)^{24}} \tag{52}$$

the same inferences are valid, that we applied above to the function $J(\omega)$. In particular, $J_1(\omega)$ always takes on the same value for equivalent arguments and in the region $G$, $J_1(\omega)$ takes every value once and only once; at the infinitely distant point of $G$, $J_1(\omega)$, just like $J(\omega)$, has the value $\infty$.

From these facts it follows that $J_1(\omega)$ and $J(\omega)$ are linear functions of each other, so that an equation of the form

$$J(\omega) = aJ_1(\omega) + b = a \cdot \frac{g_3^2}{\Delta} + b \tag{53}$$

holds true, where $a$ and $b$ denote constants.

If we write this equation in the form

$$\left[\frac{1}{12} + 20\sum \psi_3(r)h^r\right]^3 = a\left[\frac{1}{216} - \frac{7}{3}\sum \psi_5(r)h^r\right]^2 + b \cdot h\prod(1 - h^r)^{24}, \tag{54}$$

then by a comparison of the coefficients of $h^0$ and $h^1$ on both sides, we get

$$a = 27, \quad b = 1.$$

With this, equations (53) and (54) become, respectively,

$$J(\omega) = 27 \cdot \frac{g_3^2}{\Delta} + 1 \tag{55}$$

$$g_2^3 = 27g_3^2 + \Delta. \tag{56}$$

The point $B$ represents $\omega = i$.

If, however, $\omega = \frac{\omega_1}{\omega_2} = i$, then in the sum (50), the terms cancel one another by fours, as one may easily perceive; hence $g_3 = 0$. Finally from (55) follows:

*At point B*

$$J(\omega) = 1. \tag{57}$$

From the fundamental properties of the function $J(\omega)$, herewith proven, one may derive without difficulty the fact that this function provides the mapping of the positive half-plane onto an infinitely sheeted Riemann surface, whose sheets are connected at the points 0, 1, and $\infty$ in a cycle of, respectively, every three, every two, or infinitely many.[7]

## §6.  Applications to the Theory of Elliptic Functions

As is well known, the Weierstrass function $\wp(u)$ with periods $\omega_1$ and $\omega_2$ is defined by the equation

$$\wp(u) = \frac{1}{u^2} + \sum_{m_1} \sum_{m_2} \left[ \frac{1}{(u - m_1\omega_1 - m_2\omega_2)^2} - \frac{1}{(m_1\omega_1 + m_2\omega_2)^2} \right]. \tag{58}$$

This satisfies, as one may easily verify from this definition, the differential equation

$$\wp'^2(u) = 4\wp^3(u) - g_2\wp(u) - g_3, \tag{59}$$

where we set

$$g_2 = 60 \sum_{m_1} \sum_{m_2} \frac{1}{(m_1\omega_1 + m_2\omega_2)^4}, \quad g_3 = 140 \sum_{m_1} \sum_{m_2} \frac{1}{(m_1\omega_1 + m_2\omega_2)^6}. \tag{60}$$

From the values of these quantities, $g_2$ and $g_3$, one knows that the "discriminant" arising out of their composition

$$\Delta = g_2^3 - 27 g_3^2 \tag{61}$$

is different from zero.

Now it is a fundamental question for the theory of the function $\wp(u)$, whether the periods $\omega_1$ and $\omega_2$ can always be so chosen that $g_2$ and $g_3$ take on prescribed values, which satisfy only the condition that the discriminant $\Delta$, calculated from them, is not zero.

Now instead of treating this question by a discussion of the differential equation (59), as it appears within the customary presentations on the theory of elliptic functions, one could handle it in a much shorter and easier manner on the basis of the theory of the function $J(\omega)$ as developed above. In fact, let $\omega_1$ and $\omega_2$ can be determined by the equations

$$g_2(\omega_1, \omega_2) = c_2, \quad g_3(\omega_1, \omega_2) = c_2, \tag{62}$$

where $c_2$ and $c_3$ denote given values, for which

$$c_2^3 - 27 c_3^2 = c \tag{63}$$

[7] F. Klein, a.a.O., S. 121.

it not zero. The equations (62) are then and only then satisfied, if the equations

$$\frac{g_2(\omega_1, \omega_2)}{g_3(\omega_1, \omega_2)} = \frac{c_2}{c_3}, \qquad \frac{g_2^3(\omega_1, \omega_2)}{g_3^2(\omega_1, \omega_2)} = \frac{c_2^3}{c_3^2} \tag{64}$$

hold, from which the second may also be replaced by

$$\frac{g_2^3(\omega_1, \omega_2)}{g_2^3(\omega_1, \omega_2) - 27\, g_3^2(\omega_1, \omega_2)} = \frac{c_2^3}{c}. \tag{64'}$$

If we now consider

$$\omega_2 \quad \text{and} \quad \frac{\omega_1}{\omega_2} = \omega$$

as the quantities to be determined, then we rewrite equations (64) and (64') as

$$\omega_2^2 = \frac{c_2}{c_3} \cdot \frac{g_3(\omega, 1)}{g_2(\omega, 1)}, \qquad J(\omega) = \frac{c_2^3}{c}. \tag{65}$$

Now one knows that the last equation always has solutions (of which only one lies in the region $G$). If one has determined one such solution, then $\omega_2$ is given by the first equation (65) and finally $\omega_1$ by the equation $\omega_1 = \omega \cdot \omega_2$. The quantities $\omega_1, \omega_2$ can then in fact always be determined by equations (62) and, moreover, one easily recognizes from the fundamental properties of the function $J(\omega)$, that $\omega_1$ and $\omega_2$ are fully determined by equation (62) up to the linear homogeneous whole-number substitution of determinant $\pm 1$.

Zürich, 28 September 1903.

# Bibliography

Abel, N. 1965. *Oeuvres complètes*, 2 vols. New York: Johnson Reprint.

Andrews, G. 1974. Applications of basic hypergeometric functions. *SIAM Rev. 16*, 441–484.

Andrews, G., and Berndt, B. 2005–. *Ramanujan's Lost Notebook*. New York: Springer.

Archibald, T. 2002. Charles Hermite and German Mathematics in France. In Parshall, K. H., and Rice, A. C., eds., *Mathematics Unbound: The Evolution of an International Research Community, 1800–1949*, pp. 123–137. Providence, RI: AMS.

Armitage, J. V., and Eberlien, W. F. 2006. *Elliptic Functions*. Cambridge: Cambridge University Press.

Atkin, A. O. L. 1967. Proof of a Conjecture of Ramanujan. *Glasgow Math J 8*, 14–32.

Auwers, A. (ed). 1880. *Briefwechsel zwischen Gauss und Bessel*. Leipzig: Engelmann.

Bachmann, P. 1923. *Die Arithmetik der quadratischen Formen*. Leipzig: Teubner.

Barrow-Green, June. 1996. *Poincaré and the Three Body Problem*. Providence: AMS.

Bateman, P. T. 1951. On the representation of a number as the sum of three squares. *Trans. Am. Math. Soc. 71*, 70–101.

Berggren, L. J. M. Borwein, P. B. Borwein. 1997. *Pi: A Source Book*. New York: Springer-Verlag.

Berndt, B. 1985–1998. *Ramanujan's Notebooks, parts I–V*. New York: Springer-Verlag.

Berndt, B. 1992. Hans Rademacher (1892–1969). *Acta Arith. 61*, 209–231.

Berndt, B. 1993. *Theta Functions: from the Classical to the Modern*. Providence: AMS. edited by M. R. Murty. Chap. 1 Ramanujan's Theory of Theta-Functions, pages 1–63.

Berndt, B. 2005. *Number Theory in the Spirit of Ramanujan*. Providence: AMS.

Berndt, B., and Knopp, M. 2008. *Hecke's Theory of Modular Forms and Dirichlet Series*. Singapore: World Scientific.

Berndt, B., and Ono, K. 2001. *The Andrews Festschrift*. New York: Springer. edited by Foata, D., and Han, G.-N. Chap. Ramanujan's Unpublished Manuscript on the Partition and Tau Functions with Proofs and Commentary, pages 39–110.

Berndt, B., and Rankin, R. 1995. *Ramanujan: Letters and Commentary*. Providence: AMS.

Berndt, B., and Rankin, R. 2001. *Ramanujan: Essays and Surveys*. Providence: AMS.

Bernoulli, Daniel. 1982–1996. *Die Werke von Daniel Bernoulli*. Basel: Birkhäuser.

Bernoulli, Johann. 1968. *Opera Omnia*. Hildesheim: Georg Olms Verlag.

Bierman, K. R. (ed). 1977. *Briefwechsel zwischen Alexander von Humboldt und Carl Friedrich Gauss*. Berlin: Akedemie-Verlag.

Birch, B. J. 1975. A look back at Ramanujan's notebooks. *Math. Proc. Camb. Phil. Soc. 78*, 73–79.

Boole, George. 1841. Exposition of a general theory of linear transformations, Parts I and II. *Cambridge Math J 3*, 1–20, 106–111.

Boole, George. 1844. Notes on linear transformations. *Cambridge Math. J. 4*, 167–171.

Borwein, J. M., and Borwein, P. B. 1987. *Pi and the AGM*. New York: Wiley.

Bottazini, U., and Gray, J. 2013. *Hidden Harmony-Geometric Fantasies*. New York: Springer.

Brioschi, F. 1901. *Opere Matematiche*. Milano: Ulrico Hoepli.

Bruinier, Jan Hendrik, and Ono, Ken. 2013. Algebraic formulas for the coefficients of half-integral weight harmonic weak Maass forms. *Advances in Math* **246**, 198–219.

Bruns, H. 1886. Über die Perioden der elliptischen Integrale erster und zweiter Gattung. *Math. Annalen* **27**, no. 2, 234–252.

Cahen, E. 1891. Note sur un développement des quantités numériques, qui présente quelque analogie avec celui en fractions continues. *Nouv. Ann. Math* **10**, 508–514.

Cahen, E. 1894. Sur la fonction $\zeta(s)$ de Riemann et sur des fonctions analogs. *Ann. Sci. École Norm. Sup.* **11**, 75–164.

Cardano, G. 1993. *Ars Magna or the Rules of Algebra*. New York: Dover. translated by T. R. Wittmer.

Carr, G. S. 2013. *A Synopsis of Elementary Results in Pure and Applied Mathematics*. Cambridge: Cambridge University Press.

Cassels, J. W. S. 1973. Louis Joel Mordell. *Biog. Mem. Fellows Royal Soc.* **19**, 493–520.

Cauchy, A. L. 1882–1974. *Oeuvres complètes*. Paris: Gauthier-Villars.

Cayley, A. 1874. A memoir on the transformation of elliptic functions. *Phil. Trans. Royal Soc.* **164**, 397–456.

Cayley, A. 1889–1898. *Collected Mathematical Papers*. Cambridge: Cambridge University Press.

Chan, H. H., and Chua, K. S. 2003. *Papers in Memory of Robert A. Rankin*. Norwell, MA: Kluwer Acad. Press. Chap. Representations of Integers as Sums of 32 Squares, pages 79–89.

Chowla, S, D. 1934. Congruence properties of partitions. *JLMS* **9**, 247.

Clairaut, A. C. 1739. Recherches générales sur le calcul intégral. *Mém. de l'Académie Royale des Sci.* **1**, 425–436.

Clairaut, A. C. 1740. Sur l'intégration ou la construction des équations différentielles du premier ordre. *Mém. de l'Académie Royale des Sci.* **2**, 293–323.

Clarke, F. M. 1929. *Thomas Simpson and his Times*. Baltimore: Waverly Press.

Cooper, S. 2001. On sums of an even number of squares, and an even number of triangular numbers: an elementary approach based on Ramanujan's $_1\psi_1$ summation formula. *Contemporary Math,* **291**, 115–138.

Cotes, Roger. 1722. *Harmonia Mensurarum*. Cambridge: Cambridge University Press.

de Moivre, Abraham. 1707. Aequationum quarundam Potestatis tertiae, quintae, septimae, novae, & superiorum, ad infinitum usque pergendo, in terminis finitis, ad imstar Regularum pro Cubicis quae vocantur, Cardani, Resolutio Analytica. *Phil. Trans.* **25**, no. 309, 2368–2371.

de Moivre, Abraham. 1730. *Miscellanea analytica de seriebus et quadraturis*. London: Touson and Watts.

de Moivre, Abraham. 1967. *The Doctrine of Chances*. New York: Chelsea.

Dedekind, R. 1930–. *Gesammelte Mathematische Werke*. Braunschweig: Vieweg. edited by R. Fricke, E. Noether, Ø. Ore.

Deligne, P. 1974. La conjecture de Weil. I. *Pub. Math IHES* **43**, 273–307.

Dewar, Michael, and Murty, M. Ram. 2013. A derivation of the Hardy-Ramanujan formula from an arithmetic formula. *Proc. AMS* **141**, 1903–1911.

Dirichlet, L., and Dedekind, R. 1999. *Lectures on Number Theory*. Providence: AMS. translated by John Stillwell.

Dirichlet, P. G. L. 1969. *Mathematische Werke*. New York: Chelsea.

Dunnington, G. 2004. *Gauss: Titan of Science*. Washington: MAA.

Edwards, H. M. 1984. *Galois Theory*. New York: Springer-Verlag.

Eisenstein, G. 1847. *Mathematische Abhandlungen, besonders aus dem Gebiete der höheren Arithmetik und der Elliptischen Funktionen*. Berlin: G. Riemer.

Eisenstein, G. 1975. *Mathematische Werke*. New York: Chelsea.

Elfving, G. 1981. *The History of Mathematics in Finland, 1828–1918*. Helsinki: Societas Scientiarum Fennica.

Elstrodt, J. 2007. *Analytic Number Theory: A tribute to Gauss and Dirichlet*. Providence: AMS. edited by Duke, W. and Tschinkel, Y. Chap. The life and work of Gustav Lejeune Dirichlet (1905–1859), pages 1–37.

Engelsman, S. B. 1984. *Families of Curves and the Origins of Partial Differentiation*. Amsterdam: North Holland.

Enneper, A. 1890. *Elliptische Functionen*. Halle: Louis Nebert.

Euler, L. 1911–. *Leonhardi Euleri Opera Omnia*. Series I-IV A. Bassel: Birkhäuser.

Euler, Leonhard. 1988. *Introduction to Analysis of the Infinite*. New York: Springer-Verlag. translated by J. D. Blanton.

Fagnano, G. C. 1750. *Produzioni Matematiche del Conte Giulio Carlo di Fagnano*. Pesaro: Gavelli.

Feigenbaum, L. 1981. *Brook Taylor's Methodus Incrementorum: A Translation with Mathematical and Historical Commentary*. Ph.D. thesis, Yale University, New Haven.

Fine, N. J. 1988. *Basic Hypergeometric Series and Applications*. Providence: AMS.

Ford, L. R. 1957. *Automorphic Functions*. New York: Chelsea.

Fuss, P. H. (ed). 1968. *Correspondance mathématique et physique*, 3 vols. New York: Johnson Reprint.

Gårding, L. 1997. *Mathematics and Mathematicians: Mathematics in Sweden before 1950*. Providence: AMS.

Gårding, L., and Skau, C. 1994. Niels Henrik Abel and solvable equations. *Arch. Hist. of Exact Sc.* **48**, 81–103.

Gannon, T. 2006. *Moonshine Beyond the Monster*. Cambridge: Cambridge University Press.

Gauss, C. F. 1863–1927. *Werke*, vols. 1–12. Leipzig: Teubner.

Gauss, C. F. 1965. *Disquisitiones Arithmeticae*, English trans. Arthur A. Clarke, S. J. New Haven: Yale.

Gelfand, Kapranov, Zelevinsky. 1994. *Discriminants, Resultants, and Multidimensional Determinants*. Boston: Birkhäuser.

Glaisher, J. W. L. 1907a. On the number of representations of a number as a sum of $2r$ squares, where $2r$ does not exceed eighteen. *Proc. London Math. Soc. series 2*, **5**, 479–490.

Glaisher, J. W. L. 1907b. On the representations of a number as the sum of two, four, six, eight, ten, and twelve squares. *Quarterly J. Pure and Appl. Math.* **38**, 1–62.

Glaisher, J. W. L. 1908. On elliptic-function expansions in which the coefficients are powers of the complex numbers having $n$ as norm. *Quart. J. Pure and Appl. Math* **39**, 266–300.

Goldstein, Schappacher, Schwermer (ed). 2007. *The Shaping of Arithmetic after C. F. Gauss's Disquisitiones Arithmeticae*. New York: Springer.

Gowing, R. 1983. *Roger Cotes*. Cambridge: Cambridge University Press.

Gray, J. 1986. *Linear Differential Equations and Group Theory from Riemann to Poincaré*. Boston: Birkhäuser.

Green, George. 1970. *Mathematical Papers*. New York: Chelsea.

Grosswald, E. 1985. *Representations of Integers as Sums of Squares*. New York: Springer-Verlag.

Gudermann, C. 1838. Theorie der Modular-Functionen und der Modular-Integrale. *J. Reine und Angew. Math.* **18**, 1–54, 220–258.

Guetzlaff, C. 1834. Aequatio modularis pro transformatione functionum ellipticarum septimi ordinis. *J. Reine Angew. Math.* **12**, 173–177.

Gupta, H. 1935. Tables of partitions. *PLMS, series 2* **39**, 142–149.

Hardy, G. H. 1966–1979. *Collected Papers of G. H. Hardy*. Oxford: Oxford University Press.

Hardy, G. H. 1978. *Ramanujan*. New York: Chelsea.

Hardy, G. H., and Riesz, M. 1915. *The General Theory of Dirichlet Series*. Cambridge: Cambridge University Press.

Hecke, E. 1959. *Mathematische Werke*. Göttingen: Vandenhoeck und Ruprecht.

Hecke, E. 1983. *Lectures on Dirichlet Series, Modular Functions and Quadratic Forms*. Göttingen: Vandenhoeck and Ruprecht.

Hermite, C. 1905–1917. *Oeuvres*. Paris: Gauthier-Villars.

Hille, E. 1997. *Ordinary Differential Equations in the Complex Domain*. New York: Dover.

Hurwitz, A. 1962. *Mathematische Werke*. Basel: Birkhäuser.

Jacobi, C. G. J. 1846. *Opuscula Mathematica*, vol. 1. Berlin: G. Reimer.

Jacobi, C. G. J. 1965. *Mathematische Werke*. New York: Chelsea.

Joubert, C. 1858. Sur divers équations analogues aux équations modulaires dans la théorie des fonctions elliptiques. *Comptes Rendus 47*, 337–345.

Joubert, C. 1875. *Sur les équations qui se recontrent dans la théorie de la transformation des fonctions elliptiques*. Paris: Gauthier-Villars.

Kiepert, L. 1879a. Auflösung der Gleichungen fünften Grades. *J. Reine Angew. Math. 87*, 114–133.

Kiepert, L. 1879b. Zur Transformationstheorie der elliptischen Functionen. *J. Reine Angew. Math. 87*, 199–216.

Klein, F. 1921–1923. *Gessamelte Mathematische Abhandlungen*. Berlin: Verlag Julius Springer.

Klein, F. 1933. *Vorlesungen über die hypergeometrische Funktion*. Berlin: Springer. Compiled and edited by Otto Haupt.

Klein, F. 1956. *The Icosahedron*. New York: Dover. translated by George Morrice.

Klein, F., and Fricke, R. 1890–1892. *Vorlesungen über die Theorie der Modulfunktionen*. Leipzig: Teubner.

Knopp, M. 2000. *Number Theory*. New Delhi: Hindustan Book Agency. edited by Bambah, R. P., et al., Chap. Hamburger's theorem on $\zeta(s)$ and the abundance principle for Dirichlet series with fundamental equations, pages 201–216.

Knuth, Donald. 1998. *The Art of Computer Programming (third edition)*, vol. 2. Reading, MA: Addison-Wesley.

Kronecker, L. 1968. *Mathematische Werke*. New York: Chelsea.

Kronecker, Leopold. 1894. *Theorie der einfachen und der vielfachen Integrale*. Leipzig: Teubner. edited by E. Netto.

Kuhn, H. K. 1991. *History of Mathematical Programming: a collection of personal reminiscences*. Amsterdam: North-Holland. edited by Lenstra, Kan, and Schrijver. Chap. Nonlinear programming: A historical note, pages 82–96.

Kummer, Ernst. 1975. *Collected Papers*. Berlin: Springer Verlag.

Lacroix, S. F. 1819. *Traité du calcul différential et du calcul intégral*, Vol. 3. Paris: Courcier.

Lagrange, J. L. 1867–1892. *Oeuvres*, vols. 1–14. Paris: Gauthier-Villars.

Landau, E. 1906. Euler und die Funktionalgleichung der Riemannschen Zetafunktion. *Bibliotheca Math. 7*, no. 3, 69–79.

Landen, J. 1775. An investigation of a general theorem for finding the length of any arc of any conic hyperbola, by means of two elliptic arcs, with some other new and useful theorems deduced therefrom. *Phil. Trans. Roy. Soc. Lon. 65*, 283–289.

Legendre, A. M. 1792. *Mémoire sur les Transcendantes elliptiques*. Paris: Academie des Sciences.

Legendre, A. M. 1811–1817. *Exercices de calcul intégral*, 2 vols. Paris: Courcier.

Lehmer, D. H. 1937. On the Hardy-Ramanujan series for the partition function. *J. London Math. Soc. 12*, 171–176.

Leibniz, G. W. 1920. *The Early Mathematical Manuscripts of Leibniz*. Chicago: Open Court. edited by Gerhardt, C. I., translated with notes by Child, J. M.

Leibniz, G. W. 1971. *Mathematische Schriften*. Hildesheim: Georg Olms Verlag. edited by Gerhardt, C. I.

Leybourn, Thomas. 1817. *The mathematical questions, proposed in the Ladies' diary, and their original answers, together with some new solutions, from its commencement in the year 1704 to 1816*. London: Mawman.

Lindelöf, E. 1905. *Le calcul des résidus et ses applications à la théorie des fonctions*. Paris: Gauthier-Villars.

Liouville, J. 1844. Nouvelle démonstration d'un théorème sur les irrationnelles algébriques. *Comptes Rendus, 18*, 910–911.

Liouville, J. 1880. Leçons sur les fonctions doublement périodiques. *J. Reine Angew. Math. 88*, 277–310.

Lützen, J. 1990. *Joseph Liouville 1809–1882*. New York: Springer-Verlag.

Markushevich, A. 1992. *Introduction to the classical theory of Abelian functions*. Providence: AMS.

McKean, H., and Moll, V. 1997. *Ellipitc Curves*. Cambridge: Cambridge University Press.

Milne, S. C. 2002. *Infinite Families of Exact Sums of Squares Formulas, Jacobi Elliptic Functions, Continued Fractions, and Schur Functions*. Boston: Kluwer Acad. Press.

Minkowski, H. 1967. *Gesammelte Abhandlungen*. Leipzig: Teubner.

Mittag-Leffler, G. 1923. An introduction to the theory of elliptic functions. *Ann. Math.* **24**, 271–351.

Mordell, L. 1917a. On Mr. Ramanujan's empirical expansions of modular functions. *Proc. Camb. Phil. Soc.* **19**, 117–124.

Mordell, L. 1917b. On the representation of numbers as a sum of $2r$ squares. *Quart. J. Pure and Appl. Math* **48**, 93–104.

Mordell, L. 1919. On the representation of numbers as a sum of an odd number of squares. *Trans. Camb. Phil. Soc.* **22**, 361–372.

Mordell, L. 1923. On the integer solutions of the equation $ey^2 = ax^3 + bx^2 + cx + d$. *Proc. London Math. Soc.* **21**, 415–419.

Mordell, L. 1929. Poisson's summation formula in several variables and some applications in the theory of numbers. *Proc. Camb. Phil. Soc.* **25**, 412-420.

Moreno, C. J., and Wagstaff, Jr, S. S. 2006. *Sums of Squares of Integers*. Boca Raton: Chapman and Hall.

Müller, F. 1867. *De transformatione functionum ellipticarum*. Berlin: A. W. Schade.

Murty, M. R., and Murty, V. K. 2013. *The Mathematical Legacy of Srinivasa Ramanujan*. New Delhi: Springer.

Neumann, P. M. 2011. *The Mathematical Writings of Évariste Galois*. Zürich: European Mathematical Society Publishing House.

Newman, J. 1956. *The World of Mathematics*, 4 vols. New York: Simon and Schuster.

Newton, Isaac. 1959–1960. *The Correspondence of Isaac Newton*. Cambridge: Cambridge University Press. edited by Turnbull, H. W.

Newton, Isaac. 1967–1981. *The Mathematical Papers of Isaac Newton*. Cambridge: Cambridge University Press. edited by Whiteside, D. T.

Ono, K. 2002. Representations of integers as sums of squares. *J. Number Theory* **95**, 253–258.

Ono, K., and Aczel, A. 2016. *My Search for Ramanujan*. Cham: Springer.

Peiffer, J. 1983. Joseph Liouville (1809–1882): ses contributions à la théorie des fonctions d'une variable complexe. *Rev. Hist. Sci.* **36**, 209–248.

Peters, C. A. F. (ed). 1860–1865. *Briefwechsel zwischen C. F. Gauss und H. C. Schumacher*, vols. 1–6. Altona: Gustav Esch.

Pieper, H. (ed). 1987. *Briefwechsel zwischen Alexander von Humboldt und C. G. Jacob Jacobi*. Berlin: Akademie-Verlag.

Rademacher, H., and Grosswald, E. 1972. *Dedekind Sums*. Washington: MAA.

Rademacher, Hans. 1974. *Collected Papers*. Cambridge, MA: MIT Press.

Ramanujan, S. 1988. *The lost notebook and other unpublished papers*. New Delhi: Narosa.

Ramanujan, S. 2000. *Collected Papers*. Providence: AMS Chelsea.

Rangachari, S. S. 1982. Ramanujan and Dirichlet series with Euler products. *Proc. Indian Acad. Soc. (Math Sci.)* **91**, 1–15.

Rankin, R. 1946. A certain class of multiplicative functions. *Duke Math. J.* **13**, 281–306.

Rankin, R. 1962. On the representation of a number as the sum of any number of squares, and in particular of twenty. *Acta Arithmetica* **7**, 399–407.

Rankin, R. 1965. Sums of squares and cusp forms. *Amer. J. Math.* **87**, 857–862.

Rankin, R. 1977. *Modular Forms and Functions*. Cambridge: Cambridge University Press.

Remmert, R. 1991. *Theory of Complex Functions, An English translation of the second edition of Remmert's Functionentheorie I*. New York: Springer-Verlag. translated by Robert B. Burckel.

Riemann, B. 1899. *Elliptische functionen. Vorlesungen von Bernhard Riemann. Mit zusätzen herausgegeben von Hermann Stahl*. Leipzig: Teubner.

Riemann, B. 1990. *Gessammelte Mathematische Werke*. Berlin: Springer-Verlag.

Rigaud, S. P. (ed). 1841. *Correspondence of Scientific Men of the Seventeenth Century*. Oxford: Oxford University Press.

Rochat, Vecten, Fauquier, and Pilatte. 1811–12. Questions résolves. Solutions des deux problèmes proposés à la page 384 du premier volume des Annales. *Annales de Math. Pure et Appl. II*, 88–93.

Rodríguez, I. Kra, Gilman. 2012. *Complex Analysis in the Spirit of Lipman Bers*. 2nd ed. New York: Springer.

Ronan, M. 2006. *Symmetry and the Monster*. New York: Oxford University Press.

Roy, R. 2011. *Sources in the Development of Mathematics*. Cambridge, New York: Cambridge University Press.

Russ, S. B. 1980. A translation of Bolzano's paper on the intermediate value theorem. *Hist. Math. 7*, 156–185.

Scharlau, W. 1981. *Richard Dedekind 1831–1981: Eine Würdigung zu seinem 150. Geburtstag*. Braunschweig/Wiesbaden: Vieweg und Teubner.

Scheibner, W. 1860. *Über unendliche Reihen und deren Convergenz*. Leipzig: S. Hirzel.

Schwarz, H. A. 1972. *Gesammelte Mathematische Abhandlungen*. New York: Chelsea.

Serre, J. P. 1966. *Seminar on Complex Multiplication*. Berlin: Springer. edited by Borel, A. et al., Chap. II Modular forms, pages 1–16.

Shen, L. C. 1993. On the logarithmic derivative of a theta function and a fundamental identity of Ramanujan. *J. Math. Anal. Appl. 177*, no. 1, 299–307.

Shimura, G. 2002. The representation of integers as sums of squares. *Amer. J. Math, 124*, 1059–1081.

Simpson, Thomas. 1759. The invention of a general method for determining the sum of every second, third, fourth, or fifth, etc. term of a series, taken in order; the sum of the whole being known. *Phil. Trans. 50*, 757–769.

Smith, D. E. 1959. *A Source Book in Mathematics*. New York: Dover.

Smith, H. J. S. 1865. *Report on the Theory of Numbers*. N.p.: Brit. Assoc. for the Advancement of Science.

Smith, H. J. S. 1965. *Collected Mathematical Papers*. New York: Chelsea.

Smithies, F. 1997. *Cauchy and the Creation of Complex Function Theory*. Cambridge: Cambridge University Press.

Sohnke, L. 1837. Aequationes modulares pro transformatione Functionum Ellipticarum. *J. Reine Angew. Math 16*, 97–130.

Stäkel, P., and Ahrens, W. (eds). 1908. *Der Briefwechsel zwischen C. G. J. Jacobi und P. H. Fuss über die Herausgabe der Werke Leonhard Eulers*. Leipzig: Teubner.

Stalker, J. 1998. *Complex Analysis: The fundamentals of the classical theory of functions*. Boston: Birkhäuser.

Stirling, James, and Tweddle, Ian. 2003. *James Stirling's* Methodus Differentialis *An Annotated Translation of Stirling's Text*. London: Springer.

Sylvester, J. J. 1973. *Mathematical Papers*. New York: Chelsea.

Tannery, J., and Molk, J. 1972. *Éléments de la théorie des fonctions elliptiques*, 4 vols. New York: Chelsea.

Taylor, B. 1715. *Methodus Incrementorum*. London: Gulielmi Innys. Translation into English in Feigenbaum (1981).

Titchmarsh, E. C., and Heath-Brown. 1986. *The Theory of the Riemann Zeta-function*, second edition. New York: Oxford University Press.

Uspensky, J. 1928. On Jacobi's arithmetic theorems concerning the simultaneous representation of numbers by two different quadratic forms. *Trans. Am. Math. Soc. 30*, 385–404.

van der Waerden, B. L. 1975. On the sources of my book, *Modern Algebra. Hist. Math. 2*, 32–40.

Waring, Edward. 1988. *Meditationes Algebraicae*. Providence: AMS. translated by Dennis Weeks.

Weber, H. 1894–1908. *Lehrbuch der Algebra*. Braunschweig: Vieweg.

Weierstrass, K. 1894–1927. *Mathematische Werke*. Berlin: Mayer und Müller.

Weil, A. 1976. *Elliptic Functions according to Eisenstein and Kronecker*. Berlin: Springer-Verlag.

Weil, A. 1980. *Oeuvres Scientifiques*. New York: Springer-Verlag.

Weil, A. 1984. *Number Theory: An approach through history from Hammurapi to Legendre*. Boston: Birkhäuser.

Williams, K. 2011. *Number Theory in the Spirit of Liouville*. Cambridge: Cambridge University Press.

Zagier, D. 2000. A proof of the Kac-Wakimoto affine denominator formula for the strange series. *Math. Res. Letters 7*, 597–604.

# Index

Printed in the United States
By Bookmasters